Liaison Functions 2014

Core Programming Area at the 2014 AIChE Spring Meeting and 10th Global Congress on Process Safety

New Orleans, Louisiana, USA
30 March – 3 April 2014

ISBN: 978-1-63439-074-3

Printed from e-media with permission by:

Curran Associates, Inc.
57 Morehouse Lane
Red Hook, NY 12571

proceedings.com

Some format issues inherent in the e-media version may also appear in this print version.

Printed by Curran Associates, Inc. (2014)

For permission requests, please contact AIChE
at the address below.

AIChE
3 Park Avenue
New York, NY 10016-5991

Phone: (203) 702-7660
Fax: (203) 775-5177

www.aiche.org

Additional copies of this publication are available from:

Curran Associates, Inc.
57 Morehouse Lane
Red Hook, NY 12571 USA
Phone: 845-758-0400
Fax: 845-758-2634
Email: curran@proceedings.com
Web: www.proceedings.com

TABLE OF CONTENTS

354012 Possible vs. Practical: Engineers Must Lead the Development of Practical Technologies

Abstract:

Chemical Engineering is need today more than ever. There is a strong societal desire for sources of energy and feedstocks that can sustainably support a high standard of living for all. Thermodynamic, material and economic challenges stand in the way. Most of the general public does not have the basic understanding of the energy flows, chemical transformations, and the scale of the systems that support our current way of life. These concepts are fundamental to chemical engineering. Chemical engineers should be on the vanguard of separating hype from truly practical options yet occationally we let society down. This talk will present several examples where the fundamentals of chemical engineering can be used to access the practicality of proposed solutions to energy and water issues, as well as explain why the venture capital model struggles in the materials market.

352696 Hydrocarbons from Biogas Via Biogas Reforming, Syngas Cleaning and Catalytic Conversion

Abstract:

This research work focused on taking advantage of biogas as a cleaner, sustainable and renewable feedstock to produce liquid hydrocarbons. In this approach, biogas is first reformed to high quality synthesis gas (H_2 + CO) followed by Fischer-Tropsch synthesis (FTS). The Sustainable Energy Research Center (SERC) at Mississippi State University has developed and tested a continuous process over a pilot plant system to demonstrate the technology of biomass to liquid (BTL) fuels. The process includes biomass gasification, syngas conditioning and cleaning, and FTS steps. Currently, the typical gaseous product components from the gasifier are 20% H_2, 19% CO, 12% CO_2, and 47% N_2 if using wood chips as the feedstock. The N_2 and CO_2 contents are too high, the total syngas (H_2 + CO) content is only about 40%, and H_2/CO ratio of the bio-syngas is less than 1. The biomass-derived product gas is not ideal for the hydrocarbon synthesis. Exceptional product gas is produced from biogas (Table 1), it contains high concentrated syngas (CO+ H_2 > 90%), low impurities and CO_2, as well as N_2,H_2/CO ratio is between 1 and 2 which makes it perfect production of liquid hydrocarbon by FTS. The liquid products generated by FTS can be readily upgraded to transportation fuels since they are primarily composed of hydrocarbons initially.

Table 1 Product gas composition for simultaneous dry and partial oxidation (using air) of biogas.

Temperature	H_2 (%)	CO (%)	N_2 (%)	CH_4 (%)	CO_2 (%)	H_2/CO

(°C)						
750	48.96	40.82	6.33	1.75	2.01	1.2
775	49.45	41.22	6.27	1.3	1.74	1.2
800	50.06	42.2	6.05	0.96	1.12	1.19
825	50.57	42.7	6.03	0.57	0.78	1.18
850	50.75	43.11	5.93	0.4	0.25	1.18

351056 Research for an Energy Future

Abstract:

Global demand for electrical energy is increasing even as environmental concerns are changing the ways in which we produce it. Southern Company is a leading US producer of electricity and is dedicated to developing the latest technologies of pollution control, energy production, and energy efficiency, and bringing them to economic feasibility. This talk will highlight efforts in all these areas, including Southern Company's major research projects, research centers, and commercial operations. The talk will primarily focus on Southern Company's roles in developing pollution control technology, including acid gas scrubbers, selective catalytic reduction of NOx, and mercury control (including the Mercury Research Center and the High-Sulfur Pilot Baghouse Research Center). CO_2 capture will be discussed including a 25-MW demonstration plant and ongoing work at the National Carbon Capture Center, located at Alabama Power's Gaston Plant. Changing energy production methods including commercial operations of coal gasification, wind, and solar will also be discussed, as well as the new Solar Power Research Center. Efforts to develop technologies which conserve water in power production will be highlighted by a discussion of the Water Research Center located at Georgia Power's Bowen Plant. Finally, ongoing research to improve energy efficiency for our customers will be highlighted, as well as electric transportation research which would have major environmental benefits.

Sustainability and Environmental Impacts of Biofuel

Shijie Liu

Department of Paper and Bioprocess Engineering
SUNY College of Environmental Science and Forestry

1

Imagine a renewable world

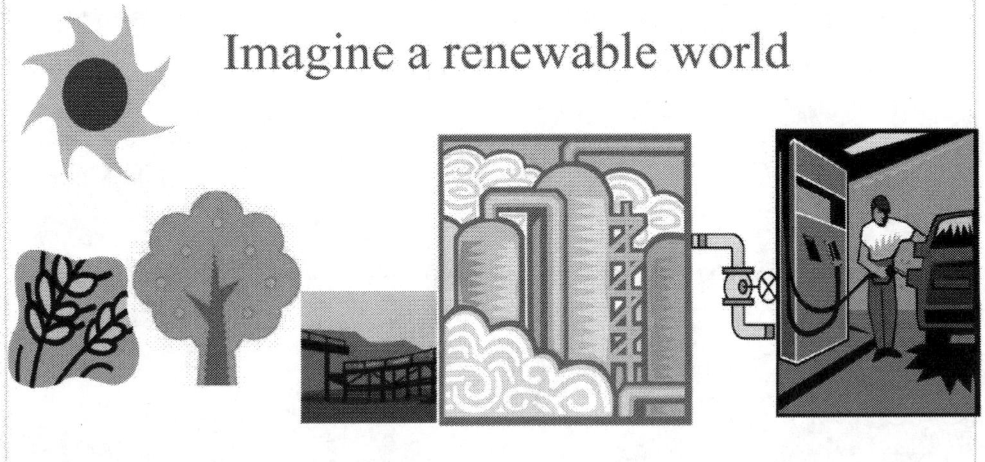

2

Outline

- **The Hype of Biomass**
- **What is Sustainability?**
- **Biomass Crop**
- **Ethanol: Good or Bad**
- **Ethanol: a Platform chemical**
- **Ethanol: Biomass conversion efficiency**
- **Conclusion**

3

The Hype in Biomass

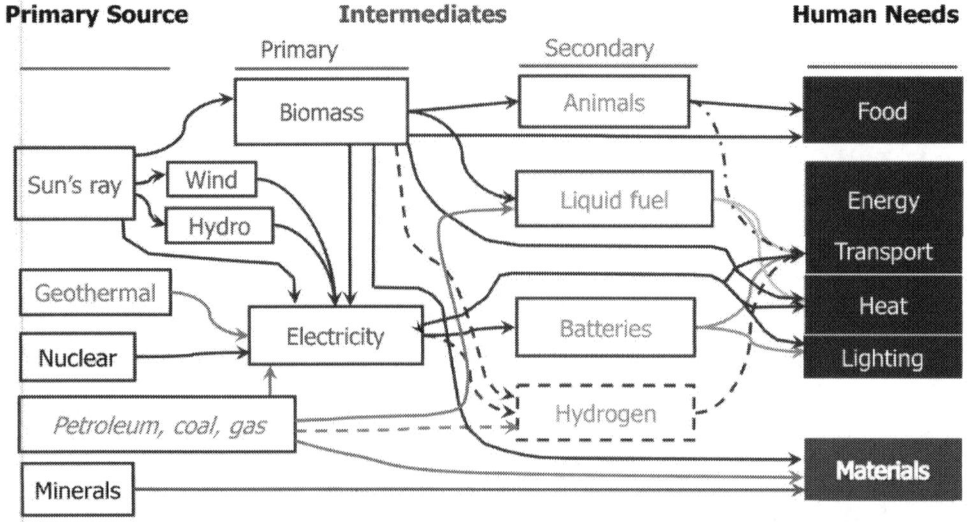

4

Where do we get our chemicals today?

- Petroleum, coal and natural gas
  - Dominant source of energy for human use
  - Materials use - plastics
- Biomass
  - Food
  - Heating use on the steep declining slope
  - Materials – dwellings, furniture
- Solar energy
  - Limited use in heating and lighting

5

What is Renewability?

- Merriam-Webster:
  - 1. Renewable – capable of being renewed
  - 2. Renewable – capable of being replaced by natural ecological cycles or sound management practices
- Point on a complete cycle

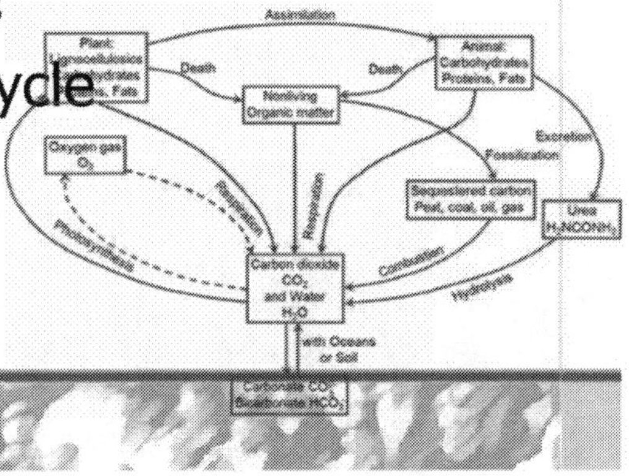

What is Sustainability?

- Merriam-Webster:
  - 1. Sustainable – capable of being sustained
  - 2. Sustainable – of, relating to, or being a method of harvesting or using a resource so that the resource is not depleted or permanently damaged.
- Non-trivial Steady State

  (S. Liu, Bioprocess Engineering: Kinetics, Biosystems, Sustainability, and Reactor Design, Elsevier, 2012)

Biomass

- Agricultural crop
  - Cereal grain; Tubers
  - Straws
  - Grass
- Forest biomass
  - Trees
- Long rotation crops
  - Fruits, nuts
  - "Short-rotation" woody biomass

Is Forest Biomass Renewable?

- Manomet Center for Conservation Sciences report "Biomass Sustainability and Carbon Policy Study" also know as the Massachusetts Biomass Report (June 2010)
  - Woody biomass utilization sending more carbon into the atmosphere than fossil
  - Equal setting of forest biomass use with fossil use will occur after 32 years IF ...
  - Forest biomass carbon neutrality will not occur after 90 years

9

Massachusetts' report

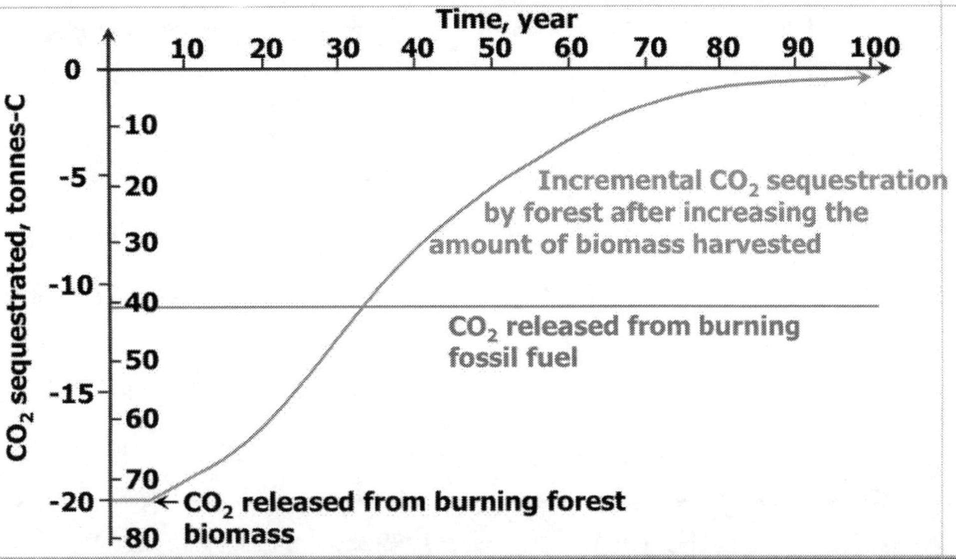

10

Biomass: Good or Bad?

- Fossil source
  - Energy dense: low CO_2 emission
- Biomass, CO_2 is CO_2
  - Less energy content (already oxygenated) - bulky
  - Not accounting: Forest disturbance – only half storage grows back when converting an untouched forest to a managed forest and equal amount underground in jeopardy
  - More CO_2 in atmosphere at sustainability
  - Hydrocarbon emission, e.g. isosprene

11

Sustainability – Biomass Production

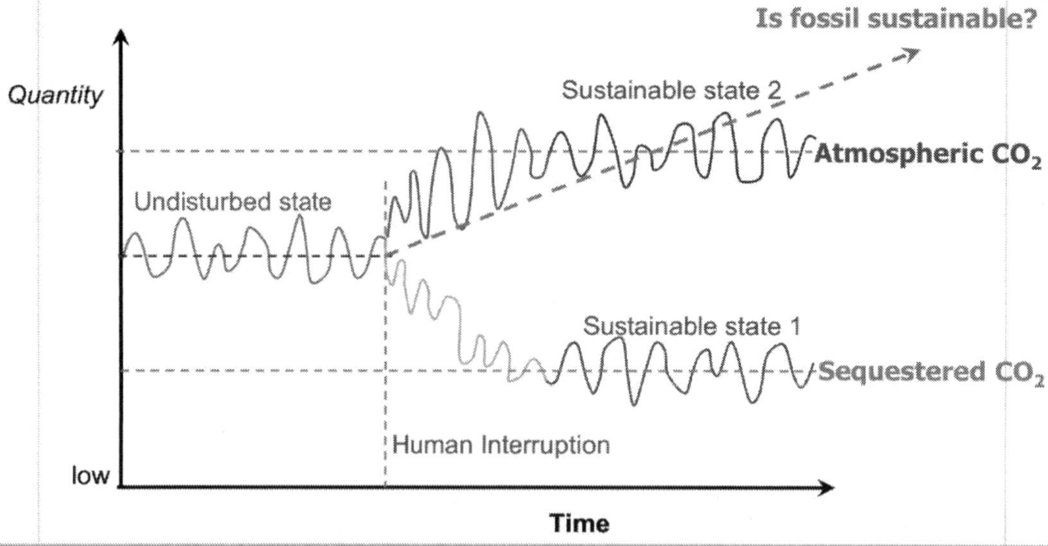

What is the Magic Biomass (Crop)?

Table 1. Recycle times and productivities of chemical/energy feedstock (Liu, 2010).

Feed stock	Recycle time	Standing Biomass tons/ha	Biomass Production tons/(ha·year)
Algae	1 month	0.9	11.25
Agricultural crops	3 month - 1 year	4.5	2.93
Temperate grasses	1 year	7.2	2.70
Savanna	1 year	18.0	4.05
Shrubs	1 – 5 years	27.0	3.15
Tropical forest	5 – 25 years	202.5	9.90
Trop. season forest	5 – 25 years	157.5	7.20
Boreal forest	25 – 80 years	90.0	3.60
Temperate deciduous	10 – 50 years	135.0	5.40
Temperate evergreen	10 – 80 years	157.5	5.85
Oil, gas and coal	*280 million years*	(38.4×10^{21} J)	(0)

Standing Biomass (~ X_∞)

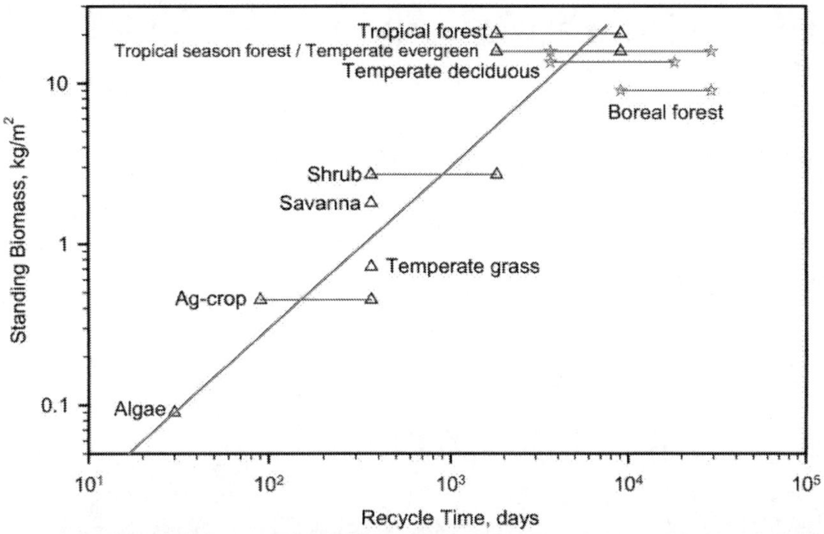

14

Biomass crop selection

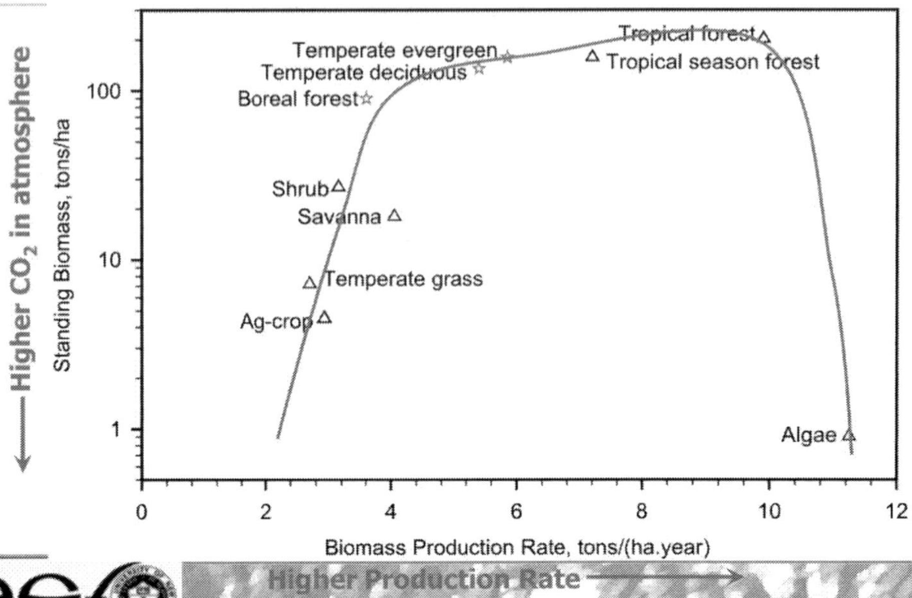

Biomass Crop Selection

- Productivity: rate of supply
- Standing biomass: CO_2 sequestration
- Water, Soil condition, Terrain
- Neighboring community
- Food needs, why "corn" is a better biomass
- Production cost
- There is no magic crop

Integrated Biorefinery:

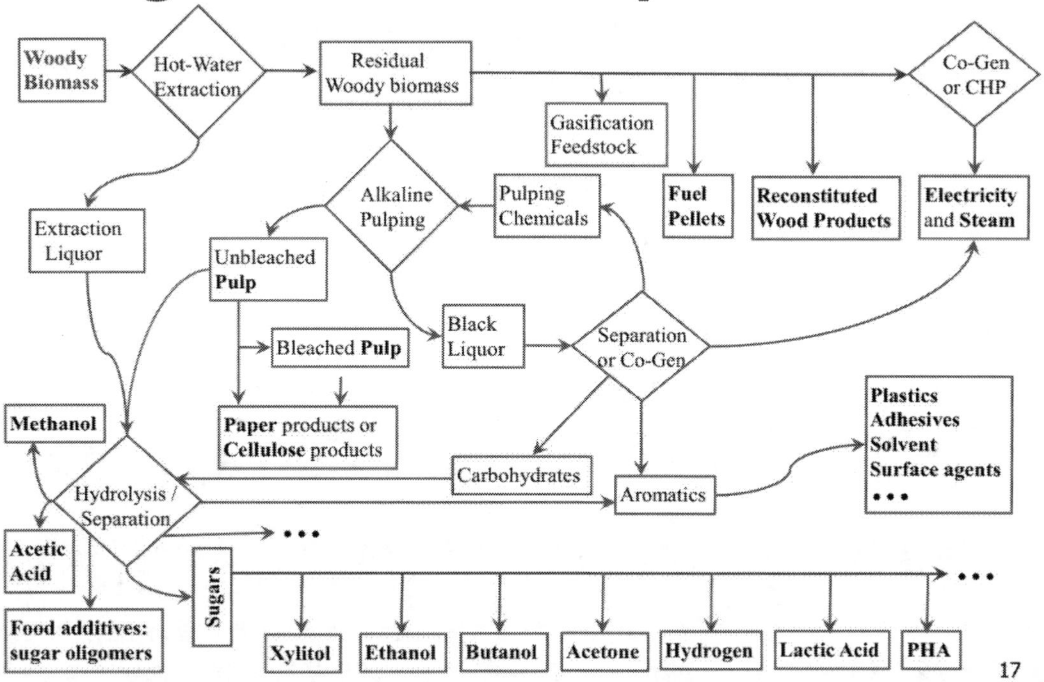

17

Ethanol: Good or Bad

- Gasoline additive: E10, going for E15
  - Commercial
  - Infrastructure established, and production facility design mature for new ones
  - Blending at distribution centers
  - Transportation
- Cereal grain use
- Physical and chemical properties

Ethanol: Platform chemical

- Jet fuel
- Bioplastics or bioploymers
- Hydrocarbons

$$CH_3CH_2OH \longrightarrow CH_2{=}CH_2 + H_2O$$
$$2CH_3CH_2OH \longrightarrow CH_2{=}CH\text{-}CH{=}CH_2 + H_2O + 2H_2$$

Wang, Y. and Liu, S. J. Bioproc. Eng. Biorefinery, 1(1), 33-43.

Ethanol: Biomass conversion

- Theoretical yield: 51% on sugars
  - Anaerobic fermentation, can tolerate O_2
  - Fermentation to high concentrations
- As compared with
- Butanol – low yield, low concentration
- PHA – low yield, high recovery cost

Fermentation: Wood extract Hydrolysate

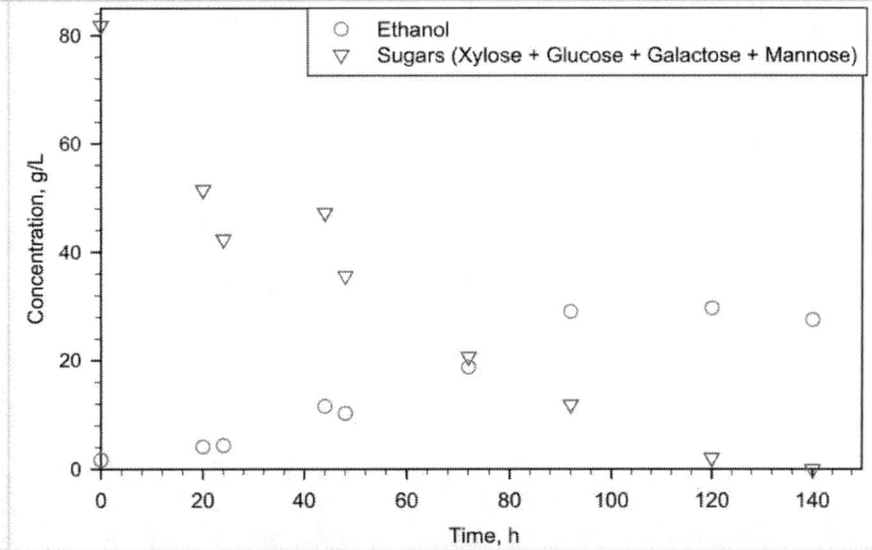

Butanol fermentation

Glucose in feed: 45 g/L

Acetone	5.2	g/L
Ethanol	1.8	g/L
Butanol	9.8	g/L
Total	16.8	g/L

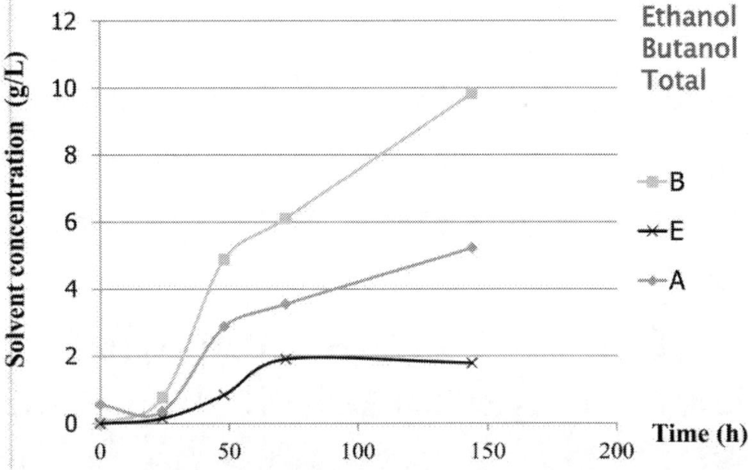

Butanol fermentation

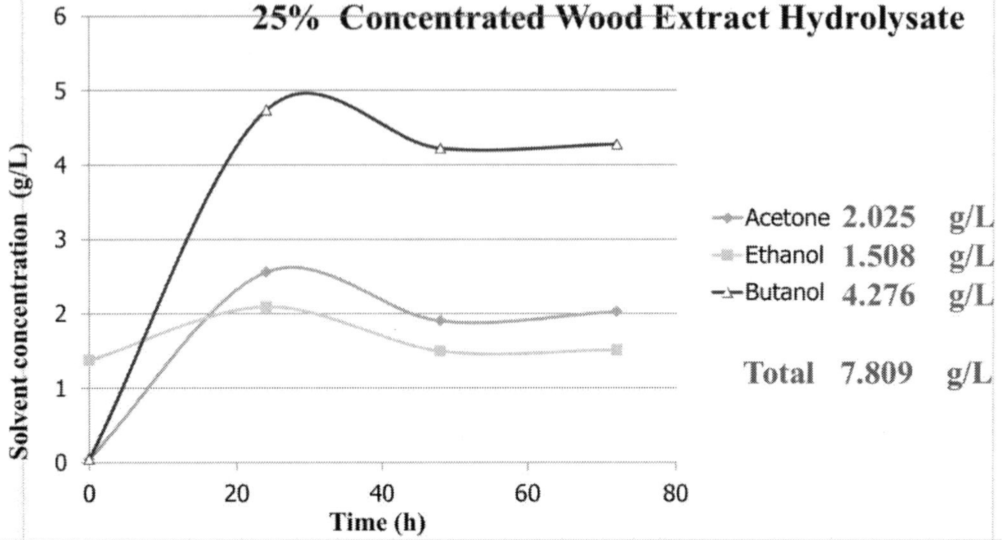

Conclusions

- Future depends on biomass, despite a hard choice to make: CO_2 is CO_2
- Biomass is both renewable and sustainable
- Ethanol is a good choice for continuing developments in renewable chemical and energy sectors
- Ethanol is a platform chemical, from which products know today from petroleum can be produced

24

Thanks!

25

Process Safety in the Classroom: The Current State of Chemical Engineering Programs at US Universities

Sean J. Dee,
Brenton L. Cox, and Russell A. Ogle
Exponent, Inc.
4580 Weaver Parkway, Warrenville, IL 60555
sdee@exponent.com, bcox@exponent.com, and rogle@exponent.com

Prepared for Presentation at
American Institute of Chemical Engineers
2014 Spring Meeting
10th Global Congress on Process Safety
New Orleans, LA
March 30 – April 2, 2014

UNPUBLISHED

Process Safety in the Classroom: The Current State of Chemical Engineering Programs at US Universities

Sean J. Dee,
Brenton L. Cox, and Russell A. Ogle
Exponent, Inc.
4580 Weaver Parkway, Warrenville, IL 60555
sdee@exponent.com, bcox@exponent.com, and rogle@exponent.com

Keywords: Process Safety, Education, Chemical Engineering, ABET

Abstract

The role of process safety in the undergraduate and graduate curriculum is a critical component of preparing chemical engineers for their future careers. In 2011, the Accreditation Board for Engineering and Technology (ABET) amended the requirements for Chemical, Biochemical, Biomolecular, and similarly named Engineering Programs. The change added a clause, indicating that the curriculum of said programs should not only include the engineering application of chemistry, physics, and/or biology to the design, analysis and control of processes, but also the hazards associated with those processes. Departments were then compelled to adjust their curricula to meet the new requirement which would go into effect during the 2012-2013 Accreditation evaluations.

Herein, we present findings related to the major milestones and challenges associated with updating the chemical engineering curriculum to include process safety components at both the undergraduate and graduate level. First, a thorough survey of publically available information regarding curriculum requirements, course syllabi, and program developments at US universities will be discussed. Next, insights from evaluating the challenges associated with adding new content to an already overloaded curriculum will be discussed. Lastly, recommendations for continued improvement in the process safety education provided to undergraduate and graduate students will be presented.

1. Introduction

The ABET certification of a degree program is a critical aspect of a university's ability to prove that it has properly educated, trained and prepared its graduates for their future careers. In the 2012 Criteria for Accrediting Engineering Programs, the curriculum criteria for Chemical Engineering programs was changed to include consideration of hazards associated with the engineering application of basic sciences. [1] Previously, safety was only addressed in outcome c of Criterion 3, which details the necessity of all engineering programs to teach students to solve design problems within various constraints including health and safety. The change in the curriculum requirement for chemical engineering programs reemphasized the importance of process safety education.

2. The Anatomy of a Chemical Engineering Curriculum

The chemical engineering curriculum requires courses in many different subjects. Specifically, ABET specifies that a chemical engineering curriculum must include a thorough grounding in basic sciences including chemistry, physics, and/or biology with some content at an advanced level. [1] ABET further elaborates that the curriculum must also include the engineering application of those sciences to the design, analysis and control of processes and the associated hazards of those processes. ABET also comments generically regarding four components that should be included in all engineering curriculum in Criterion 5. The first component specifies one year of coursework should focus on college level math and basic science. The second component calls for a year and a half of coursework bridging science and math with engineering practice and design. In both the first and second component, a year is defined as the lesser of 32 semester equivalent hours or one-fourth the total credits required for graduation. The last two components relate to a general education component that complements the technical content of the curriculum and a major design experience that allows students to apply knowledge and skills while using engineering standards and constraints.

The credit hour requirement for a chemical engineering degree varies across universities, but generally requires between 125 – 135 semester equivalent credit hours. Within a chemical engineering department, the curriculum commonly includes courses covering the following subjects:

- Thermodynamics
- Material and Energy Balances
- Transport Phenomena (Mass, Momentum, and Heat)
- Kinetics and Reaction Engineering
- Process Control
- Chemical Engineering Laboratory
- Design and/or Project
- Technical Communication

Courses covering these topics account for approximately one-third of a program's required credit hours including technical elective courses.[1] Approximately one-half of a programs required credit hours are allocated to courses in the basic sciences (physics, chemistry and math) and other engineering requirements (circuit design, computer programming, material science, etc.). General education requirements and electives outside of engineering and the basic sciences constitute the remaining credit hours.

3. Process Safety in the Undergraduate Curriculum

[1] These fractions are approximations. The one and a half year requirement set in ABET's Criterion 5 is exceeded when Chemical Engineering courses are combined with other engineering electives and core requirements. For the purposes of this paper the inclusion of a safety component class was assumed to be the responsibility of the chemical engineering department and not a general engineering course.

Finding room within the curriculum to incorporate process safety material is one of the primary challenges associated with the new language in the ABET requirements. Typically, students take between 15-18 credit hours per semester in a chemical engineering program, equating to 4-6 different courses. This leaves very little extra room in the typical student's course load, and can detract from student's opportunities to take electives and non-engineering related coursework. A completely new course requirement can result in the necessity to cut or combine other existing courses. Incorporating safety topics into existing courses can also prove difficult, specifically if professors already feel that under existing curriculum requirements, the technical material in their courses is not given the appropriate allotment of time or focus.

Including process safety in the chemical engineering curriculum can be accomplished by adding additional course(s), integrating process safety topics into existing courses, or a combination of both approaches. [2] Pintar concluded in 1999 that a combination of both approaches was the most effective way to teach chemical process safety to undergraduate chemical engineering students. Pintar argued that the effectiveness came from sensitizing students to safety issues in the dedicated course, and exposing them to practical issues through integration in courses such as unit operations. Herein, we present examples of both dedicated course topics and integration techniques observed in the chemical engineering curriculum at various ABET accredited programs.

Several universities have demonstrated the ability to find room in the curriculum for courses focused on process safety topics. Pintar noted that environmental engineering issues have significant overlap with process safety topics including toxicology, industrial hygiene, flammability, fires and explosions, dispersion models and source models, design of emergency relief systems, hazard identification, risk assessment, accident investigation, and case histories. Thus, process safety can be addressed in technical elective sequences that focus on environmental engineering issues. Several examples of this approach found by searching department websites are listed below.[2]

Technical Electives

- University of Minnesota
    - Industrial Engineering Technical Elective Sequence
        - Engineering Safety (IE 5513)
- University of California, Berkeley
    - Designated Emphasis in Energy Science and Technology
        - Nuclear Reactor Safety (NE167/NE267)
        - Safety Assessment for Geological Disposal of Radioactive Wastes (NE224)
- University of Texas, Austin
    - Technical Option Area 3: Environmental Engineering
        - Hazardous Waste Management (CE 346K)

Universities have also found ways to incorporate process safety into the curriculum with dedicated courses covering the topics suggested by Pintar as well as dust explosions, chemical

[2] These two lists are not meant to be exhaustive and are only provided as examples.

reactivity hazards, chemical plant security, and inherently safer design. [3] These courses can range from one to three credit hours. Several examples found by searching the university chemical engineering department's website are included below.

Dedicated Courses

- Georgia Institute of Technology
    - Process Safety (CHBE 4515)
- Michigan Technological University
    - Chemical Process Safety and the Environment (CM4310)
- Texas A&M University
    - Risk Analysis in Safety Engineering (CHEN430)
    - Process Safety Engineering (CHEN455)
    - Quantitative Risk Analysis in Safety Engineering (CHEN 460)
- Pennsylvania State University
    - Chemical Process Safety (CHE 452)
- The Ohio State University
    - Chemical Process Safety (CBE 5755)
- West Virginia University
    - Chemical Process Safety (CHE 475)

In addition to dedicated courses, process safety topics can be integrated into existing courses. The most common integration point is the Project or Design course, usually taken in the fourth year of the undergraduate program. In this course students are taught to apply the skills and knowledge of their chemical engineer expertise to aspects of building and designing a chemical plant, process, or product. At the Massachusetts Institute of Technology, safety issues are covered in lectures "Safety as an Industrial Concern," "Safety Assessment Models," and "Chemical Process Safety: HAZOP Analyses" in Integrated Chemical Engineering I and II. [4], [5] Pintar also noted examples of how safety topics were integrated in the Capstone Chemical Engineering Laboratory at Michigan Technological University. [6] Among the topics included were OHSA compliance, safety meetings, Job Safety Assessment Forms, lock out tag out, and management of change practices.

In addition to these two approaches, there are additional methods that can be used to reinforce process safety at the undergraduate level. In 2013, West Virginia University chemical engineering students participated in a two-day Process Safety Boot Camp, similar to the four day course offered by CCPS/AIChE. [7] In 2008, New Mexico State University integrated the Safety and Chemical Engineering Education (SAChE) program into their curriculum requiring students to obtain various certifications at different stages during pursuit of their degree. [8] As a result, NMSU has seen their chemical engineering students nearly double their scores on the Safety, Health, and Environmental portion of the Fundamentals of Engineering Exam from 2008-2011.

4. Graduate Program Considerations

Graduate and undergraduate programs fall under the same mission statement of a chemical engineering department. Generically speaking, a department mission statement usually focuses on furthering the field of chemical engineering through a combination of conducting innovative original research and teaching the department's students to become future researchers and leaders in industry, academia, and government. The emphasis in graduate level programs, specifically at the PhD level, is usually on teaching graduate students to become independent researchers. Therefore, most of the exposure that graduate students have to safety concepts within the department is usually related to laboratory safety and not process safety. This limited exposure to process safety related concepts creates both a first order and second order effect on the future of the chemical engineering industry. First, for graduate students who enter industry related jobs, core concepts of process safety have not been emphasized or reinforced during the graduate school experience. This is particularly challenging in research and development groups, because some of the largest impacts on process safety can be achieved in the initial design phases of process technologies. In these settings, concepts including inherently safety design, as well as other risk mitigation strategies can create large lasting impact on developing technologies.

The second order effect of the lack of emphasis on process safety in graduate programs is highlighted by Pintar. [2] Speaking of the undergraduate learning experience, Pintar highlights the need for process safety concepts to be embraced by faculty, and departments to develop a "safety culture" to reinforce the importance of integrating process safety concepts in the curriculum. One approach to attempt to build a "safety culture" in academia is to start enforcing the culture in graduate programs, so that the future faculty in universities have increased exposure, knowledge, and understanding of process safety issues.

The laboratory research environment provides an excellent opportunity to incorporate the core process safety concepts of hazard identification, evaluation, and control into the graduate program. In 2011, the U.S. Chemical Safety and Hazard Investigation Board (Chemical Safety Board) published a case study of a chemical laboratory explosion incident at Texas Tech University that resulted in serious injuries to a graduate student. [9] In their report, the Chemical Safety Board indicated that they had amassed information of over 120 university laboratory incidents involving fires, explosions, or toxic exposures. In response to this report, the American Chemical Society (ACS) issued a draft guidance document on hazard identification, evaluation, and control in the chemical laboratory. [10] One can easily envision the incorporation of a laboratory safety discussion into the routine project meetings between graduate students and their advisers or peers. Such routine practice would serve as a tangible mechanism for reinforcing fundamental process safety principles. It may also be possible to incorporate a similar strategy into the undergraduate laboratories.

5. Challenges Moving Forward

Several key challenges arise when attempting to incorporate process safety into the undergraduate and graduate educational programs at universities. At the outset of the program, departments must accept the notion that process safety concepts *will change* over time. This is in stark contrast to the core concepts of several key chemical engineering courses, such as thermodynamics, where a syllabus, lesson plan, and lecture material can be reused each time a professor teaches a course. Topics in process safety will likely need to be continually updated and adapted as the body of knowledge expands and matures.

The difficulty in finding time within an already full load curriculum has already been discussed. Several examples of universities that have found ways to incorporate process safety into the undergraduate program have been provided. At a graduate level, the specific focus on the research thesis makes the inclusion of a process safety component more complex. Any attempt to increase the focus on process safety in the graduate experience would likely benefit from external support from funding agencies. Specific strategies that may garner greater participation at the departmental level include tailoring requests for proposals to include a safety aspect, or asking for specific research programs to examine process safety related research problems. An allocation of research funding from government agencies for investigations relevant to process safety would likely spur greater faculty and graduate student interest.

Lastly, any improvements in process safety education at the university level cannot serve as a replacement for employer provided safety training programs. Training engineers to perform their jobs safely is a responsibility of both employers and their employees. While process safety concepts can be taught in the classroom, site and process specific hazards will always require additional considerations. However, armed with advance knowledge of process safety concepts, techniques, and approaches, the future engineers of industry can be better prepared to improve safety practices they encounter in their careers. This will be especially true if the students can learn how chemical engineering fundamentals are the foundation for process safety knowledge.

6. Conclusions and Recommendations

In response to the ABET requirements for certification, American universities and colleges have adopted a variety of strategies for incorporating process safety into the chemical engineering curriculum. It is apparent that given the diversity of curricula at these schools that there will be no single strategy that will work for every program. Each program will have to assess their unique needs and adapt a process safety pedagogical strategy that best suits them, and this strategy will likely need to change over time.

The review of process safety education in chemical engineering programs results in several potential recommendations. First, departments should consider assessing the process safety culture within their individual program to understand how the faculty, undergraduate, and graduate students value process safety as part of chemical engineering education. Programs should then consider the various approaches previously discussed to further promote the process safety culture within the department. Potential process safety components can be included in both the undergraduate and graduate programs. Finally, departments should consider tracking a metric tied to process safety education, such as relevant scores on the Fundamentals of Engineering exam or a sub-metric of the ABET Criterion 3 outcome c. After metric evaluation, programs can reassess and refine their programs to meet their process safety cultural mission and goals.

8. References

[1] ABET. *Criteria for Accrediting Engineering Programs, 2012-2013*, http://www.abet.org (accessed Jan. 6, 2014).

[2] Pintar, A.J., "Teaching Chemical Process Safety: A Separate Course Versus Integration into Existing Courses," Conference Proceedings, 1999 ASEE Annual Meeting, Charlotte, NC (June, 1999).

[3] Johnson R.W., "The Ohio State University Process Safety Education", https://chemsecurity.sandia.gov/discussion-board/RWJ%20extended%20abstract.pdf. (accessed Jan 6, 2014).

[4] Johnston, B. 10.490 Integrated Chemical Engineering I, Fall 2006. (MIT OpenCourseWare: Massachusetts Institute of Technology), http://ocw.mit.edu/courses/chemical-engineering/10-490-integrated-chemical-engineering-i-fall-2006 (accessed Jan. 6, 2014) License: Creative Commons BY-NC-SA.

[5] McRae, Gregory. 10.491 Integrated Chemical Engineering II, Spring 2006. (MIT OpenCourseWare: Massachusetts Institute of Technology), http://ocw.mit.edu/courses/chemical-engineering/10-491-integrated-chemical-engineering-ii-spring-2006 (accessed Jan. 6, 2014) License: Creative Commons BY-NC-SA.

[6] Pintar, A.J., "Capstone Chemical Engineering Laboratory Courses at Michigan Tech," Conference Proceedings, 2001 ASEE Annual Meeting, Albuquerque, NM (June, 2001).

[7] Press Release. "WVU offers first-of-its-kind Process Safety Boot Camp to chemical engineering students," http://wvutoday.wvu.edu/n/2013/04/19/wvu-offers-first-of-its-kind-process-safety-boot-camp-to-chemical-engineering-students. (accessed Jan. 6, 2014).

[8] ABET Self-Study Report for the Chemical Engineering Program at New Mexico State University, Las Cruces, New Mexico, June 25, 2012. http://chemeng.nmsu.edu/NMSUChEProgramSelfStudyReport2012.pdf (accessed Jan. 6 2014).

[9] Chemical Safety Board, Texas Tech University Laboratory Explosion, U.S. Chemical Safety and Hazard Investigation Board, Washington, DC (2011).

[10] ACS, Identifying and Evaluating Hazards in Research Laboratories, American Chemical Society, Washington, DC (2013).

350089 PSM - Just a Job, or a Calling?

Abstract:

When you wake up and look in the mirror as you prepare for work, does anything special cross your mind? Me, I look at my reflection and say to yourself, "I'm going to make a difference today; today may be the day I save a life." Being a process safety professional can be a demanding position. However, for many of us in the field we speak of it as a calling; having a passion for what we do and sharing that with the personnel in the plants where we work.

The objective of this paper is to provide a sense of passion that we experience to young professionals either starting out in or considering a career in process safety. Through sharing elemental experiences of the PSM standard, the goal is to touch at least one young professional and have them feel that they have the calling as well.

348515 Preparing and Presenting a College Workshop on Process Hazard Analysis

Abstract:

In October 2011 the ABET college accreditation organization approved a Curriculum Statement requiring Chemical Engineering programs to teach certain aspects of Process Safety Management. Engineering college professors are well qualified to teach the concepts of Process Hazard Analysis (PHA), but they may not have practical experience, creating an opportunity for safety professionals to improve preparation for the PHA responsibilities engineering students are likely to have in their careers.

The proposed paper will review a college workshop on PHA which was conducted at South Dakota Tech in 2012 and 2013. The goal is to provide industry safety professionals and college professors with information that may be useful in presenting a workshop. Topics will include what the college needs from the safety professional, how to approach the college and an outline of the workshop contents. Emphasis will be placed on examples of the practical aspects of PHA that someone with facilitation experience can provide to the students. The strategy for maximizing student participation using small teams and team advisors with industrial experience will be reviewed. The paper will be a joint effort between a college professor and a person who has facilitated PHA's in the chemical and refining industries.

353750 Introduction to Project Management Fundamentals

Abstract:

This presentation introduces the fundamentals of project management. It will be especially useful for new engineers, young professionals, and advanced students, as well as anyone who serves or will served on project teams or leads those teams. We will discuss the importance of people, the importance of setting excellent project objectives, and the importance of good planning. Working with people is perhaps the most difficult. Suggestions will be given on how to better communicate and better achieve unity with project team members. We will discuss the real meaning of the word "Communication" and how that meaning can change the whole approach we take to interacting with people in a productive way. Important ideas and tools for developing excellent project objectives will be presented, including the use of SMART objectives, the triple constraints, and tolerance diagrams. An outline of the project planning process will be reviewed. This planning process includes the use of work breakdown structures, network diagrams, and Gantt charts, as well as resource histograms. Also, we will briefly discuss doing fatal-flaw analysis.

353751 Becoming a Real Team to Accomplish Projects

Abstract:

In project management a major effort goes into becoming a real team! A team is not just a group of individuals! It is a very special group of individuals who have worked and learned to become united in their objectives and willingness to work to accomplish those objectives. This paper will discuss in depth the important principles of people interaction skills needed for people to coalesce into a working unit of synergy. We will discuss communication, what it means, and how to use it. We will talk about leading effective team meetings. We will discuss how to look at issues from other people's perspectives in order to better understand them and work with them to accomplish objectives. We will discuss listening and barriers to effective listening. All of this will go toward developing a new vision of what it means to be an effective and united team and give important meaning to the word "teamwork."

353907 Project Management:Things You Should Think About

Abstract:

"Beyond the critical fundamentals of communication, working with people, setting good objectives, and the importance of good planning, there are some other topics that are important for project managers to consider. Some of these topics are managing risks, managing conflicts, dealing with contracts, managing meetings, situational leadership, taking control of yourself, and so on. This presentation will briefly talk about many of these aspects—each one is really a major topic by itself, but we will give some key helpful hints about some things you should consider as you begin working on project teams, become a new project manager, or perhaps brush up on things to consider even as an experienced project manager. This paper will be helpful to all of these people. We'll give some important rules of effective project management, some key ideas about managing conflict, and some basics on managing risk, and hopefully aid you in your path toward effective project management!"

348790 Leadership Starts Now!

Abstract:

Often people think that leadership begins when there is a promotion to a position of authority with direct personnel responsibilities. Some believe that leadership can simply be switched on after assuming a title. Nothing could be further from the truth and to follow that prescription can be detrimental for young professionals. As new engineers come into an organization, it is essential they understand the informal opportunities and subtle behaviors that reveal their leadership aptitude.

This presentation will discuss the "hidden leadership" opportunities that are woven into your everyday roles, how to recognize these opportunities and take advantage of them. Don't wait until you are in a formal leadership position. Learn how to be a leader at every opportunity; starting now!

The main target audience is entry level engineers and professionals who are beginning their careers across all industry types.

During this presentation you will learn:

- The difference between managing and leading
- Qualities and behaviors that are recognized as leadership
- Common errors that result in missed opportunities for leadership
- Methods for leading from a non-leadership role or position
- How to recognize "hidden leadership" opportunities
  Steps you should take NOW to develop yourself as a leader

PROSYS™ Maximize Your Plant.

Operator Errors and What Can be Done to Minimize

Global Congress on Process Safety

March 31, 2014

10TH GLOBAL CONGRESS ON PROCESS SAFETY

PROSYS™ Maximize Your Plant.™

10TH GLOBAL CONGRESS ON PROCESS SAFETY

Presenter

Tom Nolan

- **Graduated from Ohio University with B.S. in Chemical Engineering**
- **24 years experience in chemical process industry in a variety of roles**
- **MAIC and DFSS Six Sigma Black Belt**

PROSYS Maximize Your Plant™

The Cost of Errors

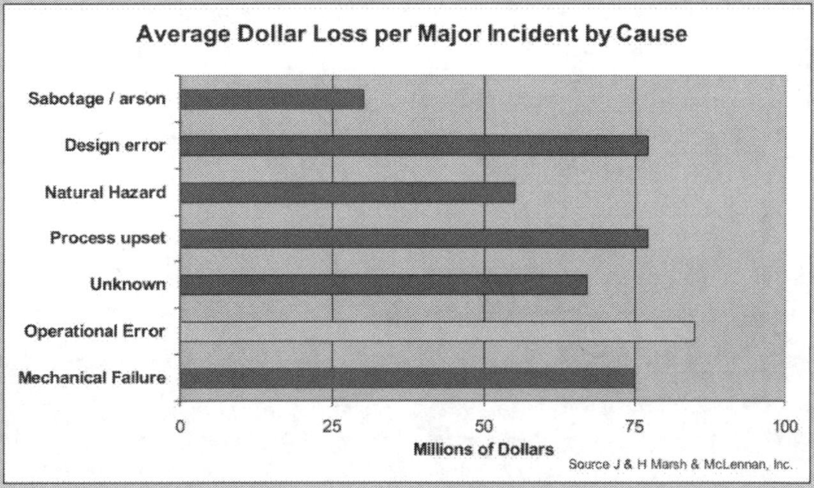

PROSYS Maximize Your Plant™

The Cost of Operator Errors

- **ASM estimates total loss due to operator error is $8B per year**
- **Chemical Safety Topical Committee - average of one chemical incident per day - cost of over $2 million per incident to comply with requirements**
- **Errors cause 42% of unscheduled shutdowns**
- **70% of process incidents occur during start-up or shutdown**

PROSYS Maximize Your Plant™

10th GLOBAL CONGRESS ON PROCESS SAFETY

Reasons for Errors

- Lack of Skill
- Lack of Knowledge
- Carelessness
- System Design
  - Operator set up to make errors by inappropriate design – built in errors
  - Errors that are predictable are preventable by better design

PROSYS Maximize Your Plant™

10th GLOBAL CONGRESS ON PROCESS SAFETY

Focus of this Discussion

- System Design
  - **Alarm Management**
  - **Operator Graphics**
  - Operator set up to make errors by inappropriate design – built in errors
  - Errors that are predictable are preventable by better design

PROSYS Maximize Your Plant™

10th GLOBAL CONGRESS ON PROCESS SAFETY

Alarm Floods

In a number of industrial incidents, alarm floods were identified as a significant contributing cause to the incident…

As found by EEMUA in 1999 and CSB

Alarm Flood defined by ISA 18.2 as -
"10 or more annunciated alarms in any 10 minute period per operator"

PROSYS Maximize Your Plant™

10th GLOBAL CONGRESS ON PROCESS SAFETY

Why do Alarm Floods Occur?

One reason is not providing dynamic alarm management

- **Alarms need to indicate abnormal situations that require operator action**
- **Processes do not operate in one state**
- **What is normal vs abnormal changes with state**
- **Alarms are typically configured for run therefore many alarms are triggered upon a Change of State: Run to Shutdown**
- **Many of the alarms are not applicable or actionable for the new state impleading the operators ability to act quickly on what is important**

PROSYS Maximize Your Plant™

10th Global Congress on Process Safety

What is Impacted by Alarm Floods?

- Product quality
- Operability or profitability of the process
- Loss of equipment
- Loss of containment – environmental releases
- Injury and loss of life in plant or community

PROSYS Maximize Your Plant™

10th Global Congress on Process Safety

What Makes Alarm Floods so Dangerous?

Can be a problem for three reasons:

- A deluge of alarms can cause critical alarms to be missed
- Floods can be a significant distraction when dealing with process upsets
- Can be an indicator of larger systemic safety issues

PROSYS Maximize Your Plant™

10th GLOBAL CONGRESS ON PROCESS SAFETY

Impacting Alarm Management Design

- Alarm rationalization is not a process to eliminate alarms - its about quality
- Good rationalization will add alarms when appropriate
- Typical (Static) rationalization is only for run mode
- Dynamic rationalization considers all plant modes

PROSYS Maximize Your Plant™

10th GLOBAL CONGRESS ON PROCESS SAFETY

Dynamic Alarm Management

- Dynamic rationalization does everything a static rationalization does plus asks "when" for each alarm
- Requires dynamic software to make changes based upon operating mode of the plant
- Eliminates redundant alarms and lowers operator loading during transitions
- Only alarms what is abnormal and actionable for the given state

PROSYS™ Maximize Your Plant™

10th GLOBAL CONGRESS ON PROCESS SAFETY

Actual Performance Metrics vs ISA 18.2

Type	Avg Alarm Rate/hr	Low/Hi of Avg Rate/hr	Peak Alarm Rate/hr	Low/Hi of Peak Rate/hr	% Time in Flood	Standing Alarms >24 hrs
Before Rationalization	30	Low=6.2 Hi=61	638	Low=152 Hi=2402	17.2%	9
ISA 18.2 Target Metrics	**6**	**---**	**<60**	**---**	**<1%**	**<5**
After Dynamic Alarm Management	2	Low=0.09 Hi=2.5	25	Low=13 Hi=42	0.25%	4

PROSYS™ Maximize Your Plant™

10th GLOBAL CONGRESS ON PROCESS SAFETY

Results After Dynamic Alarm Management

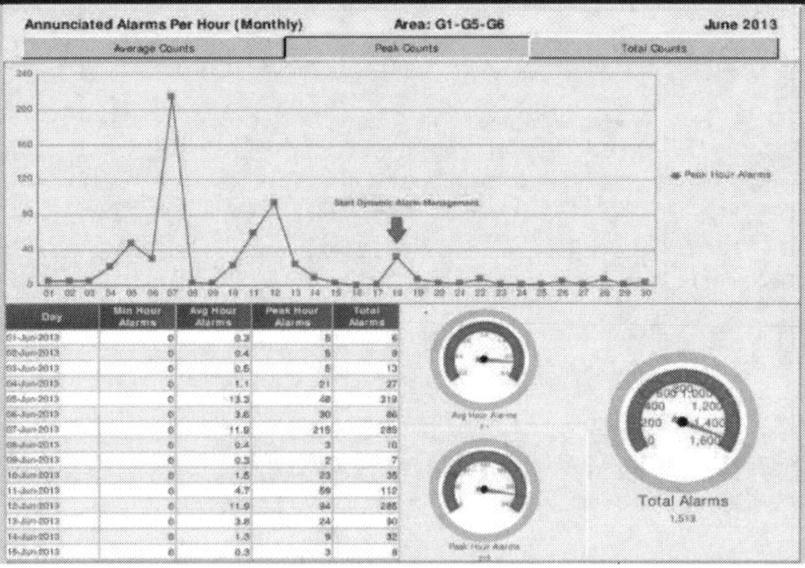

PROSYS Maximize Your Plant.

Operator Graphics – The Risk of Changing Focus

- Distraction of changing focus from process graphic to faceplate window
- Added workload from managing multiple open windows
- Increased probability of errors when changing values for un-intended tag with multiple faceplates open
- All of above become multiplied and more complex when process is transitioning from one state to another

PROSYS Maximize Your Plant.

Faceplate Operation

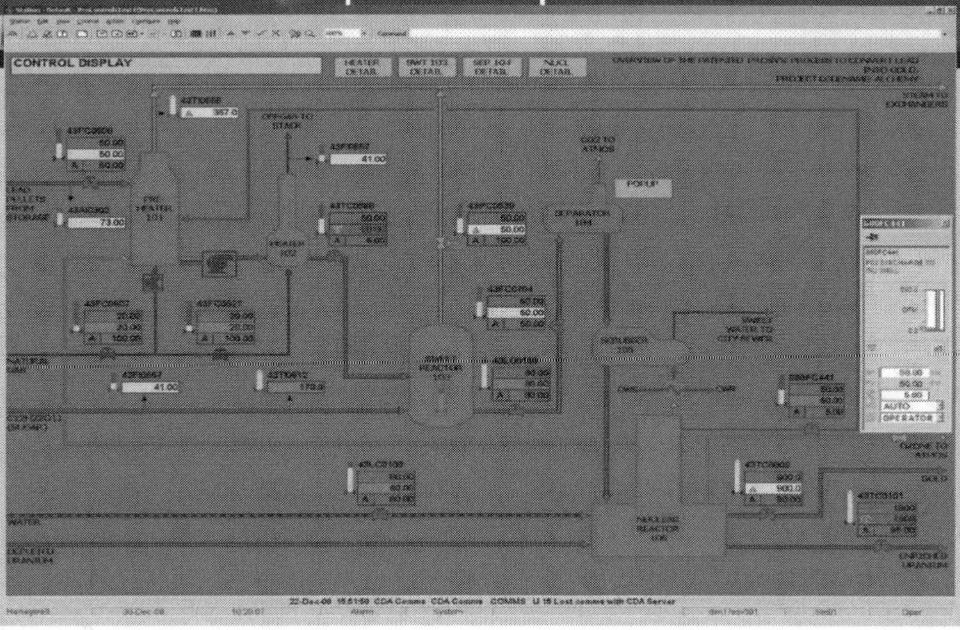

PROSYS Maximize Your Plant™

Potential for Errors

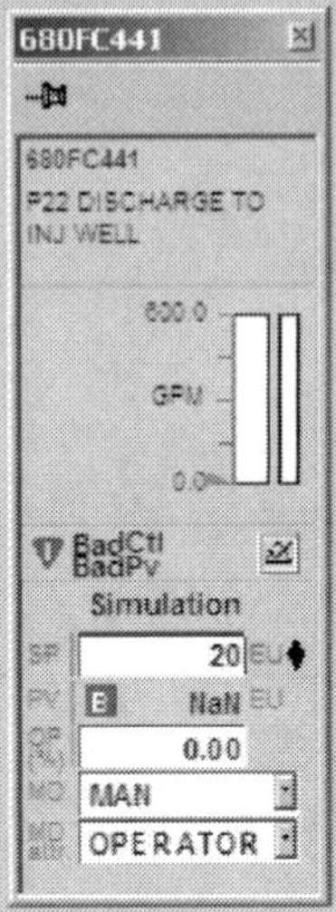

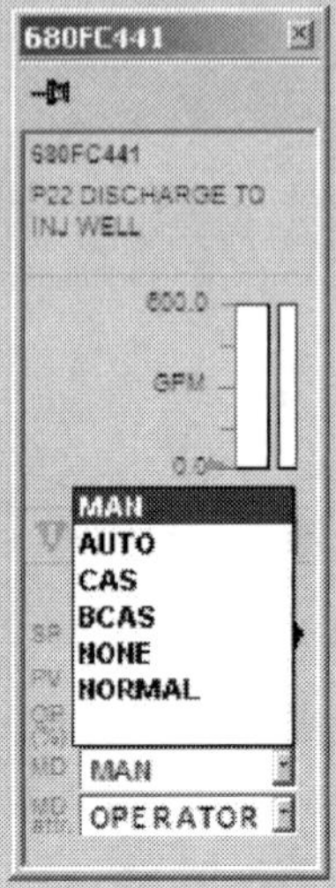

Faceplate Design -can introduce additional errors

- Mode drop down list covers SP, PV and OP values
- Operator may select wrong Mode if SP, PV and OP values are hidden from view
- Mode list offers more options than necessary

PROSYS Maximize Your Plant™

Direct Entry Operation

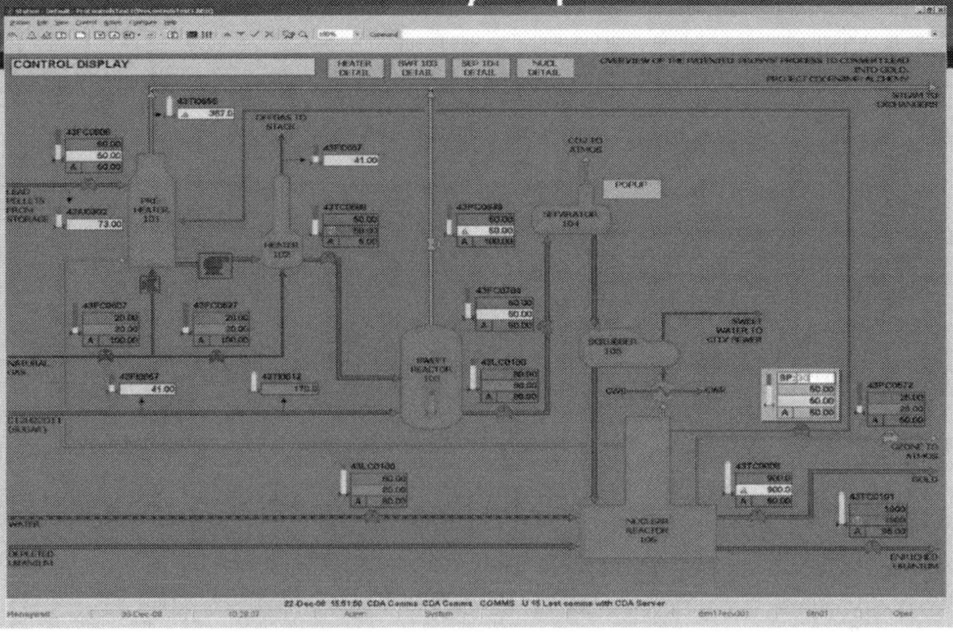

PROSYS Maximize Your Plant™

10th GLOBAL CONGRESS ON PROCESS SAFETY

Direct Entry Fields

Values can be entered by:

- Select point
- Typing value via keypad
- Pressing enter

Key Factors:

- By maintaining focus of points after value change, it improves safety for input errors, multiple clicking functions and is more time efficient
- Easily recognizable operator enterable fields

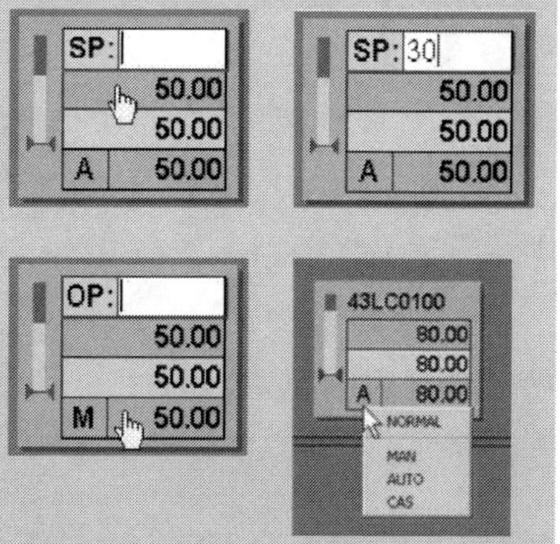

PROSYS Maximize Your Plant™

10th GLOBAL CONGRESS ON PROCESS SAFETY

Evaluation of Methods

Comparison of Faceplate to Direct Entry

- **Keystroke Level Analysis**
- **Operator Loading Analysis**
- **Risk Analysis**

Keystroke Level Model

Symbol	Time (s)	Description
K	0.28	Keystroke
P	1.1	Point to object
BB	0.2	Click on object
H	0.4	Home hands on keyboard or mouse
M	1.2+	Mental act or routine thinking

PROSYS™ Maximize Your Plant™

Estimate Execution Time Analysis

- **Evaluates execution time by an operator comparing faceplates to direct entry for SP/OP and Mode Changes**

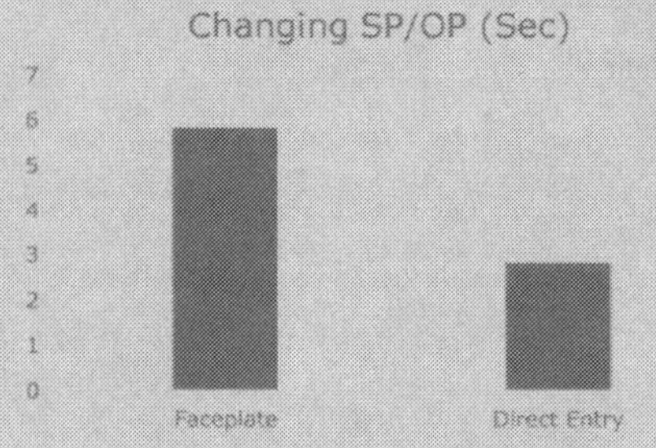

- **50 % Reduction**

PROSYS Maximize Your Plant™

10th GLOBAL CONGRESS ON PROCESS SAFETY

Estimated Operator Loading Analysis

- **Evaluates Mental and Physical Load on Operator**
  - Tasks that require very little thought are removed – such as (BB), (H)

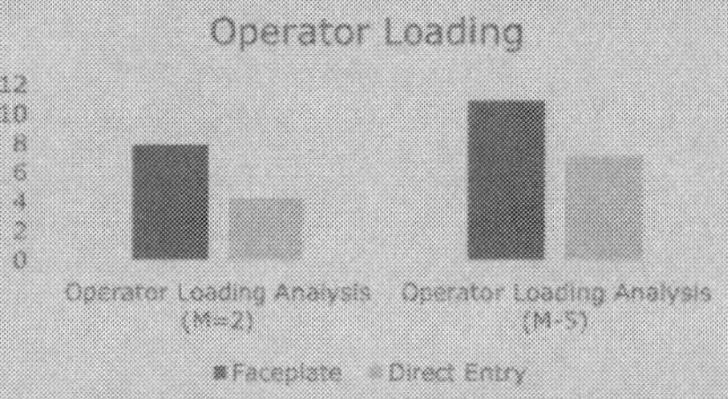

  - Mental and Physical tasks are weighted based on the amount of fatigue they cause – assigned a weight of M=2.0 to 5.0

PROSYS Maximize Your Plant™

10th GLOBAL CONGRESS ON PROCESS SAFETY

Risk Analysis

- **Identifies actions where operator entry errors can occur**

PROSYS Maximize Your Plant™

10th GLOBAL CONGRESS ON PROCESS SAFETY

Risk Analysis

Faceplate operation KLM for changing SP/OP	Potential Error
Move to shape (P)	Select wrong point
Click on shape (BB)	
Move to faceplate (P)	Point in faceplate is previous point
Click on SP/OP field (BB)	Change wrong parameter
Move hands to keyboard (H)	
Type in value and press enter (4K)	Mistype value and press enter
Move hands to mouse (H)	
Move to close faceplate (P)	
Click to close faceplate (BB)	
Potential error count	**4**

PROSYS Maximize Your Plant™

10th GLOBAL CONGRESS ON PROCESS SAFETY

Risk Analysis

Direct Entry operation KLM for changing SP/OP	Potential Error
Move to SP/OP (P)	Select wrong parameter or point
Click on SP/OP field (BB)	
Move hands to keyboard (H)	
Type in value and press enter (4K)	Mistype value and press enter
Potential error count	**2**

PROSYS Maximize Your Plant.™

10th GLOBAL CONGRESS ON PROCESS SAFETY

Direct Entry Vs Faceplate

Summary of Benefits

- **Execution Time SP/OP Changes - 51.5% less**
- **Execution Time for Mode Changes – 50% less**
- **Operator Loading – 33 % to 47% less**
- **Risk of Errors – 50% less**

PROSYS Maximize Your Plant.™

10th GLOBAL CONGRESS ON PROCESS SAFETY

Conclusion

- **Many industrial errors are a result of operators using control systems with flawed designs**
  - Poor design and performance by alarm management distracts operators and/or occludes critical alarms
  - Operator graphics with multiple open faceplates can cause a change intended for one controller to be entered into the faceplate of another
- **The two mechanisms listed above are predictable and therefore preventable through better design available today**

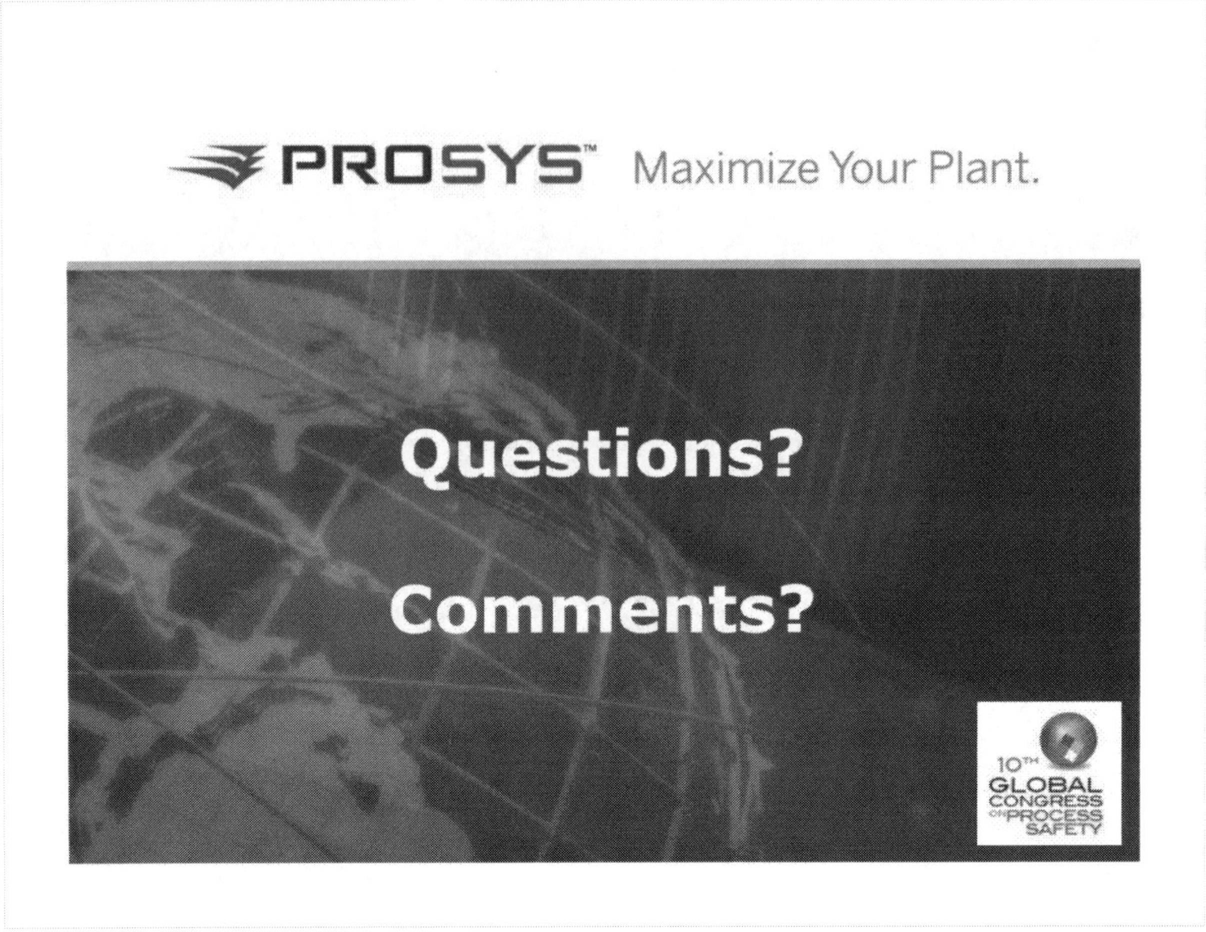
PROSYS™ Maximize Your Plant.
Questions?
Comments?
10TH GLOBAL CONGRESS ON PROCESS SAFETY

Problem Statement

visium KMS
Process Safety Management Solution

- Any organisation can improve by leveraging knowledge and lessons learned.
- Knowledge can be externally or hidden locally, in a global enterprise, where lessons learned are difficult to share beyond their initial impact.
- There is so much data, information and knowledge (opportunities) coming at your company that can make you stronger or make you look like your not paying attention.
- Not paying attention or managing the available knowledge, can sink your ship… nobody wants that!

Expert Insight Delivered 2

Osys
Part of the Rolls-Royce Group

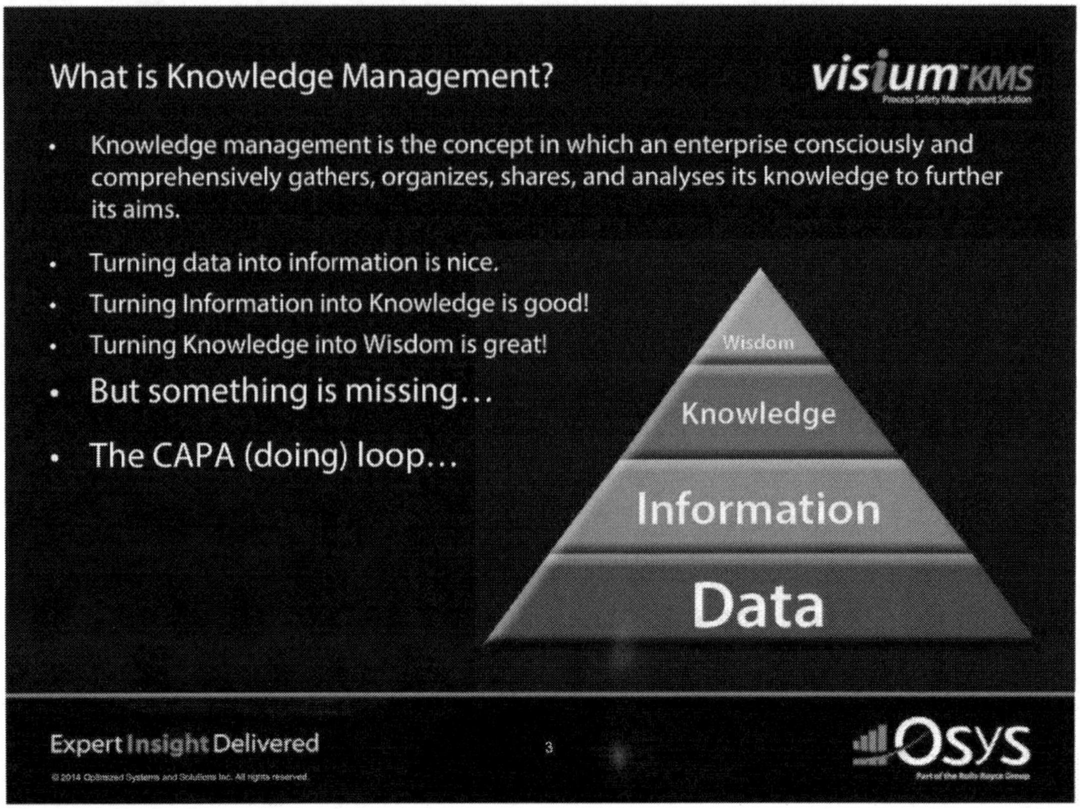
What is Knowledge Management?
visium KMS
Knowledge management is the concept in which an enterprise consciously and comprehensively gathers, organizes, shares, and analyses its knowledge to further its aims.
Turning data into information is nice.
Turning Information into Knowledge is good!
Turning Knowledge into Wisdom is great!
But something is missing…
The CAPA (doing) loop…
Wisdom
Knowledge
Information
Data
Expert Insight Delivered
3
Osys

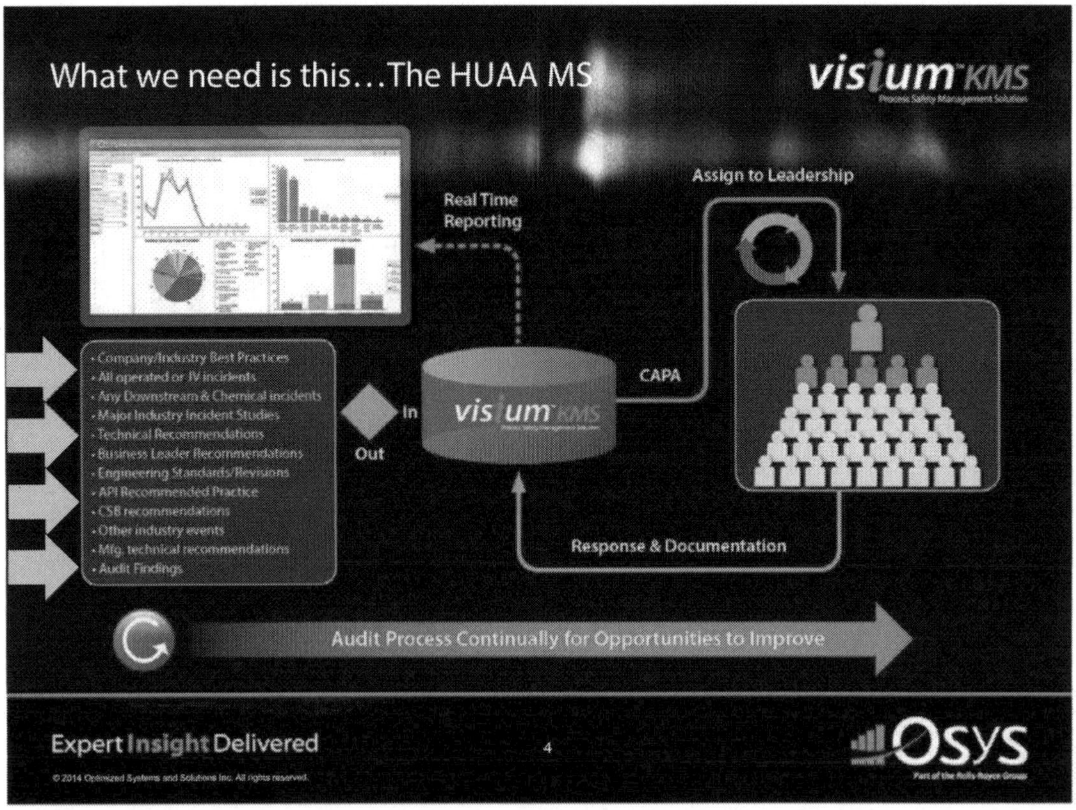
What we need is this…The HUAA MS
visium KMS
Real Time Reporting
Assign to Leadership
• Company/Industry Best Practices
• All operated or JV incidents
• Any Downstream & Chemical incidents
• Major Industry Incident Studies
• Technical Recommendations
• Business Leader Recommendations
• Engineering Standards/Revisions
• API Recommended Practice
• CSB recommendations
• Other industry events
• Mfg. technical recommendations
• Audit Findings
In
Out
visium KMS
CAPA
Response & Documentation
Audit Process Continually for Opportunities to Improve
Expert Insight Delivered
4
Osys

What is the HUAA Management System…

visium KMS
Process Safety Management Solution

HUAA – Heard, Understood, Acknowledged, Actioned.

1. Opportunities identified (heard) by your experts,
2. Captured by an electronic system,
3. Expertly Reviewed (understood),
4. Accepted or rejected (acknowledged) and
5. Assigned (actioned) to Sr. Managers for resolution and tracked to closure.

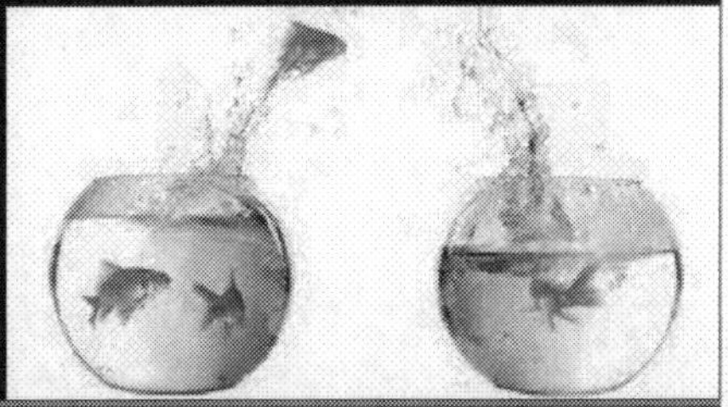

Expert Insight Delivered

5

OSYS
Part of the Rolls-Royce Group

HUAA – Management Systems - Detail

visium KMS
Process Safety Management Solution

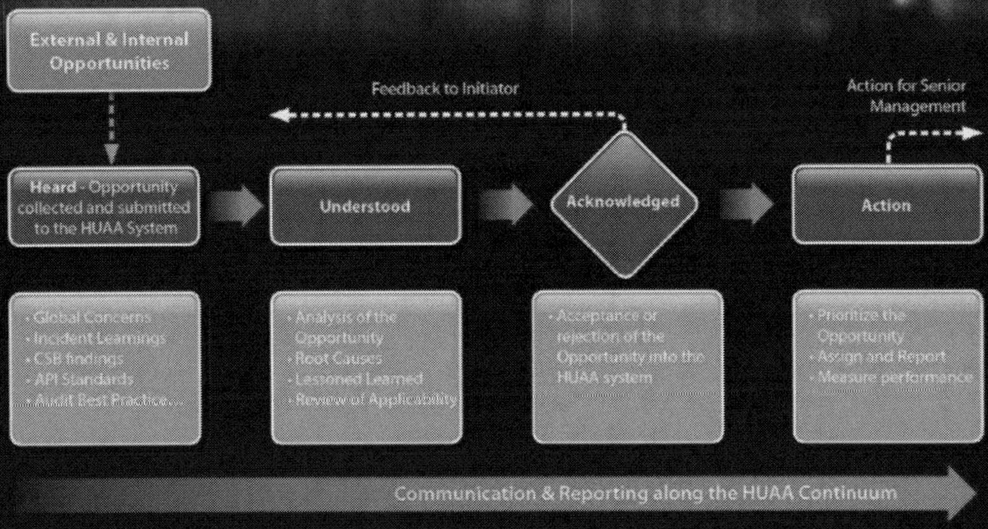

Expert Insight Delivered

6

OSYS
Part of the Rolls-Royce Group

HUAA Results

visium KMS
Process Safety Management Solution

- Efficient Collection - Important Opportunities are identified and collected from many sources (internal and external), and entered into the HUAA management system.
- Expert Vetting - These opportunities are vetted, parsed and resolved for inclusion or exclusion; documenting & communicating both.
- Accountable Assignment - Included opportunities are systematically assigned to Sr. Leaders and managed across the enterprise at SVP and C levels.
- Action - Improvement will be assured through the accountable actions of these leaders.
- Results - The organization will gain competitive advantage by active listening, learning from their experience, as well as the experience of others, and by taking ACTION!

Expert Insight Delivered 7

OSyS
Part of the Rolls-Royce Group

Sharing is not enough to keep your ship afloat!

visium KMS
Process Safety Management Solution

Expert Insight Delivered 8

© 2014 Optimized Systems and Solutions Inc. All rights reserved.

OSyS
Part of the Rolls-Royce Group

352631 Autocad - Smart Grid – Agriculture - Architecture and the United States Government

Abstract:

The computational methods, process control, process design, process safety, public policy of the smart chip for agriculture industry, and architecture industry in the United States utilizing the computational methods of AutoCAD in the structural development stages of the smart grid for device/component/equipment and geographical development and public policy for grant funding.

352747 Production, Characterization and Catalytic Studies of Biobased Carbon Materials

Abstract:

Carbon-based nano materials were produced from wood chips, lignin and bio-char. The influence of reaction conditions on the production of nano-carbons was investigated. Carbon-based nanomaterials were characterized using scanning electron microscopy (SEM), transmission electron microscopy (TEM), high resolution transmission electron microscopy (HRTEM), and X-ray diffraction (XRD). The catalytic performance of these nanomaterials as syngas cleaning adsorbents and chemical conversion catalysts for bio-fuel processes has been evaluated.

347835 Uncertainty in Sour Gas Viscosities Estimation, What Is the Effect on Your Reservoir Inflow and Tubing Performance

Abstract:

In light of the need to find an applicable, consistent and promising correlation to accurately calculating sour gases viscosity under several reservoir conditions, this study aims to compare the existing correlations and other industry standard methods, check their accuracy and range of applicability for sour/acidic gases. Moreover, investigating the impact of uncertainties in estimating sour gas viscosities on reservoir flow performance relationship (IPR), reserves and resources estimations, tubing performance, pipeline and facilities design, the design of compressors and more importantly the need to find such correlation to be used in further advancements of reservoir simulation.

354052 Chromonic Nanocarriers for Chemotherapeutics: Size Distribution and Control Release Studies

Abstract:

Conventional chemotherapy fails to release the drug in the controlled manner; rather they release the drug instantly due to which re-administration is required as therapeutic index falls. Moreover, they target both the cancer cells and non-cancer cells which make the treatment of the cancer cells highly ineffective due to excessive toxicities. Recently, the nanoparticle mediated drug delivery for chemotherapeutics has gained much attention. The present study reports the use of a novel class of materials – *Chromonics* in cancer drug delivery. Folic acid as a chromonic material has been explored to develop nanocarrier for sustained release of chemotherapeutic drugs *in-vitro*. With the help of a biocompatible polymer as a tool, folate nano-domains were formed from self-assembled liquid crystalline folate solutions. These nano-domains were crosslinked to form a stable nanoparticle system. The effect of different factors like crosslinking agent, relative polymer and folate concentration, on the size distribution was investigated in detail. Based on the observations, we have reported that not only the synthesis but also the size of the developed nanoparticles can be controlled by varying the folate: HPMC weight ratios. The present study then explored the release of drugs from these nanocarriers. We have investigated the release of folic acid as a model drug from these nanoparticles as well as release of methotrexate and cytarabine as an anticancer drug for the period of 60 days. It was observed that release of drug was nanoparticle's size-dependent, release medium dependent, pH of the release medium dependent, crosslinking agent and its concentration dependent. The significant part of the current study is the synthesis of chromonic nanoparticles, how the size can be controlled, how controlled release can be obtained as well as how the release profiles can be modified depending on the dosage forms. Therefore, we can say that folate nanoparticles can be a potent nanocarrier for the sustained release of cancer drugs.

IPEX
Committed to Excellence

Double Containment Piping Systems: A Fail-Safe, Leak-Free Solution

Darin Johnson, P.Eng.
Applications Engineer
IPEX USA LLC

AIChE – 2014 Spring Meeting
New Orleans, LA

Monday, March 31, 2014

IPEX
Committed to Excellence

Why Double Containment Piping Systems?

- December 1988 – law change within the ***Resource, Conversation, and Recovery Act*** in the United States regarding underground piping.
- U.S. Environmental Protection Agency (EPA) requirement: Code of Federal Regulations (CFR), Title 40, Part 280.
    - Petroleum Products
    - Hazardous Substances (over 3200, CFR Title 40 Part 302.4)
    - Hazardous Wastes

IPEX
Committed to Excellence

Why Double Containment Piping Systems?

- Buried tank - all buried piping connected to it must be double contained.
- Above ground tank – associated buried piping consists of 10% or more of the tank volume, it also must be double contained.

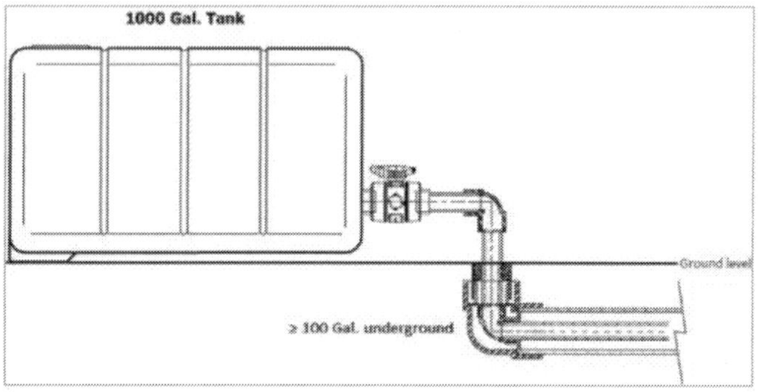

IPEX
Committed to Excellence

Why Double Containment Piping Systems?

- If the piping is less than 10% of the volume of the aboveground tank, and you choose not to double contain the buried piping, ***<u>responsibility for any clean-up or remediation resulting from a release will still fall on the Client.</u>***

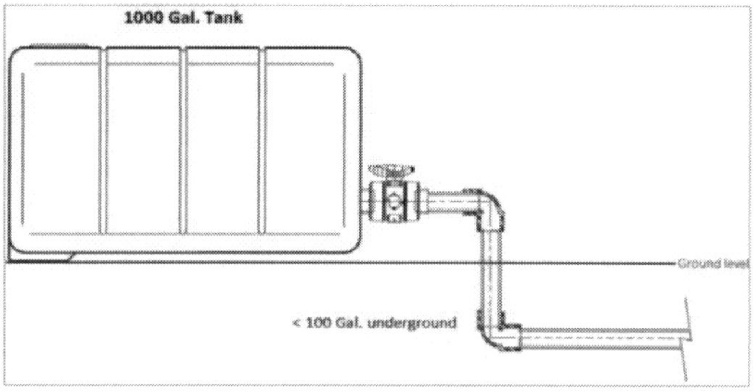

Why Double Containment Piping Systems?

- **Above Ground Piping Systems**
  - Different jurisdictions have to ability to require chemicals be double contained.

Nomenclature of Double Containment Piping

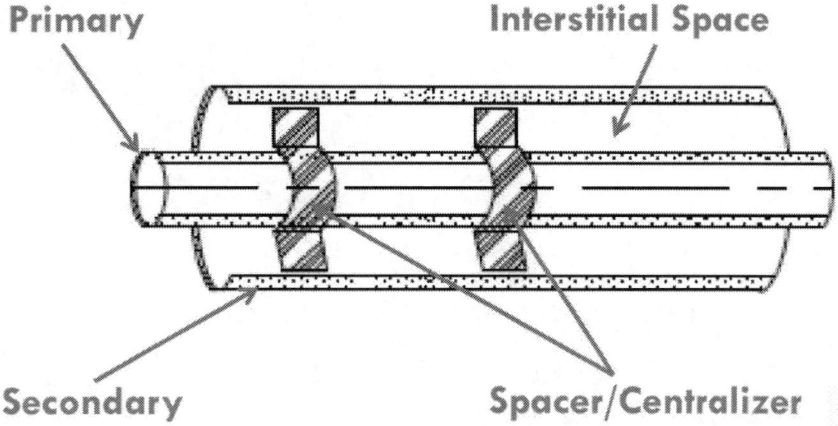

IPEX
Committed to Excellence

Double Containment Piping Systems: Material Combinations

SYSTEM Carrier	Containment	Carrier SIZE inches
PVC 80	PVC 80	1/2 - 12
PVC 80	PVC 40	1/2 - 12
PVC 40	PVC 40	1/2 - 12
PVC DWV	PVC DWV	1-1/2 - 12
PVC 80	CS SCH 10	1/2 - 12
CPVC 80	CPVC 80	1/2 - 8
CPVC 80	PVC 80	1/2 - 12
CPVC 80	PVC 40	1/2 - 12
CPVC 80	CS SCH 10	1/2 - 12
NAT PP 80	PVC 80	1/2 - 2
NAT PP 80	PVC 40	1/2 - 2
RED PVDF 80	CPVC 80	1/2 - 2
RED PVDF 80	PVC 80	1/2 - 2
NAT PVDF 80	PVC 80	1/2 - 2
ENFUSION PP	PVC DWV	1/2 - 6
S40 A53B CS BW	PVC 80	1/2 - 8
S40 A106B CS BW	PVC 80	1/2 - 8

SYSTEM Carrier	Containment	Carrier SIZE inches
S40 A53 BLK SW	FRP	1/2 - 2
S40 A53 BLK SW	PVC 80	1/2 - 2
S80 A53B CS SW	PVC 80	1/2 - 2
S10 T304L SS BW	PVC 80	1/2 - 8
S10 T304L SS BW	PVC 40	1/2 - 8
S10 T0304L SS BW	CPVC 80	1/2 - 8
S10 T304L SS BW	FRP	1/2 - 12
S10 T304L SS BW	S10 T304L	1/2 - 12
S40 T316L SS BW	PVC 80	1/2 - 8
S40 T316L SS BW	CPVC 80	1/2 - 8
S40 T316L SS BW	FRP	1/2 - 12
S40 T304L SS SW	PVC 80	1/2 - 2
S40 T304L SS SW	PVC 40	1/2 - 2
S40 T304L SS SW	CPVC 80	1/2 - 2
S40 T304L SS SW	FRP	1/2 - 2
S10 T316L SS SW	PVC 80	1/2 - 2
S10 T316L SS SW	CPVC 80	1/2 - 2

IPEX
Committed to Excellence

Double Containment Piping Systems: Design Considerations

- **Material Selection**
  - Primary Pipe
    - Chemical Compatibility
    - Pressure Capabilities
    - Thermal Properties
  - Secondary Pipe
    - Chemical Compatibility
    - Loading (Static & Live) for buried applications
    - Hanging requirements for above ground
    - Leak detection requirements
    - UV requirements

Double Containment Piping Systems: Design Considerations

- **Thermal Expansion & Contraction**
  - Two basic thermal relationships may be present:
    - When only one of the two systems experiences thermal expansion/contraction.
    - When both systems experience different magnitudes of thermal expansion/contraction.
  - The Floating System
  - The Restrained System

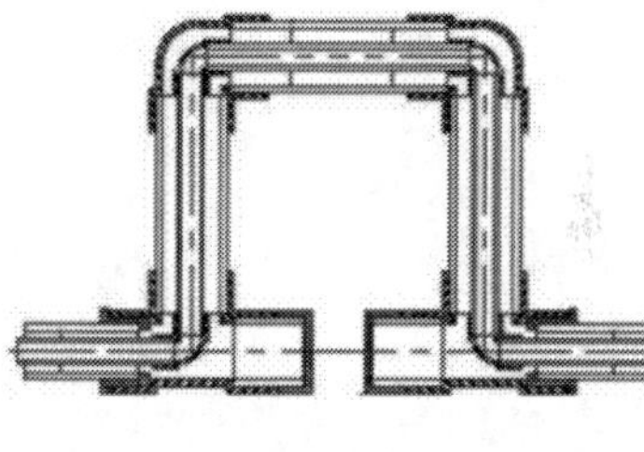

Double Containment Piping Systems: Leak Detection Requirements

- EPA requirement: (CFR), Title 40, Part 280
  - Pressurized systems require electronic leak detection that must be checked once per year.
  - Drainage systems (gravity flow) require visual leak detection with monthly manual inspection for a media leak.

Double Containment Piping Systems: Leak Detection Requirements

- The following forms of leak detection are acceptable as per regulations
  - Visual (drainage applications only)
  - Electronic low point probes
  - Continuous cable

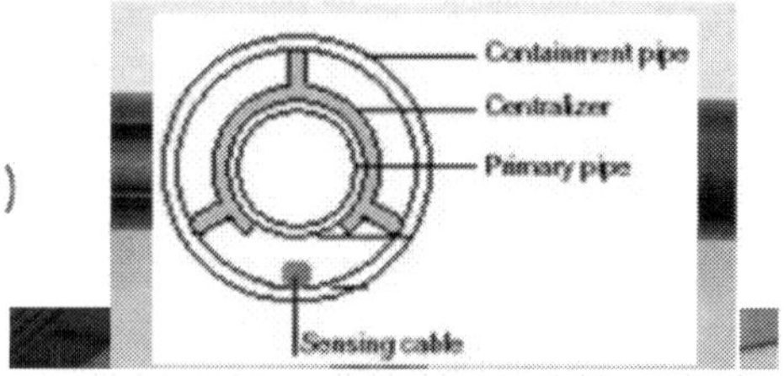

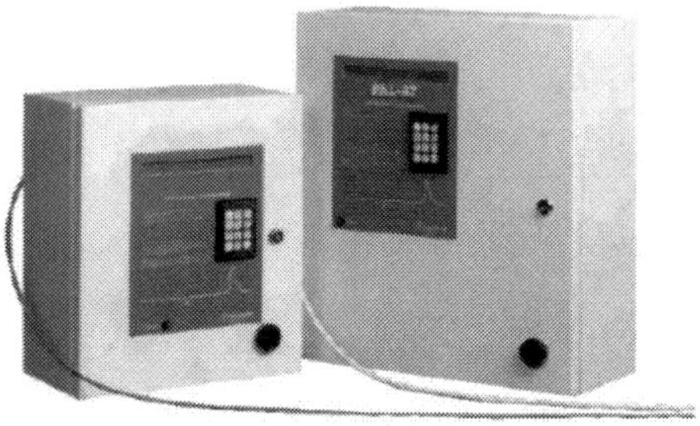

Double Containment Piping Systems: Applications

- Underground Applications
- WTP & WWTP
- Oil & Gas
- Hospitals
- Museums
- Food and Beverage
- Pharmaceuticals
- Laboratories
- Data Centers

IPEX
Committed to Excellence

Double Containment Piping Systems: A Fail-Safe, Leak-Free Solution

Questions?

Thank you!

DTP Process

"Now Ready to License"

JGC Corporation
Mitsuo Morita

2014

JGC CORPORATION

MITSUBISHI CHEMICAL

What is DTP Process ?

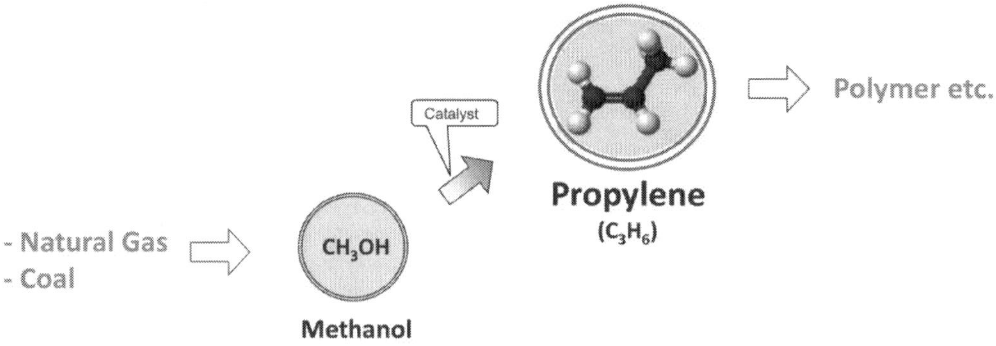

-On-Purpose Propylene Production from Methanol

JGC CORPORATION

- 2 -

MITSUBISHI CHEMICAL

What is DTP Process ?

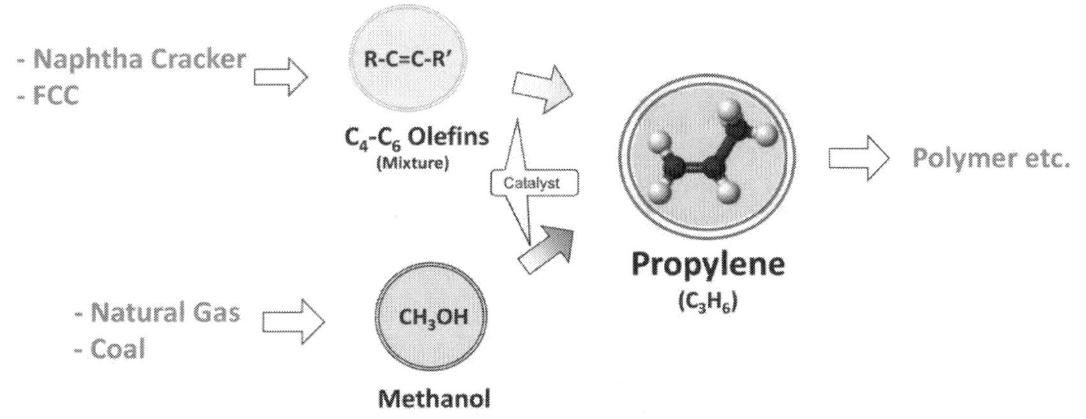

-On-Purpose Propylene Production from Methanol

- Additional Unutilized C_4-C_6 Olefins as Feedstock

Development History and Prospect

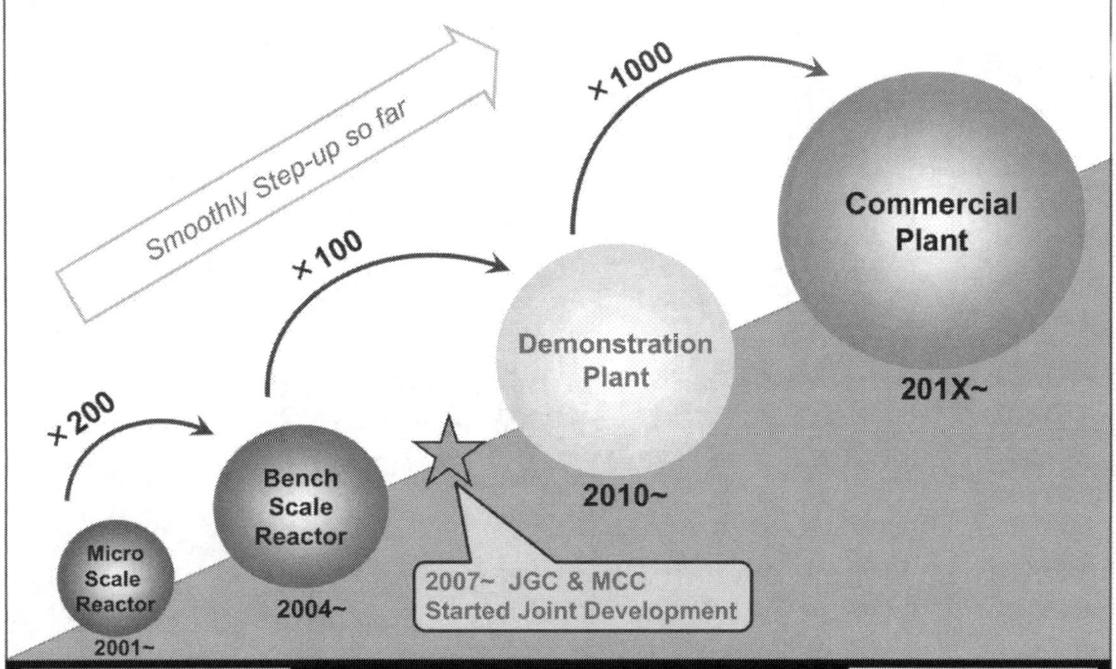

Demonstration Plant Overview

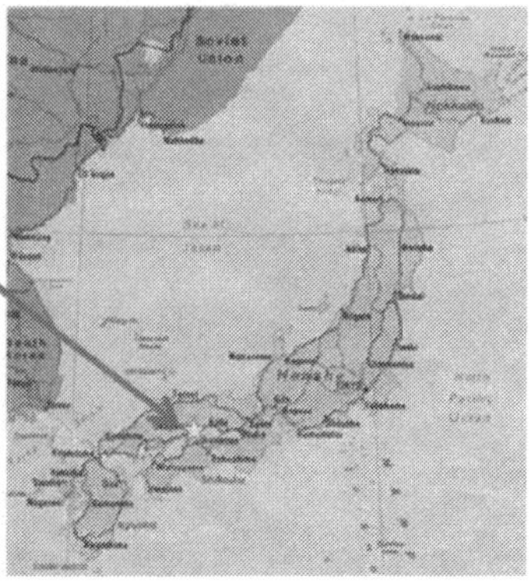

- **Located in Mitsubishi Chemical Mizushima Plant**
- **Successfully Start-up in August 2010**
- **Continuous Operation for Two Years**

Process Flow (Standalone Case)

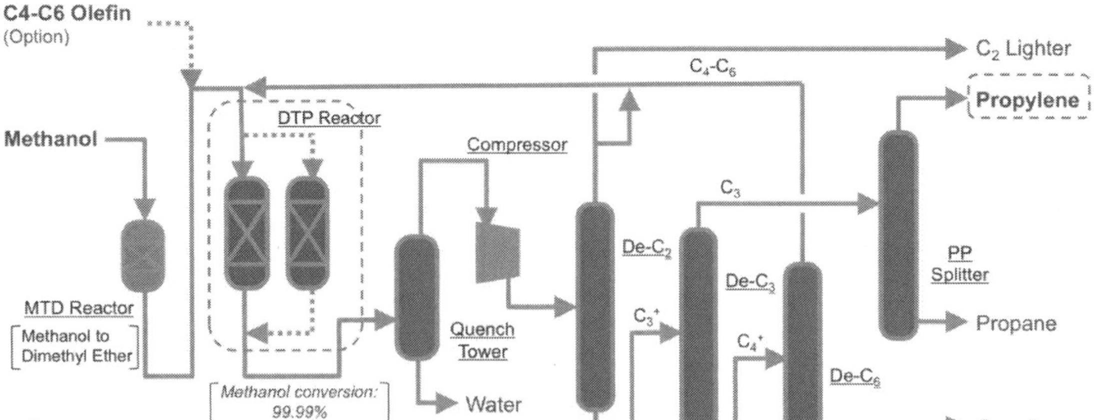

■ **Feature:**

- *High Propylene Yield: more than 70%*
- *Applied Existing Proven Technologies*

Process Flow (Integration Case)

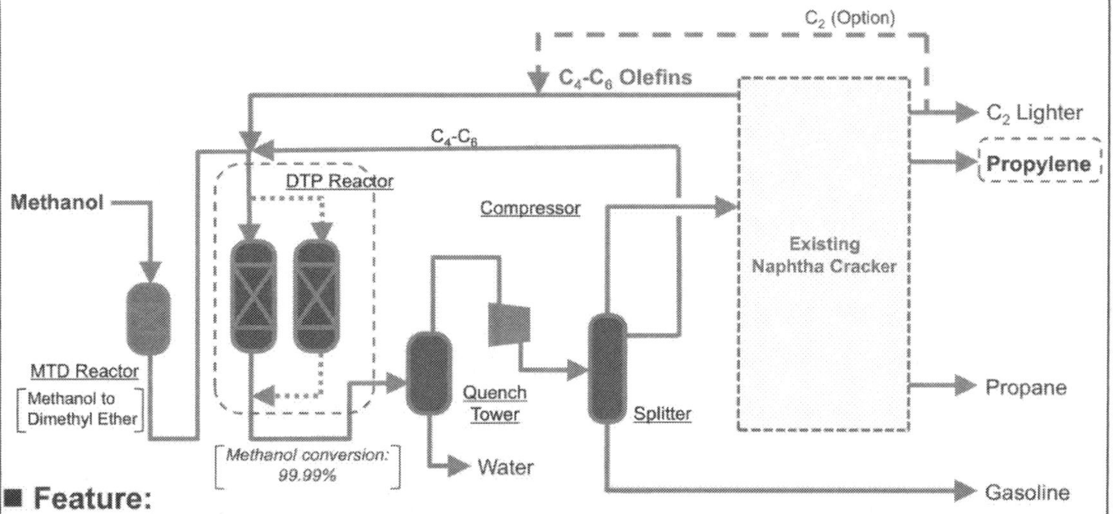

■ **Feature:**

- ***High Propylene Yield: more than 70%***
- ***Applied Existing Proven Technologies***
- ***Low Investment Cost in case of Integration with Naphtha Cracker***

Application

(1) Naphtha Cracker Integration Case (Propylene Capacity: 20KTA – 200KTA)

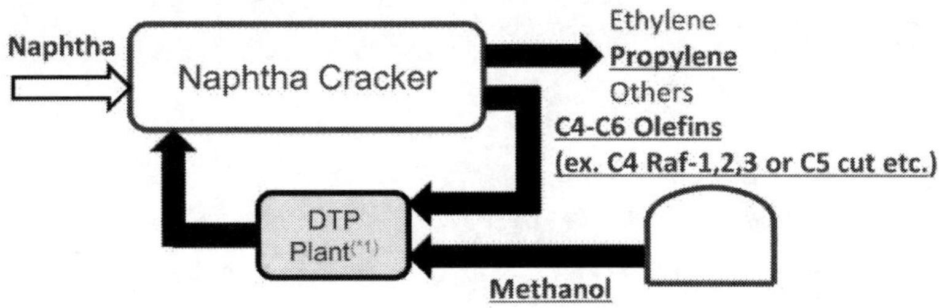

(*1) Utilizing facilities in Naphtha Cracker.

(2) Standalone Case (Propylene Capacity: 250KTA – 500KTA)

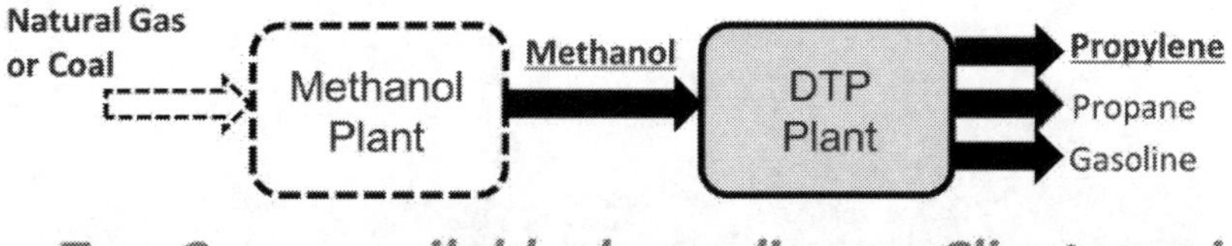

- ***Two Cases available depending on Client requirements !!***

DTP is...

- *High Yield Propylene Production Process*
- *Flexible Feed Methanol to Olefin Process*
- *Process that increase P/E Ratio for Cracker*

Thank you very much !!

Contact:
Mitsuo Morita
Market Development Dept.
Global Marketing Division
2-3-1, Minato Mirai, Nishi-ku,
Yokohama 220-6001, Japan

Phone: 81-45-682-8371
FAX: 81-45-682-8183
E-Mail: morita.mitsuo@jgc.co.jp

352533 Self-Cleaning "Bernoulli" Type Filters Used in Onshore and Offshore Applications

Abstract:

Georg Schuenemann GmbH (SAB-BREMEN) has developed a new Filter Type F-480, dubbed the "Compact Giant" for applications in onshore water intake and cooling water systems, sea water desalination plants, and in offshore projects, such as FPSO's, Drillships, and Production Platforms.

This paper will describe the essential design features of the filters, based on the "Bernoulli principle" which result in a fully automated self-cleaning filter system, and choice of materials in a sea water service with high corrosion rates.

A case study will be presented in which stringent project requirements such as substantial flow rates, adaptibility to space limitations and pipe connections had to be observed. Reliability of the system, extended run times without the need for frequent replacement of filter elements, as well as ease of monitoring and operator involvement have been achieved.

352653 Effects of Support on Sulfur Tolerance and Regeneration of Pt Catalysts Measured By Ethylene Hydrogenation and EXAFS

Abstract:

We investigated the effect of support on sulfur tolerance and regenerability under reducing environments by measuring the rate of ethylene hydrogenation, combined with hydrogen chemisorption, and extended x-ray absorption fine structure (EXAFS). Alumina and titania (P25) supports were used to prepare Pt catalysts at 1 wt.%. The catalysts were tested after treatment in H_2S/H_2 at 250 °C followed by regeneration treatments in H_2 at 250 °C, 350 °C and 450 °C. The sulfided catalysts exhibited a 20-27 times decrease in the rate of ethylene hydrogenation and a 4-6 times drop in the Pt surface area. Regenerations up to 450 °C were unable to remove all the sulfur, as shown by the presence of Pt-S bonds by EXAFS. However, a partial recovery of the hydrogenation rate per mole of Pt was observed on sulfided Pt/Al_2O_3 after reduction at 450 °C, while the induction of strong metal support interactions (SMSI) at reduction temperature above 350 °C was observed on Pt/P25, regardless of the presence of sulfur. For Pt/P25, the reversal of the SMSI state together with sulfur removal by mild oxidation suggests that sequential reduction/oxidation treatments may be more effective in restoring the S-free state of TiO_2-supported catalysts.

352799 Improved Operational Efficiency and Reliability through Insulation Materials Selection

Abstract:

This paper will explore the implications of insulation materials selection on the energy efficiency and reliability for Ethylene production units. Considerations will include insulation properties, structural support, installation methods, installation and maintenance costs.

BOILER MACT COMPLIANCE: MERCURY IN FUEL GAS

PATRICK LAINE, WES YONGE, & VIC VICKERY
PEI, HOUSTON, TX 77043

Patrick Laine, Ph.D.
Chief Scientist
Houston, TX
850-866-6686
plaine@pei-tx.com

Wes Yonge, PE
Chief Engineer
Houston, TX
713-806-8396
wyonge@pei-tx.com

James 'Vic' Vickery
Chief Technology Officer
Houston, TX
281-536-0899
vwickery@pei-tx.com

1 Introduction & Objectives

For most ethylene producers the primary objective of this sampling and analysis effort will be to verify and "establish a mercury fuel specification through which gas-fired boilers that use a fuel other than natural gas or refinery gas may be considered Gas 1 unit" as identified in 40 CFR Part 63.7521.(f).

Measurement and Monitoring Requirements: Decision Rules

❑ If Hg concentration ≤ 50% of Gas 1 specification* then no further testing required

❑ If Hg concentration > 50% and ≤ 75% of the specification* then semi-annual sampling required

❑ If Hg concentration > 75% of the specification* then monthly sampling required

*Specification is **40 µg/Sm³**

The goal will be to execute a sampling and analysis campaign that will produce compliance quality data. The key to producing and validating compliance quality data is a well-designed and executed QA/QC protocol. At a minimum, a performance based QA/QC protocol is recommended that includes the following components:

- Sorbent Trap Breakthrough Bed:
  - ❑ Sorbent traps consisting of a minimum of two sorbent beds in series so that the trailing bed can be analyzed to verify that no breakthrough occurred from the leading bed.
- Duplicate Samples:
  - ❑ The sampling system should be configured with two trains to accommodate the collection of simultaneous duplicate sorbent trap samples, in parallel, providing a means of verifying precision in the sample collection process.
- NIST (National Institute of Standards and Technology) Traceable Spike:
  - ❑ Sorbent traps with a third bed (downstream) that is spiked with a NIST traceable spiked mass of mercury serves as a matrix spike and provides a means of verifying that components in the matrix (the fuel gas) are not affecting the trapping efficiency of the sorbent trap.

2 Methods

Sorbent Trap Mercury Sampling Methods Comparison

Method	Description	Method Type: S = Sampling A = Analysis	Evaluation	Sample Collection Phase QA/QC
ISO 6978	Hg in Natural Gas by Thermal Desorption (CVAFS)	S & A	Method calls for low sample volumes can lead to non representative results.	No
EPA Method 30B	Determination of Gaseous Hg Emissions from Coal Fired Combustion Sources using Carbon Sorbent Traps	S	Requires modification (Modified M30B)	Yes
ASTM D6350	Hg S&A in Natural Gas (AFS)	S & A	Method calls for short term samples (i.e., low sample volume) which leads to poor accuracy.	No
ASTM D5954	Hg S&A in Natural Gas (AAS)	S & A	AAS is less sensitive than AFS.	No
EPA Method 1631	Measurement of Hg in Water	A	No	Yes

3 EPA Method 30B QA/QC Performance Criteria

QA/QC Test or Specification	Acceptance Criteria	Frequency
Analyzer calibration	Correlation Coefficient: 0.999	Daily
Transfer efficiency of analyzer	%Transfer Efficiency: 90%–110%	Once per sorbent trap
Initial calibration blank	< 0.5 ng	Once before each calibration
Sorbent Trap Section B breakthrough	≤ 10% of Section A Hg mass, for Hg concentrations > 1µg/dscm ≤ 20% of Section A Hg mass, for Hg concentrations ≤ 1µg/dscm	At each sample point
Spike Recovery Test	Average recovery between 75% and 125 % for Hg_0	Once per sample point
Paired Sorbent Traps Agreement	≤ 10% RD for Hg concentration > 1µg/dscm ≤ 20% RD or 0.2 µg/dscm absolute difference for Hg concentrations ≤ 1µg/dscm	At each sample point

4 Equipment

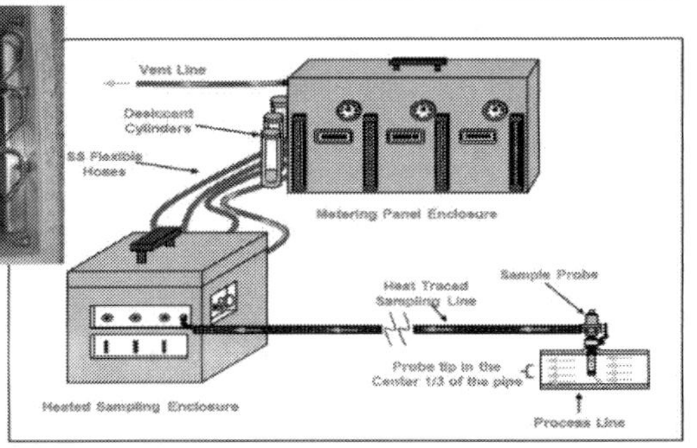

	Primary Bed	Breakthrough Bed	Spiked Bed
Flow			

QA/QC Acceptance Criteria:	99% of Total Collected Hg	≤ 10% of Primary Mass	Recovery between 75% and 125%

Gold Trap:

5 Conclusions/PEI Recomendations

❑ There are five methods accepted by the Boiler MACT rule. PEI has extensive experience and expertise employing **all five** of these methods.

❑ Operators can achieve compliance without submitting a Fuel Analysis Plan by using ASTM D6350, however, we have noticed variable results with this method.

❑ The **PEI recommendation** for obtaining the most accurate compliance quality data is to use Modified EPA Method 30B in conjunction with EPA 1631 where traps should be analyzed in a certified ISO 17025 laboratory. The recommended pathway includes a rigorous, performance based QA/QC protocol that validates accuracy and precision of all sample collection procedures.

351376 Young Professional Simulation Tutorial

Abstract:

18C00 Young Professional Simulation Tutorial

Authors: Naomi Hua and Mike Donahue, Invensys

Process simulation is a complex skill to master. Even with years of experience, a typical user is constantly facing new problems that need to be addressed. There are many issues that can lead to unexpected and unnecessary errors with the use of process simulation tools. Regardless of what simulation program you are using, these issues are common from one program to another. In this presentation, targeted for the young professional, we will discuss the "Most Common Pitfalls in Process Simulation". For each topic, we will discuss various techniques and methods these troublesome problems and prevent their occurrence.

Topics to be discussed include:

1. Erroneous data entry
2. Simulation defaults
3. Bad plant data
4. Thermodynamic method selection

5. Simulation tolerances and convergence

6. Recycle identification and handling

7. Model complexity

8. Difficult or impossible specifications

9. Assay pseudo-component development

10. Assay property blending

1. Erroneous data entry is the most common mistake made when building a process simulation model. This presentation will cover methods that can help ensure proper data entry in your process simulation model, such as checking the data you entered, particularly the units of measure, using "alternate forms" of data input such as copy/paste and data import, and reviewing all warning and error messages.

2. Defaults are methods, values, and conditions that are automatically selected for you by the program when you begin a new simulation model. These defaults will affect the results calculated by your model and every simulation program has them. Most defaults within simulation programs fall under categories such as convergence criteria, thermodynamic methods, tolerance and units of measure. The problem with defaults is unintended results from not fully understanding the effects a particular default has on the model. Take the time to understand the defaults of the program and how these defaults will affect the results. If you do not agree with the defaults selected by the program, most simulation programs will allow the user to change the default.

3. Plant data changes over time. The feed information and process conditions of the unit operations may be different. Not only does plant data change over time, but measurement information is often very inaccurate. This could be due to faulty or poorly calibrated measurement equipment, leaks in lines, sensor failures, and other issues. Because of these changes and inaccuracies over time, you always have to validate the information that is entered in your model to ensure that you are obtaining the proper results. To do so, you can utilize programs that can help with data validation and reconciliation like SimSci's ROMeo online optimization program. With it, you can connect your model to your data historian to constantly import current data measurements into your model and then reconcile these measurements to determine when problems arise within your plant and where those problems are occurring. You will also need to use your engineering judgment to determine if the data is valid. More details will be provided during the presentation.

4. The proper selection of thermodynamic methods for your simulation is the most important decision that you make when modeling a process. The method you select could have a drastic effect upon your results. Often, any number of thermodynamic methods is suitable for a given application. It is up to the user to determine the best method to use based upon knowledge, personal experience, and best judgment. Some important factors that need to be considered are the components being used and the operating conditions.

Oftentimes, the selected thermodynamic method still needs to have additional inputs such as binary interaction parameters to simulate the process accurately.

5. Many of the calculations in process simulation are iterative. When these calculations iterate, the program compares the results calculated from one iteration to the results calculated in the next iteration in order to determine

whether it is converged to a solution or needs to continue calculating. This method of comparing results is where tolerances come into play and simulation programs use tolerances to determine when a solution is reached. One of the major mistakes when modeling a process is not understanding or using the correct tolerance values. A looser tolerance may make the file easier and quicker to converge, but it also reduces the accuracy of the model. A tighter tolerance may be more accurate, but might make it impossible for the model to solve or increase the convergence time exponentially. A common practice is that when you have reached a converged solution with your model, tighten the tolerances within the model and solve it again. If you notice that your answers have changed significantly, you need to continue tightening the tolerance values until your results change little from one iteration to another. You will then know that you have reached an accurate solution for your model.

6. Recycles in process simulation occur when you return to a previously calculated unit operation to recalculate the unit due to a change made elsewhere in the model. They are very common and can often be difficult to converge. Some suggestions on troubleshooting recycle convergence include initially solving your model without the recycles to see how it behaves, providing initial estimates for recycle streams, using tighter tolerances on unit operations within recycle loops than the total recycle tolerances themselves, and ensuring that all components have a way in and out of a recycle by performing mass balances and using purge or makeup streams where necessary.

7. The general rule of thumb when modeling is to start simple and add complexity as you go. Each time you add a few streams or unit operations converge and save the model before moving on. Make sure you justify any additions that you make to the model. You can most likely model your system without including every unit from your PFD in the simulation. When modeling, think about the goal of your simulation and not about literally translating every real-world piece of equipment into the simulation model. Make the judgment call on what is most important and what the purpose is of every unit you add to your simulation. For example, if your real-life process shows a heat exchanger, a pump and a splitter, but all that is necessary is that the temperature, pressure, and separation are achieved properly, consider using a flash drum to accomplish all three.

8. Specifications within process simulation are everywhere and every stream and unit operation you place in your model contains specifications. In its simplest form, a specification is a constraint that you place on the model that the model needs to meet in order to successfully converge. For streams, these are usually temperatures, pressures, flowrates and compositions. For unit operations, they range from process conditions, like temperatures and pressures, to product conditions, like an overhead composition in a column. There are some pitfalls, however, that you can run into with specifications such as azeotropes in columns, high purity specifications, and the possibility of multiple solutions.

9. Petroleum assays are sets of distillation data that are used to characterize all of the components that reside in a given crude sample. When you enter distillation data into a process simulator, the program converts it into components that it can use when modeling your process. The components that are generated are called pseudo-components, and each pseudo-component represents all of the unknown components that would be within a given boiling point range in the crude. Some of the things you will want to ensure when properly developing your assay components are did you use the correct distillation curve basis, did you enter the data correctly, did you use enough cuts and are you using the correct characterization methods .

10. When you have more than 1 assay in a simulation model, you need to be concerned with assay blending and how it will affect the results of your file. Assay blending is the mixing of 2 or more groups of components (i.e. "C sets of cuts/pseudo-components) in order to calculate averaged component properties for cuts that are in the same boiling point range as one another. Once these new averaged component properties are determined, the program will use these averages from this point on in the simulation model any time it needs to access component

properties for a given pseudo-component. All component properties are blended, including normal boiling points, specific gravity, molecular weight and refinery inspection properties, and usually the blended component properties are calculated using a weighted average

These topics are based on experience and feedback from end-users to technical support representatives for process simulators. They often represent the most commonly asked questions or areas that most users find to be problematic. While the specific examples highlighted in each topic come from SimSci software, the intent of this tutorial presentation is to be broad-based and is particularly applicable to young professionals seeking to develop their experience with simulation.

Polymeric Sulfonic Acid Composites for Fuel Cell Applications: Effects of Casting Solvent and Sonication Time on Membrane Performance

Jimoh K. Adewole[1], *Abdullah S. Sultan*[1, *], *Amir Al-Ahmed*[2], *and S. M. Javid Zaidi*[3]

[1]*Center for Petroleum & Minerals, King Fahd University of Petroleum & Minerals, Dhahran 31261, Saudi Arabia*

[2]*Center of Research Excellence in Renewable Energy, King Fahd University of Petroleum & Minerals, Dhahran 31261, Saudi Arabia*

[3]*School of Chemical Engineering, The University of Queensland, Australia*

Abstract

Polymer blends comprising sulfonated poly (ether ether ketone) (SPEEK) and 3wt% (2-acrylamido-2-methyl-1-propanesulfonic acid) (PAMPS) were prepared by solution casting method. Two solvents (N,N-Dimethylacetamide (DMAc) and N-Methyl-2-pyrrolidone (NMP)) were uesd separately. Residual solvents in the membrane samples after drying were quantified using thermo gravimetric analysis. Effects of periods of sonication (0 to 90 minutes) and types of solvent on the membrane performance were evaluated interms of morphology and water uptake. An attempt was made to correlate membrane performance with the time of sonication in order to select the most suitable solvent and membrane casting conditions for subsequent membrane preparations. For both solvents, the water uptake was observed to first decrease at 10 minutes of sonication and then increased as the sonication time was increased. The minimum water uptake recorded for DMAc samples was 19.23% at 20 minutes sonication time while that of NMP sample was 24.24% at 10 minutes. However membrane morphology was found to deteriorate with increase in sonication time. No correlation was obtained between the membrane properties and the sonication time. In general, membrane performance was better for membranes samples using DMAc than NMP solvent. Therefore, DMAc was selected as the casting solvent for membrane preparation on which further studies will be carried out.

Keywords: morphology, water uptake, residual solvent, thermogravimetric analysis

Introduction

Water soluble polymers have received academic and industrial interests due to their potential applications in fuel cells and other electrochemical devices (Ye et al. 2012). One of the water soluble polymers that have attracted publication interest is poly (2-acrylamido-2-methyl-1-propanesulfonic acid) (PAMPS) (Florjańczyk et al. 2003). Generally, some of the major performance criteria for fuel cell materials include proton conductivity, water and methanol uptake, good mechanical properties, chemical stability, thermal stability and methanol permeability (Gil et al. 2004; Swier et al. 2005; Ye et al. 2012). PAMPS was reported to posses higher proton conductivity than partially hydrated Nafion. In addition, it has been found to be more tolerant to water fluctuation in high temperature operations. Some of the major drawbacks in the use of PAMPS are its poor mechanical properties, thermal stability, water and methanol uptake and solubility in many solvents (Florjańczyk et al. 2003). One of the approaches that have been proposed to solve these problems is development of polymer composites.

This research paper investigates the synergistic effect of combining water soluble PAMPS with sulfonated poly (ether ether ketone) (SPEEK) which is non - water soluble polymer to prepare composite membranes for application in fuel cells. Specifically, preliminary results on the performnance of the composite were presented. Performance were evaluated based on the composite morphology and water uptake. Effects of sonication time and types of casting solvent on membrane properties were also evaluated.

Material and Method

Material

Sulfonated poly (ether ether ketone) (SPEEK) purcahsed from FuMA-Tech, Germany, and poly (2-acrylamido-2-methyl-1-propanesulfonic acid) (PAMPS) purchased from Aldrich Chemical Co., are the two major polymers used in this research. The SPEEK has a DS of 0.533 and solubility of 99% in DMAc. SPEEK is derived by sulfonation of Poly (ether ether ketone) (PEEK). PEEK is a commercially successful polymer due to its excellent thermal stability and solvent resistance (Mohr et al. 1991). For sometimes, further application of PEEK have been stalled due this solubility issues. By sulfonation, PEEK can be made soluble in most commercial solvents. Importantly, the solubility of SPEEK also depends on its degree of sulfonation (DS). Since, this work involves the use of two polymers, a solvent which will disolve both the PAMPS and SPEEK must be selected. Literature review showed that PAMPS dissolves in N,N-Dimethylformamide (DMF), N,N-Dimethylacetamide (DMAc), propylene carbonate (PC), and water (Florjańczyk et al. 2003). Initial solubility studies that was carried out in our laboratory was done for N,N-Dimethylacetamide (DMAc) and N-Methyl-2-pyrrolidone (NMP). The choice of NMP was due its unique solvent properties and its structtural distiction from DMF and DMAc (Figure 1). Moreover, no literature was found where NMP was used as casting solvent for PAMPS. Thus, DMAc and NMP were used as solvents in this work. All materials were used as received from the manufacturers.

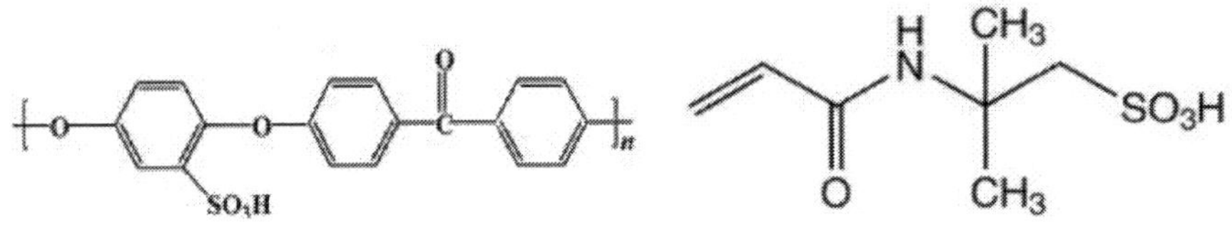

Figure 1 Structure of SPEEK and PAMPS

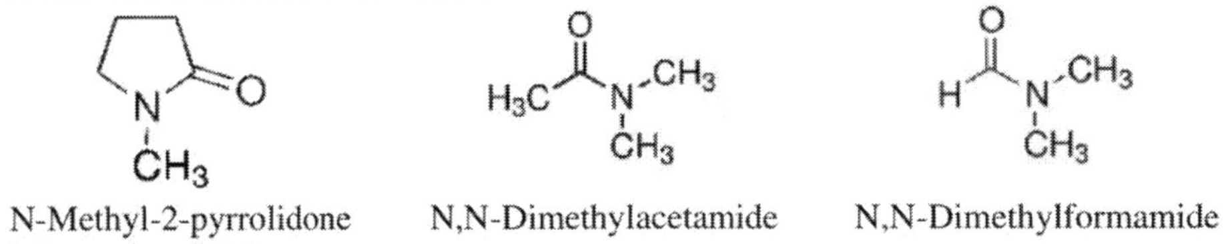

Figure 2 Structure of Solvents

Experimental Procedure

Membrane Preparation

SPEEK was first dissolved in solvents (DMAc and NMP) and stirred for 2hours at 70°C. PAMPS was then added and the blend was stirred for 24hours. The solution was sonicated for various periods of time ranging from 10 to 90minutes using an Ultrasonic bath. In untralsonic bath, the final temperature achieved for a given time of sonication depends on the tempearutre of the laboratory (summer or winter). It is therefoe genrally advisable that maximum temperature of the bath is determined and maintains (as the so - called equilibrium tempearture), when operating continuously under ambient conditions (Santos et al. 2009). During the sonication, equilibrium temperature was maitained by simply filling the bath with water that was preheated to sample temperature.

Membrane Characterization

Water uptake and morphological tests were performed. Water uptake was carried out by first vacuum drying all membrane samples at 100°C until constant weights were obtained. Dried samples were then immersed in deionized water for 24h at room temperature (25°C). Samples were taken out of the water after 24h, wipped to remove surface water and then weighed immediately. Water uptake was calculated using Eq. 1.

$$water-uptake=\frac{W_{wet}-W_{dry}}{W_{dry}} \qquad \text{Eq. 1}$$

Surface morphology of the membrane was done using scanning electron microscopy (SEM). Thermogravimetric analysis (TGA) was done from room temperature to 350°C using TA Q600 analyzer at heating rate of 20°C/min under nitrogen atmosphere with flow rate of 20mL/min using ~ 8 -10mg membranes samples.

Results and Discussion

Membrane Morphology by Scanning Electron Microspcopy (SEM)

Results of surface morphological analysis performed on membrane were shown in Figure 3 and 4. The overall pictures of this figure depicts a dense membrane with some surface defects observed for most of the sonicated samples. For example, Figure 3A showed some kind of layered arrangement with no visible pores at the surface. Also, the PAMPS seems to be well distributed throughout the SPEEK. The micrograph in Figure 3B can be observed to be rough and possesses some uneven ramdom distribution of PAMPS agglomerates indicating that the distribution of PAMPS have been disturbed by the 10minutes sonication. An increase in sonication time to 20 minutes was observed to improve the distribution of PAMPS. This was however accompanied by some cracks believed to be due to the slight increase in the viscousity of the blend solution. The vicousity increase was brought about by the excape of some fraction of the solvent during sonication. As the sonication time was increased, PAMPS was observed to become more evenly distributed and the cracks become more visible and larger in size (Figure 3D to F). SEM images of samples at sonication time of 10 and 30 minutes were reproduced in higher magnification of 10μm (Figure 4) in order to further revealed the effect of sonication on the morphology of the somposites. Similar morphological behaviour has been reported in literature, though for inorganic/organic clusters of nanoclay/epoxy composites membranes (Lam et al. 2005).

Water Uptake

Water uptake have been reported to be affected by the degree of sulfonation, membrane pretreatment, hydration, ambient relative humidity and water temperature (Xing et al. 2004; Wu et al. 2006). Figure 5 showed the effect of sonication time and solvent type on water uptake that was measured at room temperature (25°C). For both solvents, water uptake is same at zero sonication time. There was however a further decrease in the water uptake for NMP at 20 minutes sonication time before it beging to increase. Thus, the best sonication time for both solvent lied between 10 and 20 minutes with respect to water uptake. Interestingly, this result correspond to that reported by Lam (2005) on the effect of ultrasonication on microhardness of composites.

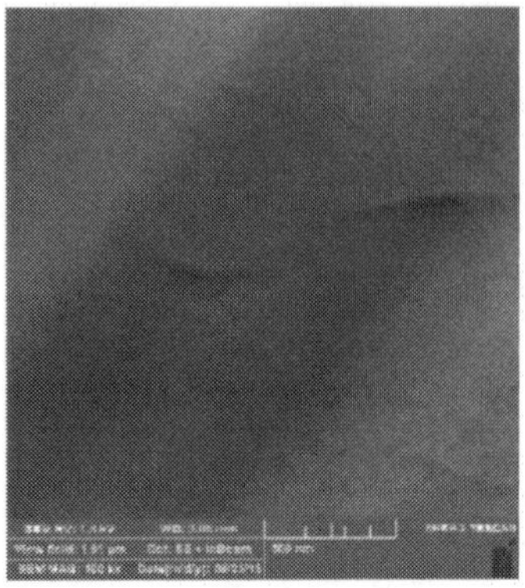

A) No sonication

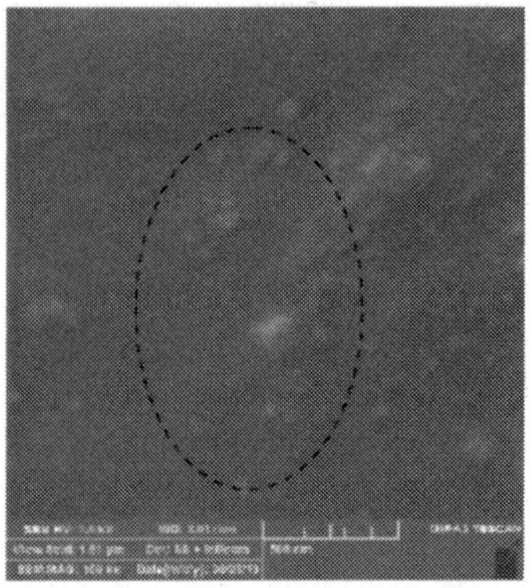

B) 10 minutes sonication

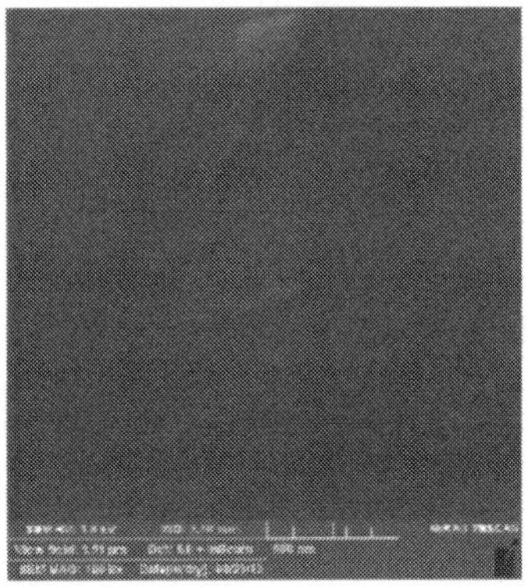

C) 20 minutes sonication

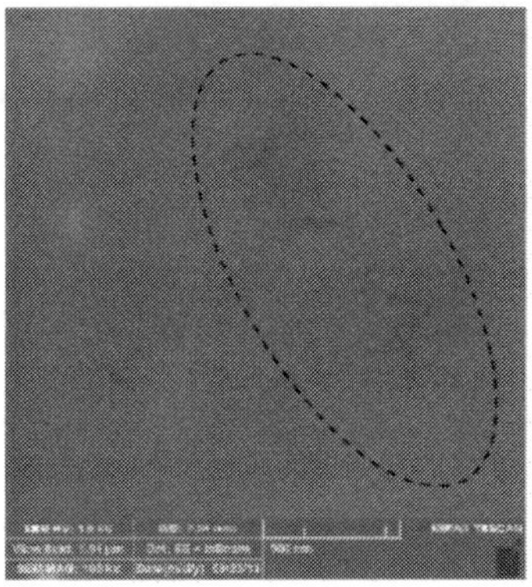

D) 30 minutes sonication

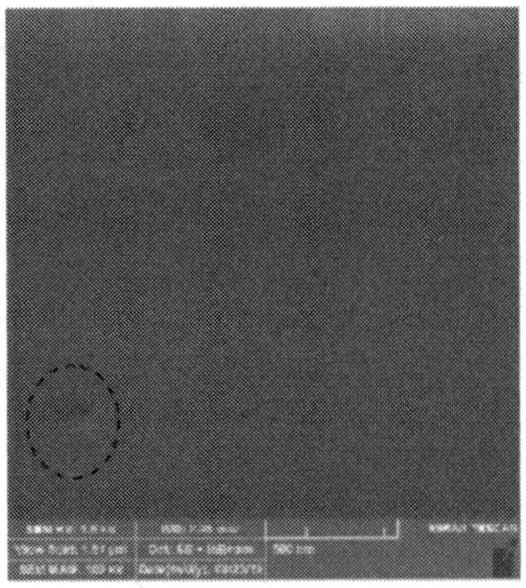

E) 60 minutes sonication

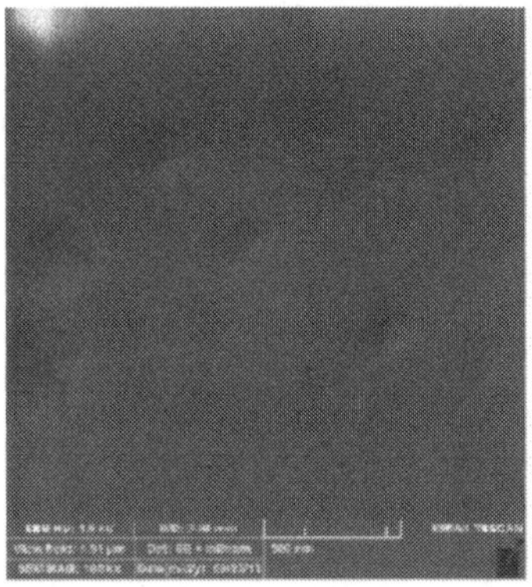

F) 90 minutes sonication

Figure 3 SEM micrograph of SPEEK-PAMS composite with various times of sonication with magnification of 500nm

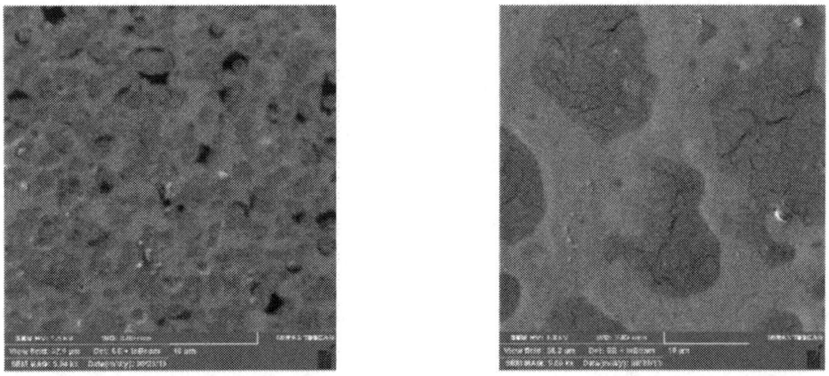

A) 10 minutes sonication B) 30 minutes sonication

Figure 4 SEM micrograph of SPEEK-PAMS composite with various times of sonication with magnification of 10μm

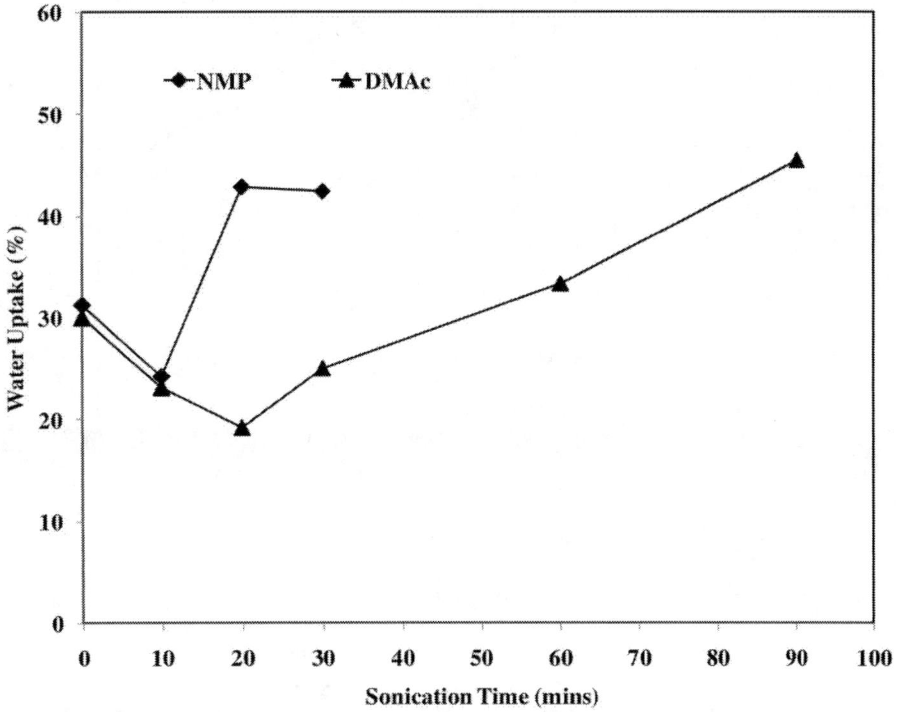

Figure 5 Water uptake of SPEEK-PAMS (3wt%) composite with variuos sonication times

Thermogravimetric analysis (TGA)

The difference in membrane performance resulting from the use different casting solvents have been attributed to the presence of residual solvents in the membrane samples after evaporation (Joly et al.

1999). TGA was thus used to quantify the residual solvent and evaluate its effect on the membrane performance as has been previously demonstrated in lieterature (Recio et al. 2007; Iqbal et al. 2008). Figure 6 shows the TGA curve for samples prepared from NMP and DMAc with different post vacuum heat treatment conditions. From this figure, weight loss can be attributed to evaporation of variety of substances, including solvents, from the samples (Rich et al. 1997). Table 2 contains the solvent and its boiling points, and weigth loss at the boiling point range that were extracted from TGA results. For sample heat treated for 24hrs, the overall percentage weight loss up to 204°C was about 8wt% of membrane sample for NMP while it is about 6.3wt% for DMAc. These will definitely include compnenents such as water, solvents and other unidentified substances. To quantify residual solvents, weight loss was estimated between 192 -214°C for SPEEK-PAMPS-NMP and between 154 and 176 for SPEEK-PAMPS-DMAc. The results obtained showed that the amount of the residual solvent in NMP is larger than that of DMAc by 1.25wt% (Table 1). To further investigate if this difference is responsible for higher value of water uptake for the NMP sample, membrane samples made from both solvents were further heat treated under vacuum for 72hrs. Residual solvent after this treatment was slightly higher in DMAc than NMP. Despite this, the value of water uptake of the NMP membrane is still higher than that of DMAc by approximately the same value as before the heat treatment (1.71wt%).

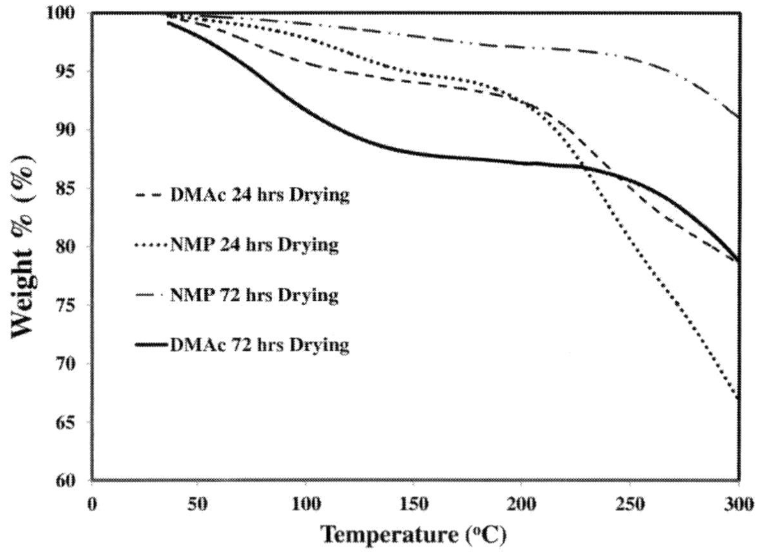

Figure 6 TGA of SPEEK-PAMPS-NMP and SPEEK-PAMPS-DMAc

Table 1 Residual solvent in SPEEK-PAMPS composites membrane samples

Membrane Sample	Solvent	B. Pt. (°C)*	Residual Solvent (wt%)	Water Uptake
DMAc 24 hrs Drying	DMAc	164.5 – 166	0.57	30.00
NMP 24 hrs Drying	NMP	202 – 204	2.85	31.25
DMAc 72 hrs Drying	DMAc	164.5 – 166	0.367	21.21
NMP 72 hrs Drying	NMP	202 – 204	0.217	22.92

*Boiling Points of Solvents

On a general note, results from Figire 1 – 5 showed that the best perfomance can be obtained for variety of sonication time depending the property of preference. Also, performance of DMAc solvent was observed to be better than that of NMP on the basis of water uptake. For this reason, DMAc was selected as the casting solvent for membrane preparation on which further studies will be carried out. Water uptake of DMAc membrane, prepared without sonication, was comapred with those reported in previuos literature. Table 2 illustrates the comparison between SPEEK-PAMPS and other relevant membranes previuosly reported in literature. In addition, this table serve to provide insight into the possible potentials of the composites. From this table, it was observed that the water uptake for SPEEK-PAMPS is within the range of values of other membranes on the table. Its water uptake is less than that of Nafion 112 which is a commercial membrane. The differences observed between the water uptake values of SPEEK – PAMPS and other membranes from the table can be explained based on three important factors - SO_3H group, the degree of sulfonation and experimental temperature (Gil et al. 2004; Shan et al. 2006; Wu et al. 2006). Water uptake of SPEEK – PAMPS is higher due to the presence of SO_3H group in the PAMPS molecules (Figure 2). SO_3H group have been reported to enhance membrane acidity and hydrophilic properties (Shan et al. 2006). Infact, the higher the SO_3H group the larger the water uptake. The degree of sulfonation has been reported to affect the water uptake of membranes. The higher the degree of sulfonation, the larger the value of water uptake (Gil et al. 2004). Thus, water uptake of SPEEK – PAMPS is lower than SPEEK -3 due their different degree of sulfonation, 53.3 and 72% respectively and the experimental temperature 24 and 60°C, respectively. The experimental temperature of SPEEK – 2 is close to that of the composite. However, the results cannot be compared in this case since the DS of SPEEK – 2 was not known. Above explanation is also true for the values of water uptake for SPEEK – 1.

Table 2 Comparison of water uptake values obtained with those from Literature (Xing et al. 2005)

Membrane	Water Uptake (wt%)	Experiment Temperature (°C)	**D S (%)	Reference
Nafion 112	22.85	N/A	-	(Xing et al. 2005)
SPEEK – 1*	16.00	30	45	(Ren et al. 2005)
SPEEK – 2*	44.90	25	N/A	(Zhong et al. 2008)
SPEEK – 3*	85.00	60	72	(Wu et al. 2006)
SPEEK– 4*	15.00	N/A	N/A	(Handayani and Dewi 2011)
SPEEK/PAMPS (3wt%)	21.21			This work

*The numbering have been included for explanation purposes.
**DS means Degree of Sulfonation

Conclusion and Recommendations

Blend of SPEEK and PAMPS was prepared by solvent casting method and its performance was evaluated with respect to its morphology and water uptake. Effect of sonication time (10 – 90minutes) and solvent types (DMAc and NMP) were also investigated. Performance of membrane prepared using DMAc was observed to be better than that of NMP. There was no correlation between the sonication time and membrane properties, but it was observed that the sonication deteriorate the membrane morphology. Membrane prepared without sonication shows better surface morphology, whereas the best water uptake was observed between 10 and 20 minutes. Thus, other properties need to be investigated in order decide the best sonication time. Therefore, further investigation such as thermal stability, methanol permeability, ion exchange capacity and gas (O_2 and H_2) permeability of the membrane are recommended for a complete evaluation of the membrane performance for direct methanol fuel cell (DMFC) application. All these investigations are in progress in our laboratory.

Acknowledgement

Authors would like to thank the National Science, Technology and Innovation Program (NSTIP) of KACST for funding this research under project number 10-ENE1374-04. Also, authors extend their appreciation to Center of Petroleum & Minearls, Center of Excellence in Renewable Energy, Mr. Dan Jerry D.Cortes of the Nanotechnology Laboratory, Chemical Engineering Department, King Fahd University of Petroeleum & Minerals, Saudi Arabis for their supports.

REFERENCE

Florjańczyk, Z., et al. (2003). "Proton conducting electrolytes based on poly(2-acrylamido-2-methyl-1-propanesulfonic acid)." Electrochimica Acta **48**(14–16): 2201-2206.
Gil, M., et al. (2004). "Direct synthesis of sulfonated aromatic poly(ether ether ketone) proton exchange membranes for fuel cell applications." Journal of Membrane Science **234**(1–2): 75-81.
Handayani, S. and E. L. Dewi (2011). "Influence of Silica/Sulfonated Polyether-Ether Ketone as Polymer Electrolyte Membrane for Hydrogen Fueled Proton Exchange Membrane Fuel Cells." Internat. J. of Sci. and Eng. **2**(2): 26-30.
Iqbal, M., et al. (2008). "Solvent effect on morphology and CO2/CH4 separation performance of asymmetric polycarbonate membranes." J. Membr. Sci. **318**(1–2): 167-175.
Joly, C., et al. (1999). "Residual solvent effect on the permeation properties of fluorinated polyimide films." Separation and Purification Technology **16**(1): 47-54.
Lam, C.-k., et al. (2005). "Effect of ultrasound sonication in nanoclay clusters of nanoclay/epoxy composites." Materials Letters **59**(11): 1369-1372.
Mohr, J. M., et al. (1991). "Gas transport properties of a series of poly(ether ketone) polymers." Polymer **32, 13**: 2387-2394
Recio, R., et al. (2007). "Gas separation of 6FDA–6FpDA membranes: Effect of the solvent on polymer surfaces and permselectivity." J. Membr. Sci. **293**(1–2): 22-28.
Ren, S., et al. (2005). "Surface modification of sulfonated poly(ether ether ketone) membranes using Nafion solution for direct methanol fuel cells." Journal of Membrane Science **247**(1–2): 59-63.
Rich, D. C., et al. (1997). "Effect of cure conditions on Probimide 32 polyamide-imide." Journal of Applied Polymer Science **63**(9): 1113-1126.
Santos, H. M., et al. (2009). The Power of Ultrasound. Ultrasound in Chemistry: Analytical Applications. J.-L. Capelo-Martinez. Weinheim, Wiley-VCH.
Shan, J., et al. (2006). "Sulfonated polyether ether ketone (PEEK-WC)/phosphotungstic acid composite: Preparation and characterization of the fuel cell membranes." Pure Appl. Chem **78**(9): 1781 - 1791.
Swier, S., et al. (2005). "Sulfonated poly(ether ketone ketone) ionomers as proton exchange membranes." Polymer Engineering & Science **45**(8): 1081-1091.
Wu, H.-L., et al. (2006). "Preparation and characterization of poly(ether sulfone)/sulfonated poly(ether ether ketone) blend membranes." European Polymer Journal **42**(7): 1688-1695.
Xing, D. M., et al. (2005). "Characterization of Sulfonated Poly(Ether Ether Ketone)/Polytetrafluoroethylene Composite Membranes for Fuel Cell Applications." Fuel Cells **5**(3): 406-411.
Xing, P., et al. (2004). "Synthesis and characterization of sulfonated poly(ether ether ketone) for proton exchange membranes." Journal of Membrane Science **229**(1–2): 95-106.
Ye, Y.-S., et al. (2012). "Water Soluble Polymers as Proton Exchange Membranes for Fuel Cells." Polymers **4**(2): 913-963.
Zhong, S., et al. (2008). "Modification of sulfonated poly(ether ether ketone) proton exchange membrane for reducing methanol crossover." Journal of Power Sources **180**(1): 23-28.

Fixed Plant Facility Siting – A Mega Project Lifecycle Framework

M.F. Haider
mech_haider@yahoo.com
mhaider@suncor.com

Prepared for Presentation at
American Institute of Chemical Engineers
2014 Spring Meeting
10th Global Congress on Process Safety
New Orleans, LA
March 30 – April 2, 2014

Fixed Plant Facility Siting – A Mega Project Lifecycle Framework

M.F. Haider
mhaider@suncor.com
mech_haider@yahoo.com

Keywords: Facility Siting, Project Management, Mega Project, Consequence Analysis (CA), Quantitative Risk Assessment (QRA), Adaptive, Change driven, Agile methodology

Abstract

Facility Siting based on tools such as Consequence Analysis (CA) and Quantitative Risk Assessment (QRA) is often times successfully applied for Brownfield operating unit projects, projects with relatively smaller footprints and for projects lasting relatively short time-periods. Challenge remains to apply and realize the full potential and value of Facility Siting for Mega-Projects. This is mostly due to project complexity and length – this paper will introduce a Management Framework to successfully apply CA and QRA as part of Facility Siting for Mega-Projects. The paper will also outline key strategies to ensure successful (i.e. value adding) implementation of facility siting, suggest a list of activities, associated sequence and pitfalls to avoid. Project Management team (e.g. Project Sponsor, Project Engineers, Managers) of both owner/operator and contractor/vendor may find this paper more useful as it provides a broad overview (i.e. when to initiate, how to plan, execute and close such assignment to maximize value) including real life examples of facility siting process over the project lifecycle.

1. Introduction

The driving force for process safety has been primarily based on catastrophic events [1]. With an increasing number of incidents, the process industry and governments started to take measures such as Facility Siting initiatives with hope to minimize loss of life and property. In the 1970s and 1980s, some of the world's most tragic industrial accidents took place. The Flixborough explosion in 1974 was reinforced by that of the Seveso tragedy in 1976 [2]. The Macondo blowout in the Gulf of Mexico on April 20, 2010, killed 11 employees and led to an uncontrolled oil spill lasting 87 days. This blowout was the most significant offshore incident in the US to-date. Finally, the Fukushima Daiichi nuclear plant incident in Japan in March 2011 drew the attention of the global process and power industries and encouraged them to incorporate natural disaster risks in a hazard analysis study [3]. As part of inherently safer design, many companies now-a-days apply CA and QRA as part of facility siting for their new projects. Due to the scarcity of information at the beginning, the studies in many cases start later in the project lifecycle for them to be effective and value adding. This paper will focus more on how to manage such initiatives for new Mega-Projects over the technical intricacies of the CA/QRA processes. The intention of the paper is to enhance the Body of Knowledge publicly available for management team in as much non technical languaga as possible to make their greenfiled Mega-Project initatives a success i.e. safer for people, environment and property.

So, what is different for industrial Mega-Projects?

- Industrial Megaprojects, as the name implies, are very large. To provide a simple and simply applied definition, we are defining a megaproject as any project with a total capital cost of more than $1 billion (U.S. dollars) as measured on January 1, 2003 [4].
- Most often engineering and construction activities run in parallel and not in sequence
- Multiple sub-projects under the 'Mega-Project umbrella' run in parallel
- Complex – design is almost 'never' complete in early stages – goes through many iterations
- Progressive elaboration is evident
- Scarcity of critical workforce with mega-project experience
- Scarcity of information – requires project team to make assumptions for CA/QRA
- Timeline: 2-5 years

2. Key Strategies

2.1 Facility Siting Charter

It is not only important or good to do but critical that the stakeholders need to identify, understand and agree on the purpose and value to conduct facility siting for their projects. The Author's personal experience over the years has been in order to maximize value of Facility Siting the project team has to start 'early' to develop project Charter and start with the end in mind. As can be seen from the following generalized Figure 1 that as the company starts to appraise opportunities, it is

Figure 1: Stages and Milestones to apply CA/QRA in a Mega-Project

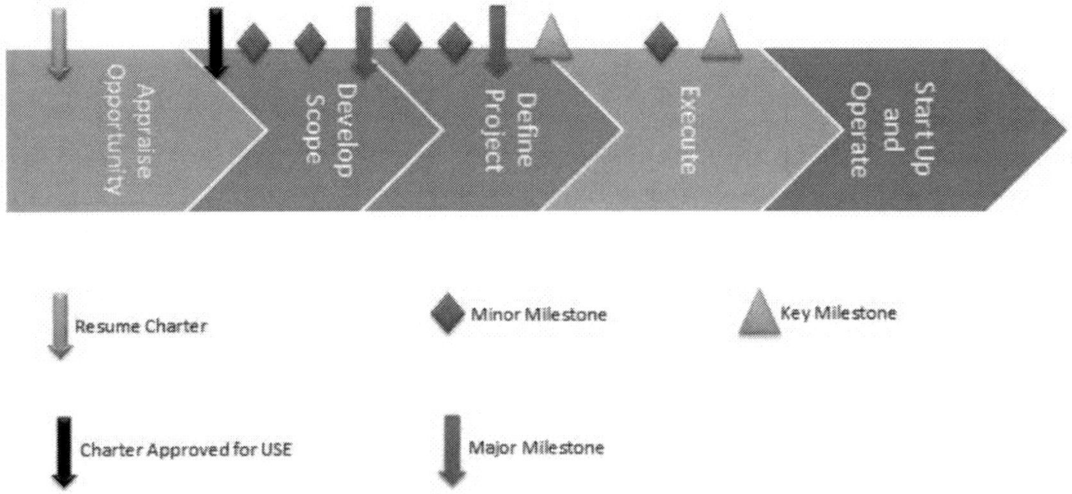

prudent to resume developing the Facility Siting Charter by bringing in the stakeholders together. Figure 1 shows a generalized project lifecycle for a typical Petro-Chemical Mega-Project. The

Figure also shows that the Key Deliverable is developed through a successive series of repeated cycles over the project life-cycle. The Milestones will be defined later in the paper. Part of starting with the end in mind is to clearly outlining ultimate goal/s, e.g. vision is to design a safer facility and utilize Mitigated QRA Report to formally document and communicate that facility is within the Risk Threshold or justification for accepting risk profile as is by Senior Management. The Vision in the Charter also provides opportunity to strike a balance between business goals and climbing the Risk Mountain – at each stage of increasing safety (decreasing risk). As part of striking a balance, the Project Engineer has to Identify options up-front if and/or when it becomes apparent that aggregate Residual Risk will not be within the Risk Threshold or acceptable so that team can put more effort in developing justification for either accepting or rejecting the risk profile by Management

Figure 2: Strike a balance between Business goal and Climbing the *expensive* Risk Mountain

2.2 Management Strategy

In order to successfully manage Facility Siting, CA and QRA for a Mega-Project, the stakeholders have to align on a management strategy. There are two primary management approaches for the purpose of managing such work:

1. Predictive or Plan driven
2. Adaptive or Change driven

The Author's past experience has been if the organization undertook very 'similar' projects in the near past (<5 years), critical experienced workforce is retained, Lessons Learned is readily available, and most importantly, the past project is still considered to be a success, it is advisable to use Plan driven approach. On the other hand, if the organization is introducing new technology, has limited past experience with similar project, involves relatively high number of stakeholders, capital cost > $1B, project life spans >2 years, it is recommended to utilize Change driven approach to maximize value. It is to be noted that it is not essential to meet all the suggested criteria mentioned above to pick one management strategy over the other.

2.2.1 Benefits of Adaptive or Change Driven Approach

In Adaptive/Change Driven Planning, project team intentionally repeats one or more project activities as the project team's understanding of the product increases ……….Also known as …agile methods …..intended to respond to high levels of change and ongoing stakeholder involvement [5]. The heart of Facility Siting involves making timely but informed decision. Some of the characteristics of this framework:

1. To add small incremental improvements that will deliver value to stakeholders
2. Milestones develop the product (e.g. Mitigated QRA Report) through a series of repeated cycles, while increments successively add to the functionality of the product
3. For each Milestone, a deliverable (read 'model') or set of deliverables will be completed. Future iterations may enhance those deliverables or create new ones. Each iteration incrementally builds the deliverables until the exit criteria for the phase are met, allowing the project team to incorporate feedback.

The biggest challenge with Mega-Project is scarcity of key information needed as inputs to develop scenarios and models. Some of the critical benefits, a project may gain by applying Change Driven approach:

1. The approach acts as Leading Indicator on Process Safety for Plant Layout - allows for multiple opportunities to make critical decisions on plant layout over the project life-cycle capitalizing on what is known at a given time. This lets the project team to use "what they know", to ensure "what they don't know" falls into the right place
2. Opportunity to make decisions over the project lifecycle not only works to save life but also helps with striking a balance between business needs and climbing the risk mountain
3. Aides design-development with progressive elaboration while producing final CA/QRA deliverable for the project

2.3 Execution Strategy

The main idea of Change Driven process starts with decomposing the known preliminary layout and process information into a set of requirements, work to be performed and their priority sequence, sometimes referred to as a *Product Backlog*.

It is advisable that the owner/operator forms a multi-use agreement (contract) with the chosen vendor. This way the plant areas under the Mega-Project umbrella can get their specific work done while the CA/QRA Project Team provides overall oversight and guidance.

At the beginning of an iteration, the team will work to determine how many of the highest priority items on the backlog list can be delivered within the next iteration. The time-length of each iteration should be less than 4 weeks. As shown in Figure 1, during "Develop Scope" and "Define Project", the length of each iteration could be even less than 2 weeks. This iterations (otherwise known as "Minor Milestone" as shown on Figure 1) will develop scenarios, assumptions and preliminary CA/QRA models. Assumptions are required due to lack of available quality information in the early stages of a project.

At the end of each iteration, the Minor Milestone/deliverable should be ready for review by the customers/stakeholders. This does not mean that the customer is required to accept delivery, just that the product should not include unfinished and/or incomplete features.

The sponsor and customer representatives should be continuously engaged with the other project stakeholders to provide feedback to Contractor on deliverables as they are created and to ensure that the *Product Backlog* reflects Sponsor/stakeholders' current needs. Below, you will find two examples using two different management strategies. For confidentiality purpose, the specific name/location is kept generic and few items were modified to suit.

2.3.1 Project X Facility Siting Using Change Driven Approach

A CA/QRA team was given the assignment to support design development during "Develop Scope" phase (Figure 1). The Sponsors already selected the site for the project. But the fixed plants were yet to be located on the site. The Vision for the initiative was for the QRA result to come under the organization Risk Threshold. Multiple EPC's were given to engineer different plant areas. The CA/QRA team selected a vendor to support them with defining the *Product Backlog* based on the preliminary Process Design Basis, intended product the facility is expected to sell, the feedstock necessary etc. The CA/QRA team had prioritized the Product Backlog based on hazard level of feedstock, preliminary PHA schedule and project schedule. This also involved obtaining feedback from different Process Engineering and Operations teams to collect necessary inputs to put together the assumptions and scenarios to be modeled within the 1st iteration.

Out of 25+ scenarios, only 7 perceived to be the most hazardous were modeled for CA that gave ample input (e.g. explosion, toxic and fire contours) for the CA/QRA team to work with the stakeholders to recommend location for specific process units on the given site. Note that, CA/QRA team integrated the potential consequence to understand local incident and global consequence events. At this time QRA was not performed. This was used as input for minor modification of preliminary plant layout. It is to be noted that the site plan was still preliminary after the first iteration but assisted different EPC accountable for their specific area to recognize where to locate the specific plants i.e. finalized key plans for the overall site plan.

With further design-development, more refined information became available from each EPC. This new information led to 130+ scenarios and was used for further CA/QRA iterations that created either new model and/or embellished the past ones. Each EPC was accountable for facility siting within their boundary while the CA/QRA team provided technical oversight and integration support for the overall project. The Sponsor/stakeholders were engaged on a regular basis to ensure that the Product Backlog reflects future modeling requirements to properly site facilities e.g. Fire Water Pump house, Tank Farms, site Medical Centre etc. Early engagement of CA/QRA team and higher management willingness to incorporate their feedback with the progressive elaboration of the project led to QRA result meeting the organization risk threshold and ultimately safer facility. Changes suggested based on CA/QRA work in earlier stage (i.e. Develop Scope, Define Project) were easier to incorporate into the design.

2.3.2 Project Y Facility Siting Using Plan Driven Approach

The sponsors for project Y selected the site for the project and decided to hold on to not perform CA/QRA work with the progressive elaboration of the project activities. This was mostly decided based on the fact that not all information was available to properly conduct the 'studies'. Fundamentally, the sponsors looked CA/QRA as deliverables requiring a more 'complete' plan. Once the Develop Scope (as referred in Figure 1) is almost complete, the sponsors decided to engage CA/QRA team and a vendor. By this time, the project had significantly high number of stakeholders. Many of the stakeholders were not very familiar with CA/QRA and the value it could bring to them. There was not much time to schedule and train the stakeholders as well to bring them up to speed. As a result, the stakeholders were unable to align on a Vision. The Vendor was asked to provide explosion, toxic and fire contour maps for specific scenarios and locations that were almost 'fixed' on the site plan at that point in time. The resultant contour maps showed significant consequence on nearby facility with over 700+ people. It was also noted that the plant area consuming the most hazardous substance would impact the nearby facilities significantly from consequence perspective. It should be noted that the result was only based on a selected few scenarios. By the time the project reached the "Execution" phase, more scenarios were added as part of the scope. But it was already too late to incorporate most of the recommended changes resulting from CA/QRA work. The project went through several area by area risk reduction workshops only to accept (by the company executives per organization risk matrix) the aggregate residual risk level which was also found to be not within the organization risk threshold.

2.3.3 Importance of Starting Early

Figure 3: Impact of variable based on project time [5]

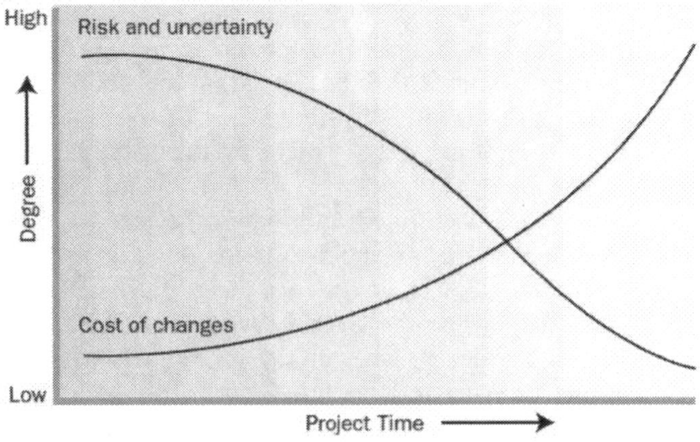

The above two examples highlight the importance of starting CA/QRA work early enough to realize and maximize the value CA/QRA brings to a Mega-Project. The high-level view provides common frame of reference for comparing projects for CA/QRA initiative, even if the projects were dissimilar. Change driven methodology allows for applying the known information at early stage in a way that supports laying out plant facilities to reduce risk and thus maximize potential value. While these characteristics remain present to some extent in almost all project life cycles, they are not always present to the same degree [5]. Change Driven framework, in particular, keeps stakeholder influences higher and the costs of changes lower especially in the early stages of the life cycle than in predictive/plan life cycles. The Figure 3 highlights a well-known fact of project management that even though risk and uncertainty remains high in the early stages of a project, cost of incorporating changes remains much lower. This translates into a key opportunity for project stakeholders to align on laying out the key areas on a site plan intelligently to reduce risk and maximize value by applying tools such as CA and QRA.

2.4 Key Decision Points

Facility Siting, from a management point of view, is all about bringing the stakeholders closer and find the 'right' spot for the facilities. Finding the 'right' spots requires making timely decisions. The challenge has always been to make timely decisions as the stakeholders more often find themselves forced to make those decisions based on information that are frequently of poor quality and at times uncertain, especially in the early stages of a project. At times, due to the scarcity of information, stakeholders rely on their past experience to make those decisions on plant layout. As a result, depending on the individual experience, two different teams for a project area may come up with two very different plant layout. CA/QRA work, if started early enough, will also depend on input that most often requires embellishment through assumptions to create needed scenarios and models. Even with this limitation, CA/QRA provides a key 'input' in allocating real estate for key facilities on a plant site at early stage of a project lifecycle. With each iteration and deliverable 'Change driven' methodology allows for multiple opportunities to make critical decisions on plant layout over the project life-cycle. It allows the stakeholders to decide when to 'sharpen the pencil' (e.g. PHAST to Computational Fluid Dynamics model), relocate facility or accept the consequence and/or risk, 'pull the plug' on further modeling/refinement and start developing justification, if required, to recommend management to accept/reject the aggregate risk profile.

2.5 Alignment with Process Hazard Analysis (PHA) Planning

CA/QRA results are a useful input to PHA (e.g. HAZOP, LOPA) activities. This is particularly true as projects move forward after establishing the plant layout on a high level and starts to conduct "Define Project" (Figure 1). PHA activities may also result in recommendations (i.e. new scenarios) that goes into the *Product Backlog* and requires project team to develop new models or refine old ones. Mega-projects have multiple sub-projects running in parallel most of the time, having an Integrated Schedule for all the reviews is a good strategy so that the right stakeholders are booked for the sessions. The integrated PHA schedule also supports prioritization of *Product Backlog* for CA/QRA activities over the lifecycle of the project.

2.6 Integration Strategy

It is critical for a Mega-Project involving multiple plant areas/silos to understand the BIG picture so that site risk can be dealt with on an aggregate basis. There is strong possibility of one area to impact other plants in the vicinity. It is recommended to review integration aspect and aggregated results on a regular basis. The following Figure 4 shows that the origin of an incident may reside in one particular plant area, while it may impact other areas in the vicinity.

Figure 4: Consequence of an incident in one area impacting multiple areas

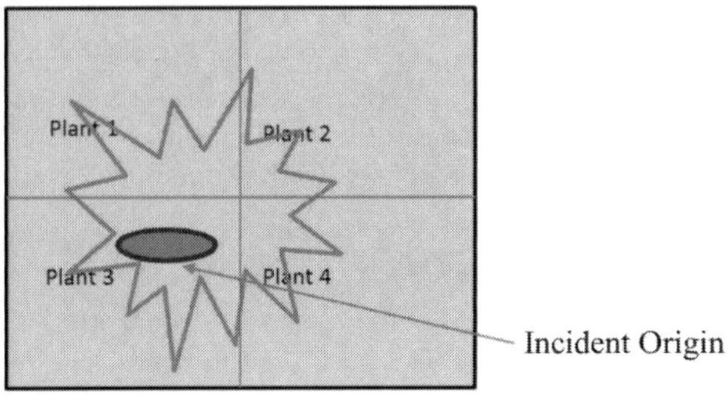

Recommended Strategy:

- Setup Strong Center - Task Force Single Point of Contact (SPOC)
- Collaborate – SPOC will promote Collaboration among Plants/Areas

3. List of Activities (suggested)

3.1 Overall Sequence

The CA/QRA work typically should start with the approved Charter as shown on Figure 1. A project Task Force i.e. SPOC is formed comprising of Project Engineer, Operations rep, Process rep and CA/QRA Technical Expert. The SPOC is ultimately accountable to produce the CA/QRA deliverable studies but also acts to support the design development and site plan finalization efforts in the early stages of the project.

The team collects 'input' such as Preliminary site Layout and design information, works with area process and operations team to identify scenarios, develop assumptions for known-unknowns and unknown-unknowns, develop modeling 'Product backlog', prioritize backlog based on PHA schedule and produce modeling output (i.e. deliverable) for stakeholder use. The stakeholder usually has the option of accepting or rejecting consequence. If the result of an iteration is acceptable within area causing the incident – communicate to outside stakeholders to ensure there is no cross boundary impact. If the result of the iteration is NOT acceptable to outside of the area – use the result as input to PHA for the areas to mitigate. Collaboration recommended between origin area and impacted area/s to identify appropriate mitigations. There is also a possibility that the consequence resulting from the iteration is not acceptable to the area causing the incident, in that case the area will initiate a PHA to mitigate the consequence. As part of developing mitigations, consider the following:

- Develop QRA if appropriate
- Perform Building Assessments for Critical Buildings
- Apply Cost-Benefit analysis to optimize risks/opportunities

- Determine Safe Haven locations and their characteristics

If Mitigation is not possible from Business perspective – perform As Low As Reasonably Practicable (ALARP) justification for Senior Management to formally accept the posed risk

- To apply mitigations on a site wide basis, review Milestone deliverables by taking on a BIG picture perspective
- Future iterations may enhance those deliverables or create new ones as new information is available. Each iteration incrementally builds the deliverables to produce the Mitigated CA/QRA report to support Facility Siting for the stage

The overall effort is expected to produce a maximum of two single reports for the overall site:

1. Mitigated version of QRA Report
2. and Consolidated CA Report

It is recommended to develop as-built, pre-start up CA/QRA reports prior to start-up. Table 1 and 2 outlines typical activities performed by SPOC and plant areas respectively once key process areas are established/finalized on Site Plan. Please note that the activities are project dependent and may warrant modifications to suit specific situation.

Table 1. SPOC Sequence of Activities to conduct Facility Siting

Initiative	Key Steps
Setup Facility Siting Team	Sponsor to support to setup a team (i.e. Project Engineer, Technical Expert, Operation, Process representative) to act as SPOC
Integration	Provide functional (technical and management) guidance to Areas
	Assess cross boundary impact/opportunity
	Communicate incremental outputs after each Area PHA/Risk Reduction
IFA versions (prior to "Define Project" stage)	Consolidated CA Report
	Mitigated and Integrated QRA Report
IFD CA/QRA (prior to "Execution" stage)	Re-validate IFA CA/QRA post PHA and vendor PHA
As-built versions	Post-oil version

3.2 Typical Area Activities

Table 2. Area (Typical) Sequence of Activities to conduct Facility Siting

Initiative	Key Steps
Plant 1 (Typical)	Perform Modeling and Calculations
	Conduct Risk Reduction Workshop (if appropriate) – Apply mitigations
	Incremental output for use based on reduced risk level
	Assess Buildings/facilities
	Identify Safe Haven/shelter Locations (if required)
	Estimate Area Project Scope impact
	Perform area scope updates/changes e.g. passive fire protection requirement
	Carry out QRA (if appropriate)
	Update Product Backlog
	Develop/update contract to support area specific activities

4. Pitfalls to Avoid

The following observations were noted by the Author during his time to manage Facility Siting scope:

1. Facility Siting does not replace good and common sense engineering practice e.g. locating a plant that uses high volume of hazardous material will significantly increase risks to other plants/areas (e.g. camp) within the vicinity
2. Evaluate effectiveness of 'Cost Reduction' philosophies as opposed to added risk e.g. different modularization techniques may decrease the plant footprint by reducing pipe lengths but makes the plant more compact and likely to increase severity of explosion consequence
3. Avoid developing models with hope to realize risk reduction
4. Sharpening the pencil for models most likely may not reduce risk
5. Avoid jumping into CFD or other similar methods without evaluating cost-benefit for undertaking such effort
6. Utilization of QRA results is much more controversial than the methodology. Watson [6] has suggested that QRA should be considered as an argument, rather than a declaration of truth. In this view – focus on accuracy over precision
7. CA/QRA Task Force should be part of Management of Change (MOC) process
8. Evaluate carefully before locating facilities e.g. tank farms close to main entrance/egress as part of Facility Siting

9. Keep the focus on VISION for the overall effort to avoid scope creep
10. Note the difference between Facility Siting as a Deliverable and tool for design-development activities

5. Conclusion

This paper presented Adaptive/Change driven framework to manage initiatives like facility siting/CA/QRA as part of Mega-Projects. Change driven framework does not mean "lack of planning or no planning". As demonstrated in the paper, this methodology starts with a vision and then develops the ultimate product both iteratively and incrementally when it is possible to define small incremental improvements that will deliver value to stakeholders. This allows making key decisions on laying out the facilities in early stages of a project rather than waiting for design to develop to a significant extent. Often times, project team wait too long for design to mature enough before applying CA/QRA to validate consequence impact and/or risk level. Most often, especially for Mega Projects, this leads to delays in executing the CA/QRA work properly at the right time to influence site layout. This delay in execution leads to scenarios when scope changes cannot be incorporated anymore. This translates into value erosion as project was unable to reap the full benefit of Facility Siting. The management framework recommended in this paper i.e. Change driven/adaptive framework allows for the stakeholders to start early in the project lifecycle to apply CA/QRA in laying out the plant facilities. The key is to start EARLY but START with the END in mind. Alignment among stakeholder on a VISION up-front is also important as this is expected to save life while striking a balance between business needs and climbing the Risk Mountain. Vision also allows avoiding scope creep while applying Adaptive/Change driven approach. It is highly recommended that stakeholders focus more on accuracy over precision. In the early stages of project lifecycle, quality information is scarce. Change driven/Adaptive framework gives you as the project stakeholder the opportunity to use "what you know" intelligently to ensure "what you don't know" falls into the right place.

6. References

[1] M. Mannan, A. Chowdhury and O. Reyes-Valdes, "A portrait of process safety: From its start to present day," College Station, 2012.

[2] M. Mannan, Lees' Loss Prevention in the Process Industries, Volumes 1–3 (3rd Edition), , 2005, Elsevier, 2005.

[3] K. McAndrews, "Consequences of Macondo: A Summary of Recently Proposed and Enacted Changes to US Offshore Drilling Safety and Environmental Regulation," in *Society of Petroleum Engineers, Americas E&P Health, Safety, Security and Environmental Conference*, 2012.

[4] E. W. Merrow, Industrial Megaprojects Concepts, Strategies and Practices for Success, New Jersey.: John Wiley & Sons, Inc., 2011.

[5] A Guide to the Project Management Body of Knowledge (PMBOK® Guide) – Fifth Edition, Newtown Square: Project Management Institute, Inc., 2013.

[6] S. Watson, "The Meaning of Probability in Probabilistic Risk Assessment," *Reliability Engineering and System Safety,* vol. 45, pp. 261-269, 1994.

348480 Global Energy and Transportation and Mobil Oil

Abstract:

Biofuels, Computational Methods, Engineering Economics, Polymers, Thermodynamics. Polymers-plastic, Thermodynamics- oil, Engineering Economics – the money it takes to finish the method of Global Engineering Research, Computational Methods – numerical input of analytical data, Biofuels – the transport fluids that utilized in more than one method from industrial usage to everyday living which enters the atmosphere. Exxon has completed a $600 million dollar deal for production of algae biofuel, amongst Mobil Oil has performed major research Global Energy and the effects of oil. Reading a most recent article on Mobil website.

348748 Characterization of Iron Phthalocyanine As the Active Material for Lithium Batteries

Abstract:

Results on the determination of the lithium-ion diffusion coefficient in the iron phthalocyanine (FePc) active material as a function of intercalation level, using a modified electrochemical impedance spectroscopy technique, will be presented. Also, developed thermodynamic functions such as the Gibbs free energy, entropy, and enthalpy changes for the lithium insertion reaction with FePc, derived from cell open-circuit voltage measurements as a function of intercalation level, are presented. Also presented are estimates of the maximum amount of lithium that can be intercalated into the FePc based on constant volume and temperature conditions.

349086 Design of a Free-Fall Reactor for Fast Pyrolysis of Waste Plastics

Abstract:

Thermal pyrolysis has earned significant reputation in the recovery of valuable chemicals, especially petrochemical based fuels from various materials, including biomass and plastics. Of many available technologies, *fast* pyrolysis was preferred in his study owing to its minimal heat and mass transfer limitations and better control on process conditions. The focus of this study is towards efficient conversion of waste plastics into gaseous products consisting primarily of alkanes, alkenes and aromatics. Lab scale experiments were performed using a micro reactor to determine favorable conditions, including reaction temperature and particle size that would result in maximum yield of ethylene. Using these results, a simple industrial size radiatively heated free-fall reactor was designed for the purpose of pyrolyzing waste plastics. The goal is to design a *fast* pyrolysis reactor that improves upon inefficiencies in traditional reactors while resulting in high yield of ethylene.

349732 Biosorptive Dehydration of Ethanol/Water Azeotropes Using Compound Starch-Based Adsorbent

Abstract:

Abstract: Recently, the interest in the production of bioethanol has greatly increased as a result of the increasing fossil fuel prices and associated environmental issues. Bioethanol is made from either grains or biomass and the broth contains 6 to 12 percent ethanol. In addition, water removal represents a serious problem as ethanol form azeotrope with water at atmospheric pressure. Distillation to water-free ethanol consumes 50 to 80 percent of energy used in the fermentation ethanol manufacturing process. Various techniques have been developed to break the azeotropic point, such as salting-out method, azeotropic distillation, extractive distillation, reactive distillation, adsorptive distillation and pervaporation. Among the methods, the adsorptive option is the most attractive, especially the biosorption, which has distinct advantages over conventional methods as it is environmentally friendly, highly selective, easily available, easy to operate, cost effective and reusable in repeated cycles in the treatment of aqueous organics.

In this work, a specially formulated compound starch-based adsorbent (CSA), which consists of corn, sweet potato and foaming agent, was developed. As an exploratory study, this research was focused on the investigation of CSA for biosorptive dehydration of aqueous ethanol and optimization of conditions for its maximum adsorption under various conditions of vapor feed flow rate, bed temperature and bed height through an orthogonal design method. Furthermore, to investigate the adsorption mechanism, dynamic saturated adsorbance of CSA was studied and characterization of the adsorbent before and after adsorption, as well as after regeneration, was performed.

In the first step, the net retention time and separation factor of ethanol and water were obtained using inverse gas chromatography (IGC), the results indicated that the net retention time of water on CSA is significantly longer than that of ethanol which attested the feasibility of this adsorptive separation purpose and low temperature could be propitious to the process which was consistent with the traditional theory of adsorption.

Then, through a batch adsorption experiment, the optimum adsorption condition was determined under vapor feed flow rate, bed temperature and bed height by orthogonal design method, which indicated that the optimization condition for ethanol was at vapor feed flow rate of 1.1g/min, bed temperature of 83 and bed height of 25cm. The packing density was 445kg/m^3. The residence time is determined by vapor feed flow rate and bed height. The residence time is more influenced than bed temperature. The longer the residence time, the more the output is. But it cannot be ignored that the adsorption selectivity may decreases. Low bed temperature is benefited to the adsorption process which agrees with the result of IGC. Five cycles experiments of adsorption-regeneration were carried out at the optimum adsorption and regeneration condition, the results showed that there is practically no change in the regeneration efficiency of CSA after the cyclic process adsorption–regeneration. It is due to the properly chosen regeneration conditions assuring complete desorption of the water adsorbed without changes in the CSA structure. This unchanged adsorption capacity after a number of cycles demonstrates reusability of the adsorbent.

To study the adsorption mechanism, experiment was conducted to study the dynamic saturated adsorbance of CSA. The water adsorption capacity of CSA has been found to be similar to some molecular sieves, which is about 0.15g/g. Besides, field emission scanning electron microscopy (FESEM) and mercury porosimetry (MP) were applied to investigate the change of surface morphology and microstructure of CSA before and after the adsorption, as well as after regeneration, as shown in Fig.1 and Fig.2. The FESEM images showed that compared with the particles before the adsorption, they became plump and the void spaces became smaller after adsorption, which indicated that the CSA was saturated with water. However, the CSA particles after regeneration became wizened like the nature particles, which demonstrated that almost all the water had been desorbed from CSA. The pore size distribution (PSD) indicated that compared with the PSD curve before adsorption, the peak at 6000nm disappeared and the peak at 17000nm became remarkably smaller after adsorption, which manifested that CSA is mainly composed of macropores which play a major role in the adsorption. However, the peaks at 6000 and 17000nm reappeared after regeneration, and the curve looked exactly the same with that before adsorption, which further demonstrated that CSA after regeneration can be totally regenerated and recovered to the original structure.

These results indicated that the starch-based adsorbent can be considered as an effective and low-cost bio-sorbent to dehydrate alcohol.

349781 Purification of 2-Amino-1-Phenylethanol Enantiomers By a New Technique Combining Distillation and Crystallization

Abstract:

Stripping crystallization (SC) is a new separation technique which combines distillation and crystallization. Basically, SC is operated at a triple-point condition, in which the liquid is simultaneously vaporized and crystallized due to the three-phase equilibrium. In essence, the process is continued until the liquid phase is completely eliminated and only the pure solid crystals remain in the feed. Thus, crystal washing is not required since only pure solid crystals remain in the feed and no impurities are adhered on the crystal surfaces at the end of SC. In the previous research, SC has been successfully applied to separate p-xylene from the mixed xylenes by Shiau et al. (2005, 2008).

Phenylethanolamine, also known as 2-amino-1-phenylethanol (2-APE), is an important basic skeleton in many adrenergic drugs and is often used as ingredients of cold medicines and anorectics. As the biological activity and pharmacokinetics of the enantiomers of a chiral drug can be different, the objective of this research is to apply SC to purify R-2-APE from the chiral 2-APE mixture. The experiments show that no crystallization occurs during SC without seeding due to subcooling of the liquid enantiomers. On the other hand, seeding can eliminate subcooling during SC, leading to the R-2-APE crystals gradually formed preferentially on the seeds. Thus, seeding with magnetic stirring can be successfully employed to purify R-2-APE from the 2-APE enantiomers.

1. Shiau, L. D., Wen, C. C. and Lin, B. S. Separation and purification of p-xylene from the mixture of m-xylene and p-xylene by distillative freezing, *Ind. Eng. Chem. Res*, **44(7)**, 2258-2265, 2005.

2. Shiau, L. D., Wen, C. C. and Lin, B. S. Separation of p-xylene from the multicomponent xylene system by stripping crystallization, AIChE Journal, 54(1), 337-342, 2008.

350841 Fabrication of Low Cost Insulating Material from Kaolin Clay for Construction Purposes

Abstract:

A new low cost, thermal insulator, construction mineral polymeric material was synthesized from local Saudi Arabia resources using Inorganic Mineral Polymerization technique (MiP). In this technique, the mineral polymer is produced from the following components: clay mineral rich in koalinite, sand, sodium hydroxide solution, and water. Inorganic polymerization experiments are conducted by mixing known amounts of clay minerals and sand in a variable speed stainless-steel mixer for some period of times until a homogenous solid mixture is obtained. Sodium hydroxide solution is added to the mixture and the polymerization reaction takes place at different mixing speeds and times. The produced mixture is divided into a number of samples of known weights. Each sample is moulded in a stainless-steel cylinder at a medium pressure. Then the moulded samples are cured in an oven for 24 hours. The curing temperature was 80°C. The mineral polymerization conditions such as the ratio of clay to sand filler, particle size of the clay, particle size of the sand, concentration and amount of sodium hydroxide, speed and time of mixing were optimized to obtain stable and hard mineral polymer. Compressive strength for different specimens of fabricated mineral polymers was tested under dry and wet conditions. Values of compressive strength with values ranging from 11.5 MPa to 27.9 MPa were obtained.

351141 Thermodynamic Study of Binary PAH (Anthracene + Phenanthrene) Solid Mixtures

Abstract:

Polycyclic aromatic hydrocarbons (PAHs) are common components of many materials, such as petroleum and various types of tars. They are generally present in mixtures, occurring both naturally and as byproducts of fuel processing operations. It is important to understand the thermodynamic properties of such mixtures in order to better understand and predict their behavior (i.e., fate and transport) in the environment and in industrial operations. The Knudsen effusion technique was combined with gas chromatography and differential scanning calorimetry to measure the vapor pressure, composition, and phase behavior of binary anthracene and phenanthrene mixtures at near ambient temperatures. Mixtures of anthracene and phenanthrene exhibited non-ideal mixture behavior: In mixtures with less than 80% phenanthrene, this more volatile component preferentially sublimed and the vapor pressures decreased as the mixtures became enriched in anthracene. The vapor pressure data do not follow Raoult's law (that is, ideal mixture) predictions. Mixtures with high phenanthrene content (> 80%) showed azeotrope like behavior, in that anthracene and phenanthrene sublimed at nearly constant vapor pressure and molecular ratio. These results confirm that the behavior of few-component PAH mixtures does not follow ideal mixture models. Hence, as tars and petroleum (and potentially other organic compound-containing mixtures) weather to few component mixtures over time, complicated, non-ideal phase behavior might be observed.

351160 How Confident Are You That Your Major Accident Risks Are Under Control?

Abstract:

Major accidents in the process industries including those at Texas City , Buncefield UK and in the Gulf of Mexico have highlighted the critical role played by senior managers and executives. Effective leadership is essential if companies are to develop a positive safety culture that remains constantly vigilant towards the risk of a major process safety incident. In the UK the HSE has made leadership a key priority for the high hazard industries and there is a cross-industry initiative to develop and deliver training focussed towards board members responsible for process safety performance.

This paper explores the leadership failings that contributed to recent major accidents, and the cost to both the company reputation and the individuals involved. Key principles are discussed that can help improve leadership in the major hazard industries. These include; attitudes and decisions of senior manager affecting the investment strategy and the safety culture of the organisation, reinforcing the importance of safety by personal example, thorough understanding of major accident hazards and key risk control systems, investigating process safety incident and near misses to find the underlying causes, developing world class safety management systems, and identifying weaknesses in these systems using targeted performance indicators.

The author bases this paper on the development of a training standard for 'Process Safety Leadership' involving a cross-industry 'expert panel'. Training of senior executives in compliance with this standard has been carried out from late 2011 and feedback from these courses and the key requirements for effective leadership will be discussed.

351236 Sil Determination: Shortcomings with the Use of LOPA

Abstract:

Proposed Topic Area: Managing Risks

For a number of years, industry has been trying to apply the principles outlined in IEC 61508. In the process sector, this has resulted in a focus on the requirements of IEC 61511 (2003) and ANSI/ISA 84.00.01 (2004). These are risk based standards. They involve the identification of potential hazardous events, the causes that could lead to each event and the necessary risk reduction for each event. Where the risk reduction involves an instrumented function (e.g. a trip or an interlock), the required performance of the instrumented function can described in terms of the average probability of failure on demand. In the standards, the performance for such instrumented functions is assigned to one of four Safety Integrity Levels. The assessment of what performance level is needed in respect to a specific hazardous event is referred to as SIL Determination. In the field of SIL Determination, IEC 61511 provides examples of a number of different approaches – these are shown in IEC 61511 Part 3. These include Event Trees, Risk Graphs, Safety Matrices, Fault Tree Analysis and Layer of Protection Analysis (LOPA).

In the first few years, following the publication of IEC 61508 Edition 1 in 2000, industry generally embraced the use of Risk Graphs as the method of choice. However, over the subsequent few years, many companies using this method found that it was (a) giving higher safety integrity level targets than really needed and (b) was less flexible than other methods and consequently more difficult to use in many real situations. This has resulted in a general move to other methods and in particular to the use of Layer of Protection Analysis. Indeed, LOPA may now be the predominant "Method of Choice" for SIL Determination. As a method, LOPA gives the impression of simplicity and appears quite straightforward. It almost invites anyone with a minimum of understanding of the method to be a practitioner.

However, there are a number of shortcomings in the use of the LOPA method that have become evident through its use across the globe.

This paper will review the application of LOPA and provide insights into what may be seen as the proper use of the method and its core principles. It will also highlight various shortcomings in use that have been found during expert reviews carried out of a number of existing studies. The paper will be of benefit both to those who are new to the LOPA method, but also of value to those who have been using the method for some time.

Shale Oil and Gas – Wellhead to Fuel

Dr. Lawrence Kremer - Baker Hughes
AIChE - Hilton New Orleans Riverside
April 1, 2014

14AIChE Spring Meeting, New Orleans, LA

BAKER HUGHES

Outline Unconventional Shale Oil

- Properties of shale oils
- Problems
  - Transportation
  - Refining

14AIChE Spring Meeting, New Orleans, LA

BAKER HUGHES

General Characteristics of Shale Oils

- API gravity typically varies 30-55+
- Very paraffinic
- Low sulfur
- Can have hydrogen sulfide and mercaptans
- Low TAN
- Very low asphaltenes
- Light crude with very little residual material
- Many samples show great quality variation

Factor	Bakken	WCS
API	39.7	21.6
S%	0.25	3.44
Light ends	2.6%	
LSR	3.5%	3%
Light Nap	12.4	3.0
Heavy Nap	9.7	5.0
Jet	10.5	10.0
LFO	27.6	12.0
Gasoil	27.2	37.0
VacTar	6.7	29.8

14AIChE Spring Meeting, New Orleans, LA

BAKER HUGHES

Shale Oil Property Comparison

Parameter	Eagle Ford	Bakken
API	52	40.8
TAN (g KOH/g)	< 0.05	0.09
Sulfur (wt %)	<0.2	0.304
Asphaltene (wt %)	0.1	0.41
Resin (wt %)	1.6	4.95
Filterable Solids (PTB)	225	76

14AIChE Spring Meeting, New Orleans, LA

BAKER HUGHES

Utica Shale Crude - Three Phase

14AIChE Spring Meeting, New Orleans, LA

BAKER HUGHES

Eagle Ford Crude Samples

14AIChE Spring Meeting, New Orleans, LA

BAKER HUGHES

Analytical Breakdown

Value	Yellow	Red	Black
API	55.0	44.6	52.3
TAN	<0.05	0.07	<0.05
Sulfur	<0.20%	<0.2%	<0.2%
Na	1.0 ppm	1.6 ppm	1.6 ppm
K	0.3 ppm	0.4 ppm	0.5 ppm
Mg	3.4 ppm	2.9 ppm	3.0 ppm
Ca	2.6 ppm	2.8 ppm	3.8 ppm
Asphaltene	0%	0%	0.1%
Resin	0.5%	3.2%	1.6%
Filterable solids	176 PTB	295 PTB	225 PTB
P	5.3 ppm	4.6 ppm	5.0 ppm

14 AIChE Spring Meeting, New Orleans, LA

BAKER HUGHES

Micrograph of Sludge

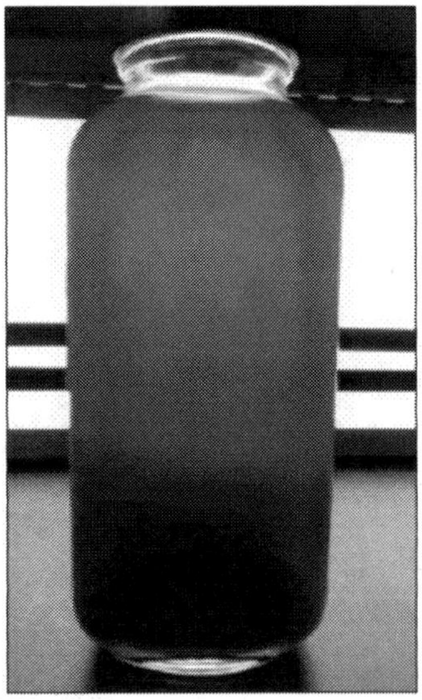

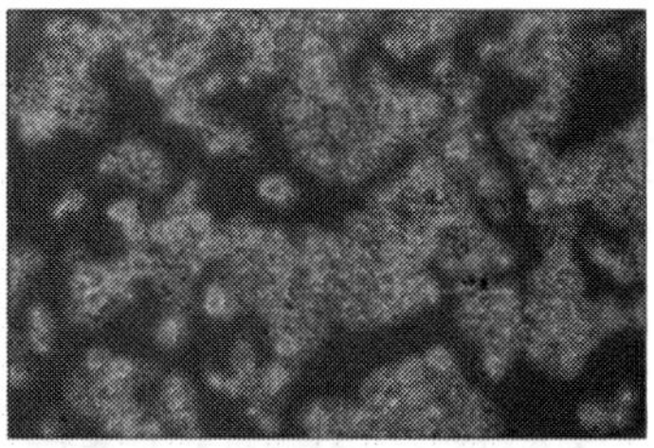

100 X

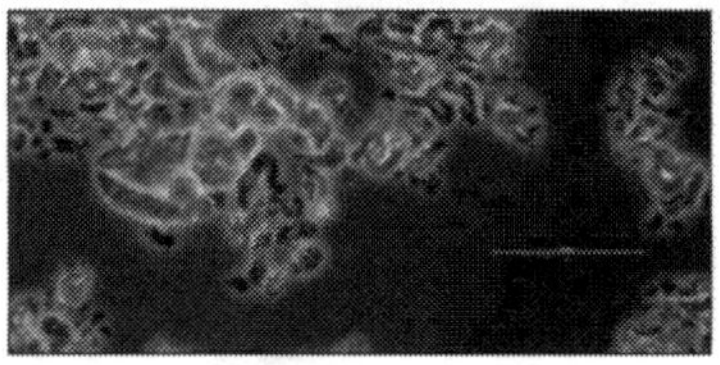

400 X

14 AIChE Spring Meeting, New Orleans, LA

BAKER HUGHES

Paraffin Analysis of Sample

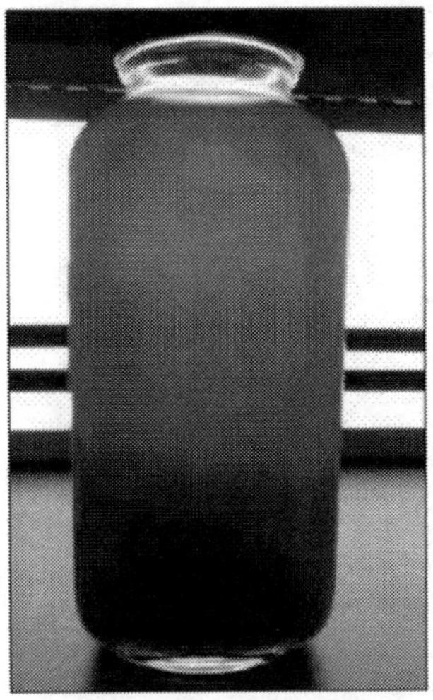

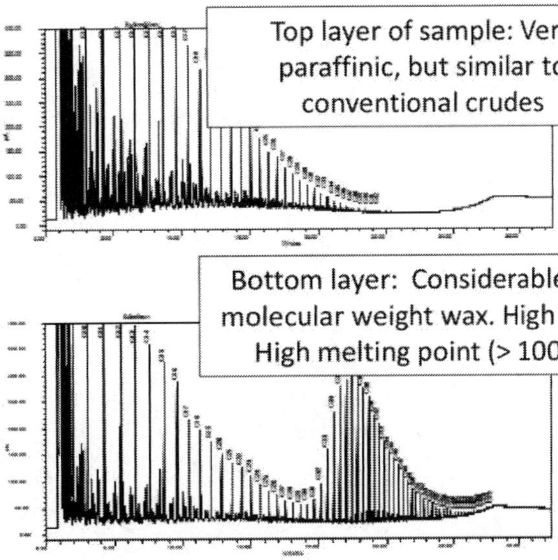

Top layer of sample: Very paraffinic, but similar to conventional crudes

Bottom layer: Considerable high molecular weight wax. High MW = High melting point (> 100 °C)

14 AIChE Spring Meeting, New Orleans, LA

BAKER HUGHES

Transport of Shale Production

- Infrastructure does not exist to meet needs
  - Most shale oil moved by rail or trucks
  - Limited pipelines expanding to meet transportation needs
- Unit trains to East Coast, West Coast, and Gulf Coast
- Problems:
  - Tankers and barges with heavy wax deposits ROB
  - Tight emulsions that look like oil or white condensate but high BS&W
  - Cars not uniformly treated for H_2S
  - Corrosion in rail cars

14 AIChE Spring Meeting, New Orleans, LA

BAKER HUGHES

Viscosity of ROB Before Treatment

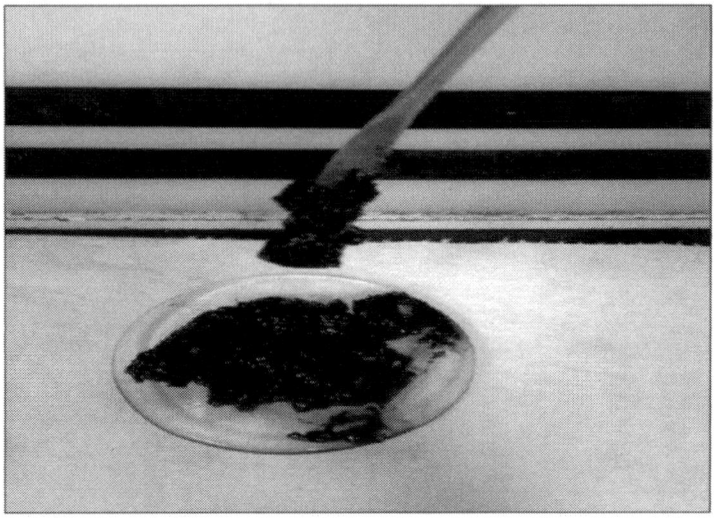

ROB from untreated compartment

14AIChE Spring Meeting, New Orleans, LA

BAKER HUGHES

Eagle Ford Shale Oil Wax Dispersion Testing

Several products were effective in dispersing the paraffin wax in an Eagle Ford shale oil sample at ambient temperature for over a month.

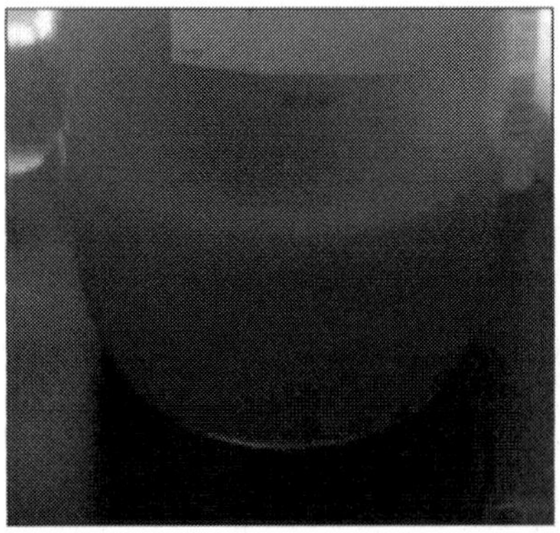

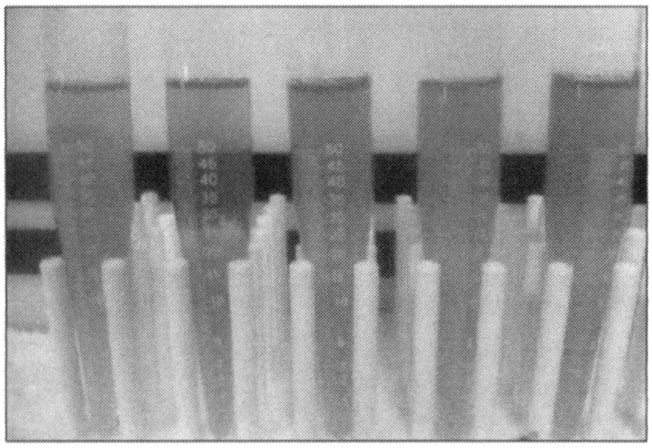

14AIChE Spring Meeting, New Orleans, LA

BAKER HUGHES

Product 'A' Effect on Wax Deposition (Utica)

Blank 200 ppm 400 ppm

14AIChE Spring Meeting, New Orleans, LA

BAKER HUGHES

Refinery Issues

- Crude Blending
- Fouling
- Corrosion
- Product quality

14AIChE Spring Meeting, New Orleans, LA

BAKER HUGHES

Crude Unit Fouling

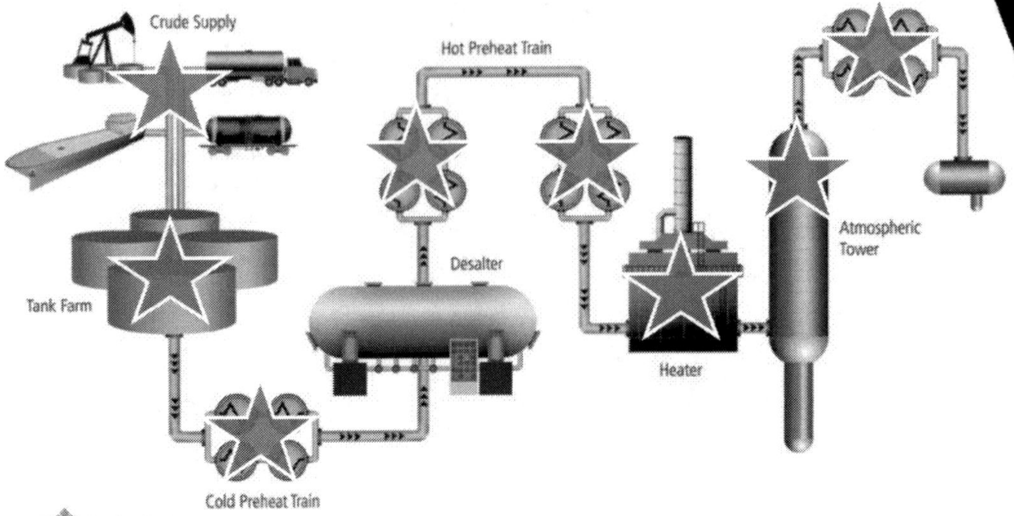

★ = Susceptible to Fouling

14 AIChE Spring Meeting, New Orleans, LA

BAKER HUGHES

Asphaltene Precipitation

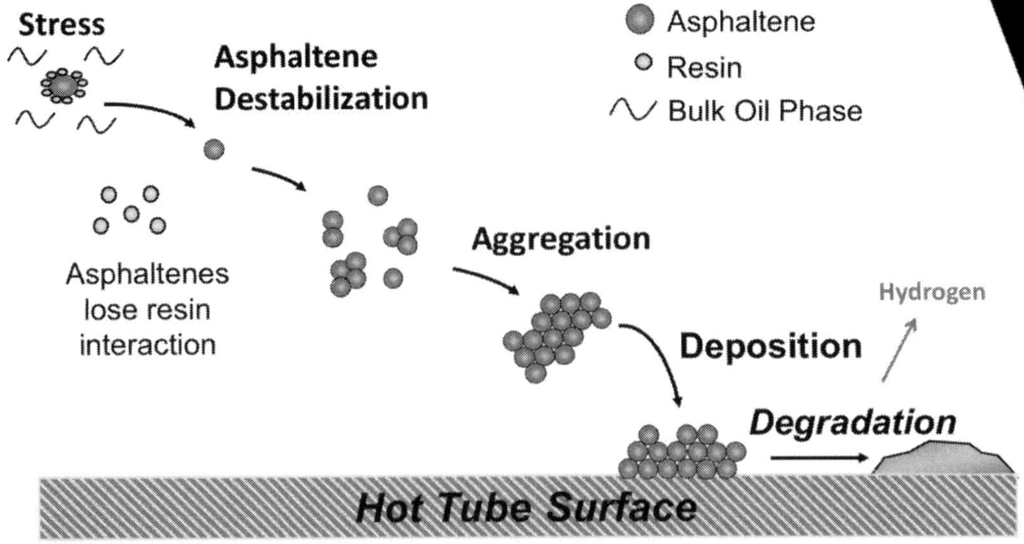

14 AIChE Spring Meeting, New Orleans, LA

BAKER HUGHES

Asphaltene Stability Index Test

Field ASIT services™ technology measures the onset of the flocculation of the asphaltenes with high accuracy by inducing the asphaltene precipitation via titration with a paraffinic solvent.

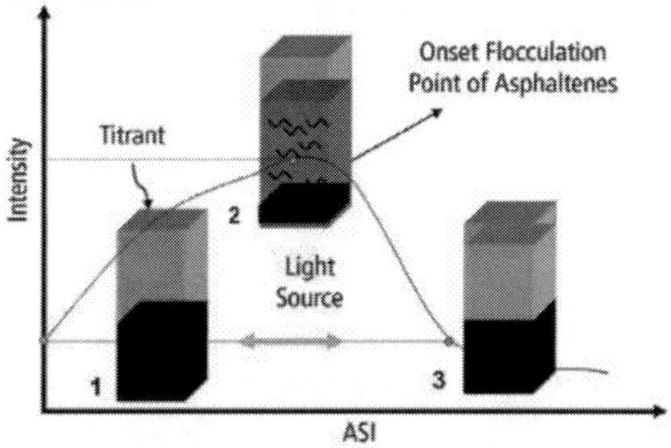

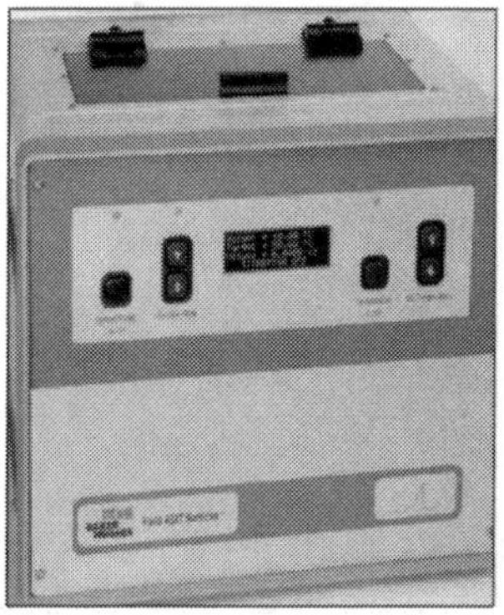

14AIChE Spring Meeting, New Orleans, LA

Field ASIT services is a trademark of Baker Hughes Incorporated

BAKER HUGHES

ASI Shows Eagle Ford Can Destabilize Crude

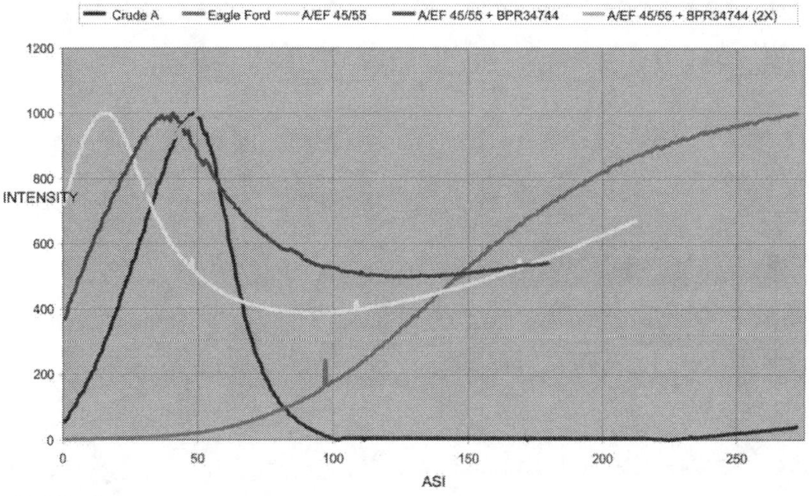

14AIChE Spring Meeting, New Orleans, LA

BAKER HUGHES

Tower Tray Fouling - Phosphorus

- Seen by refineries running Bakken, Eagle Ford, and WTI
- Operationally manifested as increasing tower dP
  - Typically in the kero/jet zone
- Tower inspection shows deposits on trays in kero/jet zone
  - Some deposition may also show a few tray levels above and/or below kero/jet zone
  - Deposit scale typically tightly bonded to tray surface restricting flow and valve function

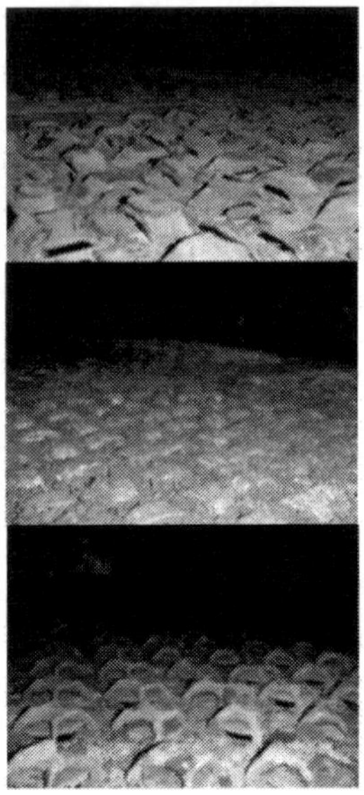

Above kero/jet zone

In kero/jet zone

Below kero/jet zone

14 AIChE Spring Meeting, New Orleans, LA

BAKER HUGHES

Shale Oil Impact – MEA Contamination

Corrosion and Fouling

- Shale oils transported by truck, rail, barge.
- Many shipments are treated with H_2S scavengers:

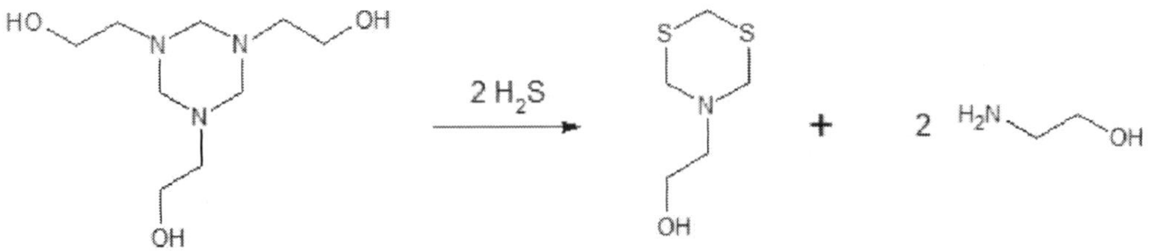

- Potential Impacts:
  - H_2S scavengers leave byproduct amine in crude oil (MEA, MA).
  - Tramp amines distill to OH system, create amine-HCl salts:
  - Corrosive to system metallurgy.

14 AIChE Spring Meeting, New Orleans, LA

BAKER HUGHES

Impact of Tramp MEA on OH Water

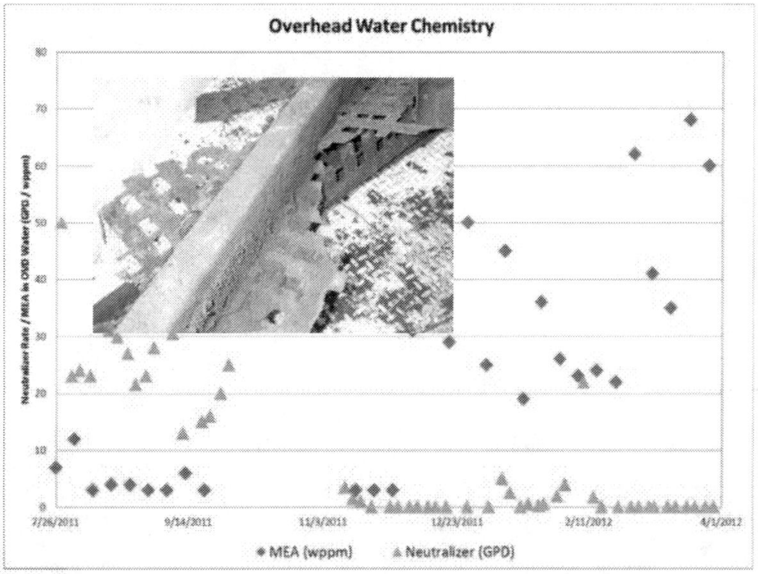

14AIChE Spring Meeting, New Orleans, LA

BAKER HUGHES

Salt Risk w/MEA Concentration

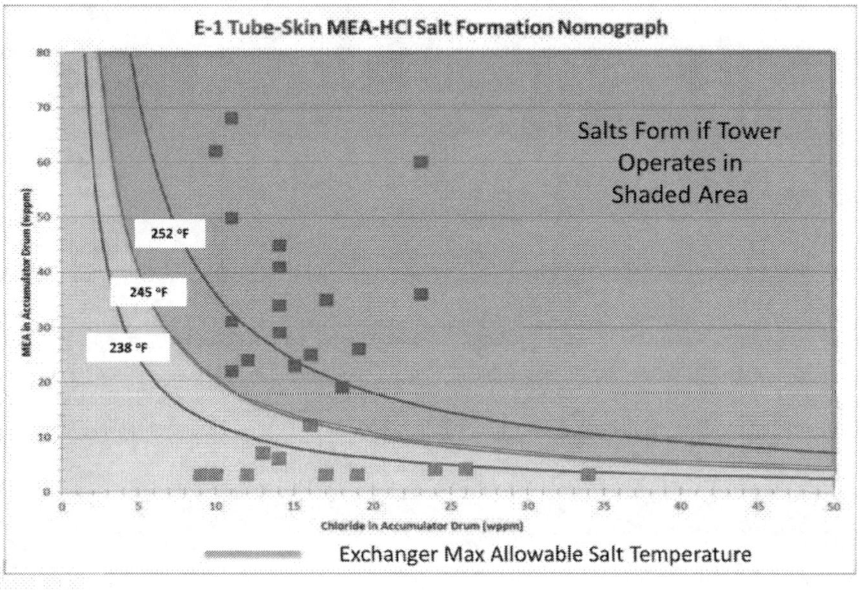

14AIChE Spring Meeting, New Orleans, LA

BAKER HUGHES

Shale Oil – Refinery Process Concerns and Monitoring

BAKER HUGHES

CHARACTERISTIC OF SHALE OIL	CONCERN IN REFINERY	PROCESS DATA	SAMPLES AND ANALYSES Samples	 Analysis	OUTPUT DATA	FREQUENCY	BHI PROGRAMS
Crystalline Wax	Tank bottom sludge	• Tank Sludge Depth	• Tank bottom	• Visual sludge • GC analysis	• Tank Sludge Depth • Crystal Wax Presence	• Monthly	• XERIC™ Pretreatment
	Cold Preheat Fouling	• Desalter Temperature			• Temp vs. Time	• Update 2x/mo	• LIFESPAN™ Wax Dispersants
	Desalter emulsions	• Emulsion layer growth	• Triline emulsions	• Visual wax • GC analysis	• Crystal Wax Presence	• After Shale Oil begins • When emulsion grows	• XERIC Demulsifiers
Asphaltene Stability in Blends	Desalter emulsions	• Emulsion layer growth	• Triline emulsions	• Precipitated Asphaltenes	• Precipitated Asphaltenes	• After Shale Oil beings • When emulsion grows	• XERIC Asphaltene Stabilizers • Field ASIT services™
	ColdPreheat Fouling	• Desalter temperature	• Crude blend	• ASIT	• Temp vs. Time • ASIT #	• Pre-Shale Oil • Update 2x/mo	• LIFESPAN Antifoulants • Field ASIT services
	Hot Preheat Fouling	• Furnace inlet temp • Furnace tube temps • Furnace fuel usage • Preheat DP	• Crude blend	• ASIT	• Temp vs. Time • Fuel use vs. Time • ASIT #	• Pre-Shale Oil • Update 2x/mo.	• LIFESPAN Antifoulants • MILESTONE™ Furnace Fouling Control • Field ASIT services
Paraffinic	Furnace Coking	• Furnace tube temps • Furnace fuel usage	• Crude blend	• BLPS	• Coking tendency	• Pre-Shale Oil • W/Shale Oil	• MILESTONE Furnace Fouling Control
Tramp MEA	Tower/OH corrosion (Early Warning)	• OH Neutralizer rate • EXCAL usage rate	• OH water • Desalter brine	• OH pH • Brine pH	• Early Sign of Amine • Early sign of Amine	• Pre-Shale Oil • W/Shale Oil	• TOPGUARD™ Overhead Corrosion Control • EXCALIBUR™ Contaminant Removal Technology
	Tower/OH corrosion (Quantitative)	• Inputs for CRM	• OH water • Desalter ww • Desalter brine	• IC Amines	• Confirm MEA in Crude • OH MEA ppm • Salt Risk Tower • Salt Risk OH System	• Pre-Shale Oil • W/Shale Oil • Update 2x/mo.	• TOPGUARD Corrosion Control • EXCALIBUR Technology • TOPGUARD Corrosion Risk Monitor (CRM)
Phosphorus	Tower P Fouling	• Tower DP			• Tower DP Increase	• 2x/mo	• LIFESPAN Antifoulants Dispersants
			• Crude	• ICP for P	• P in Crude	• 2x/mo	
			• Kerosene	• Filt Solids • XRF Filter P • ICP for P	• Incr. Kero Solids • P in Solids • P in Kerosene	• 2x/mo	
			• Kerosene • Diesel	• ASTM D86	• Incr. in Diesel IBP – Kero EP overlap	• 2x/mo	

EXCALIBUR, FIELD ASIT SERVICES, LIFESPAN, MILESTONE, TOPGUARD and XERIC are trademarks of Baker Hughes Incorporated.

How to Influence the Organization

John C. Wincek
Croda, Inc.
Mill Hall, PA 17751
john.wincek@croda.com

Prepared for Presentation at
American Institute of Chemical Engineers
2014 Spring Meeting
10th Global Congress on Process Safety
New Orleans, LA
March 30 – April 2, 2014

UNPUBLISHED

How to Influence the Organization

John C. Wincek
Croda, Inc.
Mill Hall, PA 17751
john.wincek@croda.com

Key Words: Influencing others, Effective Process Safety Implementation

Abstract

Process Safety professionals are typically not in the line management organization. Instead, they serve as engineering team members, corporate staff and facility-level advisors. Strictly speaking, they lack the authority to cause anything to happen in the operation of a process. Therefore, the Process Safety professional must know how to influence the organization to get things done. If one lacks the authority to say "we will do this," then they must be convincing when they say "we should do this." In order to influence someone, you must be able to present your idea in a way that shows the benefits outweigh the costs. This is sometimes obvious, but when it is not, or if the benefits are subtle and difficult to measure, then your credibility will have a significant impact on the decision. Credibility is not a value or a skill. It is instead a trust that is built over time. Building credibility with individuals is accomplished in different ways based upon your relationship with the person, their level in the organization, the types of work interactions you have, and both their and your past experiences. This article will discuss several different types of people with whom you may need to have credibility. It will suggest ways to build your credibility, as well strategies for influencing others.

Introduction

During over 20 years of industry experience in both staff and line management positions, the author has had the unique opportunity to have seen the grass on both sides of the fence. This paper presents lessons from the author's personal experiences. It is hoped that other Process Safety professionals can learn from these successes, failures and advice. However, much of this information is applicable to anyone in a staff position within an organization.

While an entire organization generally has the same overall goals, the daily activities, responsibilities and concerns can be vastly different between staff and line positions. While many organizations include safety-related goals for line managers, their major goals are centered on performance, production and budget. Goals for project engineering (typically a staff function) focus on completion schedules and budgets. When performance is measured primarily by these objectives, safety activities and initiatives can seem burdensome and counterproductive.

When attempting to change the organization staff safety personnel must work through the line management, influencing them to implement the things necessary for a safe working environment. The author has experienced success and failure in both line and staff positions. Some things work, and some don't. Some have worked in the past, but are no longer effective. Differences in the individuals you must influence can have a significant impact on selecting the best approach to achieve a goal. Even the timing of your efforts can affect the outcome.

Line vs. Staff Organizations

Line managers are those with direct responsibility for the business output, budget, spending and productivity, among other things. They directly contribute to the output of products or services. Line managers have the authority to change such things as what the organization does, how it functions, and what will be done when. A Shift Manager, for example, might determine which personnel are assigned to each task, what types of training are required for their employees, and how tasks are to be performed. Unit managers may determine scheduling, raw material needs, and when maintenance downtime will be scheduled.

Staff personnel are typically responsible for providing services to the line organization to assist them and the company in achieving their goals. Staff personnel include those in accounting, human resources, customer service, safety and health, research and development, and project engineering, among many others. Their customers are often members of line management, to whom they may provide information, research, expertise, advice, etc. Staff personnel typically do not have the authority, for example, to change work methods, adjust production schedules, or install safer equipment.

This difference in Line and Staff authorities is the difference between saying "we will…" and "we should..." Staff personnel can list numerous things that "we should" do. But only line management has the authority to decide which "we will" do.

Position of PSM in the Organization

Process Safety personnel are typically members of the staff organization. They may report to engineering managers, unit managers, and facility managers or, in the case of corporate-level personnel, to Technology Managers, Vice-Presidents and Directors. Like other staff personnel, their goals usually relate to assisting line managers in meeting their goals by providing technical advice and services.

The level of the organization to which process safety personnel report has a large effect on the methods used to influence what the line organization does, and how they do it. Throughout your career the people you must influence will be of different personality types, levels in the organization, and have different responsibilities. Your role in the organization will also likely change over time. Process Safety Specialists, generalists,

Subject Matter Experts and Program Champions all have different objectives. This means that each has different things they need to influence the organization to do, and thus must use different methods of influencing others.

High Consequence / Low Prob. Events

To influence others to commit to process safety and implement necessary safeguards to protect against process incidents, one must first understand the need themselves. Process Safety involves protecting against events with catastrophic consequences, including loss of life, loss of plant, and offsite effects. They can cause plants, and even entire companies, to cease to exist. Communities can be devastated, and even rendered uninhabitable. However, it is precisely because the consequences are catastrophic that the organization should expend significant resources to prevent them.

Most plants will not experience a catastrophic event over their lifespan. In terms of human experience, while these incidents can be imagined, they are experienced by very few people. It is unlikely that a manager even knows someone who has experienced a catastrophic process incident. Humans have difficulty relating process risk to their own life experiences, especially when the likelihood of an incident is several orders of magnitude less than 1. A likelihood of 10^{-5} / year equates to about one in every 1,250 human lifetimes. This can make it difficult to believe that it could happen at their particular plant.

For example, a manager who has not kept up with piping inspections likely does not believe it will cause an incident. After experiencing several minor fires due to hydrocarbon leaks, however, it may be easier to convince them to implement mechanical integrity inspections for the hydrocarbon piping. It may not be as easy to convince them that a catastrophic fire could result, or that inspections of piping systems for other types of systems should also be implemented. Adding to that the fact that experiencing several small fires can lead to the belief that fires can be adequately detected and controlled can cause a manager to lose their sense of vulnerability to major incidents

Building Personal Credibility and Integrity

Once you understand the costs and benefits of process safety, the significance of the consequences and the low likelihood of incidents, you must be able to use this knowledge to raise credible arguments for process safety. Claiming that the plant will explode as justification for every process safety improvement will not result in success for very long. It is often the case that the steps you wish the organization to take will result in reduction of risk from what is already a very small number to an even smaller one. This difference can be quite hard for individuals to understand without a firm understanding of risk. This understanding of risk often lies with the process safety personnel, who must use it to guide the thinking and understanding of the line personnel. To do this, one must have established credibility with those they are trying to influence.

The Latin origin of the word "credibility" is "credo" – which means "I believe." The Merriam-Webster dictionary defines credible as "offering reasonable grounds for being believed." The case for process safety must be credible, with adequate evidence to support its implementation. This places requirements on the messenger (Process Safety personnel), the message, and the receiver (Line Management).

The messenger must have a basic understanding of the organization and/or operation they wish to influence. For example, to credibly advise line management regarding a chemical reaction process, one should understand the basic process as well as possible. This includes:

- Raw materials used
- How chemicals are combined and reacted
- What products and byproducts result
- How the materials are stored, transported, charged to and removed from the vessels
- Process variables such as pressure and temperature
- Process utilities (heating, cooling, vacuum, nitrogen, etc.)
- How the process is controlled
- Basic knowledge about equipment location, function, size, etc.

If, for example, you wish to discuss the need for inerting a condenser receiver, you will be a much more credible advisor if you know where the equipment is located and how it functions. If you can't find the receiver in the plant, or don't know what it looks like, your understanding of the process and your message will likely be questioned.

The message you deliver must be understandable to the receiver. This can be enhanced by putting the message in the recipient's language, and talking in terms of their process. For example, while you may use the terms Hydrogen Peroxide, Sodium Hydroxide and Potassium Hydroxide, the operators might know them as bleach, caustic and potash. An operator charging what he knows by the acronym DMAPA might look at you funny if you ask him if Dimethylaminopropylamine is flammable. Similarly, discussing risk in relation to different process stages will help the operator to understand your concern. A combustible liquid would only be an ignition risk during portions of the process that raise the liquid temperature near its flashpoint. Also, ignition is not a risk after the chemical has been reacted to form a non-flammable substance.

Sometimes subtle differences in the messages being stated by two parties lead to misunderstanding. In one such instance, a rail car of solid material was heated above its melting point so that it could be transferred to a storage tank. The operator had failed to prop open the manway cover, which would prevent pressure buildup due to thermal expansion of the liquid. The error was realized after the railcar reached the desired temperature, at which time there was concern that the railcar had become pressurized. With no pressure gauge to confirm this, the only available options were to 1) allow the car to cool before opening the manway or 2) slowly loosen the manway clamps to relieve any pressure. Proponents of solution 1 (mostly managers) were concerned that if the manway clamps failed or slipped off the cover, it could open violently, spraying the operator with molten material and/or striking him with the cover. Proponents of option 2 (mostly operators) simply said "That won't happen." They felt that if the clamps were loosened with care, the pressure would be relieved through the manway seal. Who is correct?

Actually, both groups were (nearly) correct. There is no disputing that the manway clamps could fail. Everything fails at some point. But the operators had experience in slowly opening pressurized systems, allowing pressure to bleed off. They had done this many times with nitrogen hoses, water hoses, and even rail cars. Had they been permitted to implement option 2, the odds are overwhelmingly in favor of successfully relieving the pressure without incident. The messages that were intended, what was said, and what was heard are shown in Table 1.

It is often beneficial to discuss things in terms of Risk. Risk is a function of both the potential consequences of an incident, and its likelihood. No one can predict what will or will not happen. But the likelihood can be discussed. The worse the potential consequences might be, the lower the acceptable likelihood should be.

The message receiver also has an important role to play, in that they must be willing to listen to and evaluate your ideas and concerns. If they don't trust you, or don't believe you understand what you are saying, it can be difficult for them to even listen to your message. Building and maintaining credibility and trust will cause them to be open to your message, even if they disagree in the end. Remember too that communication is a two way process – you must be willing to carefully listen to the messages received from others.

Honesty and integrity are important to credibility. As shown in the previous example, being precise in your meaning, and ensuring that the message correctly conveys your thoughts can be the difference between the perception of honesty and exaggeration. Being perceived as the "boy who cried wolf" will harm your ability to make credible arguments.

Integrity also demands consistently applying the principles of Process Safety. A single concession to other priorities can destroy a one's credibility, which can be extremely difficult to rebuild. That is NOT to say the Process Safety is the ONLY priority; acting as if it is can result in others perceiving you as inflexible and unrealistic. As stated earlier, Line Managers must balance many priorities. To be effective at influencing process safety-related decisions, one must assist the line manager in balancing these priorities while ensuring that process risk is sufficiently low. This may sometimes require respectful questioning of other priorities, or suggesting concessions by production. It often requires one to be creative, suggesting innovative solutions to meet the needs of all concerned without compromising Process Safety.

Table 1 - Mixed Messages

	What was said	What was heard	What was meant
Option 1	The manway clamps could fail, causing injury.	The manway clamps are likely to (or will) fail, causing injury.	The risk of clamp failure and subsequent injury is higher than we are comfortable with.
Option 2	The clamps won't fail.	The clamps can't fail.	The risk of clamp failure and subsequent injury is minimal.

Basic Strategies

Like every employee, you must have a firm understanding of your subject area and what you are trying to achieve. In Process Safety, you should know what the minimum standards are, whether they are founded in company policy or government regulation, and the reasons for them. Some standards are prescriptive, requiring you to perform certain tasks and functions in a particular way. If so, you must understand the differences in the "have to haves" and the "nice to haves." While no one wants a process safety program that merely meets the minimum acceptable standards, you must communicate what the minimum requirements are and insure they are included in whatever decision is reached.

Other standards are performance-based, requiring that an outcome be reached while leaving the methods to the company's discretion. While this sometimes leaves the results open to interpretation, it allows much more flexibility in the practices that can be implemented. Efforts can often be tailored to best fit the organization, and its culture.

You must understand what value is provided by going beyond the minimum requirements. Providing firm benefits for the actions you wish to take will allow the line organization to weigh the costs and benefits of your proposals. Knowing what the "industry standard" is, i.e. what others in your industry are doing for Process Safety, provides a standard against which you can benchmark your own practices. Knowledge of best practices from within your company and your industry allows you to compare your proposals to others' best efforts. Proposals that match what most others in the industry are doing are easier to "sell" than those that are rarely seen in industry (even if they would represent a new best practice). This knowledge will also allow you to leverage the lessons others have learned – why rebuild something that others have already perfected?

Is your program based on risk, reducing both the frequency and severity of incidents, or does your company focus on hazards and their consequences? There are fundamental differences in the two approaches that will drive setting priorities.

In a Risk-Based program, both the potential consequences and the likelihood of the occurrence of those consequences are considered. Steps taken to reduce frequency of occurrence are proportional to the potential severity. In a hazards-based program the focus is on identifying credible potential incidents, and expending reasonable resources on mitigation with little consideration of likelihood.

Consistency

Always talk the talk, and make a point of being seen walking the walk. There is no break from Process Safety. It cannot be overlooked when it is inconvenient, such as during an upset or when trying to avoid shutting down a unit. In fact it is during these times, when rushed decisions are often made based on the judgment of one or a few people, that Process Safety can be most important. It is well known that Process Safety incidents are most likely to occur during startups, shutdowns, and other atypical operating situations. Whether a normal day or during an upset, the same systems for Process Safety must be followed.

To have credibility, you must be consistent in the advice and assistance you provide. If you are willing to take short-cuts during abnormal situations, the line managers and operators will be quick to follow. Inconsistencies in your advice in different areas of the plant or for different types of work must be avoided or, if justified, clearly explained to all involved. Process Safety must be a part of your personal culture if you are to succeed in making it a part of the company's culture.

Build Personal Relationships

Having some level of personal relationship with those you are trying to influence can be a great help. A cup of coffee or lunch with a manager to whom you provide expertise will help both of you to understand each other's communication styles. Knowing how best to communicate with each person will make discussing difficult decisions or situations easier because there is less chance for misunderstanding. Communicating in the style to which the manager is accustomed will make it easier for him or her to understand the message you are trying to convey.

Wherever possible, you should be on a first-name basis with the unit operators too. Even if the relationship is only deep enough that you say hello when you see each other, it is at least one step away from having a conversation with a stranger during a professional discussion. All the better if you spend time in the control room or in the plant getting to know the people and the jobs they do.

When called on for advice, don't be afraid to offer both advice AND assistance, even when it may not be your responsibility. If called to advise on an MOC during a process upset, offer to alert the approvers of the urgent need while the line manager writes the MOC. If information from an engineering file is needed, offer to look it up and meet them in the plant with it, instead of their coming to the office. These efforts will build a significant amount of goodwill, showing that you are a team player and willing to share the workload.

Don't make them come to you

In most cases, it is better if you go to the person you wish to influence instead of asking them to come to you. You are likely both very busy. Making this concession demonstrates a level of respect, but more importantly that you don't expect the entire burden to be on the shoulders of the line management and operators. One rarely hears the words "I GET to go to a meeting." It's much more likely to hear "I HAVE to go to a meeting." Placing yourself in the manager's shoes, the may see their day's schedule and think I "have" to go to a meeting with Process Safety. This is much less an undesirable event than it is a burden to their schedule; it is one more thing on their to-do list for the day. It can seem much less disruptive to their other duties if the Process Safety Specialist is "stopping by" this morning to discuss an issue. They can work until the moment you arrive, and resume their other obligations the moment you leave. By taking some of the burden off of them, you are seen to be less burdensome.

Remember who is in charge

Always remember who has authority over the operation. Line managers can make things happen; you can only advise that they should happen. However, the burden of compliance and performance in Process Safety is on them, as well. Your role should be to provide ideas, expertise and tools to help them meet these responsibilities to their fullest. Making demands, or speaking with the assumption that your suggestions will be implemented exactly as you present them, is a mistake from which it can be difficult to recover. Assuming you have authority over the operation for which they are responsible may be perceived as ego on your part, or that you feel you have authority over them as an individual. Most managers, and people, try to avoid working with someone of that personality type. In a busy workplace, they have neither the time nor the patience – there are more important things to be done.

Always ask for their time

While Process Safety should be an integral part of all plant operations, having a conversation about process safety is not ALWAYS the top priority at the moment for operations. Making the assumption that a manager or operator has the time to speak with you at this very moment shows a lack of respect for others, and a lack of understanding of the process and the responsibilities of others. While everyone is busy, we also perform tasks that are not very convenient to interrupt. An operator standing idle in front of a vessel may seem available to talk, but it might also be true they are watching the level of a nearly full tank to ensure that it is not overfilled. Even an operator sitting in the control room may be just minutes away from leaving to take a sample. Speaking to a Board Operator can distract him/her from monitoring the process. By expecting others to stop to talk to you anytime you wish, you are being disrespectful to them, and again showing a lack of process understanding. You will be much more respected yourself if you ask if they can stop to speak. If not, ask when they might be available. While the act of coming to speak to them shows that you value their opinion, coming back to them later, if necessary, shows an even higher interest in what they have to say.

Some professionals, in Process Safety and other fields, have used email at inappropriate times to communicate with plant personnel. Operations personnel are often out of the office for hours. Using email to send notice at 8 am for a meeting at 2 pm the same day is unlikely to be successful. You will likely be told this when you call at 2:05 pm to ask if they are coming to the meeting. Email should also not be used to issue direction on issues you have not previously discussed. While email can be a very efficient communication method, many things are best discussed in person where feedback can be seen and heard.

A similar transgression to the misuse of email is misuse of the telephone and paging system. When paging operations personnel to call you, they are often in areas of a plant remote from a phone, in mid-task. If they stop what they are doing and trek to a phone only to be asked a clearly non-urgent question, they are likely to think twice the next time you ask them to call. Radios use in this way can also damage relationships. Wherever possible, get out of the office and seek them out. At the very least, ask them to call at THEIR convenience for non-urgent matters.

Your presence should not indicate there is a problem

If people duck, hide or groan when you approach, there is obviously a problem. It may be that they only see you when there is a problem. If this is the case with you, then you are probably not spending enough time in the area. Especially in the case of operators, you should be spending time with them to understand the process, what they do and how they accomplish their tasks. Your understanding of the process increases your credibility when you discuss it. Just as importantly, you are building relationships that form a foundation on which you can work together on Process Safety. Spending time with them also provides opportunity for them to ask questions and raise topics of interest or concern to them. If they only see you when you have something to say, they don't have the chance to say what's on their mind. It is not at all unusual for comments made during these times to develop into broader discussions of what things are important, why things must be done a certain way, or alternative ideas on how to achieve an objective. This two-way dialogue puts people at ease when speaking to you, shows your interest in their opinions, and creates an environment where you work together on each other's problems, instead of them only working on your concerns. If you had to choose between working 100% of your time in an office, or in the plant, you will get greater improvement in Process Safety if you choose to spend your time in the plant.

Ask – Don't Assume

No matter how much you know about the plant and its operation, never assume you know it all. Few things are worse, professionally, than repeating someone's (misunderstood) concerns about unloading trucks at night only to be told that no deliveries are permitted at night. Different work crews have different dynamics, and sometimes differing ways of conducting their work. What you are told by one person on one crew may not be true on other crews. A person's understanding or explanation of a task may be incomplete if not actually incorrect. Making assumptions, filling in the gaps you haven't asked about, or trying to connect the dots on your own can be embarrassing during a group discussion or meeting, and even dangerous if your advice is based on your assumptions. One should always explain their understanding of the process along with their advice. For example, the statement "if the vessel is inerted with nitrogen before filling, it will remain inerted throughout the process" may very well be true. Things that you may not know, that could invalidate your statement, include:

- The flammable liquid is charged from drums by applying vacuum to the vessel, which would introduce air into the vessel.
- The vessel man way is open while filling the vessel by pump, which could introduce air to the vessel.
- When the vessel is emptied, the void volume is made up with air through the atmospheric vent instead of with nitrogen.

If you had missed, misunderstood or not been made aware of any of these situations, your process understanding and the quality of your advice could be called into question.

A good rule to follow is *Never try to change something you haven't seen for yourself.* There is no substitute for seeing something with your own eyes and hearing it with your own ears. No one can explain a process in a way that provides a better understanding than seeing it yourself. If you had observed the tank filling operation above, you likely would have realized that air was being introduced during the aspiration of liquid from drums, or that the vessel was open to atmosphere during filling. Even if an operation is completely carried out from the control room, it is often useful to observe the steps taken (e.g. opening the atmospheric vent before discharging from the vessel).

Observing a process provides the opportunity to ask questions of the operators. Discussion of the process can tease out details that might be overlooked in a verbal-only explanation. You are also building relationships during the event, as well as demonstrating the value of the operators input and advice.

Think Win-Win

As any good negotiator will tell you, think win-win. Remember that your responsibility is often to help someone else fulfill theirs. As such, whatever you propose must be seen as something that will benefit the organization (unit, plant, company) that implements it. In a company that is very explicit about its commitment to Process Safety, presenting your efforts as a way of fulfilling that objective is often successful. Where the commitment to Process Safety is less obvious, it may be beneficial to point out the business advantages of Process Safety in addition to the risk reduction[1]. A Hot Work Permitting System that decreases the risk of fire AND is more efficient than the current process will be much better received than one that increases the time required for permitting. In cases where additional time must be spent issuing permits to improve safety, the benefits to safety and the business must be clearly understood and emphasized.

Personalities

When working with others, you will discover different personality types and styles of working. Some employees are fast-paced, some singularly focused on a task, and others may have a more passive style. Some people are detail-oriented, wanting to understand every step and contingency of the process, while others focus on the big picture and prefer to leave the details to others. Many management courses distinguish between different personalities and working styles, providing advice on how to work with each.

Regardless of the location or level of your position in the organization, you must be prepared to tailor your ideas and presentation to provide the type of information each person wants and/or needs. For a high-level executive, explaining the benefits of an MOC system and an overview of how it works might be sufficient. For a Site Director and certainly for a unit manager, details such as what is included on the MOC form and which specific individuals will approve changes would likely be necessary. However, detail-oriented people become high-level executives, and persons who focus on the big picture instead of the details occupy lower management positions. Both personality type and position in the company must be taken into account when presenting your ideas. Of course, having some level of a personal relationship with them will provide this information well ahead of any presentation you might have to give.

Treat their resources as if they were your own

Even during the "good times," every line organization has limited resources, including people, money and time. Their job is to use these resources in the best way possible to produce a quality product in a cost-efficient, safe, and environmentally-friendly manner. Given this responsibility, they will not likely expend their resources without a benefit at least equal to the cost of the resources. If your recommendations require the frivolous use of their resources, they will likely not be implemented and your reputation will suffer. If you develop recommendations that conserve resources as much as possible, and justify the benefit of the resources you wish them to commit, your ideas will be much more appealing. It may even be helpful to explain where you drew the line (e.g. "we could also accomplish X, but I don't think the extra time is worth the benefit gained), to show your understanding of the value of their resources.

It is just as important to avoid under-using or undervaluing resources, especially people. A minor change due to a process upset may leave a shift manager feeling mistrusted if he

[1] The Business Case for Process Safety - CCPS

is required to obtain four, high-level approvals before implementing it. This can usually be avoided by providing a better understanding of the need for approvals, and by ensuring that a system such as MOC is not overburdened with decisions that are well within the Shift Manger's abilities to evaluate and implement safely.

Focus on Results

In every job, department, plant, business, etc. it is results that count. If the deadline isn't met or the objective not achieved, it doesn't matter how much time or effort was spent. In the same way, your efforts must focus on results. When presenting an idea or recommendation, be sure that the desired result of the effort is clear. Stating what will be achieved clearly defines the benefit that will be gained. If you can paint a picture of a better situation or circumstance than presently exists, that is both possible and attainable with reasonable resource investment, people will lead the way there for you.

Attaining results should also be your guide to supporting whatever initiatives are in place. If you are reviewing MOC system performance, you should be less focused on whether the correct boxes are checked than you are on whether the result (a thoroughly evaluated and well-executed change) has been obtained. Remember that the goal is to safely manage change, not to correctly fill out an MOC form. Some might judge the rigor of a hazard study by the number of hours spent during team meetings. Others may look at the hazards identified and the types of safeguards (inherent safety, engineering or administrative solutions) recommended. One might also judge the larger process by reviewing the recommended safeguards and evaluating why they were not included in the original design. Focusing on the result will likely always help you to find the most efficient way of achieving objectives in an effective manner.

Your way may not be the best way

When proposing an idea, one should expect there to be questions. Anticipating what questions might be asked allows for thoughtful responses. The most important questions to anticipate are those expressing dissention. You will frequently encounter individuals who have reason (valid or not) to disagree with either the goal or the methods by which it will be achieved. They may have alternative beliefs about the benefits, incident likelihood, incident severity, etc. Many of the things mentioned previously in this article will help to prepare your answers, including:

- The personal relationships you have built
- The credibility you have established
- Your knowledge of:
    - Process Safety
    - Minimum policy and regulatory requirements
    - Best practices in your company and industry
    - Standard practices within your company and industry

- Prior conversations with stakeholders who may have had the same reservations

You have undoubtedly presented the proposal you believe stands the best chance of success. However, there are always alternative methods, even if they trade effectiveness for costs, both for better and for worse. Having alternatives for which you have weighed the costs and benefits is like having an extra tool in your pocket if the primary one is failing to do the job. If your proposal is being met with significant resistance, perhaps an alternative approach may be better received. Along with having alternative plans, one must remain open to alternative points of view. A better proposal may emerge if you listen carefully to the concerns being expressed. In any case, knowing these concerns will be very helpful in proposing alternatives.

As stated previously, it is sometimes beneficial to focus on the risks instead of your proposal to mitigate them. Discussing the risk allows for alternative mitigation methods to be presented. If this does occur, the stakeholders will have a bigger investment in the results, and a greater belief in the success of actions they are to take.

Where to begin

There are those that will never propose an idea to someone who hasn't heard of it before. That is not to say that nothing new will ever be proposed. Instead, it means that if they have never heard of it before the formal presentation, then you haven't prepared well and are likely to fail. If they haven't already discussed it, then:

- They have had no opportunity to provide input
- You have not asked for their ideas
- You have not determined if they share your assessment that change is needed
- When they bring the idea to those within their organization, their reaction can't be anticipated.

Start from the bottom up so those at the lowest impacted level of the organization have heard your ideas The best place to find out if your idea will work is where it will be used and from the people who will use it. If you are considering vessel inerting practices, speak with the operators who conduct the task. If your concerns are about capital project hazard studies, talk about it with the capital project engineers. Discovering their concerns and what obstacles they perceive can help you to propose solutions that address both your and their concerns.

After discussions at that level, move up one level in the organization. Explain your concerns, your ideas, and the input from those you have already discussed it with. Always include dissenting or differing opinions offered by others in your discussion. Continue up the chain of command, incorporating all previously obtained input, until you reach the level at which the decision will be made. Discuss it informally with those at the decision-making level as well, before making a formal presentation or recommendation. This process allows you to plant the seeds of your idea beforehand, while including the thoughts (positive or negative) of those impacted by the decision. When the recommendation is presented:

- Those making the decision will have had input, provided feedback, and had the benefit of time to consider the decision.
- Those that work for the decision-makers will be able to provide informed feedback to their supervisor.
- Other stakeholders, including those that must implement the recommendation, are well informed and are invested in the process, as well.

In the author's experience, success is much more likely using the approach in the following example, where concerns have arisen regarding adequately grounding drums being filled with flammable liquid.

1. Discuss your observation with the operators performing the task. Do they think the grounding is effective? Does everyone use it? Is it ever forgotten? You might learn that they use an assured grounding device in another part of the plant and that the ground clamp is placed on an I-beam when not in use, which then indicates a good ground and could allow drums to be filled. They might tell you they'd like to try it or that it frequently fails. In any case, their feedback and opinion may change yours, or help to shape any recommendation you might make. Tell the operators on a different shift of this conversation, and ask for their thoughts.

2. Approach one or two shift managers and tell them of your conversations with the operators. Seek their thoughts and suggestions. After the conversation, a good Shift Manager will tell the operators performing the task of your conversation. If you haven't yet spoken to the operators, then the Shift Manager will be presenting your case for you – no one can describe your thoughts and concerns better than you can. If the operator tells the Shift Supervisor of his conversation with you, they can then discuss their opinions together, both having had the benefit of your technical input.

3. Finally, you can approach the Unit Manager. Tell him that you've been discussing your concern with the operators and Shift Managers, and that the group feels this action should be taken. If there has been any disagreement in your prior conversations, be honest about it. State your reasoning clearly and concisely, along with any objections raised. When the Unit Manager approaches the operators and/or Shift Manager about the issue, they have all had the benefit of receiving your technical opinion directly from you. They've also had time to consider the issue carefully.

Regardless of the outcome, you've shown respect for all three levels of the organization by being professional, discussing the issue openly and honestly, and asking for their ideas and input. They will respect you for this even if they don't accept this particular recommendation.

Escalating Issues

Despite your best efforts, you may encounter situations of imminent danger or risks that you strongly feel present excessive risk to the organization. This might be a risk that a unit manager does not seem to acknowledge, disagreement over meeting a regulatory requirement, or other differences of opinion that, in the Process Safety Specialist's opinion, have clearly not been adequately resolved. This is a very difficult situation whose handling can have huge impact on future work relationships.

You should make every effort to avoid taking an issue to managers above the appropriate decision-making level. Hopefully the information presented above will have allowed you to avoid this situation. If not, tread carefully while keeping the following in mind:

- Things are rarely either right or wrong. Both opinions likely have merit, and are based on the experience and professional judgment of the respective persons.
- Very few people choose to intentionally ignore risk. Any thoughts of intentional disregard for safety are very likely unfounded.
- A third or fourth point of view can be very helpful. The advice of a personal peer, or someone else in the line organization, may help you to better understand the other person's point of view. The more you understand the reasons for the other person's opinion, the better positioned you are to find a mutually agreeable resolution.
- Anger is always a detriment to discussion and decision-making. Never decide what to do next when under its influence.
- If the situation remains unresolved, and your still feel strongly that the risks are unacceptable, you may feel it is your responsibility to take further action. You may have a superior in the staff organization you can go to for assistance and advice. Often they can intervene in a non-threatening way to help resolve the conflict.

If, in the end, you feel you absolutely must raise the issue higher in the chain of command, respectfully advise the person with whom you disagree that you feel it is your responsibility to do so. Invite them to accompany you so that all points of view can be presented. But choose your battles wisely. Remember that is the manger's responsibility to implement the actions he/she feels are necessary for Process Safety. Only those situations that have a direct and immediate effect on Process Safety should be handled in this manner.

Conclusion

Process Safety Professionals typically occupy staff positions in their organizations. Only the line managers have the authority to cause things to happen in the organization. Because those in staff positions lack the authority to implement change, they must influence the line management to do so with regard to process safety. Effective influencers possess qualities such as credibility, integrity and good communication skills. These qualities can greatly influence success when discussing high consequence, low probability events.

Building personal relationships with those in the organization you wish to influence, and using these relationships to develop solutions, will also enhance your success. You should know the operations and maintenance employees well enough to ask about their weekend, or who won the big game last week. You should treat everyone in the line organization with respect by remembering who is in charge, asking if they have time to talk with you, and going to them instead of asking them to come to your office.

Knowledge of the process, its inputs and outputs, and how it is operated not only helps to identify workable solutions, but will also increase your credibility. Solutions are often best developed by starting with the operators and their proposed solutions. You can then refine your ideas, based on operator input, and then discuss it with the various managerial levels of the organization. Incorporate their input and ideas wherever possible to gain their ownership and acceptance.

There are times when your best efforts at influencing are unsuccessful. Remember that things are rarely black and white – both positions on an issue likely have merit. Seek the input of colleagues of independent members of line management. Instead of discussing things in terms of what might or will happen if changes are not made, develop a habit of discussing things in terms of increasing or decreasing the risk of an event occurring.

While one should never compromise your commitment to process safety, very careful consideration must be given to escalating an issue to a higher level. If you feel you absolutely must go "over their head," respectively tell them that you feel so strongly about the issue that you must do so. Propose that you meet jointly with the superior authority to discuss potential solutions.

While the information presented here is applicable to any professional in any field, it is particularly useful to staff members. Successful implementation of process safety programs often depend on it.

10TH GLOBAL CONGRESS ON PROCESS SAFETY

Blah, Blah, Blah!!! - Effective Individual and Team Communication Lead to Effective PHAs

Mr. John T. Perez, P.E.
Cognascents Consulting Group, Inc.
11777-A Katy Fwy Ste 438
Houston, TX 77079
john.perez@cognascents.com

Ms. Ashley Bourne
Cognascents Consulting Group, Inc.
11777-A Katy Fwy Ste 438
Houston, TX 77079
ashley.bourne@cognascents.com

Mr. Andrew Madewell, P.E.
Cognascents Consulting Group, Inc.
11777-A Katy Fwy Ste 438
Houston, TX 77079
andrew.madewell@cognascents.com

Prepared for Presentation at
American Institute of Chemical Engineers
2014 Spring Meeting
10th Global Congress on Process Safety
New Orleans, Louisiana
March 30 – April 2, 2014

UNPUBLISHED

Blah, Blah, Blah!!! - Effective Individual and Team Communication Lead to Effective PHAs

Mr. John T. Perez, P.E.
Cognascents Consulting Group, Inc.
11777-A Katy Fwy Ste 438
Houston, TX 77079
john.perez@cognascents.com

Ms. Ashley Bourne
Cognascents Consulting Group, Inc.

Mr. Andrew Madewell, P.E.
Cognascents Consulting Group, Inc.

Keywords: Hazard and Operability Study (HAZOP), hazard identification, Insights ® Discovery, Layer of Protection Analysis (LOPA), Process Hazard Analysis (PHA), Process Safety Management (PSM), risk assessment, risk management, team communication, team effectiveness.

Abstract

Process hazards analyses are structured, team-based exercises focused on hazard identification, risk assessment, and risk management. A competent facilitator serves a multi-disciplined team in applying the appropriate PHA methodology to identify, assess, and safeguard vulnerabilities. PHAs help a business define its overall risk profile and subsequently its overall strategy.

The various generally-accepted PHA methodologies are fairly well-defined by industry and company-specific guidance documents. PHA methodologies are processes that have evolved in response to societal norms, organizational culture shifts, and ever-changing regulatory requirements. Unlike PHA methodologies, which are typically bound by rule-sets and scope definition, multi-disciplined teams made up of diverse individuals are typically unbound when it comes to behavioral norms and personality types. Hence, as with any process, it is not the documented protocols that define the effectiveness of a process. It is the people leading, establishing, applying, and using the outputs of a process that determine its effectiveness. Hence, a PHA methodology is only as effective as the team involved.

Many factors influence team performance, no matter the task. Communication is often touted as the most critical factor for effective team performance. Hence, team communication is critical for effective application of a PHA methodology. But what is communication? More importantly, what is effective communication? Can a team's ability to communicate be assessed prior to attempting to apply a PHA methodology? If so, what safeguards can be put in place to protect against communication vulnerabilities at both the individual and team levels?

In this paper, the authors posit a team's communication capacity and potential can be assessed prior to starting a PHA. Specifically, the authors provide insight through case study and idea synthesis of potential individual and team communication vulnerabilities and safeguards. They also present tools and recommendations to enhance team communication and subsequently PHA quality, which bear directly on operational, organizational, and commercial integrity.

The target audience for this paper is anyone whose responsibilities include (1) leading within an organization that uses PHAs, (2) establishing PHA guidance documents, (3) applying PHA methodologies, and (4) reviewing PHA outputs and reports; however, anyone who works on a team will benefit from this paper's content.

1. Background and Purpose

Along with running a process safety and technical business consulting company, I facilitate brainstorming, process hazard analysis (PHA), and strategy-setting workshops. Respectively, these workshops help a business innovate and grow, identify/assess/manage risk, and set its overall vision and strategy. During a retreat in the summer of 2013, I kept turning a problem over and over trying to find a solution. The problem:

Brainstorming, strategy-setting, and PHA methodologies are fairly well-defined by industry and company-specific guidance documents. These methodologies are typically bound by documented objectives, prescriptive rule-sets, narrowed scopes, and actual business and/or design information. Despite the fairly well-defined and bounded methodologies, no two workshops are ever the same - even within the same company with the same set of team members. Some workshops are executed more effectively than others with more effective results as the outcome. To achieve excellence, inconsistent and ineffective team performance must be eliminated, prevented, and/or mitigated.

My first question was: "what drives the inconsistency?" The answer to this first question was evident - people. Unlike the abovementioned methodologies, the multi-disciplined teams utilized per methodology protocol are made up of diverse individuals who are typically unbound when it comes to behavioral norms and personality types. A methodology does not alone ensure consistent effectiveness. The people leading, establishing, applying, and using the methodologies determine their effectiveness. Hence, a brainstorm, strategy, or PHA is only as effective as the team utilized in creating it.

Farther along my thought process towards a solution, my second question was: "how can I consistently facilitate effective workshops, regardless of type?" The answer to this second question was not as evident as the first. Facilitators serve a leadership role and the teams they "lead" are often multi-disciplined, multi-generational, multi-ethnic, and multi-titular. I was essentially facing the age-old problem of how to lead teams for excellence. I was frustrated to find myself facing what seemed to be a fundamental, yet pervasive, business and organizational quandary. Let's face it...I am not alone in this dilemma. Anyone serving a team, no matter his/her role, is forced to navigate the foggy waters in search of performance excellence. I know it sounds corny and I know most of us are numb to our plight, which might explain why we tend to focus on our individual efforts and sphere of influence rather than try to achieve overall team excellence.

Nonetheless, it did not take me long to realize that developing a set of tools and recommendations focused on (1) achieving excellent team performance and (2) delivering consistent and effective workshop results would be valuable to my company, the companies we serve, and more importantly to the people we serve. Rather than tackle the elephant, I decided to focus on one of its legs - PHAs. In this paper, my colleagues and I will (1) describe PHAs and their importance, (2) describe how teams impact PHAs through their decision-making traps, (3) posit and support communication as the most critical factor to team performance, (4) present a case study using Insights® Discovery personal profiles, and (5) offer a set of tools and recommendations to improve the likelihood for consistent PHA effectiveness.

2. What Are PHAs and Why Are They Important?

2.1 PHA Methodology and Logistics Descriptions

PHAs are structured, team-based exercises focused on hazard identification, risk assessment, and risk management. The deliverables of PHAs help drive design and procedural improvements to avoid potentially catastrophic consequences to personnel, the environment, and commercial interests. Traditional PHA approaches include Checklist, What-If, What-If/Checklist, Hazard Identification (HAZID), and Hazard and Operability (HAZOP) analyses. A competent facilitator serves a multi-disciplined team in applying the appropriate PHA methodology to identify, assess, and safeguard technical design and/or procedural vulnerabilities.

Most PHAs comply with defined rule-sets, protocols, and narrowed scopes. The Checklist PHA methodology uses a prescribed element set to evaluate a proposed design against process safety and risk management guidelines. Checklist PHAs usually employ questions similar to the following:

1. Are effective mechanisms for isolation, venting, bleeding, and de-energization readily available and accessible?
2. Are the governing design cases for pressure relief devices indicated on the drawing?
3. Are safe limits for critical process parameters indicated on the drawings?

A HAZID uses guidewords and parameters (e.g. dropped objects, simultaneous operations, ergonomics) along with nodes (i.e. system sections or procedural tasks) to focus the team's discussion and scrutiny.

The HAZOP PHA methodology is another structured brainstorming exercise. HAZOPS use guidewords and parameters (e.g. high pressure, low level, no flow) to identify (1) causes of hazard scenarios, (2) consequences of hazard scenarios, (3) safeguards that eliminate/prevent/mitigate the hazard scenarios and/or their consequences. Unlike the HAZID PHA methodology, a HAZOP uses risk assessment criteria to determine Likelihood and Consequence rankings within a given risk matrix. Safeguards are not accounted for in assigning the Consequence value; however, safeguards are taken into account when determining the Likelihood value. Where the qualitative risk-ranking recognizes process safety or risk management vulnerabilities, the team crafts a recommendation to resolve the gaps.

PHAs are normally conducted via face-to-face workshops in rooms large enough to accommodate the PHA team. Remote PHA team members sometimes participate via web-meeting capability. PHAs vary in duration with some lasting one hour and others lasting months. To avoid mental fatigue, the length of a PHA workshop day is usually limited to six hours of workshop time.

2.2 PHA Inputs

In order to perform an effective PHA, team members should have the following information readily available:

1. Material safety data sheets;
2. Fire and explosion models;
3. Dispersion models of potential release scenarios;
4. Chemical/Material reactivity matrices;
5. Mass and energy balances;
6. Safe upper and lower limits for critical process parameters;
7. Operating philosophies;
8. Materials of construction;
9. Electrical area classification drawings;
10. Relief and effluent system design bases;
11. Ventilation system design basis;
12. List of design codes and standards applicable to the process;
13. Cause and effect matrices for safety controls, alarms, and interlocks (SCAI);
14. Equipment specifications;
15. Sparing philosophies (primarily used in upstream subsea asset development);
16. Electrical one-line diagrams;
17. Plot plans;
18. Safety shower/eyewash location drawings; and
19. Vendor-provided skid package information.

Some of these items may not be available due to the timing of the PHA against the project's life cycle. For example, checklist PHAs are typically performed towards the end of the Front-End Engineering Design (FEED) stage of a project when fire and blast studies may not be complete.

2.3 PHA Team Characteristics

A multi-disciplined PHA team usually includes (1) relevant engineering personnel (e.g. mechanical, process, electrical, marine, flow assurance engineers), (2) operations personnel (e.g. foremen, technicians or operators), (3) maintenance personnel as necessary, (4) third-party vendors as necessary, (5) process safety personnel (e.g. PSM manager, PSM coordinator, risk engineer), (6) a scribe and (7) a facilitator. Given current workforce demographics and business globalization, it is common for PHA teams to not only be multi-disciplined, but also multi-generational, multi-ethnic, and multi-titular. PHA team compositions are dynamic, even at units within the same refinery or assets within the same upstream business unit.

2.4 PHA Deliverables and Subsequent Utilization

PHAs identify vulnerabilities in designs and procedures. If the PHA team does not identify adequate safeguards to eliminate, prevent, or mitigate the risk to acceptable levels, then the team generates recommendations to close the risk gap. The primary goal of PHAs is to help protect people, the environment, and commercial interests by ensuring high-integrity business operations. A company's overall technical risk profile consists of the aggregate risk based on all internal independent PHAs or other risk assessments. In short, PHA teams assess a lot of information, make a lot of decisions, and generate a lot of recommendations in a relatively short of amount of time to help companies define and understand their risk profiles. Businesses then incorporate their overall technical risk profiles and their overall commercial risk profiles into their strategy-setting activities. These strategies subsequently govern resource (personnel and capital expense) allocation. Ineffective PHAs may compromise the integrity of business operations both in the short-term (via unidentified high-risk vulnerabilities and subsequent catastrophic events) and the long-term (via flawed, forward-looking, strategy-setting activities and mismanagement of resources).

3. Despite Defined Methodologies, Inconsistent PHA Effectiveness Persists Through Decision-Making Traps

As previously mentioned, despite the fairly well-defined and bounded methodologies, no two PHA workshops are ever the same. I acknowledge scopes differ from PHA to PHA and this drives some of the differences across efforts. However, the scope is simply what is to be analyzed using the designated methodology. The scope does not alone drive a PHA's effectiveness. I also acknowledge information availability differs from PHA to PHA, which can also drive differences across efforts. Once again, PHA methodologies define how to handle missing information and this factor should not drive a PHA's effectiveness. There are a number of protocol-driven elements that differ from effort to effort, but all are addressed by the various PHA methodologies and should not drive PHA effectiveness.

The truly un-controllable and undefined elements of a PHA are the team members. Humans are the elements within a PHA that change from day to day, hour to hour, and sometimes minute to minute. Humans are the elements within a PHA that change for sometimes unseen reasons. As the humans are the decision-making engines of the PHA, they drive PHA effectiveness. So, the problem evolves to leading teams for PHA excellence. In order to tackle this problem, we need to answer more questions:

1. What specifically do people do that lead to inconsistent and ineffective PHAs?
2. At what point in the PHA exercise do people exhibit the de-railing behaviors?
3. Are people aware of their de-railing actions when they occur?

John Hammond, Ralph Keeney, and Howard Raiffa, offer answers to the above questions in their *Harvard Business Review* article "The Hidden Traps in Decision Making". They assert people are inherently vulnerable to psychological (hence, invisible) "traps" when making decisions. These traps include the following:

1. Anchoring trap - letting initial information and impressions anchor minds;
2. Status-Quo trap - avoiding action that takes people out of their comfort zones and makes them vulnerable;
3. Sunk-Cost trap - perpetuating bad decisions to avoid having to admit a mistake;
4. Confirming-Evidence trap - seeking information to support a decision while ignoring contradictory information regardless of its significance;
5. Overconfidence trap - being overconfident in estimating/forecasting accuracy;
6. Prudence trap - adjusting estimates/forecasts to overly conservative values; and

7. Recallability trap - allowing dramatic past events/experiences/memories to bias estimates/forecasts.

Since these traps are ever-present during decision-making activity, people are falling into them throughout the PHA workshop...and they do not always know they have fallen into a trap. The authors mentioned above also offer tips to avoid the traps. These tips include:

1. Seeking information/opinions from a variety of people;
2. Exercising caution when communicating with others to avoid triggering a trap with your own words;
3. Challenging yourself and others to objectively evaluate all options;
4. Serving your team by supporting discipline and accountability;
5. Assigning someone to play devil's advocate to argue against the decision being contemplated;
6. Challenging the motives of evidence and data;
7. Examining how other people frame decisions and recommendations; and
8. Challenge numbers (e.g. estimates, forecasts, statistics, likelihoods, design limit values).

Effective implementation of these tips hinges on one facet of team activity - ***communication***.

4. Communication Patterns: The Most Critical Factor to a PHA Team's Performance

As promised earlier in this paper, we have (1) described PHAs and their importance and (2) described how teams impact PHAs through their decision-making traps. In this section we further posit and support communication as the most critical factor to a PHA team's performance (i.e. effectiveness). We have already noted communication as a critical factor to avoiding decision-making traps.

4.1 Communication Patterns of High-Performing Teams

According to studies led by Alex Pentland (professor at MIT and director of MIT's Human Dynamics Laboratory), patterns of communication are the most important predictors of a team's success. In the *Harvard Business Review* article, "The New Science of Building Great Teams", Pentland states that patterns of communication "are as significant as all the other factors - individual intelligence, personality, skill, and the substance of discussions." Pentland goes on to list the following as communication characteristics of high-performing teams:

1. Every team member spends about the same amount of time talking and listening;
2. Discussions consist of succinctly-delivered content;
3. Team members look at each other when speaking and use energetic gestures;
4. Team members speak to each other and not through the team leader;
5. Team members engage in "back-channel or side conversations within the team"; and
6. Team members go hunting outside of the team to gather and bring back information for discussion and application to the project.

So, we have identified communication as a necessary element to avoid decision-making traps. And, we have identified communication patterns as the most important predictor of a team's success. We have listed decision-making pitfalls as well as communication patterns of high-performing teams. But where do trust and confidence come into play? How do you get team members to have the confidence and trust to share and listen?

4.2 Critical Factors for Effective Communication in Highly Collaborative Teams

Gratton and Erickson have conducted research to identify factors that correlate with team success. Their list includes a skills factor focused on building relationships, effective communication, and conflict resolution. Highly collaborative teams have team members that know how to do these critical people-related activities. These members are the collective confidence-builders and trust-builders. Additionally, highly-collaborative teams enjoy a strong sense of community, which helps establish the confidence and trust among team members. When team members feel everyone is focused on achieving the same goals in a manner supportive of who they are, they develop the confidence to speak up, challenge, and keep an open mind. Gratton and Erickson also posit that teams should not be made up entirely of strangers. When everyone is a stranger to everyone else, it is very difficult to build the confidence and trust necessary for effective communication. The last factor worth noting is leadership. Highly collaborative teams benefit by having a leader that can focus on task-oriented activities early in the project where it is most important to get momentum and traction. As the project progresses and relationship issues arise, the leader should be able to shift to relationship-building activities to ensure collective, effective communication.

Hence, team communication and communication patterns are critical for effective application of a PHA methodology. We also now have an idea of what effective communication resembles. And lastly, we have identified other factors necessary to establish the trust and confidence for effective communication and subsequently team performance. But how does this help us? How does this help me as a facilitator? How does this help future PHA team members?

Oftentimes, I walk into a workshop without knowing some of the team members. It does not take long for me to realize that some of the team members do not know each other. It is common for representatives from several companies to be present at a single PHA. Given the short durations of PHAs, teams do not always have the luxury of time to build a sense of community that fosters trust and confidence for effective communication. So, can we assess a team's ability to communicate prior to attempting to apply a PHA methodology? If so, what safeguards can be put in place to protect against communication vulnerabilities at both the individual and team levels?

5. PHA Case Study Informed By Insights® Discovery Personality Profiles

5.1 The HAZOP of 2012

In 2012, I facilitated one of the most challenging PHAs of my career. The project involved an upstream development and the HAZOP PHA methodology was employed. There were more than 15 team members in the room at any given time, but not all of them were core. Most of the team members knew each other and had been working with each other for months before the PHA.

The first three days of the PHA were peppered with aggressive, intense, passionate, and fierce exchanges - voices were raised, fingers were pointed, and eye contact or its avoidance was strategic. These exchanges were sporadic throughout the PHA, which lasted just over two months. The number and frequency of the outbursts lessened as the PHA went on; however, the style of communication did not lend to team collaboration and effectiveness. The result was a prolonged effort to accomplish a high-quality PHA. The team averaged about one node per day.

5.2 The HAZOP of 2013

In 2013, I facilitated another PHA for the same project using the HAZOP methodology. The project had been re-run through the relevant capital deployment stage-gate process to reduce scope and cost and delivery time. Once again, there were more than 15 team members in the room at any given time, but not all of them were core.

Unlike the HAZOP of 2012, the effort was dominated by more effective communication. The heated exchanges of 2012 never flared up. The team engaged in a more constructive style of communication. The team averaged about two nodes per day.

You might expect a shorter PHA due to the reduced scope; however, the number of nodes requiring analysis did not decrease by half. In fact, the number of nodes only decreased by 10%. You might also expect more effective communication from a team comprised of members that had been through a similar exercise with each other just one year before. However, just before the 2013 PHA effort, the project was transferred to a new Engineering-Procurement-Construction (EPC) company. This transfer was accompanied by the onboarding of new EPC technical leads to the project. So unlike the 2012 effort, roughly half of the team members knew each other and had been working with each other for months before the PHA. The other half of the PHA team consisted of complete strangers to the legacy project team. So, why did the teams perform differently? We know the answer lies in the people, but what specifically was different?

5.3 Insights® Discovery Personality Profiles

Developed in 1988 in Scotland, Insights® Discovery is a tool based on the psychological-types theory of Dr. Carl G. Jung and Dr. Jolande Jacobi. Dr. Jung proposed personalities manifest through the interaction of four functions: (1) thinking, (2) feeling, (3) sensation, and (4) intuition; and two attitudes: introversion and extroversion. When combined, these functions and attitudes generate eight personality types. Dr. Jung argued that uniqueness comes from the different balances of these functions and attitudes. It is these balances that make up the Insights® 4 Color System: Cool Blue, Fiery Red, Sunshine Yellow, and Earth Green.

Figure 1: Insights® 4-Type Wheel

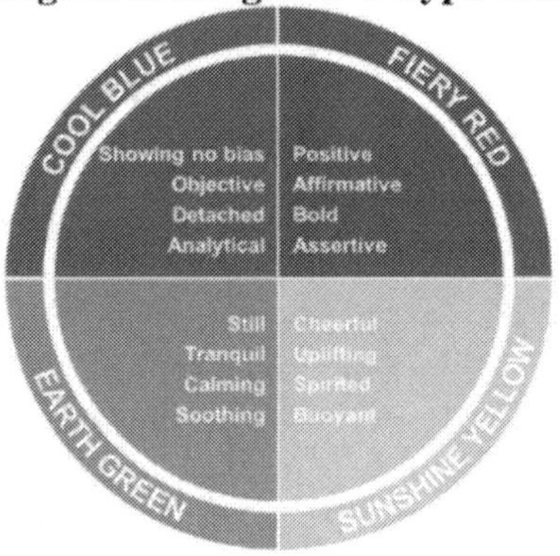

Insights® measures an individual's unique preference-set via the Discovery evaluator - a 25-frame questionnaire of 100 word pairs. The output is the Insights® Discovery Personal Profile highlighting and individual's strengths, weaknesses, communication style, and potential value to the team. In 2005, The University of Westminster published a paper outlining the case for validity and reliability of the Insights® Discovery Evaluator. In summary, strong evidence exists to support the four color measures calculated from the Insights® Discovery model being both reliable and valid. The below table presents the four colors and high-level characteristics:

Table 1. Insights® 4-Color System Quick Reference

	Cool Blue	Earth Green	Sunshine Yellow	Fiery Red
Jung equivalent:	Introverted Thinking	Introverted Feeling	Extraverted Feeling	Extraverted Thinking
Appears:	Formal, conservative	Casual, conforming	Fashionable, stylish	Business-like, functional
Work preferences:	Structured, organized, functional, formal	Personal, relaxed, friendly, informal	Stimulating, personal, cluttered, friendly	Busy, formal, efficient, structured
Style:	Slow/systematic	Slow/easy	Fast/spontaneous	Fast/decisive
Focus:	The task, the process	Maintaining relationships	Interacting/relationships	The task, the results
Fears:	Embarrassment	Confrontation	Lack of prestige	Loss of control
Under tension will:	Withdraw/avoid	Submit/acquiesce	Attack/be sarcastic	Dictate/assert
Likes:	Accuracy	Attention	Recognition	Productivity
Seeks security in:	Preparation	Close relationships	Flexibility	Control
Wants to maintain:	Credibility	Relationships	Status	Success
Help to support their:	Analysis	Warmth	Relationship	Objectives
Seeks to achieve acceptance through:	Correctness, thoroughness	Conforming, loyalty	Playfulness, stimulating the environment	Leadership, competitiveness
Likes you to be:	Precise	Pleasant	Outgoing	Brief
Wants to be:	Correct	Liked	Admired	In control
Is irritated by:	Surprises, unpredictability	Insensitivity, impatience	Boredom, routine	Inefficiency, indecision
Decisions are:	Paced/deliberate	Consideration	Spontaneous	Quick/decisive

As presented above, Cool Blues are introverted thinkers, but Sunshine Yellows are extraverted feelers. By their very natures, these two personality types represent extremes on a spectrum, which present inherent communication challenges. The same applies to Earth Greens (i.e. introverted feelers) and Fiery Reds (i.e. extraverted thinkers). When approaching decisions, Cool Blues are data-oriented, Fiery Reds are action-oriented, Earth Greens are considerate, and Sunshine Yellows are spontaneous and innovative. Hence, while it may be easier for similar personality types to communicate, they may be more susceptible to the aforementioned decision-making traps. For example, a room dominated by Fiery Red energy is more apt to make a quick decision without examining all of the evidence/information beforehand. On the other hand, a room dominated by Cool Blue energy may suffer paralysis by analysis with little action as the result.

It is important to note that most people have some degree of each color in their personality. In a sense, we are chameleons with a dominant color, but have the potential to demonstrate the personality of another color in certain situations. This multi-colored profile is what allows persons with different personalities to get along through effective communication. Understanding who you are trying to communicate with coupled with understanding your own personality profile allows you to "dial up" or "dial down" a particular color in order to adjust/improve your communication style for certain interactions. Therefore, when we use the terms "Cool Blue", "Fiery Red", "Earth Green", or "Sunshine Yellow" in this paper, we are simply referring to a personality's dominant color. A person of a specific personality type can exhibit behaviors, attitudes, and functions of other personality types, especially once the person is aware of their natural affinities and the affinities of others.

By generating the personality profiles of each team member, a color wheel depicting the team's composition can be assembled. This color wheel can be used as a communication awareness tool when assessing a team's potential for success. The color wheel can also afford action prior to conducting or during a PHA workshop.

5.4 The Insights® Color Wheels of 2012 and 2013 HAZOP Teams

After completion of the 2013 HAZOP and upon reflection of the efforts, we collaborated with Insights Learning & Development Limited to generate the personality profiles of core/critical PHA team members. The members selected represented the driving forces behind the efforts. The color wheels of both teams are presented below:

Figure 2: 2012 HAZOP Team Color Wheel

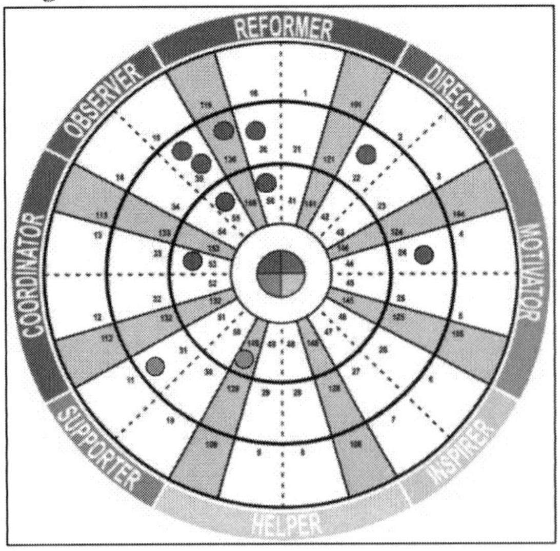

Figure 3: 2013 HAZOP Team Color Wheel

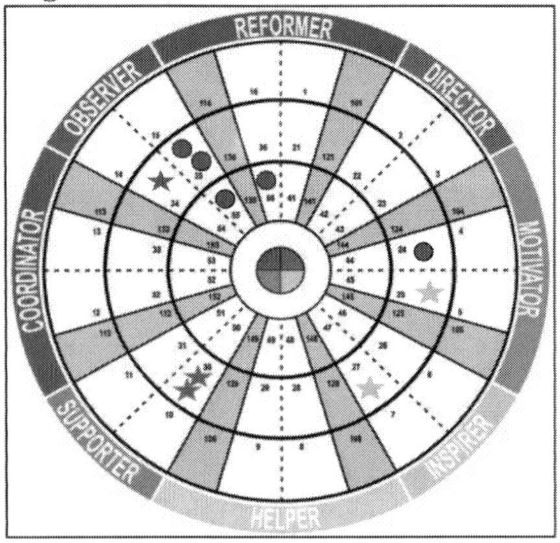

As depicted in the respective color wheels, the 2012 HAZOP team was (1) more heavily weighted on Cool Blues, (2) absent of Sunshine Yellows, and (3) potentially subjected to competing Fiery Reds. Sunshine Yellows bring energy and encourage involvement. They are outgoing and adept at building relationships. This behavior was more observable in the 2013 team where team-building outings were organized by one of the new Sunshine Yellow members.

The "★" symbol in the 2013 HAZOP team color wheel depicts new members as compared to the 2012 team. The 2013 HAZOP team enjoyed the presence of two Sunshine Yellow personalities. Two core Earth Green personalities from the 2012 team were replaced with two new Earth Greens and the 2012 directing Fiery Red dropped off the 2013 team. While these differences may seem insignificant, the 2012 Earth Greens and the 2012 directing Fiery Red were three of the players usually involved with the de-railing exchanges. A fourth player (one of the Cool Blues of 2012) usually at the heart of the heated exchanges of 2012 remained on the team for the 2013 effort. But the communication style of the fourth player seems to have been influenced in a more distracting manner by the now-departed 2012 players...or perhaps more positively influenced by the collective 2013 team.

When assessing the color wheels against potential decision-making traps, a more balanced team has the potential to avoid more traps...especially if they can evolve to a highly-effective mode of communication. With the introduction of Sunshine Yellows onto the 2013 team, a more balanced team was achieved with the potential for more effective decision-making. Additionally, with fewer core Cool Blues, a more effective balance of analysis and decision-making may have been achieved.

Please note the above analysis is not meant to assess the behavior of any one individual. The above analysis is focused on understanding communication differences across the two teams with the purpose of identifying potential communication safeguards for future efforts. Is it fair to say the 2013 team was more effective than the 2012 team? I would not make that assessment by simply looking at the above color wheels; however, the above color wheels do provide a level of awareness and insight to afford upfront and midstream communication and team performance adjustments.

5. Recommendations to Enhance Communication for More Effective PHAs

My colleagues and I have (1) described PHAs and their importance, (2) described how teams impact PHAs through their decision-making traps, (3) posited and supported communication as the most critical factor to team performance, and (4) presented a case study using Insights® Discovery personality profiles. Below, we offer a set of tools and recommendations to improve the likelihood for consistent PHA effectiveness:

1. Before Day 1
    a. Generate the personality profiles for each core team member.
    b. Generate the color wheel for the proposed team.
    c. Ensure every color is represented on the team. If the core team is missing a color, then find someone of the missing color to participate in the session. If someone of the missing color cannot be secured, then at least the team members have the insight to "dial-up" or "dial-down" colors to compensate.
    d. Ensure the PHA team is not made up entirely of strangers. This is especially critical of short PHAs as the speed to trust and confidence is greatly impacted by

healthy relationships. The facilitator should assess the team dynamics and help each team member use their strengths. The facilitator should also actively push team members to stretch their less dominant energies to overcome communication challenges.

   e. Challenge your Sunshine Yellows to serve the team with their relationship-building skills. Sunshine Yellows may need to lead the communication at first to get things rolling and then back off once enough confidence and trust has been established and others are engaged and participating freely.
   f. Similar challenges can be discussed with other colors. For example, challenges can be made to Cool Blues for data assurance and to Fiery Reds for maintaining momentum. Earth Greens may be given the task to ensure consideration is afforded to all team members and key stakeholders.
   g. Ensure the facilitator is aware of:
      i. Individual personality colors as well as the team's color wheel;
      ii. Common decision-making traps;
      iii. Communication patterns of high-performing teams;
      iv. Critical factors for effective team collaboration.
2. Day 1
   a. Designate a team member to be the decision-making trap "cop". It is his/her job to be on the lookout for offenses. This job could be split across several team members as Cool Blues have a greater affinity to challenge numbers. Fiery Reds may be more successful at enforcing discipline and accountability.
   b. Designate a team member to monitor talking/listening time. This person does not have to be a core team member.
   c. Ensure the facilitator promotes cross-talk between individuals and not through themselves.
3. Ongoing
   a. Ensure plenty of breaks early in the effort to promote and catalyze back-channel and side conversations
   b. Keep a running list of opportunities to seek information from external sources to further inform the analysis.
   c. Assign your Sunshine Yellows to coordinate inclusive, team-building activities outside of the workshop, such as ice-cream socials or celebratory lunches for achieving a significant milestone.
4. Last Day
   a. Perform a lessons learned focused on team communication and team performance. Identify best practices and opportunities for improvement. Try to recall moments of ineffective communication and decision-making deficiencies and propose future safeguards.
   b. Celebrate your success.

6. Conclusions

Brainstorming, strategy-setting, and PHA methodologies are fairly well-defined by industry and company-specific guidance documents. These methodologies are typically bound by documented objectives, prescriptive rule-sets, narrowed scopes, and actual business and/or design information. Despite the fairly well-defined and bounded methodologies, no two workshops are ever the same - even within the same company with the same set of team members. Some workshops are executed more effectively than others with more effective results as the outcome. To achieve excellence, inconsistent and ineffective team performance must be eliminated, prevented, and/or mitigated.

Humans are the primary source of inconsistent and ineffective PHA team performance. As humans, we are vulnerable to the hidden decision-making traps that dwell within our brains. In order to avoid these traps people must be effective communicators; however, few know what effective communication looks like. In addition, effective communication does not just involve body language and verbal cues, but also patterns that improve team performance. Lastly, several critical factors for team collaboration are dependent on effective communication. Given the importance of communication on a PHA team's performance, developing a set of tools and recommendations focused on (1) achieving excellent team performance and (2) delivering consistent and effective workshop results would be valuable to the energy industry and, more importantly, to the people served by the energy industry.

Moving forward, it is worth spending time and money upfront to identify and understand the various personalities on a proposed PHA team. This insight will help you predict and/or plan for the decision-making pitfalls. Knowing your team make-up will also help you assess your potential to achieve the communication patterns of high-performing teams. And lastly, determining where you stand against the factors for highly-collaborative teams will help you make adjustments for effective delivery of the PHA. In the end, organizations should achieve higher-integrity operations by safeguarding communication vulnerabilities.

7. References

[1] Center for Chemical Process Safety, *Guidelines for Hazard Evaluation Procedures*, Second Edition, Copyright © 1992, American Institute of Chemical Engineers, 345 East 47th Street, New York, NY 10017.

[2] Center for Chemical Process Safety, *Guidelines for Risk Based Process Safety*, Copyright © 2007, American Institute of Chemical Engineers, Published by John Wiley & Sons, Inc., Hoboken, 111 River Street, New Jersey 07030.

[3] Gratton, L. and T. Erickson, "Eight Ways to Build Collaborative Teams", *Harvard Business Review*, November 2007.

[4] Hammond, John S., R. Keeney, and H. Raiffa, "The Hidden Traps in Decision Making", *Harvard Business Review*, January 2006.

[5] Insights Learning & Development Ltd., "Insights® Discovery©: Validating the System", Copyright © 2008-2009.

[6] Jolande, Dr. J., *The Psychology of C G Jung*, Copyright © 1962, Yale University Press.

[7] Pentland, Alex, "The New Science of Building Great Teams", *Harvard Business Review*, April 2012.

55a
Ethylene Furnace Tutorial
Radiant Section

Trobie Thompson
Engineering Associate
Sasol North American Operations

14AIChE Spring Meeting, New Orleans, LA

Generic Furnace Schematic

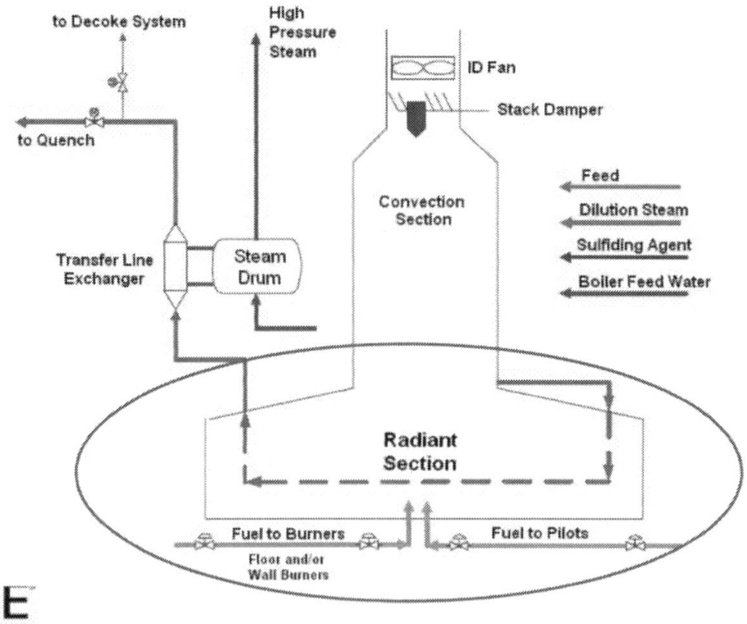

14AIChE Spring Meeting, New Orleans, LA

Furnace Performance Objectives

- Safe, Environmentally Friendly Operation
- Maximum Ethylene Production
- Minimum Energy Use
- Reliability
  - Dependable Run Lengths
  - Manageable Maintenance Costs

14 AIChE Spring Meeting, New Orleans, LA

Radiant Section Focus Areas

- Coils
- Coil Supports
- Burners
- Refractory
- Tramp Air

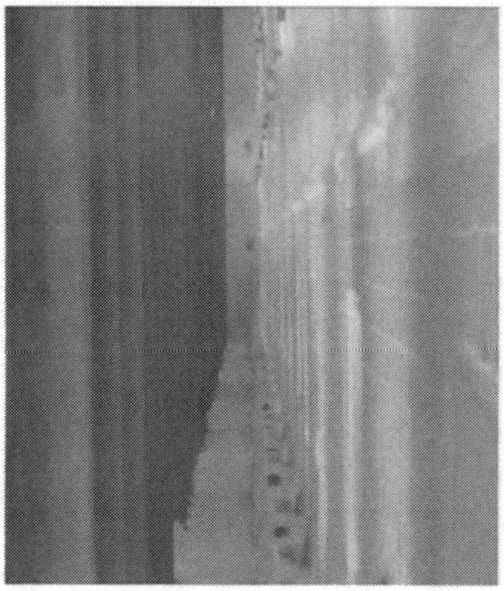

14 AIChE Spring Meeting, New Orleans, LA

Bar Stool Analogy

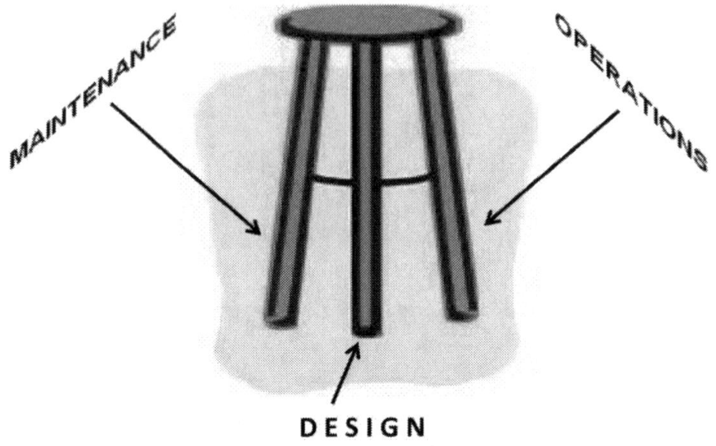

Each leg must support its equal share of the weight!!!

14 AIChE
Spring Meeting, New Orleans, LA

Coils—Design
Surface Area and Diameter

- Smaller Diameter
  - More Surface Area / More Energy Efficient
  - Better Ethylene Selectivity
  - Shorter Run Lengths
  - More Susceptible to Process Changes
- Larger Diameter
  - Less Surface Area / Higher Energy Requirements
  - Less Ethylene Selectivity
  - Longer Run Lengths
  - More Operationally Robust

14 AIChE
Spring Meeting, New Orleans, LA

Coils
Design Considerations

- Endothermic Cracking Reaction
  - Extreme Temperatures
  - Harsh Environment

14AIChE Spring Meeting, New Orleans, LA

Coils
Design Considerations

Coking

- Catalytic Coke
- Amorphous Coke

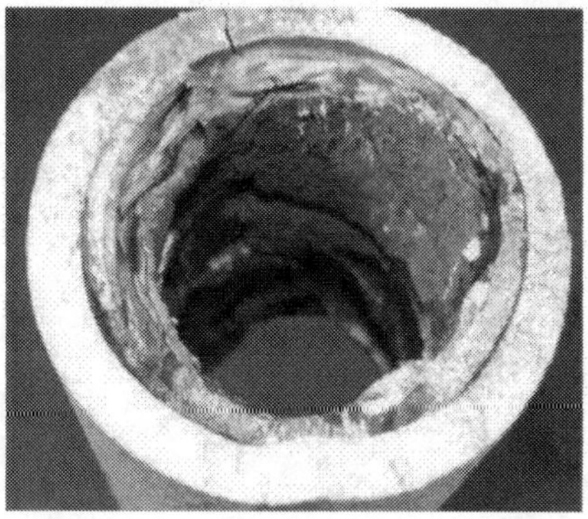

Carburization

14AIChE Spring Meeting, New Orleans, LA

Coils—Design
Creep

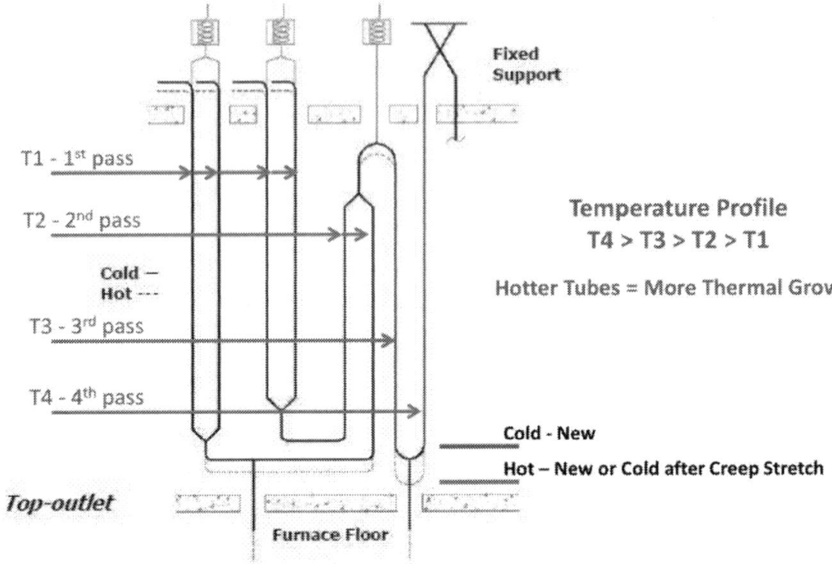

Temperature Profile
T4 > T3 > T2 > T1

Hotter Tubes = More Thermal Growth

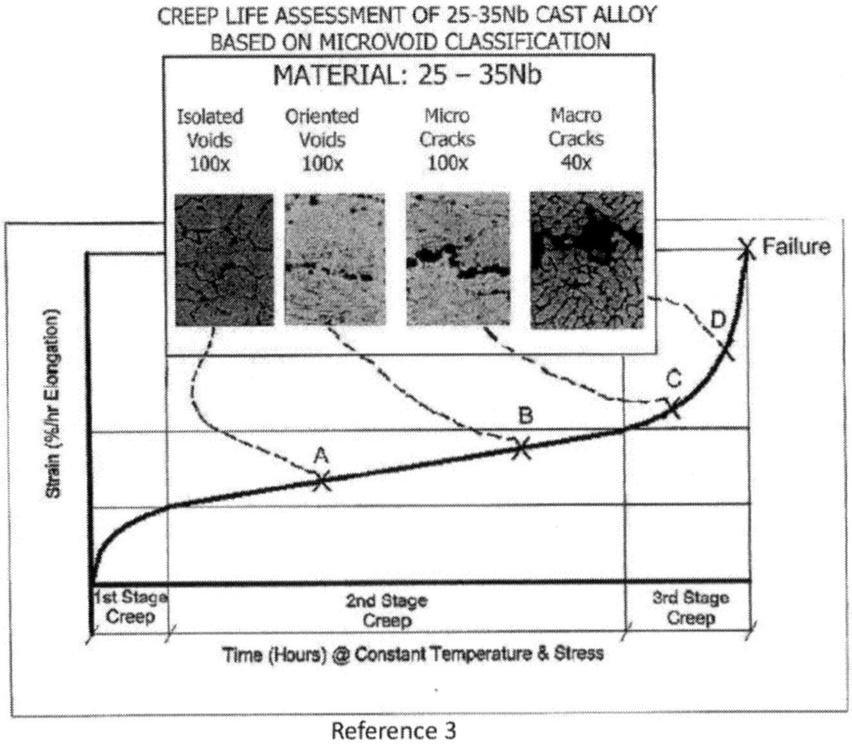

Reference 3

Coils—Design Metallurgy

- 25Cr/35Ni (25% Chromium / 35% Nickel)
- 35Cr/45Ni (35% Chromium / 45% Nickel)
- High Chromium-Containing Inner Layer
- Chromium, Nickel, Aluminum-Containing Alloys

14AIChE Spring Meeting, New Orleans, LA

Coils—Maintenance

- Why Replace Radiant Coils?
  - Bulged and bowed tubes
  - Tubes plugged with coke
  - Tube creep, hits floor of furnace
  - Erosion of 90's or 180's
  - Carburization causing short runs for gas feeds
  - Carburization leading to tube leaks

14AIChE Spring Meeting, New Orleans, LA

Coils—Maintenance
Coil Repair/Replacement Philosophy

- Run to failure
- Complete or partial coil replacement at set frequencies
  - Historical Data Analysis
  - Tube Failure Analyses
  - Statistical Techniques such as Weibull Distribution Analysis[1]
    - Based on leak frequency, planned replacement, and unplanned replacement costs

14 AIChE Spring Meeting, New Orleans, LA

Coils—Operations

- Avoid Temperature Swings
- Daily Inspections
  - Burners
  - Flame Impingement
  - Tube Metal Temperatures
- Decoking
  - Avoid Overheating
    - Potential to burn through coil
    - No Recovery from overheat damage
  - Completely Remove Coke
    - CO_2 measurement
    - Venturi Pressure Ratios
- Scheduling

14 AIChE Spring Meeting, New Orleans, LA

Coil Support Systems—Design

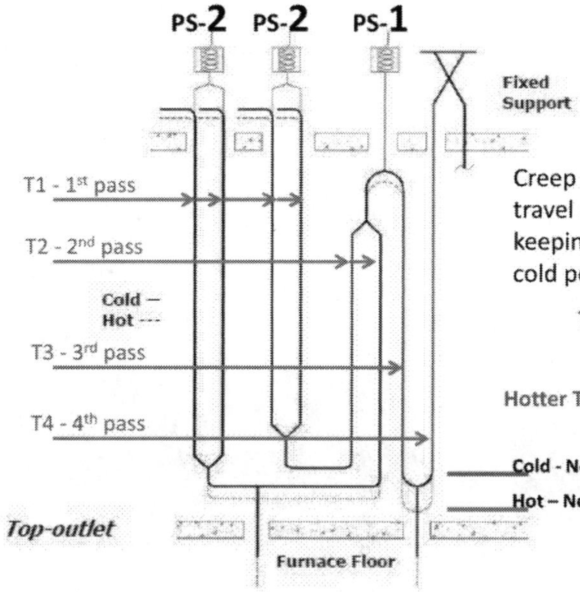

Creep Stretch reduces the effective travel range of spring hangers by keeping them from returning to the cold position!

Temperature Profile
T4 > T3 > T2 > T1

Hotter Tubes = More Thermal Growth

Coil Support Systems—Design

Must have <u>unobstructed</u> coil movement

If unable to move freely → coils bow or creep is accelerated

ultimately causes coil failure

14 AIChE
Spring Meeting, New Orleans, LA

Coil Support Systems—Design

Spring Hangers

Counter Weights

14 AIChE
Spring Meeting, New Orleans, LA

Coil Support Systems—Maintenance
Spring Hangers[3]

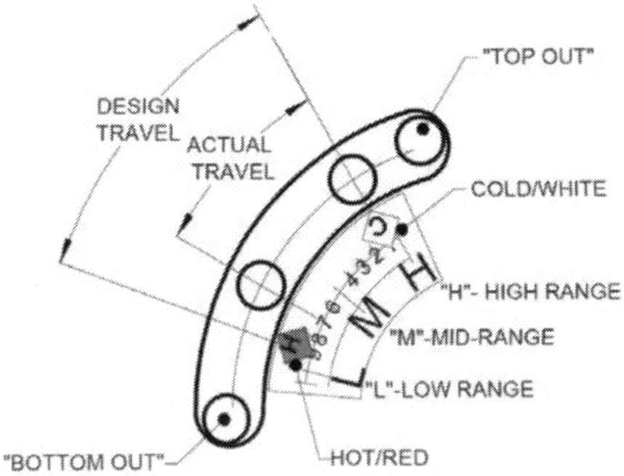

- Initial Load Adjustments
- Periodic adjustments for creep
- Record Keeping
- Re-adjusting for new coils

14 AIChE
Spring Meeting, New Orleans, LA

Coil Support Systems—Maintenance
Spring Hanger Adjustments[3]

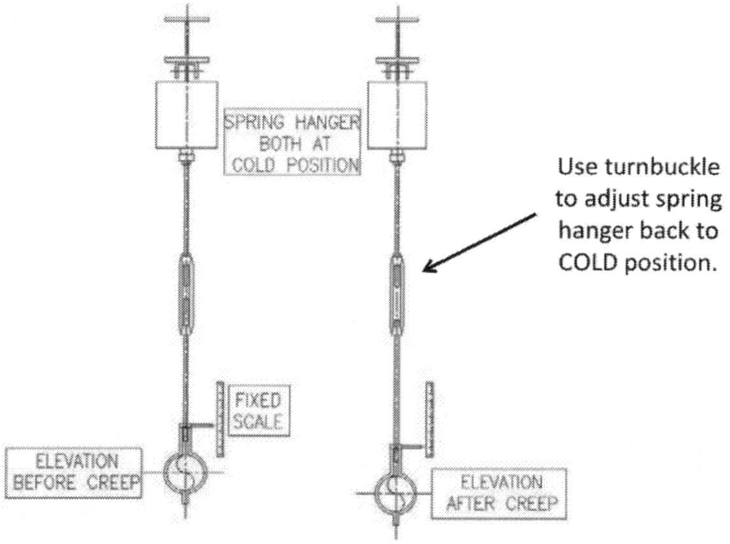

14 AIChE Spring Meeting, New Orleans, LA

Coil Support Systems—Maintenance
Counter Weight Systems[2]

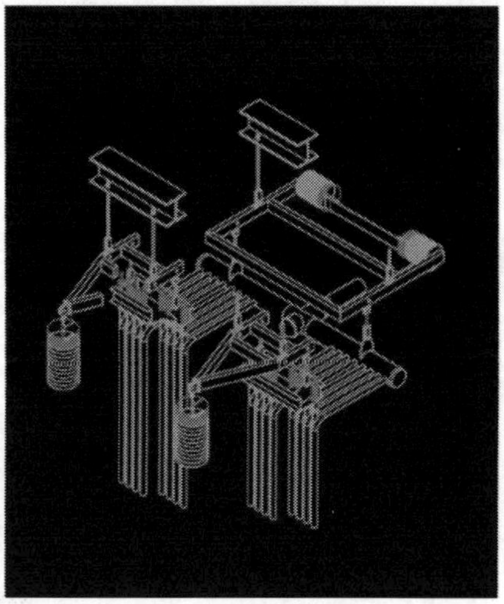

- Initial Coil Balancing
  - Field
  - Shop
- Record Keeping
- Re-adjusting for new coils

14 AIChE Spring Meeting, New Orleans, LA

Coil Support Systems—Operations

- Frequent Inspections
  - Identify "topped" or "bottomed" out supports
  - Verify *unobstructed* tube movements
  - Promptly identify issues
  - Scheduling

14 AIChE Spring Meeting, New Orleans, LA

Burners—Design

- Firing Requirements/Limits
- Emission Limits (NO_x, CO)
- Fuel Gas Composition
  - Hydrogen Impact
  - For the same heat duty,
    - 3X H_2 volume ~ CH_4 volume
- Fuel Gas Pressure *at the burner*
- Burner Type—Wall or Floor

14 AIChE Spring Meeting, New Orleans, LA

Burners—Design cont'd.
Wall Burners

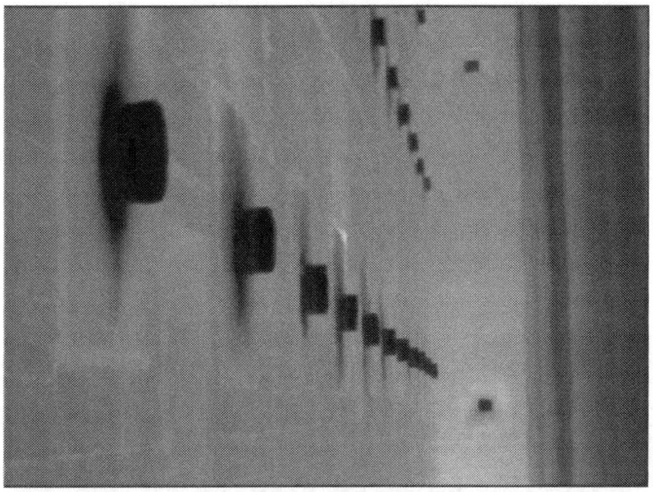

Photo Courtesy of Zeeco

14 AIChE Spring Meeting, New Orleans, LA

Burners—Design cont'd.
Typical Wall Burner

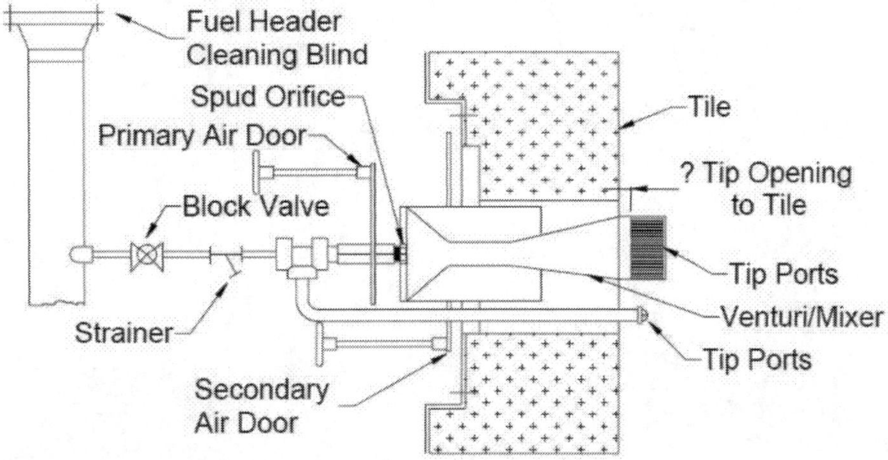

14 AIChE Spring Meeting, New Orleans, LA

Burners—Design cont'd.
Floor Burners

14 AIChE Spring Meeting, New Orleans, LA

Burners—Design cont'd.
Typical Floor Burner

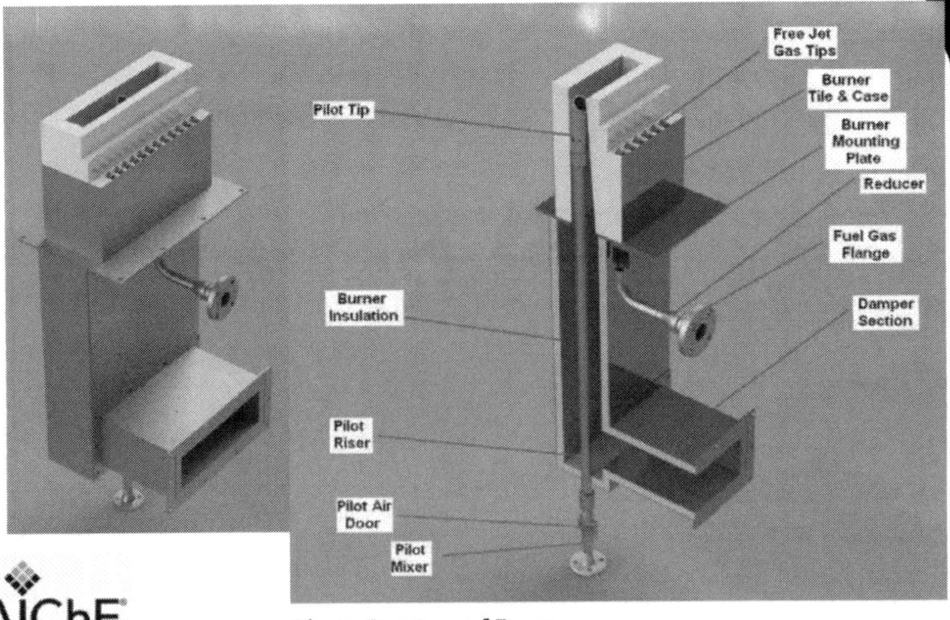

Photo Courtesy of Zeeco

14 AIChE Spring Meeting, New Orleans, LA

Burners—Maintenance

- Proper Installation
  - Wall burner tip distances
  - Floor burner nozzle orientation
- Cleaning/Repairs
  - Fuel Gas Lines
  - Venturi's
  - Spud/Nozzle Orifices
  - Air Openings
  - Lighting Ports
  - Tip Slots
- Burner Management Program
  - As Needed
  - Set Frequency

14AIChE Spring Meeting, New Orleans, LA

Burner—Operations

- Fuel Gas Conditioning
  - Coalescer
  - In-Line Strainers
- Daily Firebox Inspections
  - Prompt identification of dirty/broken burners
  - Burner firing within design/environmental limits
  - Primary/Secondary Air
  - Staged Fuel
- Scheduling

14AIChE Spring Meeting, New Orleans, LA

Refractory—Design

- Brick
- Castable
- Ceramic Fiber

Photo Courtesy of Thorpe

14 AIChE Spring Meeting, New Orleans, LA

Refractory—Design Brick

Photos Courtesy of Thorpe

14 AIChE Spring Meeting, New Orleans, LA

Refractory—Design
Castable

14AIChE Spring Meeting, New Orleans, LA

Photos Courtesy of Thorpe

Refractory—Design
Ceramic Fiber (CF)

14AIChE Spring Meeting, New Orleans, LA

Photo Courtesy of Thorpe

Refractory—Design

Basic
Construction

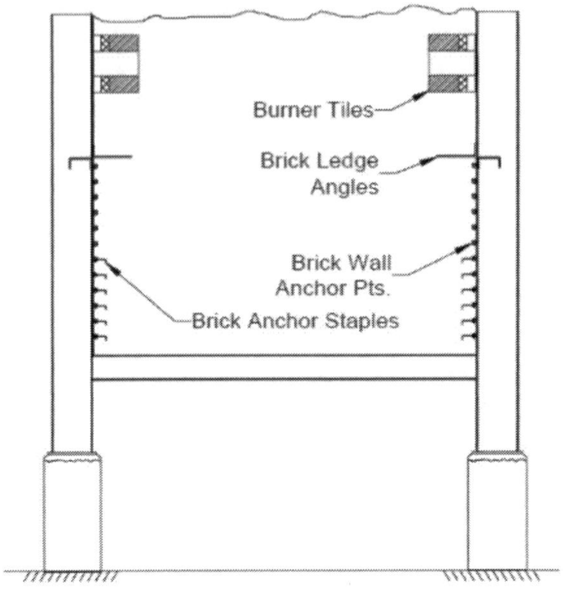

14AIChE
Spring Meeting, New Orleans, LA

Refractory—Design

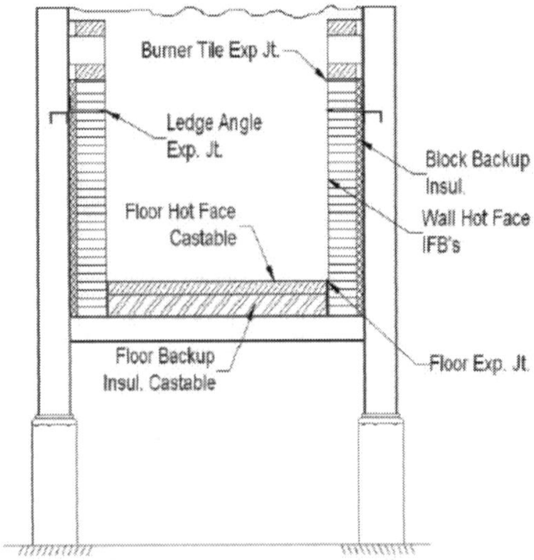

14AIChE
Spring Meeting, New Orleans, LA

Refractory Common Issues

- Shrinkage
- Anchor Damage
- Erosion
- Overheating Material Loss
- Mechanical Damage
- Water Damage

14 AIChE
Spring Meeting, New Orleans, LA

Refractory—Maintenance

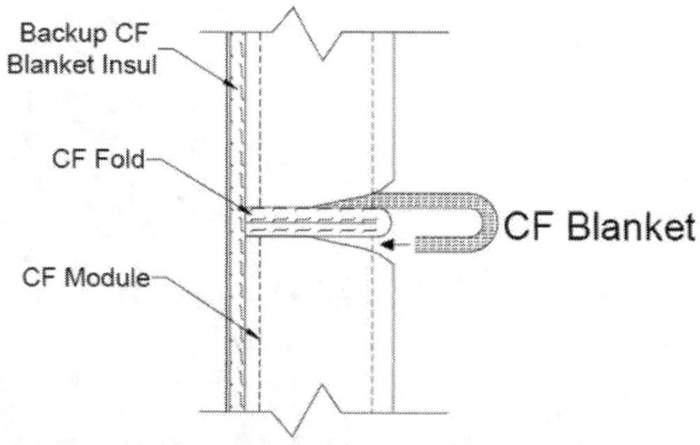

Tuck-And-Roll Method
To Address Initial Shrinkage in CFMs

14 AIChE
Spring Meeting, New Orleans, LA

Refractory—Maintenance

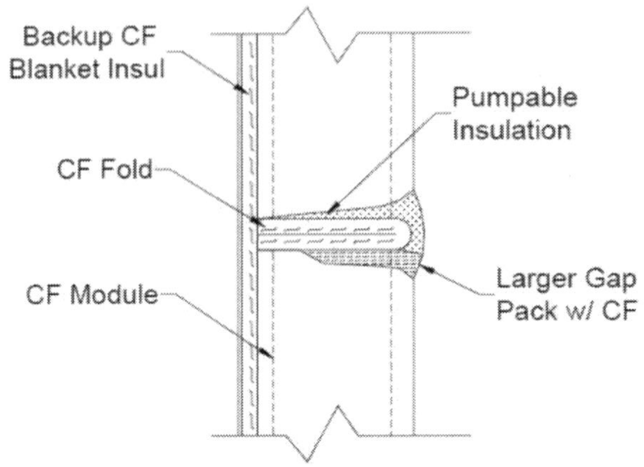

Pump And Pack Method
To Address Initial Shrinkage in CFMs

14AIChE
Spring Meeting, New Orleans, LA

Refractory—Operations

- Minimize Temperature Swings/Thermal Cycles
- Daily Inspections
- Infrared Camera Scans
  - Radiant Roof
  - Tube Penetrations
  - Erosion-Prone Areas
    - Decoke Return Nozzles
    - Burner Tiles

Radiant Roof Tube Penetration

Photo Courtesy of Thorpe

14AIChE
Spring Meeting, New Orleans, LA

Tramp Air/Leakage Points/Efficiency

- Tube Penetration Areas
- Peep Sites
- Access Doors

80%

- Too Much Excess Oxygen
  - 2 to 3% Excess Oxygen typical
  - Draft Set Points and Control

14 AIChE Spring Meeting, New Orleans, LA

Summary

Furnace Performance Objectives

- Safe, Environmentally Friendly Operation
- Maximum Ethylene Production
- Minimum Energy Use
- Reliability
  - Dependable Run Lengths
  - Manageable Maintenance Costs

14 AIChE Spring Meeting, New Orleans, LA

Summary

- Extreme Temperatures / Harsh Environment
- Radiant Components must:
  - Withstand firebox environment
  - Maintain heat in the firebox
  - Protect personnel/equipment outside firebox
- Equal balanced needed among all three legs
  - Design
  - Maintenance
  - Operations

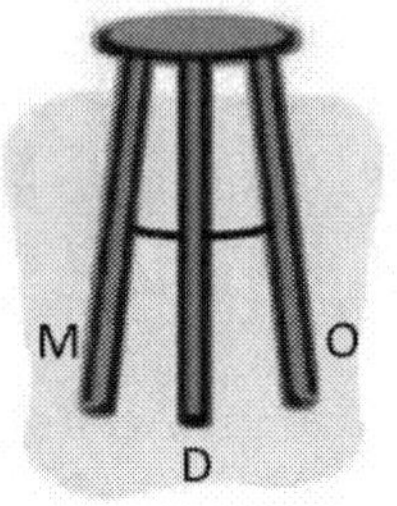

14 AIChE
Spring Meeting, New Orleans, LA

References

1. Cleavinger, Jimmy, "Optimum Furnace Tube Replacements" Presentation, 2009 NPRA Conference.
2. Karrs, Mark and Cannon, T.C., "Cracking Heater Radiant Coil Balancing," March 13, 2002, EPC Topical Conference, 2002 AIChE Spring Meeting.
3. Karrs, Mark and Webb, Ken, "Radiant Coil Spring Hanger Maintenance Considerations," April 26-30, 2009, EPC Topical Conference, 2009 AIChE Spring Meeting.

14 AIChE
Spring Meeting, New Orleans, LA

(55b) Ethylene Cracking Heater Decoking Tutorial

Brian K. Sullivan
Manager of Technical Support
CB&I

Prepared for Presentation at the 26th Ethylene Producers' Conference
AIChE 2014 Spring Meeting
New Orleans, LA, April 1, 2014

14 AIChE Spring Meeting, New Orleans, LA

CB&I

Overview

- Introduction
  - The what, where, why and how of coke
- Fundamentals of radiant decoking
  - Steps
- Points of emphasis
  - Issues, difficulties and possibilities
- Other considerations
  - Polishing
  - TLE
  - Automation
- Summary

14 AIChE Spring Meeting, New Orleans, LA

CB&I

Introduction

14 AIChE Spring Meeting, New Orleans, LA

CBI

Introduction

What is coke?

- Normal by-product of the steam cracking reaction
- Typically ~ 98% carbon and small amount of hydrogen
- Coke properties, amount and location depend upon:
  - Feed
    - Ethane cracking – very hard, brittle, difficult to remove coke
    - Liquids cracking – softer coke, more prone to crumble
  - Cracking severity
  - Length of run
    - Shorter run typically easier to decoke
  - Condition of radiant coil

14 AIChE Spring Meeting, New Orleans, LA

CBI

Introduction

Where does coke deposit and accumulate?

- Convection tubes
- Radiant furnace tubes
  - Greater on outlet/hottest tubes
- Primary transferline exchangers (TLEs)
  - On tube walls for liquids crackers (condensation coke)
  - On inlet tube sheet for gas crackers (catalytic and accumulation)

14 AIChE Spring Meeting, New Orleans, LA

CBI

Introduction

Why do we care about coke deposition in cracking heaters?

- Increases thermal resistance to heat transfer between firebox and process fluid inside radiant tube
  - Reach mechanical temperature limit for radiant coil
- Increases process pressure drop
  - Raises hydrocarbon partial pressure = poorer yields
  - Limits capacity at end-of-run

14 AIChE Spring Meeting, New Orleans, LA

CBI

Introduction

How is coke removed from a cracking heater?

- Coke is gasified using a controlled burn with steam and/or air
- Decoking reactions

$$C + O_2 = CO_2$$ → Major reaction

$$C + CO_2 = 2CO$$

$$C + 2H_2 = CH_4$$

$$C + H_2O = CO + H_2$$ →

$$CO + H_2O = CO_2 + H_2$$

$$H_2 + 1/2\,O_2 = H_2O$$

Steam only decoke

- Much slower
- Very high temperature required (1000 °C+)
- Effect on gasoline fractionator or quench tower
- Online possibility

14 AIChE Spring Meeting, New Orleans, LA

CB&I

Introduction

The objective of radiant decoking is to:

- rapidly
- safely, and
- completely

remove the deposited coke from the radiant coil and form a continuous oxide layer

Improper radiant decoking can lead to:

- Short run-lengths (incomplete decoke)
- Surface damage (decoke too aggressive)
- Higher carburization/shorter coil life

14 AIChE Spring Meeting, New Orleans, LA

CB&I

Fundamentals of Decoking

14AIChE
Spring Meeting, New Orleans, LA

CB&I

Fundamentals of Decoking

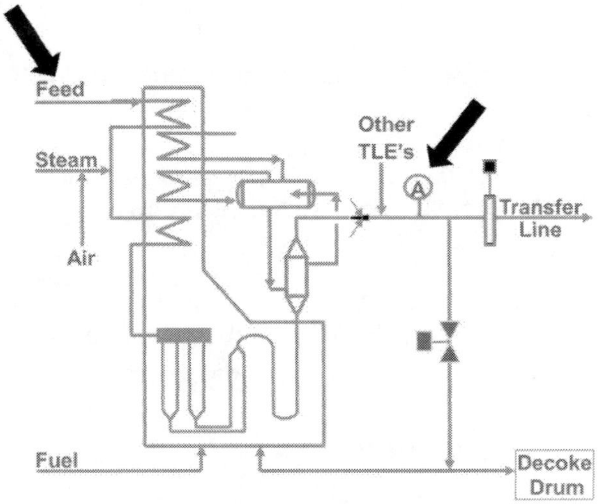

Block in effluent sampler

- Avoid carryover of steam condensate

Preparation

- Withdraw hydrocarbon feed and increase dilution steam
  - Minimize high residence time coking
  - Minimize thermal shock of radiant coil by limiting coil outlet temperature (COTs) excursions (minimize coke spalling)
  - If applicable, maintain venturis choked to prevent mal-distribution of flow and potential tube plugging

14AIChE
Spring Meeting, New Orleans, LA

CB&I

Fundamentals of Decoking

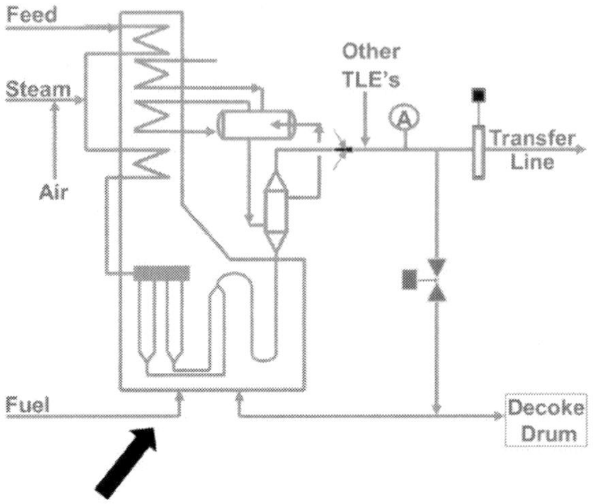

Preparation (continued)

- Adjust or shut off burners as necessary to reduce heat load
  - Establish balanced heat flux profile
  - Avoid low fuel pressure limit
  - Heat steam/air mixture to coke combustion temperature as quickly as possible
  - Establish standard decoke burner pattern

14AIChE Spring Meeting, New Orleans, LA

CBI

Fundamentals of Decoking

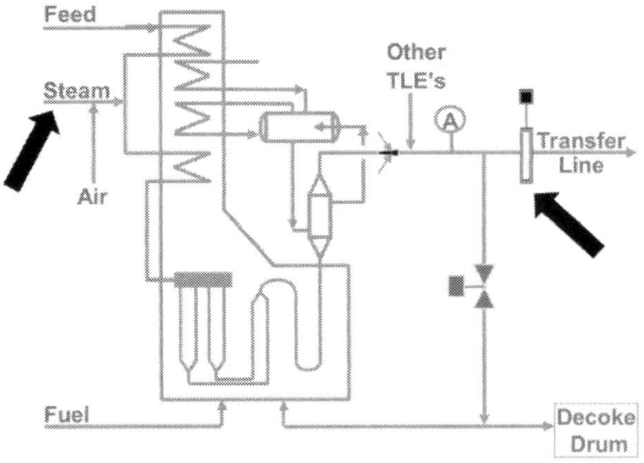

Preparation (continued)

- Switch furnace effluent from transfer line to decoking header (decoke drum or firebox)
  - Purge for ~ 15 minutes prior to switch to ensure no hydrocarbon remains in the coils or piping
- Set initial steam flow rate
  - Provide sufficient heat load

14AIChE Spring Meeting, New Orleans, LA

CBI

Fundamentals of Decoking

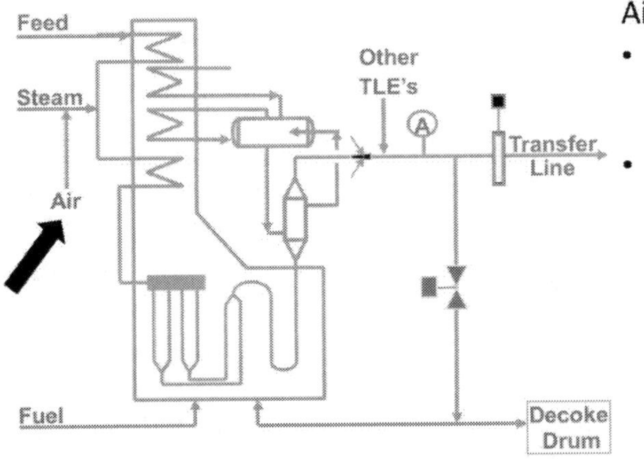

Air introduction

- Connect air hoses to piping or remove the blind from the air line
- Introduce air
  - Initially use small, slow increments
    - Initial air flow may be below range of air flow meter
    - Spalled coke can burn rapidly
  - Duration of initial steps based upon response or time schedule

14 AIChE Spring Meeting, New Orleans, LA

CB&I

Fundamentals of Decoking

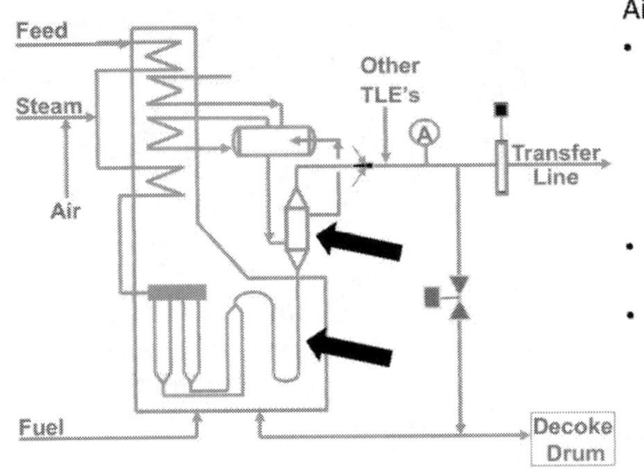

Air Introduction (continued)

- Visually monitor coils for hot spots >1000 °C (1830 °F) or plugged tubes (red hot due to lack of flow)
  - Reduce air rate if overheating is observed
- Check pressure drop across TLE to ensure no plugging
- If some tubes are severely coked or partially blocked, then air introduction may need to be more conservative

14 AIChE Spring Meeting, New Orleans, LA

CB&I

Fundamentals of Decoking

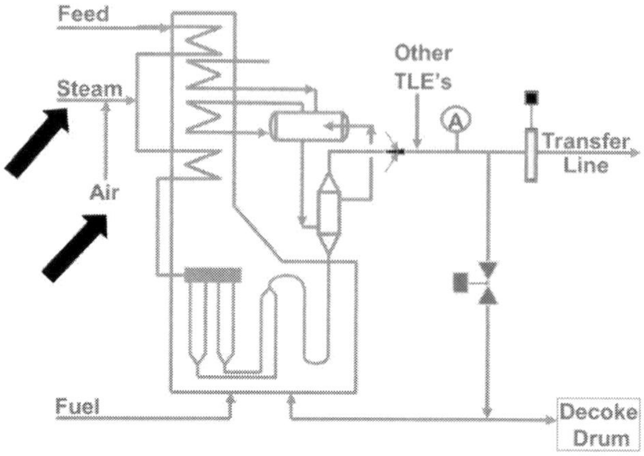

Air and steam adjustments

- Air is increased on time schedule or as temperature allows
  - As initial coke is burned, it becomes safer to proceed at a faster rate
- Steam can be reduced as air is increased
  - Increase air/steam ratio (oxygen partial pressure)
  - Typical terminal target is 1:3 to 1:1

14 AIChE Spring Meeting, New Orleans, LA

CB&I

Fundamentals of Decoking

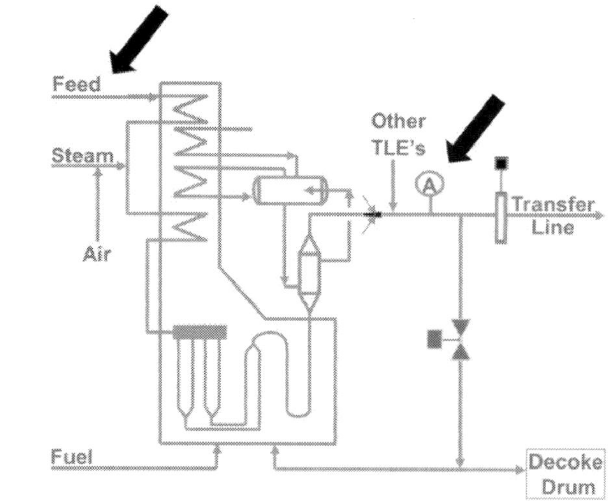

Completion of decoke

- Various criteria are used
  - CO_2 concentration in furnace decoke effluent
  - Venturi critical pressure ratios
  - Time schedule
- Some plants perform a proof burn to confirm completion

14 AIChE Spring Meeting, New Orleans, LA

CB&I

Fundamentals of Decoking

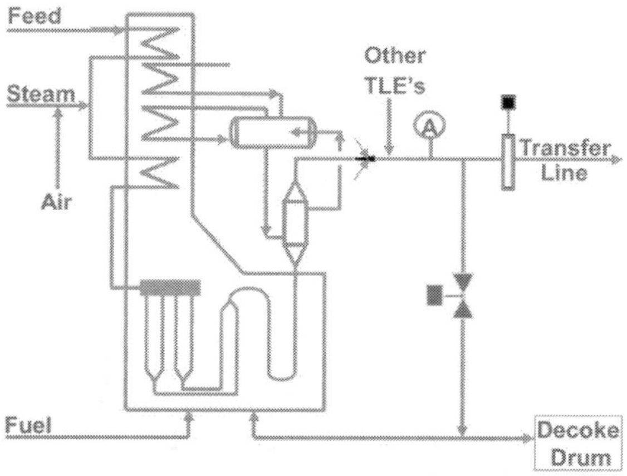

Completion of decoke (continued)

- Return furnace to pre-operation dilution steam flow rate
- Remove air lines from furnace and blind
  - Automatic interlock or logic typically prevents furnace restart while air lines are connected

14 AIChE Spring Meeting, New Orleans, LA

CBI

Points of Emphasis

Suggestions for Common Problems/Issues

With experience, each plant adjusts the general procedure to reflect its experience and specific configuration

14 AIChE Spring Meeting, New Orleans, LA

CBI

Point of Emphasis - Feed Withdrawal

How Do I Minimize Coke Spalling?

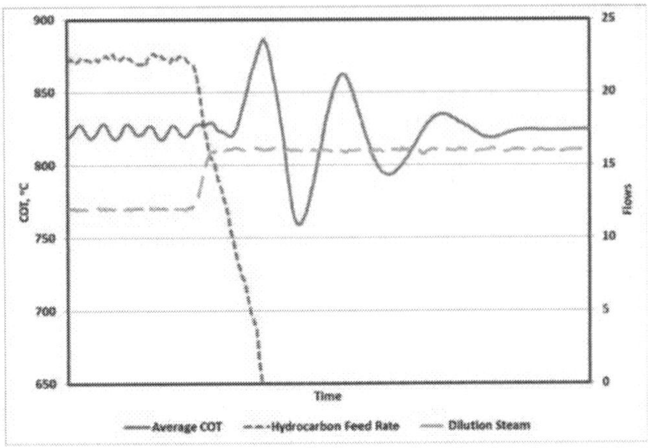

- Rapid changes in COT reflect rapid changes in TMT which expand/ contract tube radially
- TMT will drop significantly as a result of feed withdrawal
- Dynamic response of COT controller lags feed disturbance
  - Anticipate firing requirement manually

Poor temperature control during feed withdrawal (>100 °C swing)

- Loss of heat of cracking and manual burner adjustments = difficulty
- Target is +/- 20 °C (+/- 35 °F)

14 AIChE Spring Meeting, New Orleans, LA

CB&I

Point of Emphasis - Feed Withdrawal

Due to different coefficients of expansion for coke and tube, stresses are induced in coke that can fracture it -- or the tube!!

Minimize temperature swings to reduce coke spalling prior to decoke

14 AIChE Spring Meeting, New Orleans, LA

CB&I

Point of Emphasis - Burner Pattern

How Do I Obtain Thorough Coke Burn?

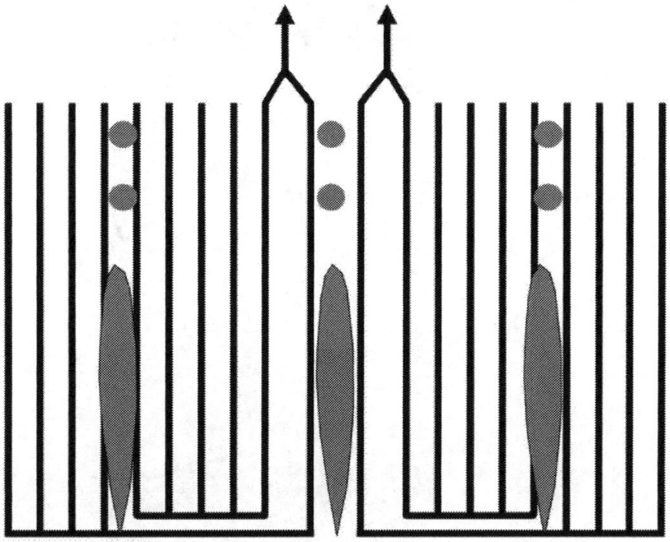

Example of normal burner pattern for firebox with hearth and wall burners

14AIChE Spring Meeting, New Orleans, LA

CBI

Point of Emphasis - Burner Pattern

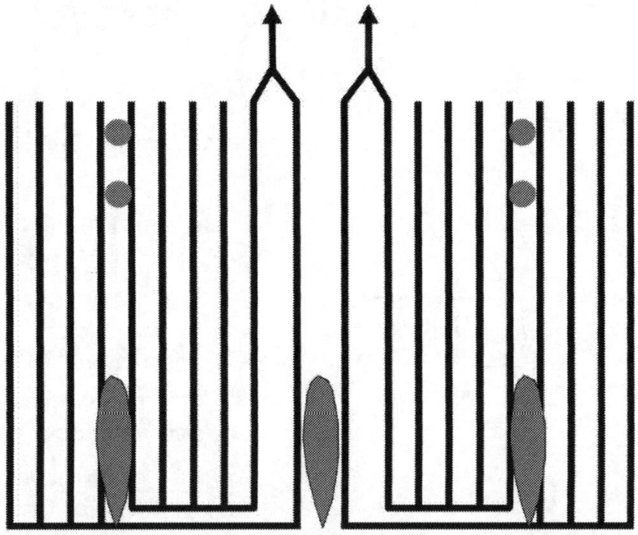

Raise temperature of steam/air mixture at entrance to coil to increase combustion rate

Typical target is up to 700 °C crossover temperature (XOT), subject to metallurgy limits

Once optimal burner pattern is determined, it should be sufficient for subsequent decokes

14AIChE Spring Meeting, New Orleans, LA

CBI

Point of Emphasis - Burner Pattern

Decoke nozzles can impact burner pattern when decoke effluent is sent to firebox

- Compensate for cooling effect of decoke effluent to firebox
- Route decoke steam to firebox gradually

14 AIChE Spring Meeting, New Orleans, LA

CBI

Point of Emphasis – Damper Settings

How Do I Prevent Convection Section Overheating?

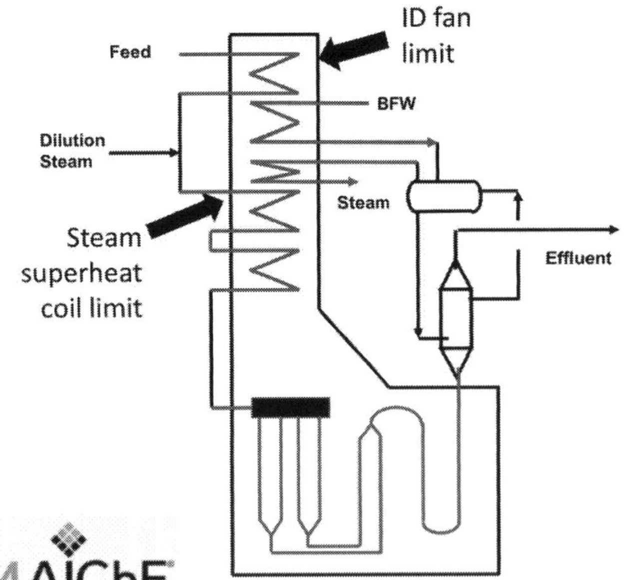

- Minimize excess air to firebox
  - Close air dampers on burners out of service
    - Wall: keep primary open and close secondary
    - Hearth: air registers (minimize opening)
  - Minimize draft while maintaining control

14 AIChE Spring Meeting, New Orleans, LA

CBI

Point of Emphasis - Steam/Air Flow & Velocity

How Do I Select Air Rate?

Steam and air flow rate schedule

- Maximum steam + air velocity set by erosion limit
  - Suggested limits
    - Radiant tubes < 125 - 150 m/sec
    - Decoke lines < 100 ~ 110 m/sec
- Minimum steam + air flow rates set by
  - Load to support balanced firebox firing/heat distribution
  - Maintain choked venturis (if applicable)
    - Different backpressure before and after
      - TLV/DV swap
      - Coke removal
  - Time requirements

14 AIChE Spring Meeting, New Orleans, LA

CB&I

Point of Emphasis - Steam/Air Flow & Velocity

Steam and air flow rate schedule (continued)

- Maximum mass ratio of air to steam typically 1:3 to 1:1
  - Limit partial pressure of O2 to limit combustion rate
- Other considerations
  - Air compressor capacity and steam availability

14 AIChE Spring Meeting, New Orleans, LA

CB&I

Point of Emphasis - Coke Burn

How Do I Ensure Coke is Burning?

Confirmation of burning can be seen as:

- Increase in COTs
  - Coke burning with air at constant fired duty will raise COT
  - Allows monitoring to confirm coke burn in individual coils
- Increase in CO_2 in furnace effluent
  - Initially rises above 10 vol%
  - Rapidly drops off to a lower value
  - Large or rapid increases in CO_2 at low initial air rates are expected and do not indicate excessive coke burning
  - Not as good as COTs/does not ensure burning in all coils

14 AIChE Spring Meeting, New Orleans, LA

CB&I

Point of Emphasis - Coke Burn

Do not be alarmed by spikes in CO_2 when air rate is low

- Measurement on dry basis
- Relative measure, not absolute measure
- At low air rate, coke burn and related heat release is low even if CO_2 reading is high

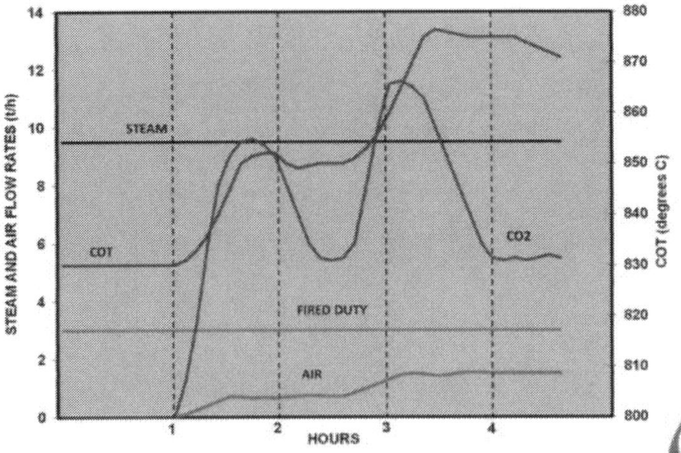

14 AIChE Spring Meeting, New Orleans, LA

CB&I

Point of Emphasis - EOR Criteria

How Do I Know When Decoke is Complete?

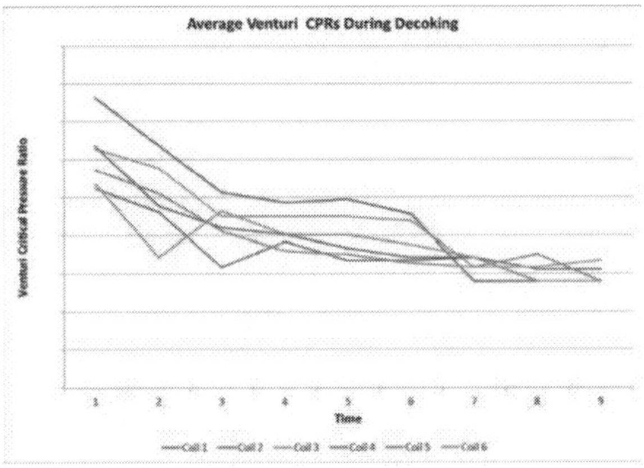

- Venturi CPRs may start at very different values
  - Coils are coked in different amounts
- Venturi CPRs will slowly converge as decoke proceeds
  - Wait until venturi CPRs for all coils converge within a reasonable tolerance and reach a target value as another indication of completion

14 AIChE Spring Meeting, New Orleans, LA

CB&I

Point of Emphasis - EOR Criteria

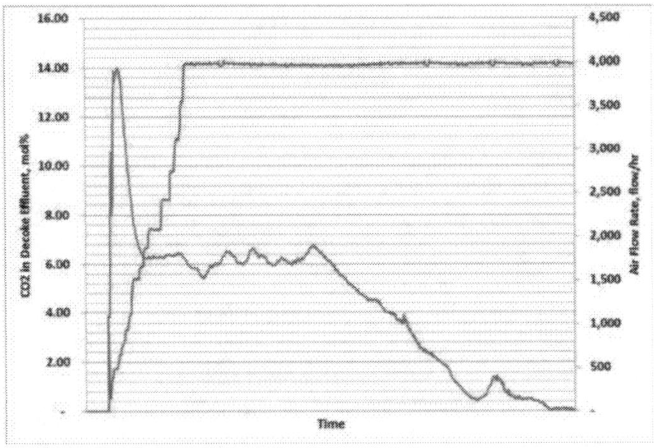

- CO_2 will spike early in decoke
  - Not a concern when air rate is low
- CO_2 will slowly taper off as decoke proceeds
  - Wait until CO_2 < 0.1vol% as one indication of completion

14 AIChE Spring Meeting, New Orleans, LA

CB&I

Point of Emphasis–Decoke Effluent Safety

Are There Safety Concerns While Decoking?

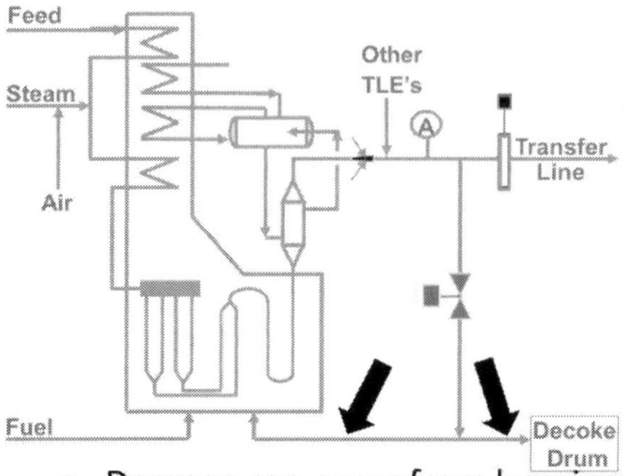

Safety Issue

- Must not contaminate the dilution steam
  - Quench water/gasoline separation interface instrumentation must be working properly

- Damage can occur from burning hydrocarbon in decoke effluent
- Potential hydrocarbon release and fire when discharging effluent to decoke drum

14 AIChE Spring Meeting, New Orleans, LA

CB&I

Point of Emphasis – Fuel Firing Safety

Decoking Safety Report – Operation without automatic COT control

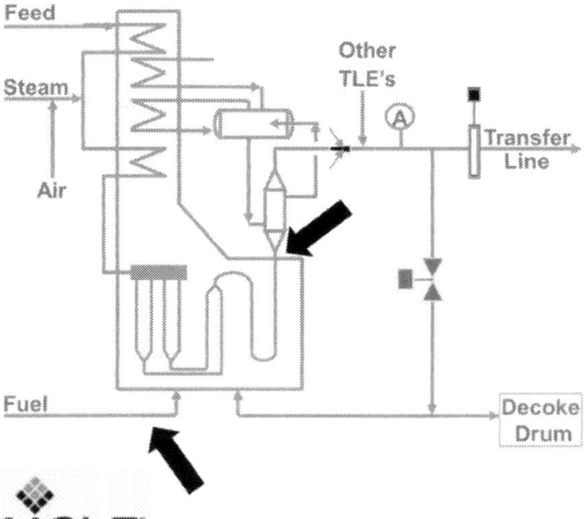

Incident

- Furnace on steam only
- Fuel heat duty control in manual
- Upset caused heavy backup fuel to enter the fuel gas system
- At constant valve opening and denser fuel, heat released increased significantly (i.e., doubled)
- Operator did not notice for 20 minutes
- Result = damage to radiant coils and convection section

14 AIChE Spring Meeting, New Orleans, LA

CB&I

Point of Emphasis – Fuel Firing Safety

Decoking Safety Report – Operation without automatic COT control

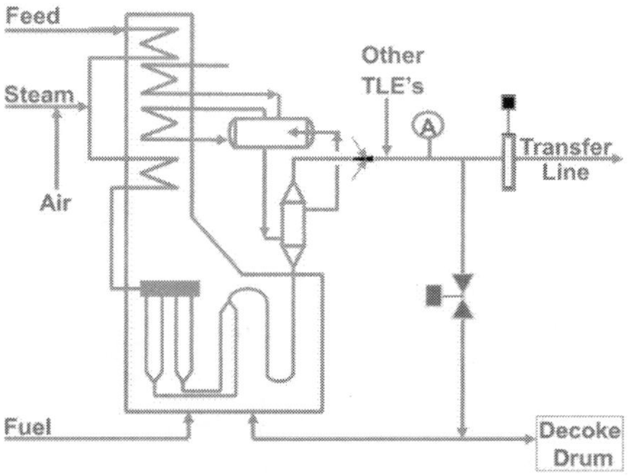

Recommendation

- Use of a Wobbe meter or molecular weight analyzer allows control of heat released from burners
- Operators should be educated on the consequences of shifting to backup fuel
- Must diligently monitor the fuel system when the analyzer is offline
- Alarm for back-up fuel

14 AIChE Spring Meeting, New Orleans, LA

CB&I

Other Considerations

14 AIChE Spring Meeting, New Orleans, LA

CB&I

Polishing

Some plants prolong the final air step to re-establish the Cr_2O_3 layer on tube surface

- Theory is Cr_2O_3 improves coking and carburization resistance
- Cr_2O_3 layer is damaged during operations and decoke
- Another theory - a prolonged air step can create a Cr_2O_3 layer that is too thick and brittle, which can lead to coke spalling

14 AIChE Spring Meeting, New Orleans, LA

CB&I

TLE Decoking Procedure

On-line TLE decoking

- No cold furnace shutdown required for mechanical cleaning
- Extension of furnace decoke procedure
  - Modified steam and air rates
  - Burn coke from TLE
- Possibility of Cr_2O_3 layer that is too thick

14 AIChE Spring Meeting, New Orleans, LA

CB&I

Automation & Control

Startup and Shutdown Automation Issues

- Additional instrumentation and control
  - Individual pass air flow control
  - Online decoke effluent CO_2 analysis
  - Burner damper control
  - Adaptive gain for COT control
- Economically impractical to fully automate due to safety requirements surrounding shutting and lighting burners
- Difficulty with integration of continuous controls, sequenced logic and manual operations (i.e., lighting of burners, etc.)
- Several alternatives (e.g., advisory, partially automate, etc.)

14AIChE Spring Meeting, New Orleans, LA

CB&I

Summary

Improper radiant decoking can lead to:

- Short run-lengths (incomplete decoke)
- Surface damage (decoke too aggressive)
- Higher carburization/shorter coil life

Key points of emphasis address problematic issues

- Minimize feed withdrawal temperature swings
- Set burner pattern for optimal coke combustion
- Set air dampers to prevent convection overheating
- Select air/steam flow rate to minimize erosion/maximize firing load
- Confirm initial coke burn with COT rise and CO_2 analysis
- Ensure complete coke removal using appropriate criteria
  - Time, CO_2, venturi critical pressure ratios
- Incorporate appropriate safeguards

14AIChE Spring Meeting, New Orleans, LA

CB&I

352606 Ethylene Furnace Convection Section Tutorial

Abstract:

Cracking furnaces generate a lot of waste heat. The furnace convection sections are designed to recover this heat from flue gas by pre-heating and vaporizing process feed prior to the radiant section, pre-heating boiler feed water, superheating steam streams and, in some cases, pre-heating air for combustion. The efficient recovery of this heat in the convection section is critical to the economics of the process.

This ethylene furnace tutorial will cover the objective of the convection section, typical layouts, internal components and their failure modes, inspection areas and repairs.

Ethylene Furnace Tutorial
Steam System

Mark Karrs
Engineering Advisor
Becht Engineering

14AIChE
Spring Meeting, New Orleans, LA

Overview

- Steam System Functionality
- Design Considerations
- Operational Considerations
- Failure Mechanisms
- Inspection Focus

14AIChE
Spring Meeting, New Orleans, LA

Steam System Schematic

- Process Waste Heat (reaction effluent) recovered in a Transferline Exchanger (TLE), aka Primary Quench Exchanger (PQE) via a thermosyphon loop with an elevated Steam Drum.
- Flue Gas Hi Temp Waste Heat recovered by superheating generated steam
- Flue Gas Lo Temp Waste Heat recovered by BFW preheating in an economizer coil.
- Steam generated powers rotating equipment via turbine drives, typically Cracked Gas and Refrigeration compressors.

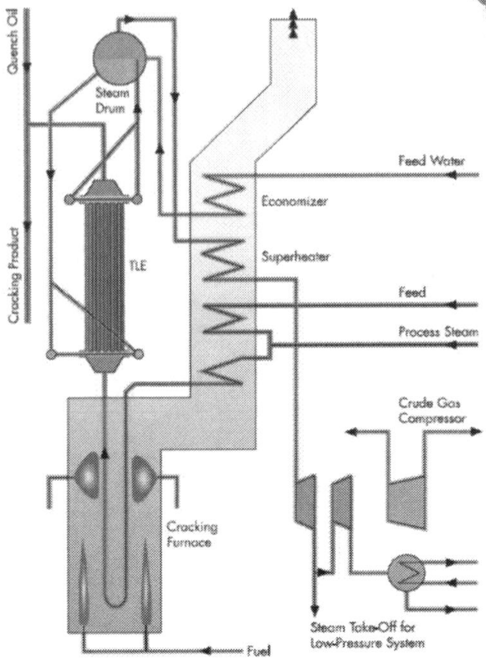

14 AIChE
Spring Meeting, New Orleans, LA

Steam Pressure

- Normally controlled at CG turbine inlet
- Drum pressure floats on header + SSH dP
- 1200 to 1800 psig level for contemporary design
- Vent + Silencer provided for startup
- Non return valve in outlet line to header

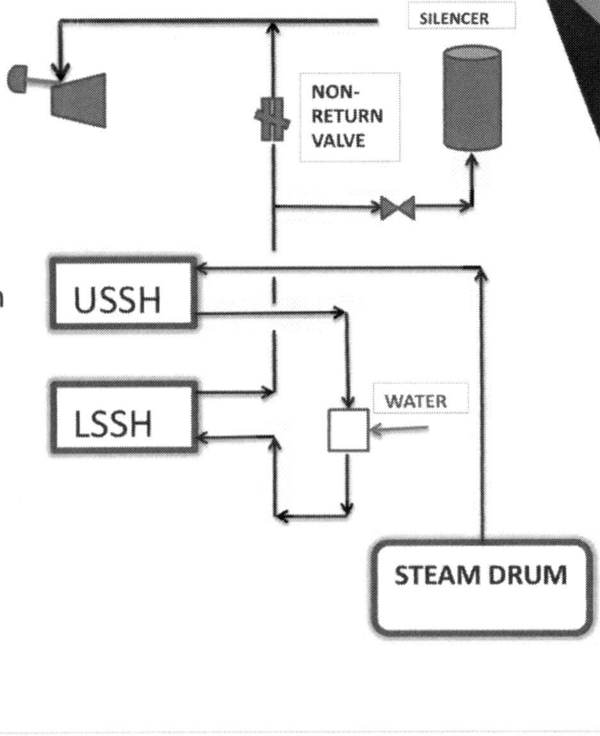

14 AIChE
Spring Meeting, New Orleans, LA

Design Codes

- Steam Superheat (SSH) Coil and Boiler Feedwater (BFW) Coil usually designed to ASME Sect I (Power Boilers) and B31.1 (Power Piping)
- Steam Drum and Transferline Exchanger (TLE) Steam side can be designed to ASME Sect VIII Div 1 (Pressure Vessels)
- If stamped to ASME Sect I, annual inspection requirements of that code will apply.
- State Boiler Authority has jurisdiction.

14AIChE
Spring Meeting, New Orleans, LA

Steam Desuperheater

- Superheat is req'd to maintain turbine efficiency
- Superheat Q a function of firing rate only, whereas steam generation Q also a strong function of process flow
- Intermediate desuperheater provides for control of SSH outlet temp to header
- Water to desuperheater must be free of non-volatile components, treatment chemicals

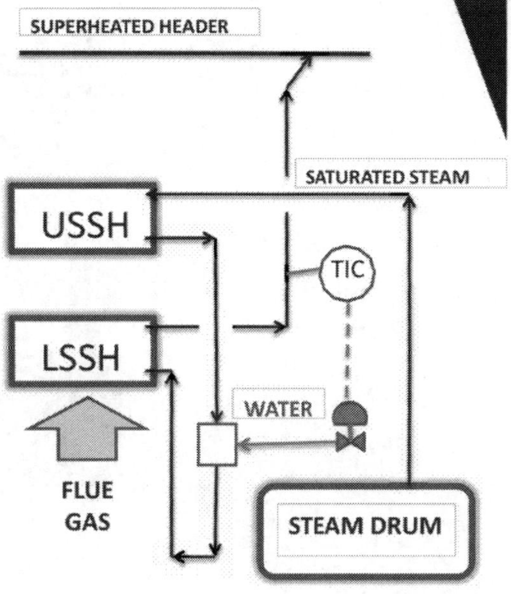

14AIChE
Spring Meeting, New Orleans, LA

Start-of-Run vs. End-of-Run

- Radiant Coil fouling increases the firing rate, increasing superheat Q toward EOR
- TLE fouling decreases steam generation Q, decreasing steam flow toward EOR
- Lower steam flow and higher SSH Q will cause desuperheat water flow to increase over a typical run
- TLE fouling most severe for heavier feedstocks, two levels of desuperheat can be required

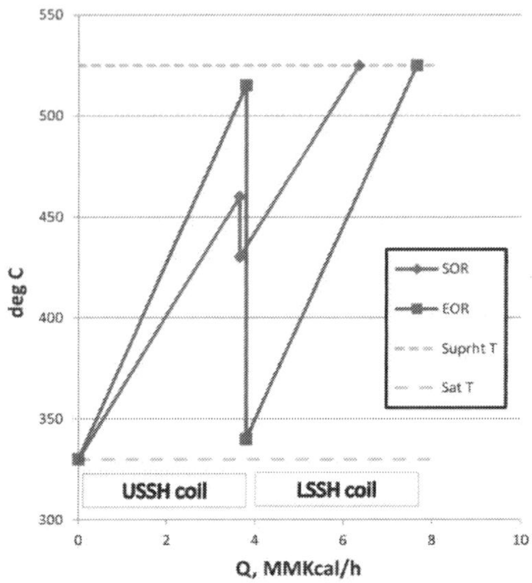

14 AIChE Spring Meeting, New Orleans, LA

Expansion Considerations

- Steam system coils and piping must be free to expand from ambient to the hot position
- Springhangers required to support heavy wall pipe during large movements, to avoid overstress due to expansion

Photo courtesy NOVA Chemicals

14 AIChE Spring Meeting, New Orleans, LA

Expansion Considerations

- Furnace coils must expand freely into the headerboxes
- Even number of passes desirable
- Desuperheater Loop will expand between coil inlet/outlet headers

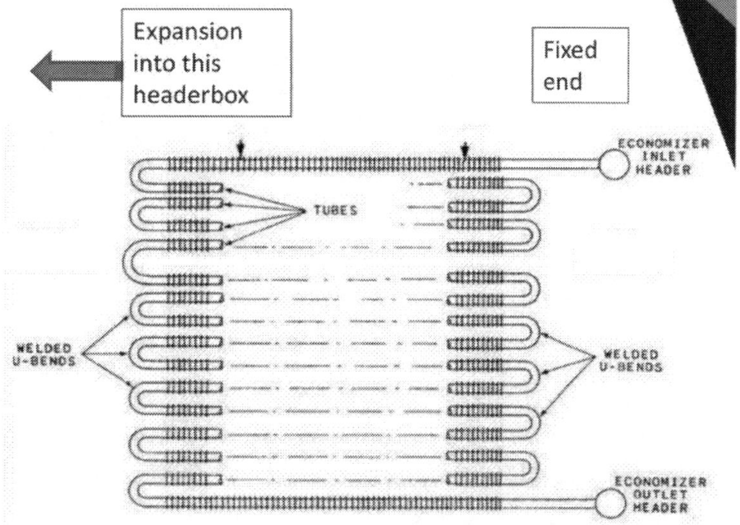

14 AIChE
Spring Meeting, New Orleans, LA

Expansion Considerations

- Downcomers/risers must expand between Steam Drum and TLE (fixed supports) – flexibility is critical
- TLE fixed support is near inlet tubesheet
- Riser flexibility must accommodate:
  - Riser expansion, plus
  - TLE expansion, plus
  - Steam Drum movement

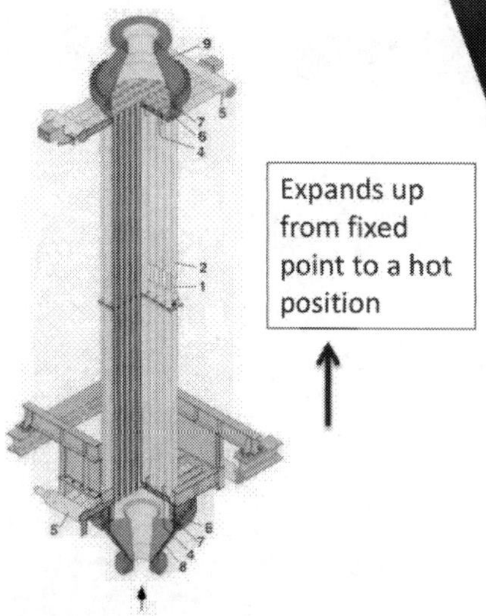

Graphic courtesy ALSTOM Power

14 AIChE
Spring Meeting, New Orleans, LA

Water Quality is critical

- Water treatment should include pH control, dissolved O2 scavanging and silica limitation
- Continuous and Intermittent blowdown from the Steam Drum to limit TDS

Photo courtesy BASF

14AIChE Spring Meeting, New Orleans, LA

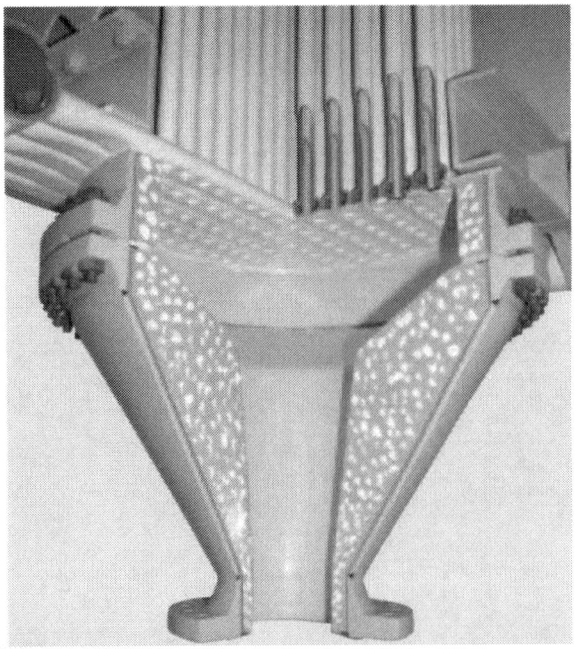

TLE Inlet tubesheet

- Low point in the steam system, subject to solids accumulation, blowdown is critical

- High temp, high velocity erosive flow on the process side

14AIChE Spring Meeting, New Orleans, LA

Intermittent TLE Blowdown

- Drain connections on downcomer headers (low points of low velocity)
- Blowdown of inlet headers <u>should be done offline</u>, as this can starve water from the annular space and result in localized overheat and stress

Photo courtesy ALSTOM Power

14 AIChE Spring Meeting, New Orleans, LA

Heat Input Balance

Startup and SOR

- Clean radiant coil and TLE result in low firing and high steam make
- High process waste heat recovered and low flue gas waste heat
- Low desuperheat water req'd , even zero

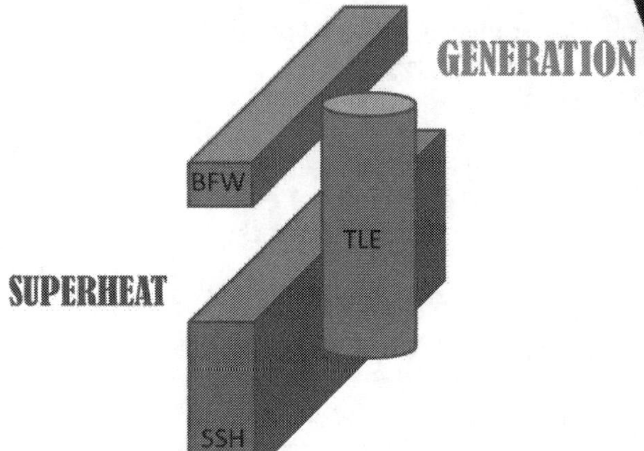

14 AIChE Spring Meeting, New Orleans, LA

Heat Input Balance

EOR and Decoke

- Fouled TLE can result in low process waste heat recovered, low steam make
- Fouled radiant coil will result in high firing and high flue gas waste heat
- Max. desuperheat water req'd

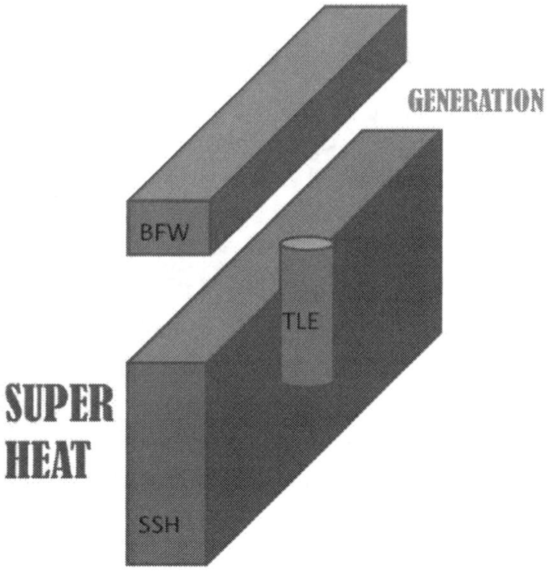

Decoking

- Radiant coil flow switches from highly endothermic (absorbing heat of cracking reaction) to slightly endothermic (heating of steam/air mass flow, offset by exothermic heat of combustion of the coke)
- Radiant coil flowrates can vary widely over the decoke sequence
- Key process variable needed to balance steam system duties is the combustion air flow (unmetered)

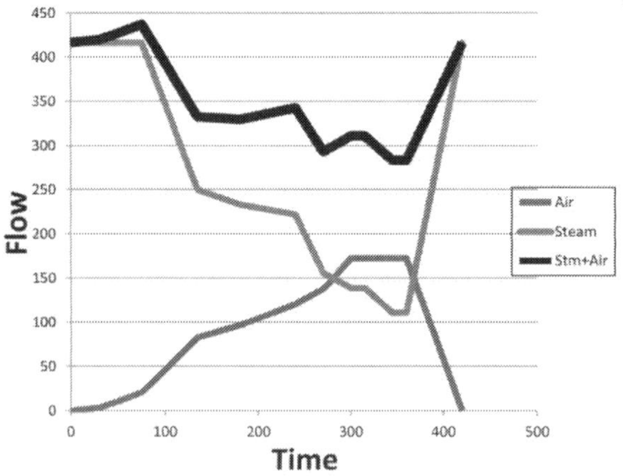

14AIChE
Spring Meeting, New Orleans, LA

Failure Mechanisms

- Erosion
- Corrosion
- Stress
  - Overheat
  - Inhibited Expansion/Movement
  - Dis-similar Welds

14 AIChE
Spring Meeting, New Orleans, LA

Stress on tubesheets

- Plugged tube causes stress at tubesheet due to differential temperature between adjacent tubes
- Plugged tube gets no flow, cools and contracts

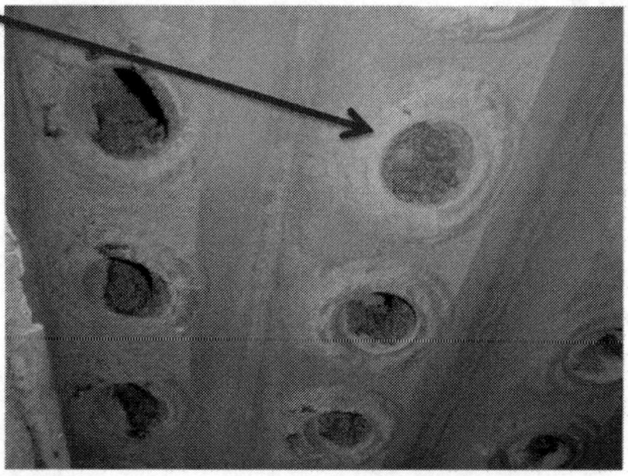

Photo courtesy NOVA Chemicals

14 AIChE
Spring Meeting, New Orleans, LA

Erosion of tubesheets

- Erosion of the inlet tubesheet by coke fines (mainly during decoking)
- Erosion protection of the tubesheet can be provided

Photo courtesy ALSTOM Power

14 AIChE Spring Meeting, New Orleans, LA

Erosion protection

- Erosion protection shield device mounted in the inlet channel refractory
- Avoids welding to the inlet tubesheet

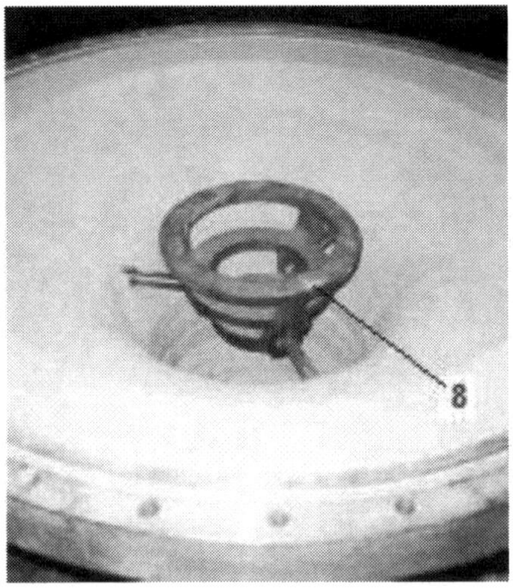

Photo courtesy BORSIG GmbH

14 AIChE Spring Meeting, New Orleans, LA

Erosion of Desuperheater Loop

- Erosion of the piping downstream of the desuperheater can result from overquench
- Resulting droplets will be accelerated to vapor velocity
- Piping walls and elbow directly downstream of the desuperheater outlet can see erosive impingement
- If overquench reaches hot coil, thermal fatigue and failure

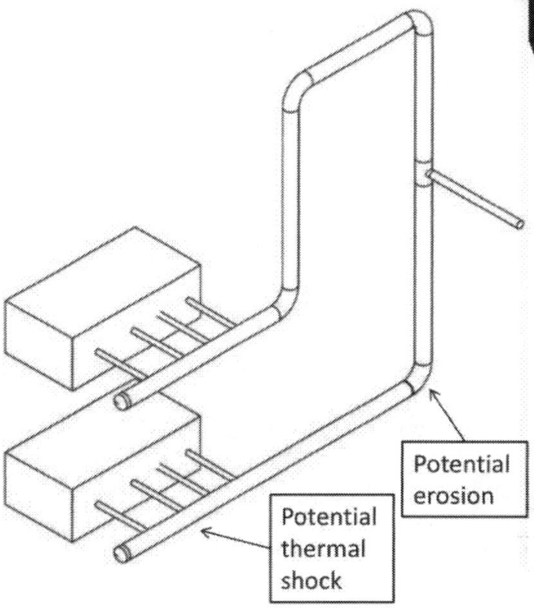

14AIChE Spring Meeting, New Orleans, LA

Stress due to overheat

- Tube temp excursion can be caused by firing upset, high flue gas temps
- Steam piping temp excursion can be caused by desuperheat control failure, reaching limits of desuperheater system
  - ➢ Reduced strength can result in deformation, bulging, wall thinning, failure
  - ➢ Overexpansion can result in stress imposed by mechanical limits

14AIChE Spring Meeting, New Orleans, LA

Stress due to inhibited expansion

- Expansion beyond the design temp typically not accommodated in the support design
  - Return bends hit the headerbox covers
  - Shoes may fall off their supports
  - Springhangers reach their limits of travel
  - Piping contacts structure or platforms

14 AIChE
Spring Meeting, New Orleans, LA

Dis-similar welds

- BFW coil and steam drum typically CS
- SSH coil can be Cr-Mo or austenitic SS
- Steam Header piping Cr-Mo
  - ➢ Dis-similar welds will be present in the system
  - ➢ These welds will be under stress due to differential expansion between alloys, even in the absence of external stress
- Welds at lugs, shoes, trunnions will also be subject to thermal stress due to temperature gradient, even if they are same material

14 AIChE
Spring Meeting, New Orleans, LA

Steam system dis-similar weld

Cracking of a dis-similar P11-P91 socket weld at steam manifold connection

Removable insulation at all dis-similar weld locations should be considered to facilitate regular inspection

Photo courtesy BASF

14 AIChE Spring Meeting, New Orleans, LA

Inspection Focus

- TLE tubes / tubesheets
- Coil welds / connections
- External piping / valves
- BFW control valves
  - BFW inlet control (reliability critical)
  - Desuperheater water control (shutoff critical)
- Support Points
- Steam Drum internals

14 AIChE Spring Meeting, New Orleans, LA

Boroscope inspection of inlet tubesheet

BORSIG "Tunnelflow" TLE

Waterside inspection of inlet tubesheet using the inspection nozzles furnished on the blowdown header

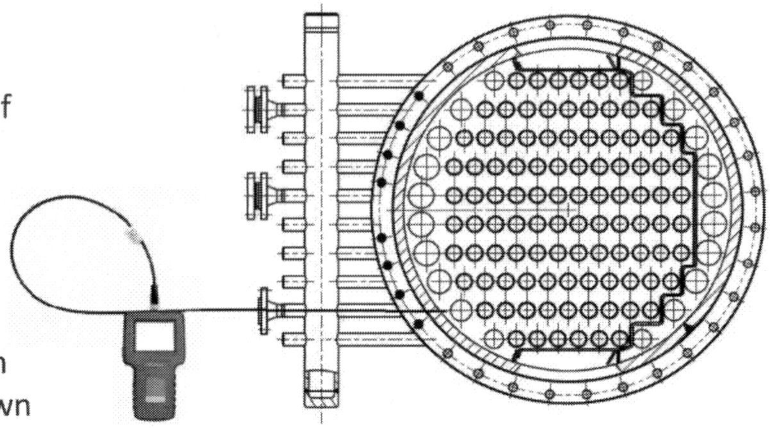

"Tunnelflow" graphic courtesy BORSIG GmbH

14 AIChE Spring Meeting, New Orleans, LA

TLE inner tube corrosion

Loss of protected magnetite layer at locations close to the inlet tubesheet can result in accelerated corrosion.

Waterside inspection of external surface of inlet tubes will reveal this type of damage.

Tube replacement is required to correct this condition, and water quality should be scrutinized.

Photo courtesy BASF

14 AIChE Spring Meeting, New Orleans, LA

Isolation of leaking TLE tubes

Welding of blanks at inlet and outlet of individual leaking tubes can isolate these tubes in case of waterside leakage.

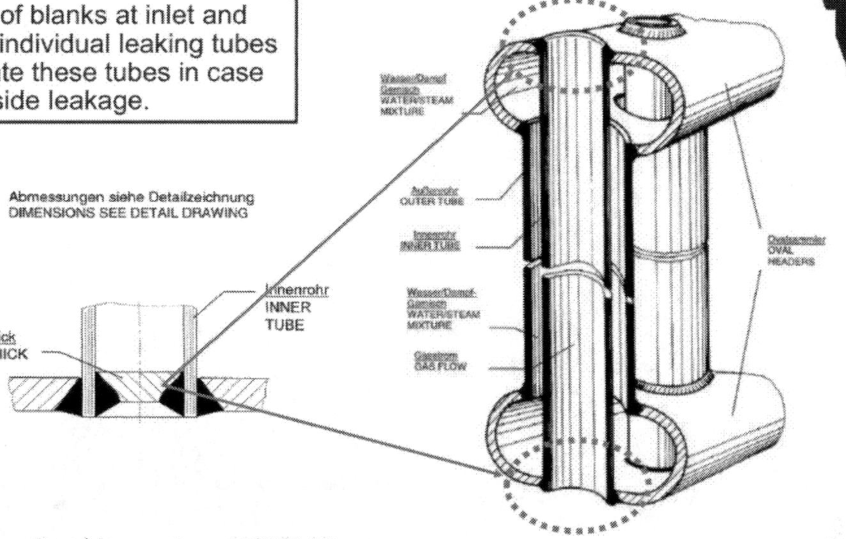

Graphic courtesy ALSTOM Power

14 AIChE Spring Meeting, New Orleans, LA

Replacement of leaking TLE tubes

Cutting with special jig allows removal of tube wall and weld at inlet and outlet of individual leaking tubes.

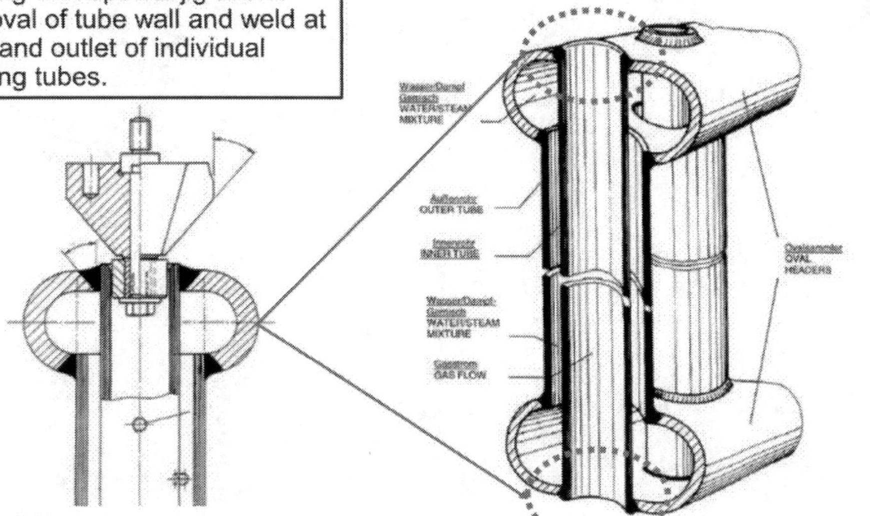

Graphic courtesy ALSTOM Power

14 AIChE Spring Meeting, New Orleans, LA

Removal of tube to tubesheet weld

Photos courtesy ALSTOM Power

14AIChE Spring Meeting, New Orleans, LA

Replacement of inlet tubesheet

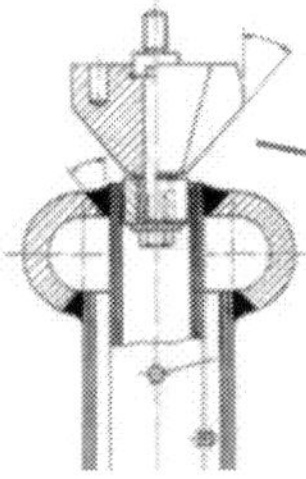

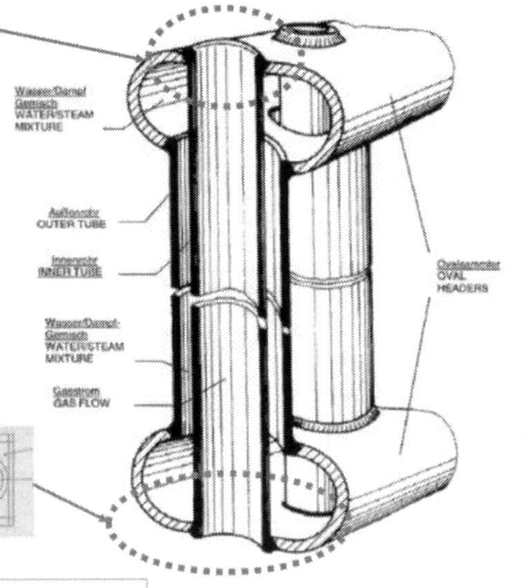

Longitudinal cutting of inlet oval headers can allow replacement of lower tubesheet together with inner tubes.

Graphic courtesy ALSTOM Power

14AIChE Spring Meeting, New Orleans, LA

Replacement of lower tubesheet

Clearance limitations below the TLE make this operation non-feasible without removal of the TLE from the furnace structure.

Photos courtesy ALSTOM Power

14 AIChE Spring Meeting, New Orleans, LA

Schmidt'sche® Transfer Line Exchangers

Double tube / oval header design
Basic element of all types of Schmidt'sche® TLEs

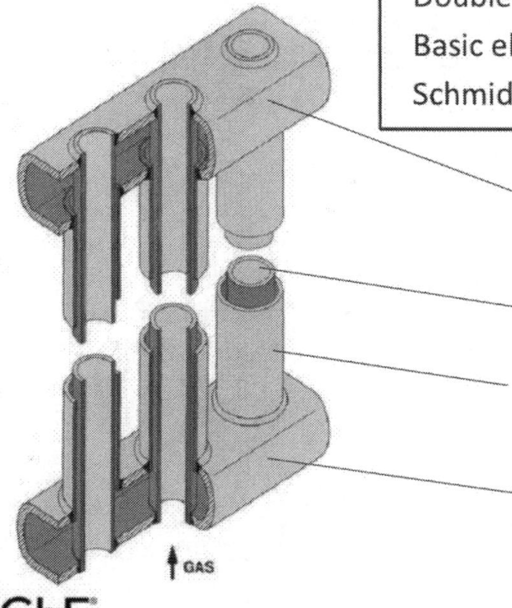

Graphic courtesy ALSTOM Power

14 AIChE Spring Meeting, New Orleans, LA

Linear Quench Exchanger, LQE
Transition Pieces

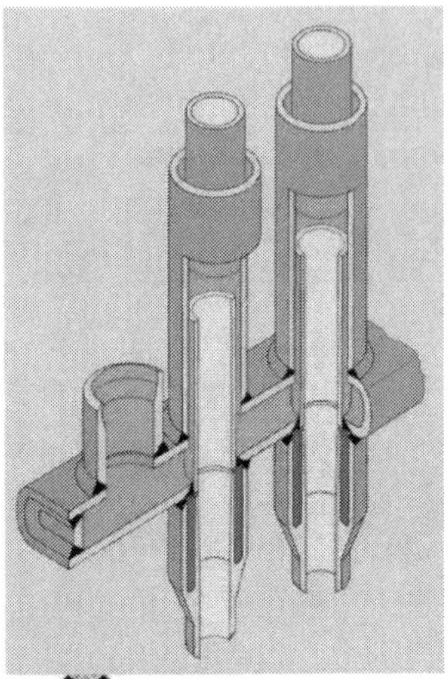

Photos courtesy ALSTOM Power

14 AIChE Spring Meeting, New Orleans, LA

SLE inlet transition
Temperature profile

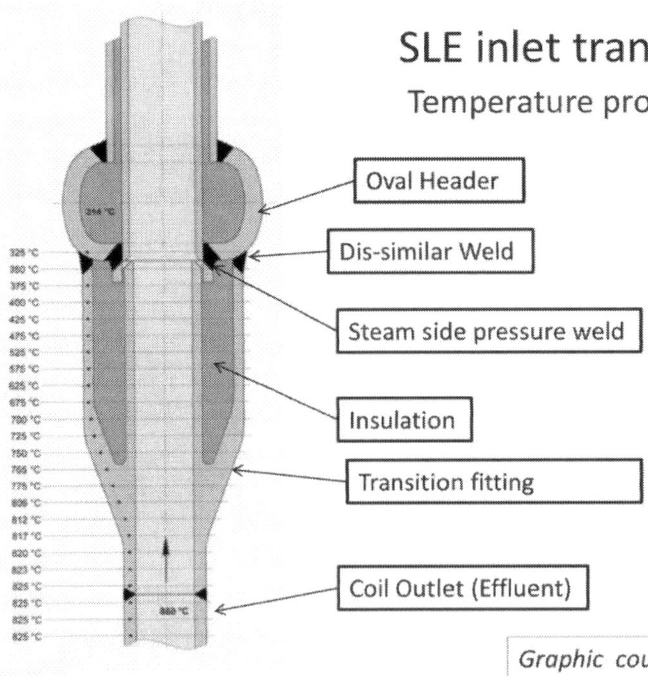

Graphic courtesy ALSTOM Power

14 AIChE Spring Meeting, New Orleans, LA

Temperature profile at SLE Outlet fitting

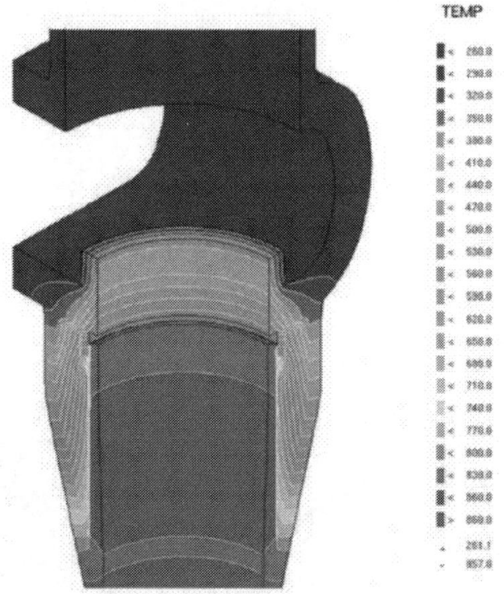

FEA of the temperature profile along the inlet transition piece of a Schmidt'sche® Linear Close-Coupled TLE

Graphic courtesy ALSTOM Power

14AIChE
Spring Meeting, New Orleans, LA

Borsig "Turboflow" Arrangement

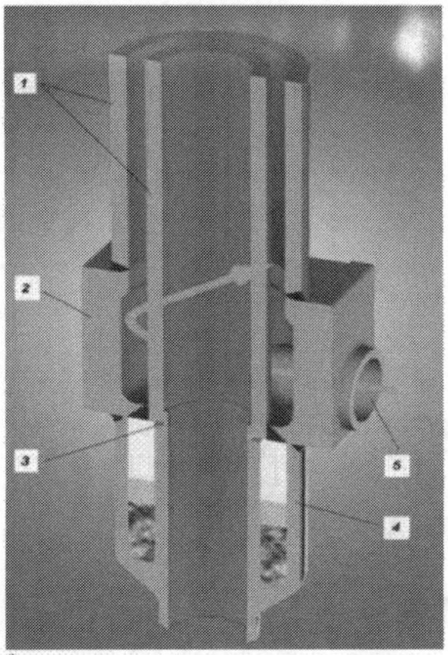

The BORSIG "Turboflow" concept uses a tangentially arranged water inlet nozzle at the water chamber which produces a rotating water flow (turboflow) around the process gas pipe, in the lowest region near the hot gas inlet.

The BORSIG "Turboflow" concept:
1. Double pipe
2. "Turboflow" chamber
3. Metal sealing ring
4. Gas inlet head
5. Water inlet nozzle

Graphic courtesy BORSIG GmbH

14AIChE
Spring Meeting, New Orleans, LA

Borsig Linear Quench (BLQ)

Typical BLQ arrangement:
1. Outlet chambers
2. Inspection/Clean out nozzles for Hydro jetting
3. Gas collection header
4. Riser nozzle
5. Riser header
6. Guides
7. Double pipes
8. Supporting brackets
9. Supporting frame
10. "Turboflow" inlet chambers
11. Downcomer header
12. Downcomer nozzle
13. Gas inlet nozzle
14. Blow down header

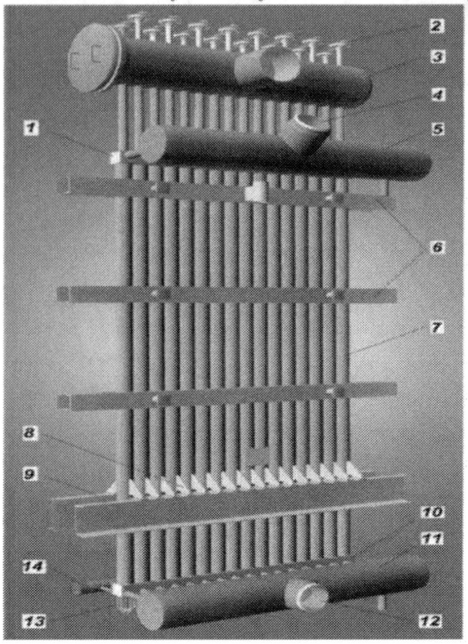

Graphic courtesy BORSIG GmbH

14 AIChE
Spring Meeting, New Orleans, LA

Boroscope inspection of inlet chambers

BORSIG Linear Quench "BLQ"

Waterside inspection of inlet turbo water chambers using the inspection connections furnished on the blowdown header

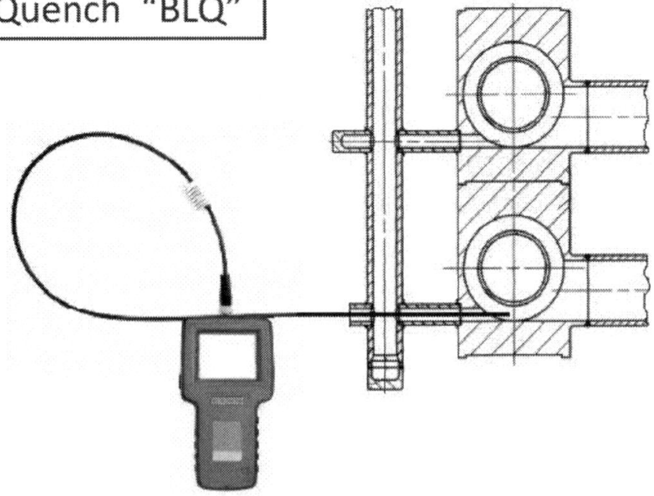

"BLQ" graphic courtesy BORSIG GmbH

14 AIChE
Spring Meeting, New Orleans, LA

Steam Drum Internals

Inspect for:

- Mechanical Integrity
- Protective Magnetite layer
- Scale and solids deposits

Photo courtesy NOVA Chemicals

14 AIChE Spring Meeting, New Orleans, LA

Corrosion Under Insulation

- CUI at Steam Drum saddle attachment to Steam Drum
- Potential stress concentration point

Photo courtesy NOVA Chemicals

14 AIChE Spring Meeting, New Orleans, LA

Operate, Inspect and Maintain

- Minimize operational upsets – this will minimize stress on the components
- Implement a Risk Based Inspection (RBI) program to focus resources
- Inspect the most vulnerable areas, thoroughly and on a regular basis
- Maintain critical components and replace or repair before they fail

14 AIChE
Spring Meeting, New Orleans, LA

Innovation from Beginning to End

Cosponsored By:

Career Guidance Committee Liaison (04H)

Education (04)

This is a workshop, with invited speakers, designed to overview the entire research process from conception to project management completion

Co-Chair

Eldon R. Larsen

Professor of Engineering:

Marshall University Graduate College

100 Angus E. Peyton Drive

Engineering

South Charleston, WV 25303

Email: Larsene@marshall.edu

Co-Chair

Jack Hipple

Principal:

Innovation-TRIZ, Inc.

Tampa, FL 33647

Phone Number: 813-994-9999

Fax Number: 858-434-9311

Email: jwhinnovator@earthlink.net

352529 A Study on Exergy-Based Thermodynamic Analysis and Process Synthesis of Mixed-Refrigerant Systems for Ethylene Plants

Abstract:

Refrigeration system holds an important role in most chemical/petrochemical processes. A refrigerant system transfers heat from low-temperature sources to high-temperature sinks through vapor-compression cycles at the expense of mechanical work, magnetism, laser or other means. In many industrial processes, the refrigeration systems are employed for chilling and freezing the final liquid/solid products or liquefying the gas intermediates/products for the downstream cryogenic separation. Optimal synthesis of these systems is attractive not only because of their economic benefits but also because they incorporate most of energy transfer operations in widespread use in the chemical industry. The traditional cascade refrigeration system used in ethylene plants includes multiple refrigerants and multiple temperature and pressure levels. Recently, mixed-refrigerant system (MRS) has shown some advantages comparing to the single-component one. A MRS uses a mixture as the refrigerant instead of several pure components as in conventional cascade refrigeration systems. This system operates within smaller temperature differences at the lower limit; therefore, it leads to a smaller increase in entropy and consequently a smaller loss in energy. Besides, an MRS also requires a simpler unit configuration and fewer maintenance problems.

In this study, MRS used in ethylene plants is studied based on in-depth exergy analysis, a powerful method to explore system energy loss within a large temperature range crossing the ambient temperature due to various compositions of the employed mixed refrigerant. An exergy-based mixed-integer linear programming (MINLP) model is developed to maximize the energy efficiency of each system, where multiple refrigerant combinations and their recycling loops are simultaneously addressed and all of cooling demands are fulfilled as inputs for a heat exchanger network synthesis. The efficacy of the developed methodology is demonstrated a case study of a mixed-refrigerant system used in a typical ethylene plant.

352748 Plant-Wide Simulation of Ethanol Oxidation Process for Acetic Acid Production

Abstract:

Acetic acid is an important chemicals that is widely used in manufacturing for many products, such as medicine, photographic film, synthetic fiber, food additives, and paint. The annual global demand of acetic acid is about 6.5 million tons. There are three major processes for acetic acid productions: i) low-pressure carbonylation of methanol synthesis process, which employs methanol and carbon monoxide reaction to produce acetic. It is currently most popular method for acetic acid production. ii) Ethylene oxidation process, where ethylene is oxidized to acetaldehyde first, and then further oxidized to acetic acid. iii) Ethanol oxidation process, where acetaldehyde is prepared by ethanol oxidation, and then acetaldehyde is oxidized to acetic.

In this study, a plant-wide model for ethanol oxidation process to produce acetic acid is developed by using Aspen Hysys. The modeled process is divide into two main sections: the ethanol oxidization to acetaldehyde process and the acetaldehyde oxidization to acetic acid process. In the first section, the raw feeds of ethanol and water is mixed, evaporated, and directed to a reactor for oxidation. Besides the major product of acetaldehyde, there are some byproducts of CO,CO_2, methane, tr-Crotonal, small quantity of acetic acid as well as unreacted materials. After the reaction, the outflow goes to an absorber system to separate the insoluble gases from the liquid. Next, the liquid effluent enter two distillation columns to separate unreacted ethanol for recycle and the byproduct of formaldehyde, as well as obtain the high-purity acetaldehyde. In second manufacturing section, the enriched acetaldehyde, air, and the catalyst of manganese acetate are sent into a distillation column for oxidation reaction and separation. then the bottom outflow goes to a distillation column to get the product of acetic acid, while the light component goes to another two distillations to separate oxidation byproduct of lipid and recycle unreacted acetaldehyde. The recycled acetaldehyde goes back to the oxidation column. We simulate the entire plant with these both process in steady-state model first, and obtain the simulation results agreeing well with real plant data. Then some critical sections are further examined by dynamic models to examine the process control ability. Some valuable process and control knowledge have been disclosed by our simulations, which can provide more technical support for plant future manufacturing.

351249 Dynamic Modeling and Optimization of Ring Opening Polymerization: Modeling Challenges and Computation Improvements

Abstract:

Recent work has focused on modeling and optimizing semi-batch reactors for ring opening polymerization. Dynamic models have been developed, using first principles (mass balance, heat balance, reaction kinetics, etc.) one based on population balances and another using the method of moments. These models have been validated with historical plant data, matching reactor pressure profiles. The moment model has the added benefit of increased computational efficiency. When using these models to optimize the reactor temperature and monomer feed rate, reductions in polymerization time are seen. Because of the promise from initial product recipes, these simulations and optimizations should be applied to many similar polymerizations.

A reformulation procedure has already been applied to the above system, to separate the fast dynamic model from the differential equation, modeling the fast kinetics algebraically in their quasi-steady state form. However, further challenges with a large, nonlinear system make it difficult for simulations and optimizations to converge. The most realistic system inputs must be used, and the best initial conditions for the optimization are critical. From the chemistry of polymerization, heuristics were developed to determine these initial conditions. Here we present rules of thumb, and general assumptions that enable good initial conditions to be calculated, given a realistic polymerization system. With these informed initial guesses, improved computation time is seen for these optimizations.

POLYMERIZATION IN A SPRAY DRYER – DESIGNING A PRE-REACTION BEFORE ATOMIZATION TO BOOST THE REACTION IN THE SPRAY

Magnus Tewes, Technische Universität Bergakademie Freiberg, Germany
Professor Dr.- Ing. Urs A. Peuker, Technische Universität Bergakademie Freiberg, Germany

Introduction

Powder polymers with well defined size and a spherical shape are commonly generated by the use of spraying devices, e. g. spray dryer. Here polymers in solution or suspensions of polymers are atomized to small droplets and dried to particles. These particles are mostly used in a wide field of industrial applications like pharmaceutical, cosmetics and chemical industries. Especially water soluble or water absorbing polymers are present as additives in drugs, hair styling products or cleaning agents. Polymers as powders instead of gels or solutions are in advantage concerning transport and storage or further processing because of their flowing characteristics and stability. But the production of these powders is complex. After the polymerization in different reaction steps for dissipating the high amount of exothermic heat formation, processes are required like drying, milling and agglomeration. The idea to combine all the single steps in a so called one step process is obvious and theoretical described in U.S. patent from 1993 [1]. Krüger adopted this new process concept, designed a plant and called the process spray polymerization. The plant is a vertical heated pipe reactor with an ultra sonic droplet generator that creates droplets in a row in the sizes of 90–220 µm. He investigated the generation of superabsorbent polymer particles [2]. Previous works show that efficiency and yield of the spray polymerization is a key problem. Most of the educts evaporate during the drying and aren't available for the reaction consequently. The modification of the spray polymerization process and a development of an economical concept is the motivation of this research.

The new concept is based on a common spray dryer, the main aggregate, where the liquid is atomized by a high velocity gas and dries by a hot gas in a co-current stream. The dryer itself combines three different sub-processes, atomization of the liquid to small droplets, drying and solid formation of the dispersed droplets. Finally the dried particles are separated from the gas stream. The advantage of a spray dryer is the fast process time (< 10 s) because of the small droplets and the high convective heat transfer of the drying gas. It is used to generate solid particles from a liquid rapidly but gently. Even a high drying temperature causes low surface temperature of the droplets, so glass transition temperature of the polymer will not be reached. A new process will be created if additional to these three operations a reactive system is sprayed. In this case polymer reactions are applied, so the idea is a one step process from a liquid monomer solution to a ready powder polymer.

The reaction in a droplet is compared to a small batch reactor, which is heated by the drying gas in the spray dryer. If the monomer is volatile, it will evaporate during the heating and the concentration of monomer in solution changes. With the heating the reaction accelerates and the concentration of monomer decreases. Polymer molecules are formed and change the viscosity of the liquid in the droplet. In addition to that, the polymer will tend to build a skin on the surface of the droplet and change the heat and mass transfer from the inner droplet to the outside. If the monomer isn't volatile, only the solvent evaporates and the concentration increases first, but then the reaction decreases the concentration again. It becomes clear now that the process becomes more complex by the competition of all four sub-processes. The advantage of the reaction in the droplets is that no safety problems will occur because of the small volume. A runaway of the reaction, as the worst case, will only cause unintended particle morphologies. But the advantage of the small droplets and the fast drying times of a spray dryer is also a bottle neck of the spray polymerization.

Due to the low retention time in the spray dryer the process requires a very intensive reaction with fast polymerization kinetics. Therefore the polymerization is a radical reaction of acrylic acid and its acrylates. This system is also used by Krüger because of the fast growing rates [2]. The acrylates are nonvolatile and nonflammable compared to the acid and therefore, easier to use without taking account of explosion protection.

Although the monomer system is very fast and has been tested for a polymerization in a droplet, first considerations about the reaction kinetics show that the retention time is still too low. A pre-treatment of the reactive liquid has to be done to start and properly progress the reaction before it gets atomized. The focus of this research is on the pre-reaction before the atomization in order to guarantee an effective one step process and production of well defined particles in size and shape. The kinetics of the pre-reaction is also important to figure out the impact of the four different sub-processes in the droplet during the spray drying.

The start of the polymerization is mainly depended on the decomposition kinetics of the starter or of the monomers. The decomposition rate is a function of temperature and concentration of the educts. Radicals activate monomers and macro molecules are formed. These oligomers or molecules with higher polymerization degree effect the atomization. In earlier works the influence of molecular mass of a polymer solution has been investigated. It has a big influence on the atomization if molecular mass is high enough to cause a complex rheology with non-Newtonian viscoelastic behavior of the polymer solution. Then filaments instead of droplets are formed [3].

The pre-reaction is necessary to overcome the problem of the short retention time. If the start of the reaction is before the atomization, the properties of the liquid will change depending on the amount of pre-treatment. Therefore a special nozzle is designed to investigate the effect of the retention time of the reactive liquid within the nozzle. This nozzle is modular designed to change the retention time in the nozzle and to heat the nozzle inside. The next chapter is dealing with the design.

Because of the fact that the rheology has a big influence on the atomization, a new approach has been developed to measure the rheokinetics or function of viscosity over the time during the polymerization. Here results are presented and used to demonstrate the progress of reaction within the nozzle. A big challenge is to handle the experimental data, but a more ambitious work is to generate kinetic data that are able to describe the chemistry of polymerization in the nozzle and in the droplets.

Designing the pre-reaction

The reason for developing a pre-reaction is to enlarge the process time the reactive liquid has to polymerize. This pre-treatment will be done before the atomization in a special nozzle. In order to have a compact process and keep the idea of a one-step-process a first idea was to design a new nozzle, a so called 3-fluid nozzle. Here the mixing of the monomers with a starter in solution was intended to take place within the nozzle. Due to mixing problems within the nozzle this concept is frozen by now. The alternative nozzle is therefore a 2-fluid nozzle with special properties.

One special design feature is the double jacket that is passed through by a heating liquid to heat up the feeding monomer/starter solution. The radical starter is a thermal initiator and therefore the temperature has a high impact on the decomposition rate. Because of safety aspects the

monomer/starter solution is fed at a low temperature of 5 to 15 °C. Consequently a rapid heating to the reaction temperature is necessary.
The nozzle has a modifiable length. Variation of the length influences the retention time of the reactive liquid in the nozzle at a constant flow rate. A constant flow rate is set in order to have no effect on the atomization later. The variation of the pre-reaction time is available by changing the length. An optimum of the geometry will be needed where, on the one hand the temperature for reaction is reached and, on the other hand the progress of the reaction is sufficient. The more the reaction is progressing the better the efficiency of the spray polymerization, but the higher the degree of polymerization at the nozzle tip. With the length of the polymer chain the viscosity raises. This can hinder the droplet formation or block the nozzle.

For atomization the second fluid, nitrogen, is used at a high pressure. The gas is fed through the double jacket and concentrical to the tip of the nozzle. Through thin channels and with a tangential direction to the liquid's exit the gas is swirled and expands outside the nozzle. High velocity gas atomizes the reactive liquid compared to standard twin fluid nozzles. After the atomization the reaction progresses in the droplet. This process is comparable with a small batch reactor, whereby the nozzle is more a laminar flow reactor.

Another advantage of the nozzle is that its design is like a lance. So every inlet is on the bottom of the nozzle including the outlet of the double jacket. The outlet of the spraying fluids is at the tip of the nozzle. The application of this special nozzle in a common spray dryer is without any problems. The constructed nozzle uses a nozzle tip from Buchi® Corporation with a liquid diameter of 0.7 mm and a gas diameter of the annular gap of 1.5 mm.

Measuring the effect of the rheokinetics

A free radical polymerization is mostly an exothermic reaction, because the combination of small monomer molecules to a polymer chain means losing of degree of freedom. The amount of reaction heat becoming free is a safety problem, especially if very fast reactions are processed. These fast reactions are used for polymerizations in a spray dryer.

The kinetics of the polymerization are generally described in three different steps; initiation, propagation and termination. Monomers are inactive under certain conditions so an initiator is used. The initiator is decomposed and forms radicals that activate monomers if an initiator radical is connected with a monomer. The polymerization starts and the radical monomers form a polymer chain by connecting more and more monomers. This chain growing step is called propagation. If the growing is interrupted and the chain becomes inactive the reaction is at its last step, the termination. Often the combination or disproportionation of two active polymer chains ends the reaction.

The kinetic of the reaction's beginning is normally described by the following equation (see eq. 1). The rate of the propagation reaction rate is described as the change in concentration in the time interval by the use of some simplifications [4].

$$r = \frac{dc_M}{dt} = \sqrt{\frac{k_i}{k_t}} \cdot k_p \cdot \sqrt{c_i} \cdot c_M \tag{1}$$

The indices *i*, *p* and *t* shows the influence of the three different steps. All constants *k* are depended on temperature and the material and describes by the Arrhenius law.
In view of the complete transformation of monomers to polymers the radical polymerization is more complex. Regarding this, the reaction is described by the help of statistic values, like the distribution or mean value of the molecular weight. The molecular weight will be measured with special

chromatography. This analysing technique often uses light scattering or viscosity measurement to determine the volume of the polymer coils. The intrinsic viscosity is measured in dilute solution and is a reciprocal concentration which can determine the mean molecular weight. Equation 2 shows the relation of mean molecular weight $\overline{M}$, intrinsic viscosity η and concentration c and was developed by several scientists; Staudinger, Fikentscher, Houwink, Huggins, Hess and Sakurada [5-9].

$$\eta = K \cdot c^b \cdot \overline{M}^a \tag{2}$$

The idea of measuring the viscosity over time during a polymerization and therefore be able to describe the kinetics is first described in detail by Malkin and Kulichikhin and called rheokinetics. The aspiring target of this method is to generate a semi empirical description. For the initial stage of the radical polymerization of acrylamide in water they find equation 3. This function is developed with the use of equation 1 and 2 and describes the viscosity η over the time t [10].

$$\eta(t) = K' \cdot c_{M,0}^{a} \cdot c_{I,0}^{b-a} \left(k_p \sqrt{\frac{k_i}{k_t}}\right)^{a+b} \cdot t^b \tag{3}$$

The parameter K' takes into account the dependency of the viscosity with change of temperature. It is like the kinetic parameters k calculated with the Arrhenius law.

In addition the knowledge of the function of viscosity over the reaction time is of great benefit in terms of the atomization. The results of former investigations on the atomization of polymer solutions are helpful optimizing the retention time of the reactive solution in the nozzle. To measure the rheokinetics a modification of a rheometer is necessary, as it is seen in figure 1.

The standard rheometer is a HR-1 by TA Instruments and has a cone-plate geometry. The 2° cone with 40 mm diameter is connected to a motor. The plate is coupled with a Peltier temperature control. The Peltier element controls a temperature range between -40 and 200 °C with a controlled heating rate of up to 10 °C/min.

The modification is done by covering the sample with two semi caps. They are connected to a nitrogen supply and at a certain step the sample gets purged with nitrogen during the measurement. Here the nitrogen is when turned on, starting the reaction, while oxygen that inhibits the polymerization is displaced rapidly. The geometry of the caps fits with no contact to the moveable cone. Through small gaps the gas is forced from the caps over the holes and sample between caps and cone to the outside. Several holes are drilled in an additional plate and arranged in a circle around the upper cone. This modified plate has two functions. It fits with the Peltier plate and the two are assembled together. The holes in it are filled with water, so during the measurement the nitrogen flows over the water and is saturated. Thus, evaporation of the sample is suppressed because of a saturated atmosphere.

With heating up the sample to a temperature of 75 °C the procedure to get the time dependence of the viscosity starts. The sample is a solution of monomer, water and initiator and its volume is 525 µL. A damage of the rheometer because of the high reaction enthalpy is therefore excluded. Then nitrogen flow is turned on (0.1 L/min) and the lack of an inhibition of the radicals will start the polymerization immediately. Normally an inhibitor is present in monomers to avoid a self-ignition, but it works in combination with oxygen only [11]. At a constant shear rate of 300 1/s and a sampling rate of 0.5 Hz the function of viscosity during the polymerization is plotted.

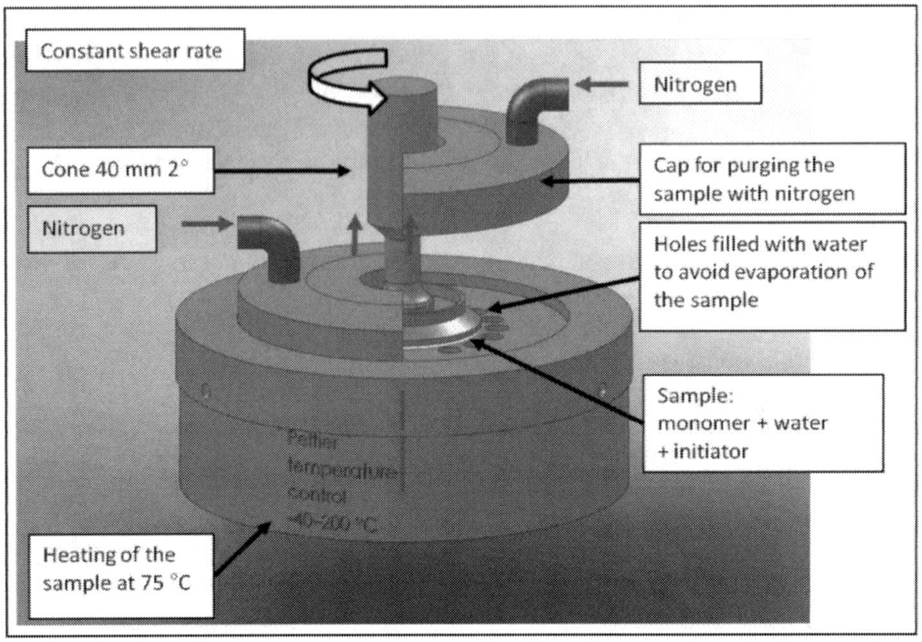

Figure 1. Modification of a rheometer to measure the viscosity over time during a polymerization

Results of the polymerization in the rheometer

The reactive system used is acrylic acid and its acrylates. A work of Krüger shows that a polymerization with this material in a droplet during a free fall is successful [2]. The monomer is applied by BASF SE and contains an inhibitor. With a vapor pressure of 4.3 hPa the acid is less volatile than water. But acrylic acid is flammable and explosion protection has to be guaranteed if used in spray dryers. That is why the salts of the acrylic acid, acrylates, are less harmful and to be preferred. They are produced by neutralization of the acid e.g. with sodium hydroxide in a solution.

The results of the rheokinetics have two benefits. On the one hand the viscosity function is measured and an optimal nozzle design is possible. On the other hand it is an ambitious target to use the method of Kulichikhin and Malkin to get information about the kinetic of the reaction as it is mentioned in equation 3. But this equation has a big disadvantage because of the dimensions. For the initial boundary, $t = 0$, the viscosity would also zero if b is not equal to zero and is consequently not logical.

All following results are done with the same testing procedure as described before and with a minimum of three repetitions. In order to explain the analyzing, four different measurements of acrylic acid (AA) and sodium acrylate (SA) with two initiator ratios, 0.8 mmol/mol (dotted lines) and 1.2 mmol/mol (plain lines) related to the monomer's molar amount, and two different concentrations, 2.8 and 3.5 mol/L, are used as an example. In figure 2, left, the raw data of the measurement are shown. The shear viscosity is a function of time, which, after a plateau, is getting a high gradient after a characteristic time, called zero time or time of the viscosity´s increase t_0. It can be directly captured by the raw data for acrylic acid. Sodium acrylate does not show such a rapid increase and an overall threshold has to be defined.

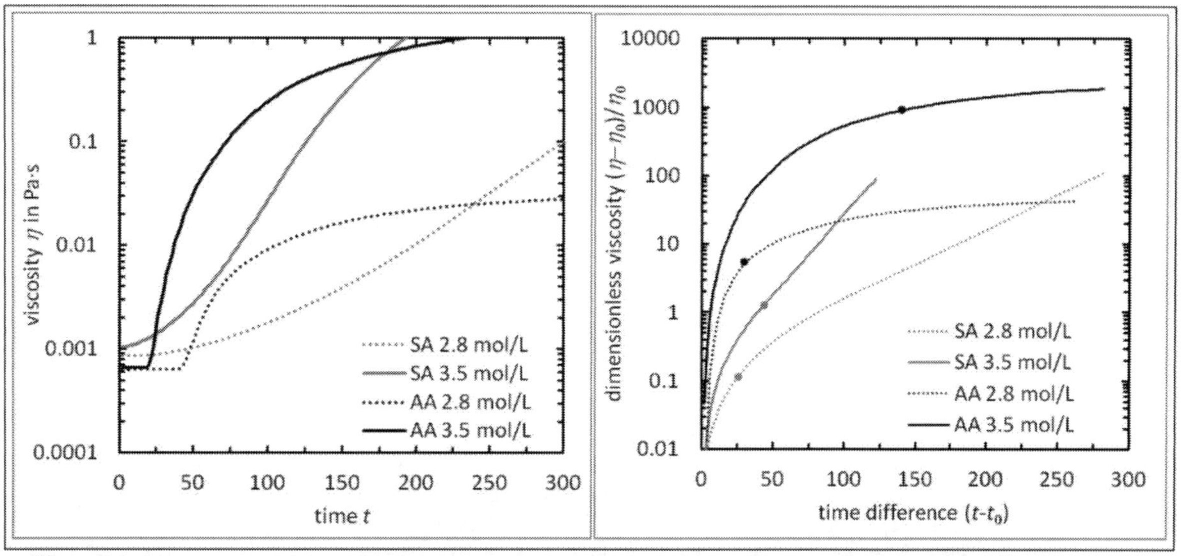

Figure 2. Working with rheokinetics – l. data by rheometer, r. dimensionsless view for the analysis

The next information that is necessary to describe the function $\eta(t)$ is the gradient after passing the zero time. In order to work with the data a transformation to dimensionless quantities is necessary. Therefore the viscosity is calculated by the quotient of the difference between start and current time and the starting point. To focus the gradient easily the Savitzki and Golay smoothing is an efficient tool to smooth the data first and calculate the gradient with the first derivation of the smoothing [12, 13]. Then a fitting to a power function and an exponential function is done mathematically with a linear regression. The breaking point between the power and exponential function is marked in figure 2 with a dot. This intersection gives a good basis to discuss the differences between the polymerization of acrylic acid and the acrylates. Although sodium acrylates show an earlier rise in the viscosity the gradient is lower. The polymerization kinetic is therefore different and the only reason results from material properties. It seems that the pH value, which differs from 1 (acrylic acid) to 7 (acrylates) is the main influence, because it affects the solubility and morphology of a polymer in the solution. Another difference is the exponent of the exponential fitting. Acrylic acid has a negative exponent that decreases the gradient of the viscosity but the acrylates' gradient is still increasing and positive.

The result of the zero time is presented in figure 3. Here acrylic acid and first results of sodium and lithium acrylate are plotted with view of the concentration of the initiator in the liquid. Different analyses have shown that only the amount of starter is determining the resulting zero time. And it is obvious, that the more initiator is present the more radicals are getting free during the decomposition and consequently the zero time is decreasing. But the comparison between acrylic acid and the acrylate is not significant by now. It seems that the zero time of the acrylate is less than for acrylic acid. For every concentration the zero time can be calculated by using an exponential regression. A first simple rule of optimizing the nozzle could be, that the retention time has to be smaller or equal to the zero time. Then the viscosity of the liquid is low enough (compared to water) for excluding problems concerning the atomization.

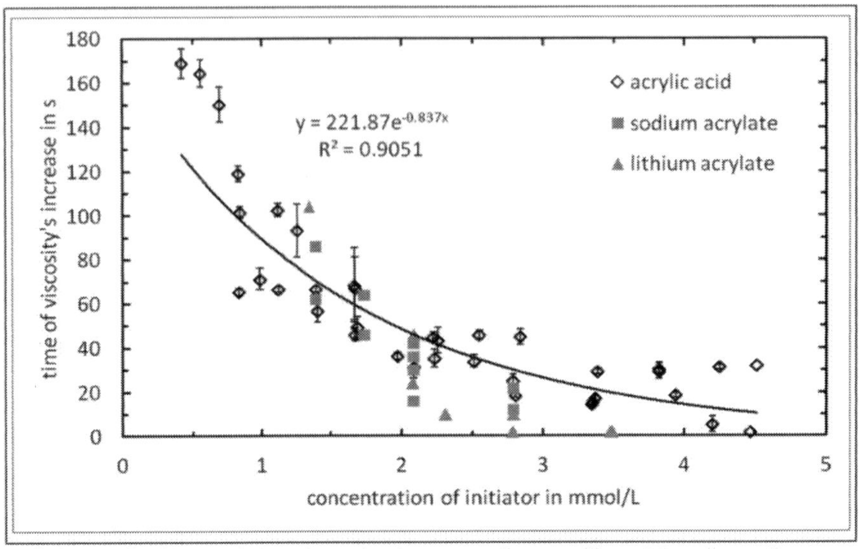

Figure 3. Time of the viscosity increase for acrylic acid and acrylates

Figure 4 is presenting the missing information to describe the function of viscosity over the whole time when only the power function describes the increase. Taking into account the work of Kulichikhin and Malkin the equation 3 can be simplified by combining all constants and get a simple equation. This is done in equation 4. Now the viscosity is only a function of time and two constants, a factor β and an exponent b.

$$\frac{\Delta\eta(t)}{\eta_0} = \beta \cdot (\Delta t)^b \tag{4}$$

As it is seen in figure 4 left, the factor β is a function of the molar fraction of the monomer and of the initiator at the beginning when $t=t_0$. This dependence is given by a product of two power functions and a factor. The result shows that the influence of monomer molar fraction is very high compared to the amount of the initiator. In addition the range of values is very broad, so it has to be noticed that although the square of the correlation coefficient is near to one the variance of the power function is high. There is a good agreement with the power function of equation 3.

But on the right of figure 4 another mathematical correlation is in no agreement with the results of Kulichikhin and Malkin. Here a similar power function is correlated with other coefficients. The exponent b is therefore also depended on the monomer and initiator molar fraction at the reaction's start. Here the influence of the monomer is only the power of 3 higher than the influence of the initiator. Thus, this result is in contrast to the rheokinetic of equation 3, because no influence of the initiator concentration is mentioned. The meaning is that during the polymerization the influence of the monomer and initiator start concentration is present at all time. The influence of the monomer is, at the start of the viscosity´s increase, higher than afterwards.

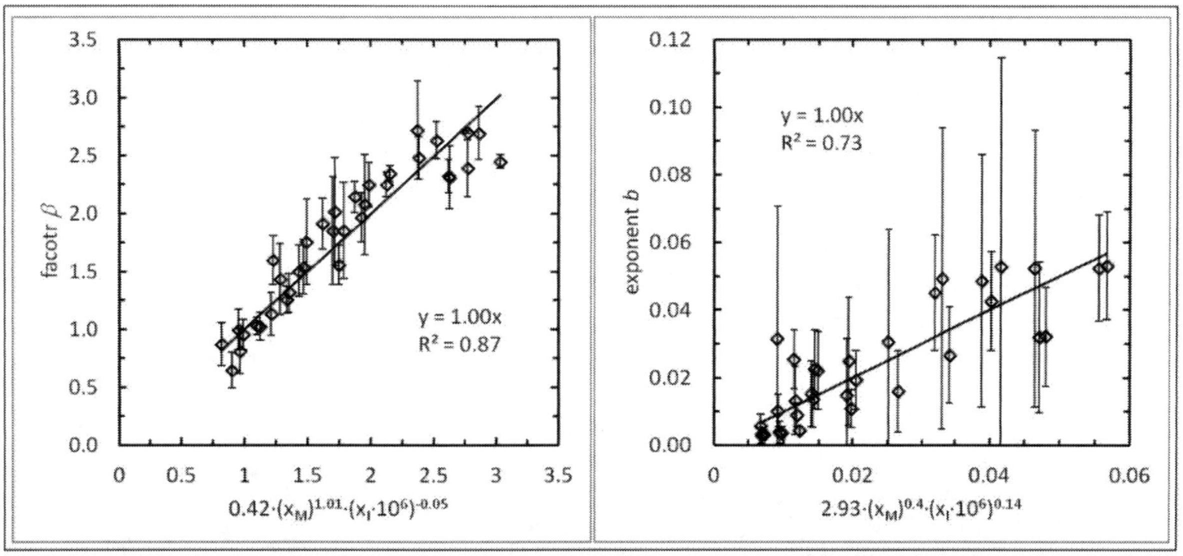

Figure 4. Presenting the values for calculating the viscosity gradient in the nozzle over time

In figure 5 a good agreement to the power function is shown. The biggest different between the calculated and measured values is due to the small differences in the zero time. The table in figure 5 demonstrates the good agreement of the power function and a relative error of 7 % between the exponents.

The presented results are able to extend the following theoretical model of the progress of the polymerization within nozzle. That model is based on a simplified heat and mass transfer for a plug flow reactor. Now the function of viscosity with time will be integrated. The extended modeling will generate base points for the optimization. For validation of the theoretical model plenty of tests with the nozzle have to be arranged. In an intermediate device, where the nozzle is processed in a dripping mode and no drying is applied, results for temperature distribution are generated. Now the focus is on the theoretical model.

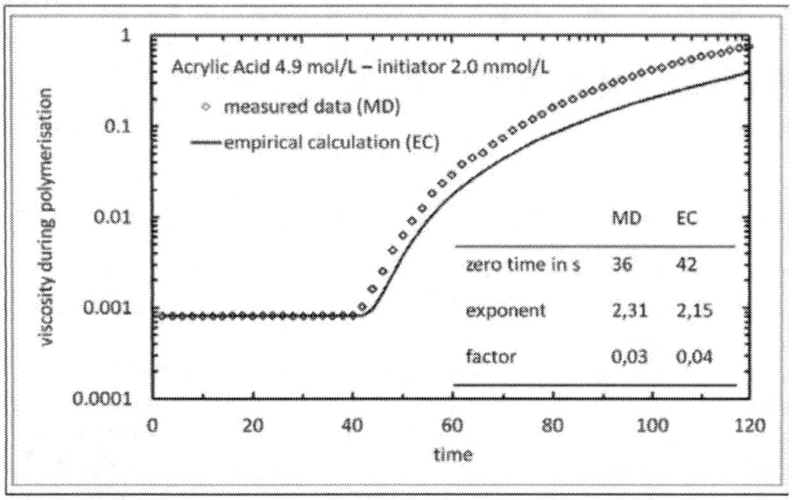

Figure 5. An example of the calculation of the viscosity function during polymerization

Modeling the progress of the polymerization within the nozzle

Before testing the reactive system, theoretical considerations are necessary. The first reason for the consideration is that with a good model the number of tests can be reduced. But the more important reason is the fact that the tests deal with a fast reactive monomer, which will probably cause safety problems. A runaway of the 3-fluid nozzle is possible, because the high reaction enthalpy will accelerate the reaction. The pressure within the nozzle will rapidly increase, if the nozzle gets blocked because of the resulting viscosity´s increase.

The modeling is based on the heat and mass transfer equations. These differential equations are presented in equation 5 and 6. As it is seen a transformation to a dimensionless form is made. On the one hand the mass transfer is the quotient of the conversion and the change of the dimensionless length d*z*/*L*. It is calculated with the use of the Damköhler number. The rate *r* describes the reaction kinetic and was introduced in equation 1. The heat transfer is given as the quotient of the change of the dimensionless temperature and the dimensionless length. The heat transfer is depended on the mass transfer, making a simultaneous solution of both equations necessary.

$$\frac{dU}{dz/L} = \frac{\tau \cdot r_0}{c_{10}}\left(\frac{r}{r_0}\right) = Da_0\left(\frac{r}{r_0}\right) \tag{5}$$

$$\frac{d\vartheta}{dz/L} = Da_0\left(\frac{r}{r_0}\right)\frac{\Delta T_{ad}}{T_0} - \frac{K_W A_W}{\dot{V}\rho\hat{c}_p}(\vartheta_W - \vartheta) \tag{6}$$

To solve the problem of the simultaneous solution easily the differential quotients are approximated as difference quotients. By choosing a small increment for the control variable *z*, length of the pipe, the solving is done numerical by several iterations.

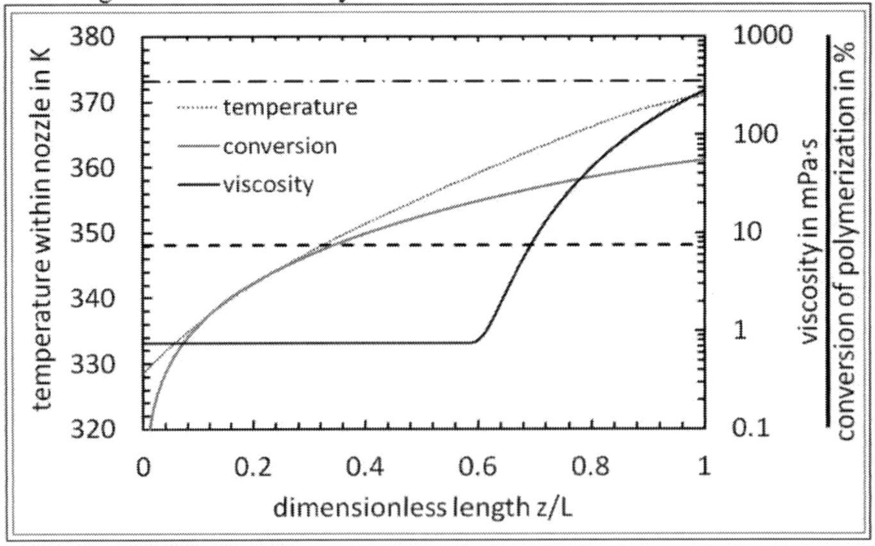

Figure 6. Calculating the temperature, conversion and viscosity in the nozzle for 5.6 mol/L acrylic acid in water and an initiator concentration of 2.3 mmol/L during polymerization

A result of the calculation is shown in figure 6. Here the temperature and conversion is plotted over the dimensionless length. It is an example for a high conversion within the nozzle when using a 5.6 mol/L acrylic acid solution in water with an initiator ratio of 0.6 $mmol_{initiator}$ / $mol_{monomer}$. The dashed line marks the reaction temperature of 75 °C. As an upper limit the dash-dot line marks the boiling temperature of water at 1 bar. At this point the solution overcomes the boundaries of constant pressure and one-phase state. It is shown that the temperature and the conversion increase until the end of the nozzle at 400 mm. That is obvious because of the enthalpy of the reaction. After a high rise of the conversion its gradient becomes lower because the polymerization consumes the monomers; the maximum of 100 % is the theoretical limit. The lower gradient of the conversion at the end damps the temperature gradient because of the linking of heat and mass transfer. If only the temperature and the conversion would be considered, the safety of the pre-reaction seems to be guaranteed, because the temperature is below the boiling point of water. The high conversion should increase the efficiency of the process very much as no reactive mass is evaporated.

But the result of the viscosity development is overcritical. The function of viscosity is calculated with the results presented before. The start of the function is when the temperature crosses the first mark at 75 °C. After the plateau ending with the zero time the viscosity function is calculated by the use of the power function. If the rheokinetics can be transferred to the behavior within the nozzle and a critical viscosity for the atomization is known, the model is able to find optimal parameters for a safe reaction. In the example above a viscosity of 324 mPas would cause the blocking of the nozzle sooner or later. Because it is just a simple model, a testing of the forecast is absolutely necessary.
In the intermediate device, described before, tests with different temperatures are done and the heating up within the nozzle are measured with thermo elements and finally validate the model concerning the energy balance.

Conclusions

The presented work shows the development that has been done to design a spray polymerizing process. The focus is on the pre-reaction before the atomization in order to improve the efficiency of the process as only low retention times are present. First a special nozzle has been mentioned and explained in detail. In order to do first practical tests it was said that there is still a big lack of information concerning the right use of a high reactive radical polymerization. Therefore a new measuring technique is presented to study the influence of the polymerization, especially the influence on the viscosity. The effect is called rheokinetics. Results of acrylic acid and acrylate are presented. A semi empirical function of the starting point of the viscosity increase is only depended on the initiator concentration. The function of the viscosity increase of acrylic acid is determined and it is demonstrated that here the influence of the concentration of monomer is higher than of the initiator.

In addition a modeling of the heat and mass transfer during the pre-reaction in the nozzle is presented. A model calculates the temperature and conversion distribution along the axis of the nozzle. With the combination of the viscosity function the model is able to find basic points to plan further experiments with a reactive solution and avoid a runaway of the reaction and consequently risks.

Finally the outlook work is to test the nozzle with a reactive solution in the intermediate device. Then the step to work in a spray dryer with atomization is not very far away. Another research is the linking of the function of viscosity to kinetics of the polymerization to have a better rheokinetic approach for acrylic acid and its acrylates.

Acknowledgements

We would like to thank the German research foundation for the financial support and the BASF SE for donating the materials. A special thank is to all the co-workers in the SPP 1423 and to our team at our institute, especially to my students.

Nomenclature

Latin

a	exponent of power function
A	area of heat transfer
b	exponent of power function
c	concentration in mol/L
$\tilde{c}_p$	heat capacity at constant pressure in J/(kg·K)
Da	Damköhler number
k	reaction constants in L/(mol·s) or 1/s
K	empirical constant
K	overall heat transfer coefficient in W/(m²·K)
l	length in mm
L	length in mm
M	molecular mass in g/mol
r	rate of polymerization in mol/(L·s)
t	time in s
T	temperature in K
U	conversion
x	molar fraction
z	length in mm

Greek

β	factor of power function
η	viscosity in Pa·s
ρ	density in g/L
ϑ	dimensionless temperature
τ	retention time in s

Indices

0	initial condition
ad	adiabatic
i	initiation
I	initiator
ideal	ideal
M	monomer
p	propagation
t	termination
W	wall

References

1. Levendis, Y. A.; Panagiotou, T. and Flagan, R. (1993), Patent U.S. 5,269,980.
2. Krüger, M. (2003) "Sprühpolymerisation," PhD thesis, Hamburg.
3. Tewes, M. and Peuker, U. A. (2012), "Polymerization in sprays – one step from a monomer to a powder polymer," *12th Triennial ICLASS*, Germany, September.
4. Stickler, M. (1979), *Makromol. Chem*, 180, pp. 2615.
5. Fikentscher, H. and Mark, H. (1929), *Kolloid Z.*, 49, pp. 135.
6. Staudinger, H. (1930), *Kolloid Z.*, 51, pp. 71.
7. Hess, K. and Sakurada, I. (1931), *Ber. Dtsch. Chem. Ges.*, 64, pp. 1183.
8. Huggins, M. (1938), *J. Phys. Chem.*, 42, pp. 911 (1938)
9. Houwink, R. (1940), *J. Prakt. Chem.*, 157, pp. 15.
10. Kulichikhin, S. G. and Malkin, A. Y. (1997), *Polym. Eng. Sci.*, 37, pp. 1331.
11. Becker, H. (2003), PhD thesis, Darmstadt.
12. Savitzky, A. and Golay, M. J. E. (1964), "Smoothing and differentiation of data by simplified least squares procedures," *Analytical Chemistry*, 36, pp. 1627.
13. Gorry, P. A. (1990), "General least-square smoothing and differentiation by the convolution (Savitzky-Golay) Method," *Analytical Chemistry*, 62, pp. 570

UTILIZATION AND BENEFITS OF A NANOPARTICLE EXTRACTION PROCESS IN A DROP COLUMN FOR PRODUCTION OF HIGH QUALITY ORGANOSOLS

Jacqueline V. Erler[1]; Tom Leistner[2]; Urs A. Peuker[1]
[1]TU Bergakademie Freiberg, Freiberg, Germany
[2]Helmholtz-Institute Freiberg of Resource Technology, Freiberg, Germany

Abstract

The emphasis of this research is the development of a continuous particle extraction process at the interface of two liquids to produce stable colloidal and functionalized particles. To demonstrate the phase transfer process we designed a miniplant using a drop column as transfer device. For process development, the investigation of the miniplant in various operating methods is required. The necessary experimental setup, materials and relevant parameters for a stable process behavior are presented. The process behavior within the drop column is characterized by a particle-free phase transfer, whereby the measured total organic carbon (TOC) is only dependent on the used surfactant. Based on the modeling of the system the transfer behavior can be characterized, if the transfer fluxes of the surfactant are evaluated. Furthermore, the mass flows and the yield of transferred magnetite are determined by ICP-OES measurements. The additional investigation of interfacial tensions using the pendant-drop method provides conclusions regarding to the impact of interaction of the elementary processes and procedures at the phase boundary. Moreover, the transfer kinetics is specified and compared with calculated theoretical values resulting from a kinetic approach. As a consequence of this, the colloidal stability of the resulting organosols is estimated and compared with the results of previous investigations using a laboratory centrifuge as transfer device. In both devices different forces affect the phase transfer so different organic solvents are applied. The differences in colloidal stability are characterized by a dispersion analyzer as well as theoretically with the use of solubility distances by Hansen.

Introduction

The establishing of high quality organosols requires the production of stable colloidal and functionalized particles with a minimum amount of stabilizing surfactant in a liquid organic medium. The magnetite nanoparticles used have special magnetic properties and therefore, they have a great technological application potential. Thus, they are very interesting for a broad range of research areas for example in the field of functional materials and magnetic fluids applied in low friction dynamic systems, vibration dampers and tweeters (1, 2). Furthermore, possible areas of application are coatings and composite materials (3, 4) used in chemical reactors as magnetically separable catalyst material (5) as well as in the biomedical sector (6, 7, 8). Particularly in combination with polymers it is necessary, that the magnetite nanoparticles are available as stabilized colloids in the organic phase (9, 10, 11). But on an industrial scale the nanoparticles are synthesized mainly in the aqueous phase. The transfer of the particles into the organic phase is conventionally done by strategies based on filtration with subsequent drying and redispersion steps. But the properties of magnetite nanoparticles are limited, because they tend to oxidation and agglomeration due to their increased surface area/volume ratio (12).

Hence, a new type of particle extraction process is of great importance, which includes the transfer of magnetite particles from the aqueous into an immiscible organic phase directly through the liquid-liquid phase boundary. Further advantages are no exposure of the particles to a gas atmosphere and with it the avoidance of sintering by capillary forces and a high particle concentration in the receiving phase as well. The engineering concept is a miniplant, where the continuous phase transfer can be realized using a drop column. In 1968 the first liquid-liquid extraction experiments of ultrafine particles were conducted by Lai and Fuerstenau. For the separation of a mixture of alumina, water and oil they used a separatory funnel to put on top of each other the two phases (13). In our study in all operating methods a drop column is used in the simplest form without internals. The fundamental steps of the phase transfer process mechanism are at first the partly agglomeration of the synthesized magnetite nanoparticles caused by the instable conditions in the aqueous phase (14). After injecting the organic phase as drops through a distributor into the magnetite suspension in the drop column, the sedimentation is prevented by the stirring effect of the rising liquid drops. The organic phase represents the disperse phase of lower density and the magnetite suspension demonstrates the continuous aqueous phase, whereby collisions of particles and drops can be realized. However, the first step is necessary for the laboratory centrifuge as transfer device, because through sedimentation the agglomerates are transported into the organic phase. At the liquid-liquid interfaces the surfactant molecules from the organic phase interact with the hydrophilic particle surface, which leads to a chemical adsorption (15, 16). Thus, the formed partially nonpolar magnetite surface implicates the hydrophobization and functionalization of the particles as well as the wetting of the modified particle surface by the organic phase, which allows the phase transition of magnetite. Stable organic colloids are formed by deagglomeration of the particles as well as physicochemical dispersion due to the strong repulsive potential of the adsorbed surfactant molecules (17). The assumption of an efficient particle extraction process requires a high yield of transferred particles in the organic phase, low process times and a stable product. Therefore, a stable process in the drop column is necessary characterized by a stable drop formation and upward movement with a sufficient coalescence rate, which leads to a clearing off of the column.

Materials, Methods and Experimental Setup

By using a co-precipitation reaction the magnetite nanoparticles are synthesized. The procedure is widely applied and specified in (18) as well as the pH-conditioning of the aqueous phase. The used surfactants are ricinoleic acid and oleic acid, which are unsaturated fatty acids (FA) at the 9th C-atom with a carbon chain length of 18 C-atoms. In addition, ricinoleic acid has a hydroxyl group at the C-atom C12. Both surfactants are of technical grade with 90% purity from Sigma. The study presents the results with the parameters $x_{surf} = 1.4$ mass-% (mass fraction of surfactant in the solvent iso-octane as organic phase) and a specific amount of surfactant per magnetite $X_{S/M} = 0.2$ g/g. In (18) the methods are described in detail.

The drop column has a length of 700 mm in the partial recirculation operation. In the simple multi-level circuit as well as in the continuous circuit of both phases the length of the drop column is 920 mm caused by additional glass components for the implementation of the operating method. The internal diameter of the drop column of 25 mm is kept constant. Thus, a volume of 300 ml of the magnetite suspension is utilized at the first mentioned operation and 600 ml for the other one. The volume used corresponds to a mass of 6 g or rather 12 g magnetite nanoparticles. The dispersing system at the bottom of the transfer device is a single metal capillary with an inside diameter of 3.2 mm centrally mounted in a perforated plate. Both supply system and removal system consist of a peristaltic pump (Ismatec Reglo Analog) and a Tygon-tube with an inside diameter of 3.2 mm. In addition, the removal system has an automated sampling system with a pneumatic 3-way valve, manufactured by Swagelok, which is controllable by measurement software. The sample volume is 2.9 ml. To keep the volume of the disperse phase constant, the same volume is added simultaneously as organic phase in the receiver tank. The volume flow rate is 29 ml/min in the partial recirculation operation and in the simple multi-level circuit of the miniplant. To control the continuous circuit of both phases a volume flow rate of 14.5 ml/min is used.

Previous investigations using a laboratory centrifuge (Hettich Universal 30F) as transfer device are carried out with dichloromethane (DCM) as solvent, because the aqueous phase is placed on top of the organic phase as described in (14).

Results and Discussion

As mentioned for the development of a continuous particle extraction process the miniplant is investigated in different operating methods. The necessary parameters for a stable process behavior, which are the mass fraction of surfactant in the organic solvent x_{surf} as well as the specific surfactant amount to magnetite $X_{S/M}$, are determined using the miniplant in a partial recirculation operation (Figure 1a). This means, that the continuous phase is stationary and the disperse phase is in recirculation. Figure 1b represents the miniplant in continuous circuit of both phases, which is regarded later. Figure 1c outlines the procedures in the drop column during the extraction process. Thus, the disperse phase is pumped from the receiver tank through the capillary into the drop column, which is already filled with magnetite suspension, the formed drops rise through the column in the continuous phase, due to their lower density. This leads to collisions of particles and drops, whereby a phase separation in an organic phase, containing the extracted magnetite particles, and an almost particle-free aqueous phase takes place. Hence, the clearing off of the column and a black color of the organosol can be observed. Furthermore, a coalescence area is formed at the surface of the continuous phase (interface layer). The enrichment at the interface is a pickering emulsion consisting of droplets of the organic phase in water with only partial hydrophobized particles. If the coalescence is prevented, the particles can not be extracted in the organic phase resulting in a dispersion ribbon (19).

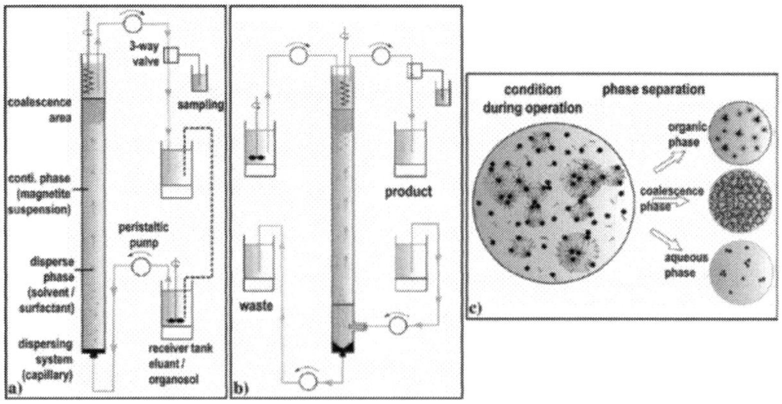

Figure 1. Experimental setup of the miniplant in a partial recirculation operation (a) as well as in continuous circuit of both phases (b) and illustration of the occurring procedures during the particle extraction process (c).

For the calculation of the corrected mass concentration of transferred magnetite particles in the receiver tank $\beta_{magn,rt}(t)$ a validation of the system is essential, due to the fact that after each removed sample volume the amount of magnetite is reduced. In consideration of the input and output to the receiver tank and the residence time distribution Equation (1) can be applied for the evaluation:

$$\beta_{magn,rt}(t) = \left(\beta_{magn,rt,0} + \frac{a_i * V_{rt}}{\dot{V}_{disp}} - b_i \right) * e^{-\frac{V_{rt}}{\dot{V}_{disp}} t} + a_i t + b_i - \frac{a_i * V_{rt}}{\dot{V}_{disp}} \tag{1}$$

where $\beta_{magn,rt,0}$ is the initial magnetite mass concentration of the system, V_{rt} is the volume in the receiver tank, $\dot{V}_{disp}$ is the volume flow of the disperse phase and a_i as well as b_i are formed by linear interpolation between the removed samples.

For the characterization of the process behavior within the drop column in dependence from surfactants used the method of a particle-free operation is applied. Thus, it was possible to observe how the organic phase immediately disintegrates in drops directly above the opening of the metal capillary. With ricinoleic acid as surfactant a multitude of very small droplets as well as a turbidity in the column is formed. In contrast to oleic acid the noticed larger and significantly less drops without turbidity are similar like the mechanism of a single periodic drop formation (20). As a consequence of this, the available phase interface is smaller for the particle extraction process. The turbidity occurred is distributed in the whole column within a few minutes caused by the low buoyant force of the droplets. Owing to the additional hydroxyl group ricinoleic acid is also soluble in the salts and ammonia consisting aqueous phase, whereby the spontaneous formation of an emulsion is generated. As a result the small emulsion droplets with a median value of 5-10 μm measured by laser diffraction spectroscopy are iso-octane droplets, which are stabilized of ammonium ricinoleates existed at the interface by the dissolving of ricinoleic acid in the aqueous phase. This leads to additional available phase interface for the transfer process. This connection is also proved by the implementation of the experiments using a pH-indicator bromothymol blue in the continuous phase, whereby the formation of the surface active ammonium ricinoleates is observed resulting in color changes.

The measurements of the TOC content at defined time points of sampling in the aqueous phase (every 5 minutes till 45 minutes) confirmed these observations. By using ricinoleic acid as surfactant, a significant increase of the TOC content can be observed. Finally, after 5 minutes process time the aqueous phase is reached an average value of 2.0 g/l of dissolved ricinoleic acid. Whereas oleic acid is used, the poor ability of dissolving within the aqueous phase is expressed considerably, since the TOC content is on average 0.02 g/l. Thus, it can be recognized that in contrast to oleic acid approximately the 100-fold amount of ricinoleic acid dissolved in the aqueous phase irrespective of time.

Furthermore, the measurements of interfacial tensions on the particle system using the pendant-drop method (Figure 2) are demonstrated that the preferred dissolution of ricinoleic acid in the aqueous phase has a higher influence on the interfacial tension in contrast to the oleic acid. The interfacial tension between the aqueous and organic phase determines the drop size and the available phase interface for the particle extraction process. Thus, with ricinoleic acid as surfactant the interfacial tension is reduced from 31 mN/m to 8 mN/m and only for 2 minutes a stable drop can be formed. After this time the drop disintegrates caused by the dissolution into the aqueous phase resulting in a fast mass transport through the phase boundary. In the presence of oleic acid in the system an interfacial tension of 11 mN/m is reached and the time for a stable drop, which also means the equilibration at the interface amounts to 30-45 minutes. Due to the comparability a dimensionless representation of the time is chosen.

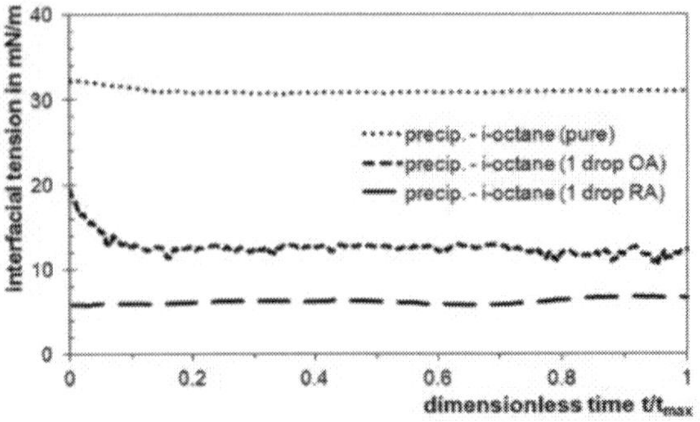

Figure 2. Demonstration of the interfacial tensions dependent on the surfactant used.

Consequently, the both surfactants used have a different process behavior with varying consequences to the particle extraction process. The determination of the yield of transferred magnetite in dependency of the process time indicates that the phase transfer with ricinoleic acid takes place rapidly. Because the increase of transferred magnetite to around 80% after 10 minutes is nearly twice the amount compared to oleic acid. The longer process times for the extraction of magnetite with oleic acid are caused by the lower dissolution in the aqueous phase resulting in a smaller available phase interface. This also can be recognized during long-time trials. Already after 15 minutes the yield of transferred magnetite with ricinoleic acid as surfactant amounts nearly 90% and is kept constant over the process time up to 165 minutes. In contrast with oleic acid after 270 minutes processing time approximately 86% of the magnetite particles have been extracted into the organic phase. Moreover, the clearing off of the column with ricinoleic acid is achieved after nearly 6 minutes, in contrast with oleic acid it can be observed after 160 minutes.

For the characterization of transfer kinetics, the reaction of 1st order is assumed. The experimental and theoretical calculated relative change in mass concentration conforms only for oleic acid, so that the particle extraction process can be estimated as a 1st order reaction. This means that the dissolution of oleic acid from the organic into aqueous phase does not influence the process kinetics, whereby the surfactant mass transfer is negligible. In contrast to ricinoleic acid another approximation represented by a sigmoid function in terms of a dose-response-curve has to be applied. Thus, the mass transport of the surfactant by dissolving in the aqueous phase strongly influences the kinetics of the particle extraction process within the column. This can be demonstrated as the dose for the process and is prevalent especially in the first minutes of the extraction process. This is the case, if the equilibrium of ricinoleic acid concentration in either phase is not reached. Furthermore, the response is represented by the functionalization of the particles resulting in the following extraction of them. Therefore, it can be assumed that the particle extraction process with ricinoleic acid as surfactant is controlled by process imbalance.

The colloidal stability of organosols is investigated using the calculation of the solubility distances $D_{FA-solvent}$ between the surfactant and the solvent (Equation 2) by the determination of the Hildebrandt parameter δ_t. This parameter can be calculated with the three Hansen solubility parameters in consideration of the dispersive δ_d, polar δ_p and hydrogen δ_h bonding components (Equation 3). Thus, the interaction between the used surfactants as pure substances and capped onto the magnetite surface can be characterized (17, 21). The calculated values are presented in Table 1 and it applies the following correlation: The smaller the value, the more soluble is the surfactant in the solvent. Ricinoleic acid has a lower solubility in iso-octane caused by the polar and hydrogen bonding parameters and therefore a better dissolution into the aqueous phase resulting in a more rapid particle extraction process. In contrast, oleic acid is more soluble in iso-octane, whereby more stable drops are formed during the phase transfer. Therefore, the colloidal stability of the organosol is increased. Compared with the produced organosols in the transfer device lab centrifuge using the solvent DCM the colloidal stability with ricinoleic acid is increased. Due to the very good solubility of ricinoleic acid in DCM, all surfactant molecules are very fast available at the phase interface to disintegrate the agglomerates in primary particles. Therefore, the result is a stable colloidal organosol.

$$D_{FA-solvent} = \sqrt{4*(\delta_{d,FA} - \delta_{d,solvent})^2 + (\delta_{p,FA} - \delta_{p,solvent})^2 + (\delta_{h,FA} - \delta_{h,solvent})^2} \quad (2)$$

$$\delta_t = \sqrt{\delta_d{}^2 + \delta_p{}^2 + \delta_h{}^2} \quad (3)$$

Table 1. Summary of the calculated Hansen solubility parameters and solubility distances.

substance —	δ_t in MPa$^{1/2}$	δ_d in MPa$^{1/2}$	δ_p in MPa$^{1/2}$	δ_h in MPa$^{1/2}$	$D_{FA\text{-}iOCT}$ in MPa$^{1/2}$	$D_{FA\text{-}DCM}$ in MPa$^{1/2}$
iso-octane	14.30	14.30	0	0	–	–
DCM	20.20	18.20	6.30	6.10	–	–
RA	19.68	16.84	8.32	5.89	11.39	3.39
OA	18.31	17.30	5.73	1.78	8.49	4.71
RA@Fe_3O_4	19.40	17.27	6.89	5.55	10.66	2.03
OA@Fe_3O_4	17.90	17.68	2.81	0	7.32	7.11

For the analysis of the product stability a dispersion analyzer (Lumisizer) is applied, whereby the integral transmission as a function of time can be determined. By the increase of integral transmission, conclusions regarding to the colloidal stability can be drawn. These experimental observations are confirmed with the calculated solubility distances using the Hansen solubility parameters.

Due to the interesting process behavior and short processing times with ricinoleic acid in the partial recirculation operation of the miniplant as mentioned before, further operating methods are developed by experiments using ricinoleic acid as surfactant. Therefore, the simple multi-level circuit of the miniplant can be generated. This means that both phases are in circuit, but the removal and supply of the phases as well as the production of organosol are discontinuously. Due to stable processing control and yields of transferred magnetite about 80-90%, the continuous circuit is developed. Therefore, the organosol can be produced continuously including the simultaneous formation of waste and product as shown in Figure 1b. Yields of transferred magnetite are reached of about 90% (Figure 3a). Furthermore, the miniplant is utilized for a selective flotation process of magnetite and quartz particles in the partial recirculation operation. In Figure 3b the recovery of transferred magnetite dependent on the process time is demonstrated. Including the transferred particles of the interfacial layer (IL) a recovery of extracted magnetite particles of 76% is reached. This process engineering application presents, that the miniplant can be utilized for separation processes of two minerals for example.

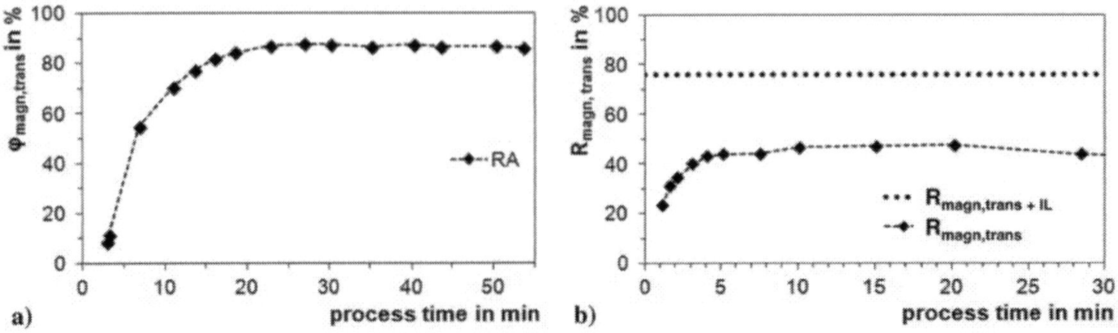

Figure 3. Demonstration of the process time dependent yield of transferred magnetite in continuous circuit of both phases using ricinoleic acid as surfactant (a) and the recovery of transferred magnetite by the implementation of a separation process in partial recirculation operation of the miniplant (b).

Conclusions

The study has been presented that the particle extraction process can be performed in different transfer devices. As a consequence of this, the interaction between the surfactants as well as the organic and aqueous phase used has to be considered, because this leads to different procedures at the phase boundary. A continuous particle extraction process with a drop column as transfer device was developed. Furthermore, the process behavior, transfer kinetics and colloidal stability were investigated and the process engineering application for a selective separation process was proved.

Acknowledgments

We give thanks to the Deutsche Forschungsgemeinschaft (DFG) for financial support (project no. PE1160/6-3). Additionally, we would like to thank Andre Rieger and the laboratory technicians from Institute of Thermal Process Engineering, Environmental and Natural Products Process Engineering of the TU Bergakademie Freiberg for the TOC measurements as well as Christian Strötges from Institute of Physical Chemistry of the TU Dortmund for the measurements of interfacial tensions.

References

1. Scherer, M. and Figueiredo Neto, A. M. (2005), "Ferrofluids: properties and applications", *Brazilian Journal of Physics*, Vol. 35, No. 3A, pp. 718-727.
2. Hurlebaus, S. and Gaul, L. (2006), "Smart structure dynamics", *Mechanical Systems and Signal Processing*, Vol. 20, No. 2, pp. 255-281.
3. Dallas, P., Georgakilas, V., Niarchos, D., Komninou, P., Kehagias, T. and Petridis, D. (2006), "Synthesis, characterization and thermal properties of polymer / magnetite nanocomposites", *Nanotechnology*, Vol. 17, pp. 2046-2053.
4. Teja, A. S. and Koh, P.-Y. (2009), "Synthesis, properties and applications of magnetic iron oxide nanoparticles", *Progress in Crystal Growth and Characterization of Materials*, Vol. 55, pp. 22-45.

5. Hickstein, B. and Peuker, U. A. (2009), "Modular process for the flexible synthesis of magnetic beads - process and product validation", *Journal of Applied Polymer Science*, Vol. 112, pp. 2366-2373.
6. Banert, T. and Peuker, U. A. (2007), "Synthesis of magnetic beads for bio-separation using the solution method", *Chemical Engineering Communications*, Vol. 194, No. 6, pp. 707-719.
7. Laurent, S., Forge, D., Port, M., Roch, A., Robic, C., Vander Elst, L. and Muller, R. N. (2008), "Magnetic iron oxide nanoparticles: synthesis, stabilization, vectorization, physicochemical characterization and biological application", *Chemical Reviews*, Vol. 108, No. 6, pp. 2064-2110.
8. Mahmoudi, M., Sant, S., Wang, B., Laurent, S. and Sen, T. (2011), "Superparamagnetic iron oxide nanoparticles (SPIONs): development, surface modification and applications in chemotherapy", *Advanced Drug Delivery Reviews*, Vol. 63, pp. 24-46.
9. Banert, T. and Peuker, U. A. (2006), "Preparation of highly filled super-paramagnetic PMMA-magnetite nano composites using the solution method", *Journal of Material Science*, Vol. 41, pp. 3051-3056.
10. Kirchberg, S., Rudolph, M., Ziegmann, G. and Peuker, U. A. (2011), „Nanocomposites based on technical polymers and sterically functionalized soft magnetic magnetite nanoparticles: synthesis, processing and characterization", *Journal of Nanomaterials*, Article ID 670531.
11. Rudolph, M. and Peuker, U. A. (2011), "Coagulation and stabilization of sterically functionalized magnetite nanoparticles in an organic solvent with different technical polymers", *Journal of Colloid and Interface Science*, Vol. 357, pp. 292-299.
12. Machunsky, S. and Peuker, U. A. (2007), "Liquid-liquid interfacial transport of nanoparticles", *Physical Separation in Science and Engineering*, Article ID 34832.
13. Lai, R. W. M. and Fuerstenau, D. W. (1968), "Liquid-liquid extraction of ultrafine particles", *Transactions of the American Institute of Mining, Metallurgical*, and Petroleum Engineers, Vol. 241, pp. 549-556.
14. Erler, J., Machunsky, S., Grimm, P., Schmid, H.-J. and Peuker, U. A. (2013), "Liquid-liquid phase transfer of magnetite nanoparticles – evaluation of surfactants", *Powder Technology*, Vol. 247, pp. 265-269.
15. Zhang, L., He, R. and Gu, H.-C. (2006), "Oleic acid coating on the monodisperse magnetite nanoparticles", *Applied Surface Science*, Vol. 253, pp. 2611-2617.
16. Rudolph, M., Erler, J. and Peuker, U. A. (2012), "A TGA-FTIR perspective of fatty acids adsorbed on magnetite nanoparticles – decomposition steps and magnetite reduction", *Colloid and Surfaces A: Physicochemical and Engineering Aspects*, Vol. 397, pp. 16-23.
17. Rudolph, M. and Peuker, U. A. (2012), "Phase transfer of agglomerated nanoparticles – deagglomeration by adsorbing grafted molecules and colloidal stability in polymer solutions", *Journal of Nanoparticle Research*, Vol. 14, No. 7, Article 990.
18. Erler, J. V., Leistner, T. and Peuker, U. A. (2014), "Application of a particle extraction process at the interface of two liquids in a drop column – consideration of the process behavior and kinetic approach", *Advances in Chemical Engineering and Science*, doi:10.4236/aces.2014, in press.
19. Blaß, E. (1988), "Bildung und Koaleszenz von Blasen und Tropfen", *Chemie Ingenieur Technik*, Vol. 60, pp. 935-947.
20. Räbiger, N. and Schlüter, M. (2006), "Bildung und Bewegung von Tropfen und Blasen", *VDI Wärmeatlas*, Hrsg. von VDI Gesellschaft, Springer, Lda1-Lda15.
21. Hansen, C. M. (2007), "Hansen solubility parameters", A User´s Handbook, Second Edition CRC, Boca Raton.

350108 Process Safety Performance Management – a Strategic Approach for Sustainable Improvement

Abstract:

The process industry has over 20 years experience in using a formal management systems approach to managing process safety. Yet, data and experience reveal that few companies have mastered it, to achieve and sustain high levels of HSE performance over long periods. Some companies/facilities have tried to implement PSM but have had either cyclic performance or ineffective implementation. Many other companies/facilities that have tried to implement PSM and have had reasonably good performance, but they have plateaued and have been unable to really continuously improve.

Benchmarking of companies in a variety of industries has revealed numerous PSM implementation failure modes and success practices. One of the reasons that companies have had inconsistent Process Safety performance is because they have not embedded an effective learning process into their PSM/HSE system, including systematic diagnosis of performance problems from ALL sources of learning and employ tangible, practical ways to pursue continuous improvement.

This paper will describe a systematic approach for **Process Safety Performance Management** that addresses the following issues:

- Methods and information sources for early recognition of PS performance degradation
- Ways to diagnose the technical, management practice, behavioral, and cultural causal factors for the performance issues

- Ways to monitor long-term slow degradation of organizational HSE performance

This paper will highlight case studies of three company efforts to ensure a robust PSM system, avoid past PSM system failure modes and monitor organizational culture warning signs for future potential process safety performance.

Systematic Analysis and Learning from Process Safety Incidents

Stephen James
Department of Fire Protection & Safety
Oklahoma State University, Stillwater, OK 74078
jstepht@orangemail.okstate.edu

Fazle Rabbi
Department of Fire Protection & Safety and School of Chemical Engineering
Oklahoma State University, Stillwater, OK 74078
fazle.rabbi@okstate.edu

Dr. Qingsheng Wang
Department of Fire Protection & Safety and School of Chemical Engineering
Oklahoma State University, Stillwater, OK 74078
qingsheng.wang@okstate.edu

Prepared for Presentation at
American Institute of Chemical Engineers
2014 Spring Meeting
10th Global Congress on Process Safety
New Orleans, LA
March 30 – April 2, 2014

UNPUBLISHED

Systematic Analysis and Learning from Process Safety Incidents

Stephen James
Department of Fire Protection & Safety
Oklahoma State University, Stillwater, OK 74078
jstepht@orangemail.okstate.edu

Fazle Rabbi
Department of Fire Protection & Safety and School of Chemical Engineering
Oklahoma State University, Stillwater, OK 74078
fazle.rabbi@okstate.edu

Dr. Qingsheng Wang
Department of Fire Protection & Safety and School of Chemical Engineering
Oklahoma State University, Stillwater, OK 74078
qingsheng.wang@okstate.edu

Keywords: Systematic analysis, incident analysis, process safety, lessons learn

Abstract

Numerous incidents have occurred in the process industries and caused hundreds of fatalities and injuries. The US Chemical Safety Board (CSB) has conducted incident investigations as one element of the Process Safety Management (PSM) program and provided final reports for over 77 incidents. This paper is to look into all those final reports and summarize the findings. The type of accidents is identified as vapor explosion, dust explosion, reactive chemical explosion, and toxic chemical release. The main type of accidents is found to be vapor explosion that resulted in nearly 40% of all the types of accidents. Through the comprehensive analysis of PSM violations, the results show that mechanical integrity, process safety information, and process hazard analysis are the top three types of violations while about 20% of the PSM violations are due to mechanical integrity. The results also show that about 80% of these incidents are somehow due to human error. Detailed human error analysis for all these incidents is performed and the results show that lack of training is the main contribution to incidents. Recommendations are given to each type of incident that occurred based on the analysis results. The purpose of this research is to develop learning from previous incidents and therefore to improve future process safety.

1. Introduction

Chemical Process Industries (CPI) has gone through major changes in the last two decades. Severity and complexity of the processes have significantly increased to create quality products to meet market demand. On the other hand, risk associated with these processes has also increased due to the complexity of the processes. Hence, to save lives, environment and to avoid financial damages, process safety has drawn public attention which is driven by the growing scale of the modern plants [1]. Process safety management has become more matured in last decades. Now, accidents hardly happen in petrochemical industries due to unknown hazards [2]. However, most of the incidents which occurred could be related to organizational safety culture and lack of knowledge and awareness.

Enrico et al. has presented a short history of major chemical process incidents that occurred over the last century which changed the landscape of the industries viewpoint on process safety. These major incidents not only took many people's lives but also damaged the financial status of the companies. Mannan et al. have studied the financial impacts of catastrophic events in the process industries [3]. To avoid these incidents, many researchers have emphasized the importance of learning from previous incidents [4]. Statistical studies based on previous incidents have pointed out recurring causes [5].

Most common incidents in process industries which may result into fatalities are fire, explosion or toxic chemical release. Pekalski et al. has claimed that explosions, fire and toxic releases are responsible for 75%, 20% and 5% respectively of the total loss. Human errors are mainly responsible for majority of the industrial disasters [6]. Responsible human factors and remedial actions have been studied [7]. Process safety knowledge gained from previous incidents could be utilized to identify the root causes more efficiently and apply process safety management effectively to achieve operational excellence without affecting people and environment.

This study examines the root cause of various incidents which occurred within last two decades. Statistical analysis has been carried out to find the weightage of different type of incidents, major elements of process safety management which were violated. This study also depicts the role of human errors in studied incidents and contributions of different factors which fall within human error. Analyses and findings of this study would help to invest attention of researchers and concerned management into key safety features to implement process safety management more effectively.

2. Systematic Analysis of Previous Incidents

Learning from previous incidents gives safety researchers and safety professionals a scope for reviewing or modifying current safety practices. Advancement of technology has improved chemical processes but hazards associated with it could be minimized. However, many severe incidents still have been observed. Systematic analysis of the incident data is required to identify the responsible factors and based on the findings process safety management system could be updated. In this study, incident data of 77 incidents recorded in Chemical Safety Board website has been collected [8]. First, all the incidents which took place in USA within last 20 years have been sorted out according to their type. Then, contribution of different PSM violations has been found for those incidents. Finally, contribution factors that are under the category of human error have been studied.

2.1 Types of Incidents

Totally 77 process safety incidents from 1999 to 2012 in the United States were reviewed. Abbasi et al. have recapitulated some of the major accidents which occurred during 1926-1997. From his study it could be depicted that almost all the incidents were related to explosion, fire and toxic gas/vapor release. And it was also observed that about 70% of all the incidents were gas/vapor cloud explosion. This study covered the time frame of 1998-2012. Figure 1 shows that vapor cloud explosion, dust explosion and reactive chemical explosion contributed 39%, 9% and 18% of all the incidents that took place in the USA process industries. Totally about 66% of all the incidents are related with explosion, which is consistent with previous studies. Vapor explosion that resulted in nearly 40% of all the types of accidents and therefore we should pay more attention to vapor cloud explosion. Many mitigation methods can be proposed, for example, how to prevent vapor/gas release, how to limit ignition sources, and how to detect gas release.

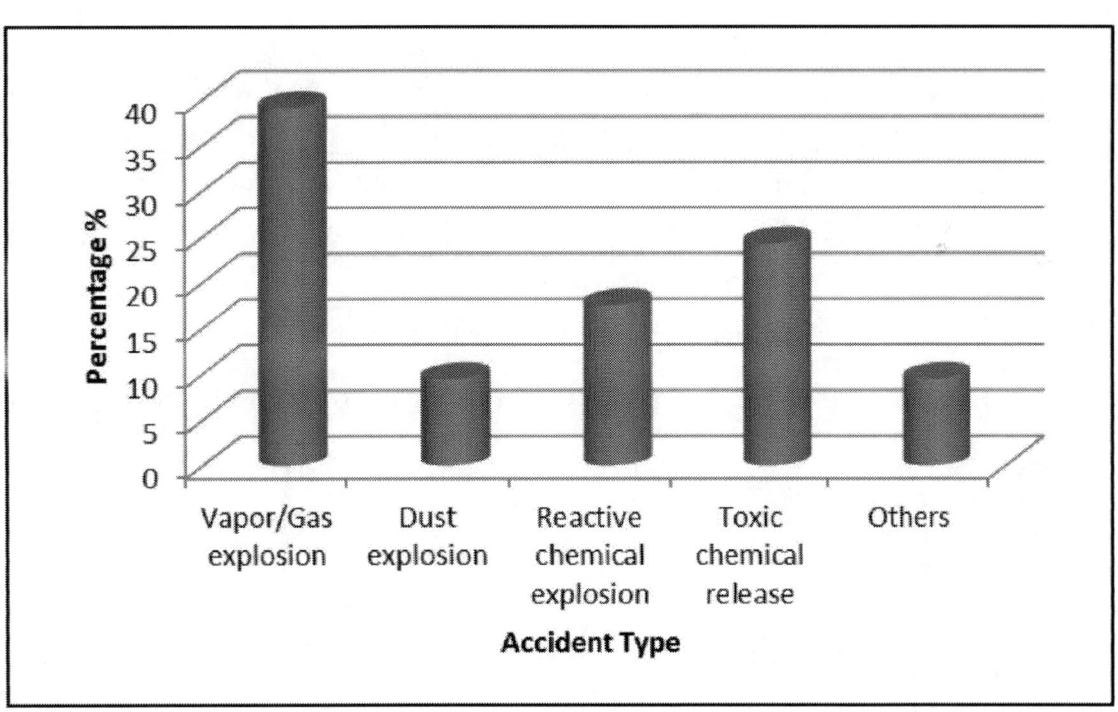

Figure 1. Weightage of Process Safety Incidents between 1998 and 2012

2.2 *Process Safety Management Violation Analysis*

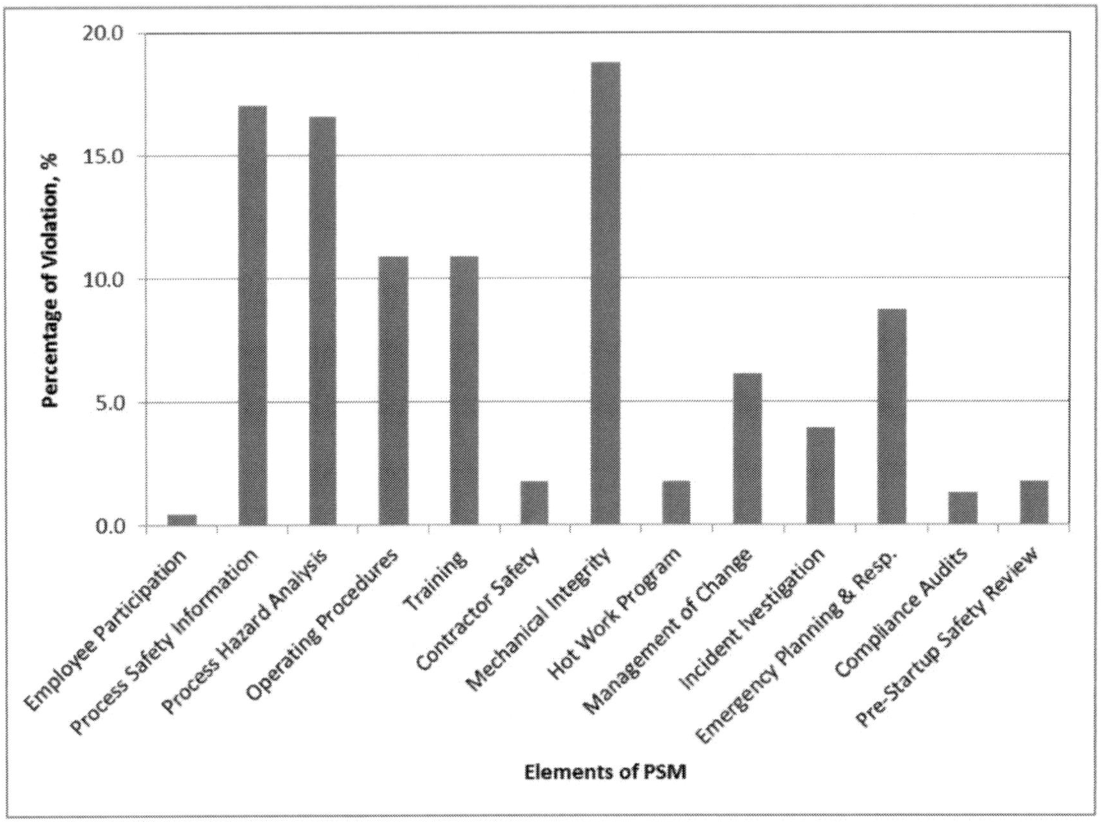

Figure 2. PSM Violation Analysis for the Process Safety Incidents

Katizima has defined PSM as the management system which manages prevention of, preparedness for, mitigation of, response to and restoration from catastrophic chemical releases or energy from any chemical process. Process Safety Management standard 29 CFR 1910.119 has fourteen elements as follows:

- Employee Participation
- Process Safety Information (PSI)
- Process Hazard Analysis (PHA)
- Operating Procedures
- Training
- Contractors
- Pre-startup Safety Review (PSSR)
- Mechanical Integrity
- Hot Work
- Management of Change(MOC)

- Incident Investigation
- Emergency Planning and Response
- Compliance Audits
- Trade Secrets

Figure 2 depicts that Mechanical Integrity, Process Safety Information (PSI) and Process Hazard Analysis (PHA) are the elements which have the significant frequency of violation among other elements of PSM. Contributions of these three elements are 19%, 17% and 16.6% respectively. Operating procedures, training of personnel, management of change (MOC) and Emergency planning and response are other four elements of PSM which were violated in 11%, 11%, 6% and 9% of total number of violation. Mitigation plan should consider these mentioned elements which have been violated, especially mechanical integrity.

2.3 Factors Contributing Human Errors which Lead to Incidents

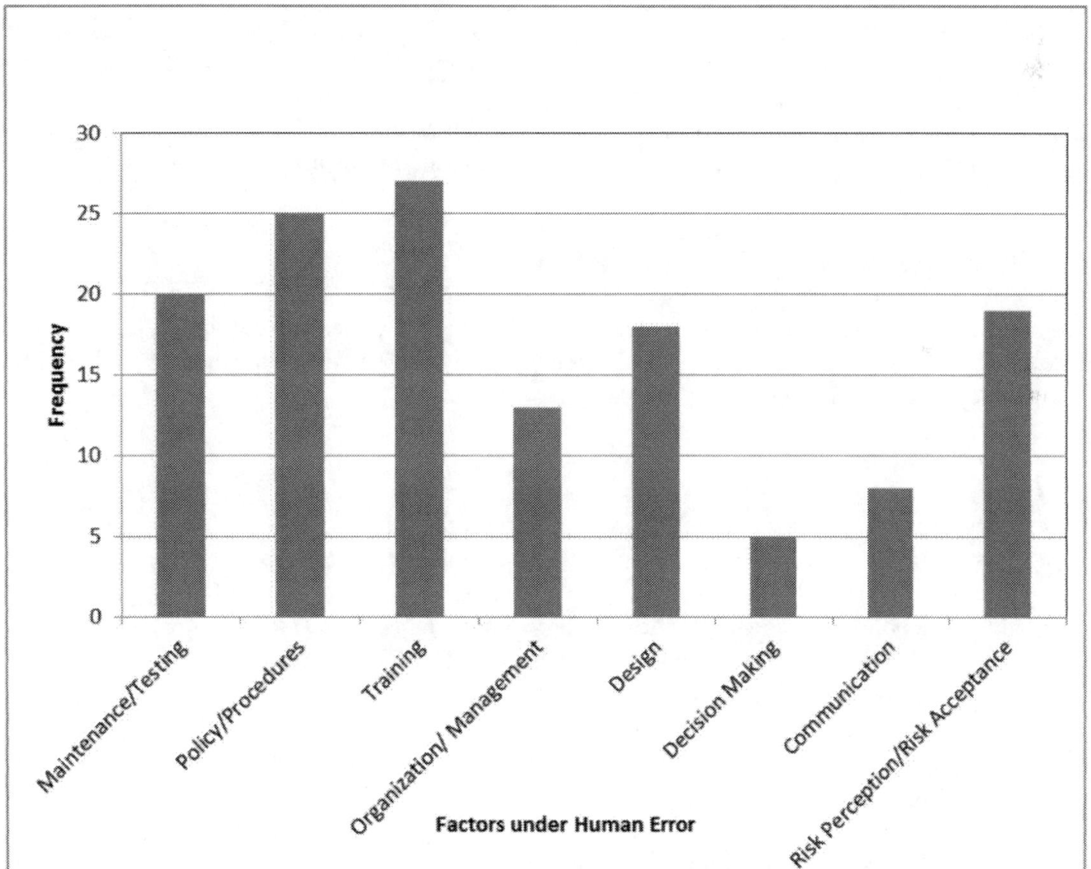

Figure 3. Factors Analysis under Human Errors Leading to Incidents

This study reveals that nearly 80% of the incidents were directly or indirectly caused by human errors. Eight factors under the category of human error have been identified by Wang et al. from the Macondo well blowout incident. In this work, all eight factors were considered for those 80% of the incidents [9]. These factors are included as follows:

- Maintenance/Testing
- Policy/Procedures
- Training
- Organization/Management
- Design
- Decision Making
- Communication
- Risk Perception/Risk Acceptance

The results are shown in Figure 3. Detailed human error analysis shows that lack of training is the main contribution to incidents, which is responsible for 27% of the incidents which are caused by human errors. Lack of training for the personnel, poor maintenance of the equipment, wrong design of equipment and lack of risk perception are the key factors which lead to majority of the incidents. Lack of training, poor maintenance, and risk perception are all related to safety culture. Figure 3 can help the safety professionals and management level to improve the safety culture and thus minimizing the human errors which may lead to disasters.

3. Lessons Learned and Recommendations

Although the cause of many of the incidents covered in this paper are due to unsafe practices. Lessons learned have been systematically studied and recommendations for individual four types of incidents have been proposed.

3.1 Vapor/Gas Explosion

- Development of practical guideline of vapor cloud explosion [10].
- To avoid pipeline explosion, use of inherently safer fuel gas pipe cleaning methods are required.
- Implementing inherently safer design for process plants [11].
- Emphasizing on Mechanical Integrity element in PSM.
- Building appropriate layer of protections.
- Proper gas testing for hot work operation.

3.2 Dust Explosion

- Understand of fundamental of dust explosion [12].
- Development and implement combustible dust hazard awareness training.
- Combustible dust hazard identification in the audit protocols.
- Effective emergency response plan should be developed.

3.3 *Reactive Chemical Explosion*

- Comprehensive mechanical integrity programs to be implemented, emphasizing on thorough inspections of critical equipment.
- Frequency of preventive maintenance of equipment could be revised.
- Create a culture where leaders encourage operations and maintenance personnel to maintain process equipment more effectively.
- Near misses and spills should be investigated and reported properly.

3.4 *Toxic Chemical Release*

- Personnel should be aware of the toxic materials and their impact on human health and environment.
- LOP should be implemented.
- Develop guidance for effective design of HVAC system and other necessary control room components to protect employee and equipment in the event of release of hazardous materials.
- Effective emergency response plan should be in place. Safety drills should be carried out regularly to enhance emergency response [13].

Based on the detailed human error analysis, several lessons learned were summarized to minimize human errors during process operation and maintenance.

- Effective training for personnel is required, even if the personnel are contractors or short service employee.
- Proper maintenance plus management approved standard operating procedure should be in place in order to achieve operational excellence.
- HAZOP, PHA should be in place to identify hazards and communicate potential hazard and impact of these hazards to personnel.
- Design of critical equipment should be reviewed and simulated to predict the impact of scaling up or down operations.
- Behavior based safety should be adopted [14].
- Effective leadership would effective to communicate the hazards to personnel.
- OSHA permit-required confined spaces rule should be amended for general industry to establish a maximum permissible percentage substantially below the LEL for safe entry and maximum occupancy load in confined spaces [15].

4. Conclusion

This study reveals that during 1998-2012 several industrial disasters caused 199 fatalities and left 650 people severely injured as recorded in the Chemical Safety Board website. The research shows that 66% of the incidents recorded during the mentioned time period are due to explosions. This follows the trend from the previous studies. This study also revealed that 80%

of the incidents were caused by human error. Error in maintenance, lack of standard operating procedures, insufficient training for personnel, mismanagement, wrong design of equipment and inadequate risk perception are the most significant causes contributing to these industrial disasters.

Process safety management plays a vital role in operational excellence. Violation of PSM elements has led to most accidents in the past. This study shows that out of 14 elements of PSM, process safety information, PHA, operating procedures, training, mechanical integrity and emergency response and planning are frequently violated. Most common amount of process safety management violations is the mechanical integrity program. Effective maintenance program and well trained workers dedicated to safety will help to prevent these accidents to continue to occur.

Acknowledgement

The research was supported by the National Science Foundation (NSF) through OK-LSAMP and Boots & Coots Center for Fire Safety and Pressure Control within the Department of Fire Protection & Safety at Oklahoma State University.

References

1. E. Zio and T. Aven, Industrial disasters: Extreme events, extremely rare: Some reflections on the treatment of uncertainties in the assessment of the associated risks, Process Saf Environ Prot, 91 (2013), 31–45.
2. B. Knegtering and H.J. Pasman, Safety of the process industries in the 21st century: A changing need of process safety management for a changing industry, J Loss Prevent Proc, 22 (2009), 162–168.
3. B. K. Kim, J. Krams, E. Krug, M. Leaseburge, J. Lemley, A. Alkhawaldeh, R. A. Mentzer, and M. S. Mannan, Case study analysis of the financial impact of catastrophic safety events, J Loss Prevent Proc, 25 (2012), 780-787.
4. R. E. Sanders, Chemical process safety learning from case histories, 3rd ed., Elsevier Butterworth Heinemann, 1999.
5. K. Kidman and M. Hurme, Statistical analysis of contributors, Chem Eng Technol 36 (2013), 167-176.
6. A. A. Pekalski, J. F. Zevenbergen, S. M. Lemkowitz, and H. J. Pasman, A review of explosion prevention and protection systems suitable as ultimate layer of protection in chemical process installations, Process Saf Environ Prot, 83 (2005), 1–17.
7. F. I. Khan and S. A. Abbasi, Major accidents in process industries and an analysis of causes and consequences, J Loss Prevent Proc, 12 (1999), 361-378.
8. Chemical Safety Board reports, available at: www.csb.gov
9. P. Smith, H. Kincannon, R. Lehnert, Q. Wang, and M. Larranaga, Human error analysis of the Macondo well blowout, Process Saf Prog 32 (2013), 217-221.
10. Guidelines for vapor cloud explosion, pressure vessel burst, BLEVE and flash fire hazards, Center for Chemical Process Safety (CCPS), 2nd ed., Wiley-AIChE, 2010.
11. T. A. Kletz and P. Amyotte, Process plants: A handbook for inherently safer design, 2nd ed., CRC Press, 2010.
12. P. Amyotte, An introduction to dust explosions, 1st ed., Elsevier Butterworth Heinemann, 2013.
13. Q. Wang, T. Ma, J. Hanson, and M. Larranaga, Application of incident command system in emergency response, Process Saf Prog 31 (2012), 402–406.
14. O. Wirth and S. O. Sigurdsson, When workplace safety depends on behavior change: Topics for behavioral safety research, J Saf Res 39 (2008), 589–598.
15. P. Wilson and Q. Wang, Development of a protocol for determining confined space occupant load, Process Saf Prog (2014), in press.

"All In" Means "All the Time"

Jennifer F. Mize
Principal Chemical Engineer
Eastman Chemical Company
P.O. Box 511, Building18
Kingsport, TN 37662
jfmize@eastman.com

Prepared for Presentation at
American Institute of Chemical Engineers
2014 Spring Meeting
10th Global Congress on Process Safety
New Orleans, LA
March 30 – April 2, 2014

UNPUBLISHED

"All In" Means "All the Time"

Jennifer F. Mize
Principal Chemical Engineer
Eastman Chemical Company
P.O. Box 511, Building18
Kingsport, TN 37662
jfmize@eastman.com

Keywords: Process safety; near misses; management of change; conduct of operations; operational discipline

Abstract

Process safety must be the responsibility of all technical and operations personnel. As such, this responsibility is not limited to attendance at Process Hazard Analysis (PHA) revalidations; rather, it must be an ongoing, integral part of both operating and supporting manufacturing facilities. Compartmentalization of process safety knowledge and practice can hinder proper identification of potential safety consequences of change. Near misses and incidents can occur due to lack of understanding about controls (both administrative and engineering) necessary to ensure continued safe operation.

This paper will present examples of actual near misses that have occurred due to misunderstanding, miscommunication, loss of institutional knowledge, and improperly managed personnel changes that lead to failures in understanding and properly evaluating the consequences of changes made to support daily operations.

I. Introduction

Process safety is an essential element of the day-to-day operations of a chemical facility. Engineers, managers, and operators make decisions each day to ensure continued safe production in their manufacturing areas. If not properly managed, these "routine" changes can lead to unanticipated negative safety consequences. As facilities age and/or require upgrades to meet current and future operating needs, changes associated with debottlenecking efforts, equipment replacement and maintenance, and modifications to operating conditions must be properly evaluated to ensure that all potential hazards are adequately identified and controlled.

In many companies, process safety expertise is concentrated in a centralized organization which has responsibility for multiple facilities and/or manufacturing sites. In others, this expertise may be obtained from contractors to support specific projects or PHA studies. This compartmentalization of process safety knowledge and practice can prevent the personnel responsible for the day-to-day technical and operational support of the facility from effectively identifying the hazards and potential safety consequences that can result from seemingly routine decisions. All personnel responsible for supporting manufacturing operations should have sufficient process safety knowledge to allow them to critically evaluate their decisions and proposed changes to determine if there could be potential safety consequences. They must be able to effectively perform an initial safety review (typically included as part of the Management of Change process) to determine if additional process safety expertise is needed to fully evaluate the impact of the proposed change. Attendance at a PHA revalidation every five years, as required for OSHA PSM-covered processes, is insufficient. Process safety must be an ongoing, integral part of operating and supporting manufacturing facilities. It is the responsibility of the employer to determine the level of process safety knowledge required for each job and to ensure that the people in those jobs get the training they need.

Efforts to achieve zero injuries at chemical facilities must also include a commitment to preventing process safety incidents. PHA team members must identify potential hazards based on What-If or checklist questions. The successful identification of what could go wrong is highly dependent upon the expertise and experience of the team members and the quality of the process safety information. OSHA-recordable and lost-time injuries are much easier to see and track than potential process safety incidents. Process safety and personal safety are both critical elements of a comprehensive safety program dedicated to being "All In" for Safety.

The following examples document near misses that have occurred due to misunderstanding, miscommunication, loss of institutional knowledge, and improperly managed personnel changes that lead to failures in understanding and properly evaluating the consequences of changes made to support daily operations. In most cases, a process safety expert was not consulted or involved in the decisions that lead to the near-miss. These examples have been selected from a wide variety of industrial applications and processes.

II. Managing change during capital projects

The best opportunity to effectively implement inherently safer design concepts occurs during conceptual design. When inherently safer alternatives are not available, engineering controls such as relief systems and safety instrumented Systems (SIS) can then be implemented to address the hazards. The process safety engineer can utilize each of these options to optimize the safety design [1]. The conceptual safety design is the basis for many of the engineering controls that will be documented in the preliminary PHA. Any changes that occur after the preliminary PHA should be communicated to the process safety engineer to determine if they impact the process safety design. Recognizing these changes late in the project can result in significant re-work of the process safety design and/or implementation of additional controls that could impact cost and schedule.

2.1 Process changes that resulted in additional safety concerns

A capital project with an aggressive schedule was initiated to provide a new distillation column for an existing process. This column was intended to provide additional solvent recovery capability. The reboiler was originally designed to use low pressure steam, which was readily available and less expensive than medium pressure steam. Due to the compressed timeline for installation, the equipment was ordered very early in the design process to ensure timely delivery. The process safety engineer provided a conceptual design for the relief system for the column and associated equipment, including recommended pressure ratings.

After the preliminary PHA, the process design and staff engineers determined that the low pressure steam would not provide sufficient steam chest pressure to generate the vapor required to operate the column. They determined that medium pressure steam would need to be used instead. The reboiler was already on order and provided significantly more heat transfer area than was needed when using medium pressure steam. This change was not communicated promptly to the process safety engineer, who had to update the relief evaluation for the column. The change to medium pressure steam introduced a new relief scenario, tube rupture in the reboiler, along with an increased required relief capacity for abnormal heat input. These cases were not adequately relieved by the relief system that was originally sized.

Since all the equipment was already ordered and the original relief system would not provide adequate relief for the new operating conditions, the process safety engineer had to determine how to address the relief issues. Operations did not want to install an SIS at the time. The required operating pressure in the reboiler was significantly less than the source header pressure for the medium pressure steam supply. The steam control loop was designed to limit the maximum steam pressure such that the maximum temperature in the reboiler was less than the boiling point of the components in the column system at the system rated pressure. A new relief valve was installed downstream of the steam control valve to insure that the steam pressure was limited in the event of an automatic control failure [2]. Delivery of the new relief valve had to be expedited to meet the project schedule. With this relief valve in place in the steam supply, operations personnel were limited in their capability to adjust the steam chest pressure, which ultimately lead to operational difficulties.

2.2 Improperly managed changes during detailed design

A relief review conducted during a PHA revalidation revealed that the relief calculations generated several years earlier during the detailed design phase of the facility had not been updated when design changes were made. The feed tank to the process is a large API 620 storage tank. The existing relief documentation was based on a spherical tank with significantly higher pressure rating relieved by a large relief valve. A handwritten note in the margin indicated that someone had "updated" the relief sizing for the change in wetted area for the cylindrical tank. Unfortunately, the person doing the update failed to update the sizing calculation for the new, lower MAWP. Additionally, at the lower MAWP, the subcritical flow calculations needed to be used. Updating the relief calculations revealed that there was excessive pressure drop in the discharge piping on the low pressure vents installed on the tank. Piping modifications were required to provide adequate emergency pressure relief.

III. Inadequate vendor and contractor oversight

Many companies rely on vendors and/or contractors to provide expertise that they do not possess or maintain internally. This is an effective way to manage resources that are not routinely required by the organization. However, inadequate oversight of vendors and contractors can lead to undesirable results, including inadequately addressed safety concerns. Operations personnel have valuable process knowledge and expertise that should be effectively utilized to monitor and critically review the work of vendors and contractors. It is essential to determine the scope of expertise that can be provided by a given vendor or contractor; a vendor who specializes in the design of packaged skid units may not be an expert in Safety Instrumented Systems, relief system design, or PHAs. Similarly, contractors who provide general process engineering expertise may not be experts in all elements of process safety. It is the responsibility of the operating company to ensure the limitations of the selected vendor and/or contractors are recognized and understood, and to provide sufficient oversight to ensure that all safety concerns are identified and addressed.

3.1 Skid unit designed by the vendor

A skid unit was purchased from a vendor recognized as a process expert in this technology. General process engineering was provided by a contractor routinely used by the operating company. The vendor sized and provided all the relief devices; the contractor routed and installed the discharge piping and effluent handling system. The responsibility for each group was well-defined. The operating company did not review the relief system design, relying on the vendor to provide the skid unit and all associated equipment. The contractor had no responsibility to verify the adequacy of the relief design or the impact of installation of the discharge piping on the relief system. The initial PHA was conducted by the operating company. Relief documentation received from the vendor was archived without any review.

A relief design review conducted as part of the PHA revalidation lead to a recommendation to update the relief documentation received from the vendor to evaluate the impact of the installation of the discharge piping on the relief systems. Review of the existing documentation provided by the vendor revealed that they had no understanding of the facility siting issues, since the relief devices were all sized for external fire exposure. In fact, the facility in question had no source of liquid fuel that could sustain a pool fire. Additionally, relief valves were sized to provide relief for external fire on gas-filled vessels. Evaluation of the piping systems revealed concerns with inlet and outlet pressure drop, which were most likely the result of the third-party piping design with no verification and the fact that the vendor sized the relief devices based only on an orifice calculation. The vendor also failed to recognize inherently safer design opportunities provided by the pressure rating of the equipment. With the elimination of pool fire as a credible relief scenario, numerous vessels for this installation could then be effectively protected by system design as documented in ASME Section VIII, UG-140(b) [3].

3.2 Contractor sizes relief for aqueous process for pool fire

A contractor was given full responsibility for relief system design for a process containing only water and a solid. The contractor sub-contracted the work to a subsidiary outside of the United States. The subcontractor was told to treat the process fluid as water and to use the standard API RP 520 relief cases. They were not told that there was no solvent or any other fuel source in the area. A full set of calculations was submitted to the customer without any internal review. When a relief expert was able to obtain the calculations for review, they found that pool fire was the governing case in almost all of the designs. The customer had to pay the contractor to update the calculations to redesign the relief systems for the remaining credible relief cases.

3.3 Contractor applies ASME Section I

A contractor sized relief valves per ASME Section I for a boiler, steam drum, and superheater. Two relief valves were sized for the steam drum and a third was sized for the superheater as specified directly in Section I [4]. However, the contractor used saturated steam conditions to size all three valves, failing to recognize that the superheater contains superheated steam. Significant superheating reduces the capacity of a given relief valve as compared to relief of saturated steam. Additionally, the contractor failed to evaluate the installation of inlet and outlet piping on the three relief valves, instead basing their sizing evaluation on an orifice calculation only. The only documentation of the relief system design was a table on the vessel drawing showing the results of the orifice area calculations.

Operations requested that the system be evaluated to determine if they could add a spool piece to the inlet of each relief valve to increase the height above the nearby platform to provide improved access for maintenance. The evaluation revealed that the superheater relief valve was undersized due to the original sizing at saturated steam conditions. It also revealed that all three valves had excessive backpressure with the installed discharge piping, which was never evaluated by the contractor. All three valves had to be replaced and the discharge piping redesigned to reach a safe discharge location.

IV. Understand all sources of material into your system

Heat and material balances and operating procedures focus on the normal, steady-state expected flows into and out of a unit operation. PHA teams, by comparison, necessarily focus on upset conditions including unintended and intermittent flows, reverse flow, start-up and shut-down conditions, and other means for introducing material where it is not normally expected to be. Small flows can be overlooked as a means for tank overfilling and overpressure, particularly under extended shutdown conditions, where flows are assumed to be zero. This can lead to a significant release of flammable or toxic material with the potential for a large process safety incident such as a hazardous vapor release or pool fire.

4.1 Seal flushes lead to tank overfilling

New pumps were installed in an existing process. The pump seals required flushes that needed to be directed to a safe location. An in-process tank was selected as the destination for the pump seal flushes, which were supplied from a solvent header at a pressure greater than the rating of the tank. The tank outlet valve was closed during operation leading to an increase in level in the tank and activation of the high level interlock, which shut off the process flows into the tank. Seal flushes continued to be directed into the tank since this source of liquid was not included in the high level interlock. A release of solvent occurred through the relief system on the tank. The relief system design had not been evaluated as part of the Management of Change process for the introduction of solvent flushes into the tank because they were not recognized as a potential source of overpressure. Fortunately, the flowrate was small enough that the relief system had sufficient capacity to relieve the liquid entering the tank from the seal flushes.

4.2 Thermal expansion valve bypasses level interlock

A new thermal expansion valve was added on a transfer line that had been experiencing gasket leaks due to heating of blocked-in liquid between transfers. Since the transfer line contained flammable solvent, the relief effluent had to be directed to a safe location. The transfer line was owned by one operating department and the destination tank was owned by a different operating department. The department that owned the transfer line was responsible for the new relief valve. It was decided to discharge the relief valve into the destination tank by tying the discharge piping into the existing inlet piping after the block valve. Under the procedures in place at the time, the installation of this new relief valve did not require formal approval by the operating department that owned the tank since it did not directly touch their equipment.

The tank experienced high level and the level interlock responded to close the inlet block valve. The transfer pump continued to run per normal operating practice. The thermal expansion relief valve lifted and allowed flow to continue into the tank since the relief valve discharge piping bypassed the level interlock. The operators were not monitoring the tank level because they knew it had already reached the high level trip point and the level interlock was activated. They did not understand that there could be an additional source of liquid flow into the tank through the leaking relief valve. The tank overfilled leading to a significant release of the solvent. The over-fill continued for several hours undetected by operations. This incident resulted in a procedure change which now requires both departments to approve changes to the system.

V. Upset conditions during shutdowns

Routine shutdowns are not typically considered to be a change. Detailed procedures are used to ensure that equipment is properly emptied, cleaned, and isolated so that maintenance can be conducted safely. However, changes that occur to the equipment need to be evaluated to determine if they can impact how the equipment should be safely shutdown. In preparation for an upcoming shutdown, the shutdown procedures should be reviewed and compared to changes that have been made since the previous shutdown to determine if the procedures should be updated.

5.1 Vessel damage due to vacuum from a sample pump

During a routine shutdown, a vessel was isolated using the existing lock-out/tag-out procedure. The procedure did not include shutting down the small sample pump used to capture an off-gas sample during operation. The pump continued to run during the shutdown. Three days into the shutdown, the vessel experienced damage due to vacuum.

5.2 Vessel overpressure due to plugging

A low pressure storage tank contains a molten material with a freezing point significantly above normal ambient temperatures (the material is a solid at room temperature). Steam is utilized to heat the tank contents (via an internal coil), the conservation vent (via a jacket), and the nitrogen supply (via tracing). Plugging in the conservation vent and/or nitrogen supply could lead to ineffective pressure and vacuum relief. During a shutdown of the associated manufacturing complex, the steam was isolated at the main header per the existing shutdown procedure. The shutdown procedure had not been updated when this tank was put into service for bulk storage of the molten material. Loss of heating on the nitrogen supply lead to freezing in the remote sense line. The nitrogen regulator responded by continuously flowing nitrogen into the tank to attempt to reach the set point. The tank pressure increased to the set pressure of the conservation vent over a 12-hour period and vapors were relieved through the conservation vent. Although some plugging was observed in the conservation vent due to loss of steam to the jacket, the vent functioned properly to protect the tank. However, if the shutdown had occurred during the winter, it is possible that the conservation vent could have been plugged completely, resulting in vessel failure due to overpressure by the nitrogen supply.

VI. Incident investigations with overly-narrow focus

Incident investigations are a critical part of a comprehensive safety program. The goal of any incident investigation should be to identify all the causes of the event. Near miss investigations are also essential to ensure that potential hazards are properly controlled to prevent a serious incident from occurring. All incident and near-miss investigations should look for potential process safety incidents in addition to personal safety issues. For PSM-covered processes, 29 CFR 1910.119(m) requires investigation of each incident that resulted in or could have resulted in a catastrophic release of a highly hazardous chemical [5]. One of the causes of an event or near miss could be hazards that were not previously identified in a hazard analysis. Another potential cause could be a change that was made without adequate risk assessment resulting in the introduction of a hazard that resulted in the incident. The results of all incident and near-miss investigations should be reviewed as part of the PHA revalidation process.

6.1 Failure to look for the process safety concerns

An incident occurred when unloading a truck trailer containing caustic soda. The incident investigation focused on the release of caustic which resulted in injuries due to exposure. A brief description of the incident was provided to the entire manufacturing site to promote learnings across multiple operating areas. This description included the statement that the operator had mistakenly connected the unloading line to a pump used to transfer acid. The operator recognized the mistake and the release of caustic occurred when the unloading line was disconnected. The original incident investigation did not address the potential process safety incident that could have resulted from unloading caustic into a tank containing acid. A process safety expert reading the brief description of the incident contacted the operating department to ask them to address the potential process safety concern.

VII. Loss of institutional knowledge – back to the future

Personnel changes may occur during the five-year PHA revalidation cycle that lead to the person responsible for implementing the team's recommendations not having attended a single PHA meeting. In some cases, multiple personnel changes occurring in a manufacturing area can lead to significant loss of institutional knowledge. It is essential to effectively document not only the recommendation but the reasons for that recommendation to ensure that essential process safety knowledge does not reside only with the members of the PHA team. This information must be accurately and concisely communicated to those responsible for implementation. Additionally, OSHA issued a letter of interpretation in April 2009 specifically to address management of organizational change with respect to the requirements for Management of Change in 29 CFR 1910.119(l) [5]. An MOC may be required to address personnel changes, including changes in staffing levels, staff experience, or contracting out tasks that directly impact PSM-covered processes.

A process to refine a thermally sensitive material produces two types of waste, one that is reactive and one that is non-reactive. The reactive waste contains the thermally sensitive material and a water-reactive material. In order to generate the non-reactive waste, the reactive waste is mixed with water in a controlled manner. This waste, once rendered non-reactive, is loaded into dumpsters for disposal. Due to capacity limitations in the waste recovery system, it became necessary to load some reactive waste into dumpsters as well. For many years, the same dumpsters had been used for both reactive and non-reactive waste streams. The wastes were not loaded into the dumpsters at the same time, as this had been recognized as a safety concern due to the reaction that could occur in the dumpster, and procedures had been written accordingly. The procedure called for an "empty" dumpster to be hooked up to the appropriate waste stream. However, dumpsters were returned from the waste disposal complex with a heel of material, which could be either reactive or non-reactive (water-containing).

Several near-miss incidents had occurred where dumpsters of the reactive waste were observed to be heating up. Temperature indication on the dumpsters showed that an exothermic reaction was occurring. Investigations concluded that the water-reactive material in the reactive waste was most likely reacting with water from the heel of non-reactive waste in the dumpster. Fortunately, the small amount of water in the heel limited the amount of heat generated due to the reaction and the dumpster contents did not reach the onset temperature of the thermally sensitive material.

A study team was formed to determine appropriate emergency response practices for a reacting dumpster. The team expanded its focus to evaluate appropriate practices for handling both wastes to prevent reaction in the dumpsters. This team included process safety experts, emergency responders, and representatives from the manufacturing area and the waste disposal complex. The team recommended that dedicated, clearly marked dumpsters be used for each waste and that those dumpsters have unique connections to ensure that they could not be connected to both waste streams. Recommendations included appropriate signage and color-coding on each set of dumpsters to facilitate placement of the correct dumpster at each waste loading location by the truck driver. These visual cues would also assist the operators at the waste disposal complex with identification of each waste stream. Both the operating area and the waste disposal complex updated their procedures accordingly. An emergency response plan was also developed.

Four years after the study team made its recommendations, the PHA revalidation was conducted for the process. The PHA study was updated to include the administrative controls that were in place to ensure that the two waste streams were segregated, including the use of dedicated dumpsters for each waste. The procedures were reviewed and it was noted that they adequately documented the operating practices. The staff engineer who served as the team leader for the

PHA revalidation was new to the operating area and had not been part of the study team. However, she was made aware of the concerns and recommendations through the PHA discussion. The operations manager was also new to the area. He did not serve as a member of the PHA team. A relief design review conducted as part of the PHA revalidation lead to a recommendation to generate a relief system evaluation for both sets of dumpsters. This relief evaluation was completed as part of the resolution phase for the PHA revalidation. In order to provide adequate emergency pressure relief for a reacting dumpster, it was recommended that the existing dumpsters used for the reactive waste be replaced with dumpsters with a significantly higher pressure rating.

The staff engineer started a project to acquire three new dumpsters with the recommended pressure rating; the existing dumpsters were determined to be acceptable for use with any other waste stream than the reactive waste. The staff engineer left the company prior to completion of the project. A new staff engineer, with no previous experience in the operating area, was tasked with completing the project and all other open PHA action items from the revalidation. Once the three new dumpsters were commissioned and put into service, the PHA action item was closed. Subsequently, a new operations manager, also with no previous experience in the operating area, was named.

Four years after the PHA revalidation, an incident occurred at the waste disposal complex. A dumpster containing reactive waste reacted, leading to a pressure increase and release of material through the rupture disk. The incident investigation uncovered that the three new dumpsters were being used for both reactive and non-reactive waste, which was a return to the practices that were in place prior to the formation of the study team. The dumpsters had proper signage for each waste stream, but the placarding was removable and interchangeable. Evaluation of the existing procedures revealed that only the environmental procedure had been updated at the time the new dumpsters were commissioned. It clearly stated that the new dumpsters were approved for both reactive and non-reactive waste. However, this practice was not supported by the current operating procedure, the emergency response plan, the PHA documentation, and the recommendations made by the study team.

The purpose of replacing the existing dumpsters with new dumpsters with a higher MAWP was to be able to provide adequate relief for runaway reaction. However, adequate relief for runaway reaction was never intended to make the practice of mixing reactive and non-reactive waste acceptable, as was evident in the PHA which documented the need for both adequate relief and the administrative controls recommended by the study team. This intent was not clearly understood by the current operations personnel, who were not in their positions at the time of the study team and/or the PHA revalidation. Lack of institutional knowledge of the history of incidents and safety concerns around the dumpsters lead to an assumption that the "new and improved" dumpsters superseded the administrative controls.

VIII. Conclusion

Process safety is essential to ensure continued safe operation of chemical facilities. All decisions and changes must be critically evaluated for potential hazards before implementation. Personnel involved with day-to-day operations must have sufficient process safety knowledge to initially evaluate changes and to request additional process safety expertise as needed. It is the responsibility of the employer to determine the level of process safety knowledge required for each job and to ensure that the people in those jobs get the training they need. Efforts to achieve zero injuries at chemical facilities must also include a commitment to preventing process safety incidents. Process safety and personal safety are both critical elements of a comprehensive safety program dedicated to being "All In" for Safety. Being "All In" means focusing on process safety "all the time".

IX. References

[1] Mize, Jennifer F., "Process Safety: From Concept to Commissioning", Proceedings of the 9th Global Congress on Process Safety, April 28 – May 1, 2013

[2] Mize, Jennifer F. and J. Wayne Chastain, " Instrumented Protective Systems for Distillation Operations", Proceedings of the 6th Global Congress on Process Safety, March 22-24, 2010

[3] American Society of Mechanical Engineers, Boiler and Pressure Vessel Code (BPVC), Section VIII, Division 1, Subsection A, Part UG

[4] American Society of Mechanical Engineers, Boiler and Pressure Vessel Code (BPVC), Section I

[5] Occupational Safety & Health Administration, U.S. Department of Labor, Occupational Safety and Health Standards: 1910.119, Process safety management of highly hazardous chemicals

Typical Elements of Process Safety Management

Flávio L B Diniz
Det Norske Veritas - Rua Sete de Setembro 111, 12º andar
Centro, Rio de Janeiro, RJ, 20050-006 - Brazil
Email: flavio.diniz@dnvgl.com

Nilda Maria Visco Vieira
Email: nilda.visco@dnvgl.com

Tatiana Barboza Coelho Cordeiro
Email: tatiana.cordeiro@dnvgl.com

Prepared for Presentation at
American Institute of Chemical Engineers
2014 Spring Meeting
10th Global Congress on Process Safety
New Orleans, LA
March 30 – April 2, 2014

UNPUBLISHED

Typical Elements of Process Safety Management

Flavio L B Diniz
Det Norske Veritas - Rua Sete de Setembro 111, 12° andar
Centro, Rio de Janeiro, RJ, 20050-006 - Brazil
Email: flavio.diniz@dnvgl.com

Nilda Maria Visco Vieira
Det Norske Veritas

Tatiana Cordeiro
Det Norske Veritas

Keywords: Management, process safety, management system

Abstract

This paper will present a comparison of key elements that will provide grants for companies to optimize their process safety management system, which is a great support tool to enhance company's performance.

The end result of this work will be the consolidation of the elements that comprise the typical composition of a process safety management system with cover information from several Brazilian and international legal requirements as well as guidelines and best practices for Brazilian and international technical-scientific entities. Among the sources consulted we can highlight: CCPS (Center for Chemical Process Safety), DNV (ISRS8), PETROBRAS, BRASKEM, ANP (National Petroleum Agency), CETESB/SP (Society of Environmental Sanitation Technology / São Paulo), FEPAM-RS (Foundation State Environmental Protection Henrique Luiz Roessler Rio Grande do Sul), OSHA (Occupational Safety and Health Administration) and API (American Petroleum Institute).

1. Introduction

It is possible to find several definitions for the process safety management, but according to the Center for Chemical Process Safety (CCPS), created by the American Institute of Chemical Engineers (AIChE) on March 25, 1985 after the Bhopal accident in India, process safety management is a management system that is focused on prevention, preparedness, mitigation, response and restoration of catastrophic releases of chemicals or energy from a process associated with an ease. Today CCPS has over 135 members companies.

With to aim to support companies in developing a process safety management appropriate, this paper presents a consolidation of the typical elements for the composition of a management system, from Brazilian and international legal requirements as well as guidelines or practices recommended by international technical and scientific entities.

Note: Do not add page numbers. Do not refer to page numbers when referencing different portions of the paper

2. Typical Requirements

Among the major systems of safety management processes employed nationally (Brazil) and internationally we highlighted the following references: [1] CCPS, [2] OSHA, [3] API, [4] ANP, [5] CETESB/SP, [6] FEPAM/RS, [7] DNV/ISRS8, [8] PETROBRAS and [9] BRASKEM.

2.1 CCPS

Center for Chemical Process Safety (CCPS) was created by the American Institute of Chemical Engineers (AIChE) on March 25, 1985 after the Bhopal accident in India, with the aim of contributing to the development of process safety in the chemical, pharmaceutical and oil industries. The CCPS brings together companies in the industry, government agencies, consultants and academics.

Those associated with the CCPS work at project subcommittees, defining and developing useful guidelines and that have practical applications in industry. The project topics cover a range of important areas for process industry and range from human factor issues for qualitative risk analysis to simulations to determine the vulnerability of major accidents.

CCPS and its members are committed to protecting employees, communities and the environment through the development of practical engineering and management to prevent or mitigate catastrophic releases of chemicals, hydrocarbons and other hazardous materials.

The management system suggested by CCPS consists of 20 elements listed below:

1. Process Safety Culture
2. Compliance with Standards
3. Process Safety Competency
4. Workforce Involvement
5. Stakeholder Outreach
6. Process Knowledge Management
7. Hazard Identification & Risk Management
8. Operating Procedures
9. Safe Work Practices
10. Asset Integrity & Reliability
11. Contractor Management
12. Training & Performance Assurance
13. Management of Change
14. Operational Readiness
15. Conduct of Operations
16. Emergency Management
17. Incident Investigation
18. Measures & Metrics
19. Auditing
20. Management Review & continuous Improvement

2.2 OSHA

OSHA is part of the United States Department of Labor. With the Occupational Safety and Health Act of 1970, Congress created the Occupational Safety and Health Administration (OSHA) to assure safe and healthful working conditions for working men and women by setting and enforcing standards and by providing training, outreach, education and assistance.

The OSH Act covers companies and their respective employees either directly through federal OSHA or through an OSHA state program that has been approved. State programs must meet or exceed federal OSHA standards for safety and health in the workplace.

OSHA has developed a Process Safety Management Guidelines for Compliance (OSHA3133 – 1994) with the major objective to prevent unwanted releases of hazardous chemicals especially into locations that could expose employees and others to serious hazards. This federal program introduces 14 requirements for process safety, listed below:

1. Employees
2. Information on process safety
3. Hazard analysis process
4. Operational Procedures
5. Training
6. Contracted
7. Review of pre - operational safety
8. Mechanical integrity
9. Hot Work Permit
10. Change management
11. Incident Investigation
12. Emergency Planning
13. Auditing
14. Confidentiality

2.3 API

American Petroleum Institute (API) is an association that was founded in the time of World War I. API members are dedicated to continuously improving the compatibility of its operations with the environment and economically developing energy resources to supply quality products and services to consumers.

The American Petroleum Institute (API) is the only national trade association that represents all aspects of America's oil and natural gas industry. American National Standards Institute (ANSI)/API RP 754 identifies leading and lagging process safety indicators.

The API provides a recommendation of process safety management system consists of 12 elements, listed below:

1. Leadership and Commitment
2. Safety and environment
3. Hazard Analysis
4. Change management
5. Operational Procedures
6. Safe work practices
7. Training
8. Ensuring the quality and mechanical integrity of critical equipment
9. Review of Pre –startup
10. Control and emergency response
11. Incident investigation
12. Audit security management program and environment

2.4 ANP

The Brazilian National Agency of Petroleum, Natural Gas and Biofuels (*Agência Nacional do Petróleo, Gás Natural e Biocombustíveis* - ANP) is the federal government agency linked to the Ministry of Mines and Energy responsible for the regulation of the oil sector.

In order to establish criteria for the conduct of exploration and production according to internationally recognized standards, ANP launched the operational safety regulation, prioritizing offshore facilities, due to the complexity and risks inherent to it.

The main objective of ANP supervision is to audit the Operational Safety Management System (OSMS) of offshore drilling and oil and natural gas production installations, and where effective implementation and compliance with the requirements established by resolution n^0 43/2007 is checked.

The 17 OSMS requirements are spread over safety management practices which are required in drilling rigs, production, stockpiling and transfer, such as:

1. Safety Culture, Commitment and Management Responsibility
2. Involvement of Personnel
3. Qualification, Training and Personnel Performance
4. Work Environment and Human Factors
5. Selection, Control and 3rd Party Management
6. Performance Monitoring and Continuous Improvement
7. Audits
8. Information and Documentation Management
9. Investigation of Incidents
10. Project, Construction, Installation and Deactivation
11. Critical Elements of Operational Safety
12. Identification and Risk Analysis
13. Mechanical Integrity
14. Planning and Management of Major Emergencies
15. Operational Procedures

16. Management of Change
17. Safe Work Safety and Procedure of Control for Special Activities

2.5 CETESB/SP

The Environmental Company of São Paulo (Companhia Ambiental do Estado de São Paulo - CETESB) is the State Government agency responsible for the control, supervision, monitoring and licensing of pollution-generating activities, with the primary concern of preserving and restoring the quality of water, air and soil.

The CETESB in addition to maintaining the function of inspecting agency and licensing of potentially polluting activities considered, also license activities involving cutting of vegetation and interventions in areas considered permanent, environmentally protected conservation.

The licensing process for companies which produce, store, use or transport hazardous substances, within the CETESB, requires the preparation of a Management System, as described at the standard Accidental risk from technological origin (P4.261) with the following elements:

1. Characterizations of the development and the surrounding
2. Identification of hazards
3. Study Review Risk Analysis or the identification of hazards
4. Operating procedures
5. Modifications Management
6. Maintenance and integrity assurance
7. Training of human resources
8. Investigation of incidents and accidents
9. Emergency Action Plan (EAP)
10. Auditing the Risk Management Program

In addition to the 10 elements presented, companies should also consider the Law 13.577 of July 8, 2009 of the State of São Paulo, which has additional requirements for protection against soil contamination by harmful changes, to be followed by the company in the development of a management system.

2.6 FEPAM/RS

State Environmental Protection Henrique Luis Roessler (Fundação Estadual de Proteção Ambiental Henrique Luis Roessler – FEPAM), is the institution responsible for environmental licensing in Rio Grande do Sul (Brazil) since 1999 FEPAM is linked to the Department of Environment - SEMA.

Similar to CETESB is São Paulo State, FEPAM is the institution responsible for environmental licensing in Rio Grande do Sul (State located in South of Brazil)

FEPAM developed an Handbook of Industrial Risk Analysis that provides a systematic for reference for internal procedures FEPAM in licensing activities and/or facilities capable of causing harm to people and/or the environment, external to the premises points, due to accidental releases of hazardous substances and/or energy in an uncontrolled manner, within a context of analysis of industrial hazards. At this document is possible to find some elements to establish a Risk Management Program for activities controlled and must contain at least the following elements of management:

1. Allocation of Responsibilities
2. Information on process safety
3. Risk Analysis
4. Operating procedures
5. Training
6. Modifications Management
7. Ensuring integrity of critical equipment and systems protection
8. Emergency Planning

2.7 DNV/ISRS8

DNV – Det Norske Veritas is a leading, independent provider of services for managing risk with a global presence. International Safety Rating System (ISRS) developed initially in 1978 by Frank Bird describes best practice in safety and sustainability management and represents over than 30 years of accumulated best practice experience in safety.

The ISRS management system has been implemented on thousands of sites worldwide and has become a global benchmark for excellence in safety management. This type of management system helps the companies to: ensure business operations are safe and sustainable; ensure systematic and effective risk management; save money by reducing accidents, process stoppages, and other loss events; provide assurance to stakeholders that external standards are being met; establish or confirm regulatory compliance; benchmark performance against industry peers and world class; highlight current strengths and weaknesses in the management system; identify any improvements required and monitor their implementation; establish optimized work processes using industry best practice; develop employee skills and foster the right workplace behaviors and culture; improve systems leading to accreditation in relevant certification standards and establish a single integrated management system to drive continual improvement.

Nowadays, the ISRS is in the version 8 (including PSM issues) is structured based on 15 elements:

1. Leadership
2. Planning and Administration
3. Risk Evaluation
4. Human Resources
5. Compliance Assurance
6. Project Management
7. Training and Competences
8. Communication and Promotion
9. Risk Control
10. Asset Management

11. Contractor Management and Purchasing
12. Emergency Preparedness
13. Learning from Events
14. Risk Monitoring
15. Results and Review.

2.8 PETROBRAS

The biggest company in Brazil, PETROBRAS has around of 100 production platforms, 16 refineries, 30,000 kilometers of pipelines and more than 6,000 service stations. 7th biggest energy company in the world. The company owns oil refineries, oil tankers, and is a major distributor of oil products. PETROBRAS is a world leader in development of advanced technology from deep-water and ultra-deep water oil production.

The management system of Safety, Environment and Health of PETROBRAS, aims to improve their results in these areas. Guidelines and requirements are elements beacon for managers and their teams operate and properly conduct issues related to SMS.

The safety management system is composed of 15 HSE guidelines listed below:
Guideline 1 – Leadership and Accountability
Guideline 2 – Regulatory compliance
Guideline 3 – Risk evaluation and management
Guideline 4 – New Projects
Guideline 5 – Operation and maintenance
Guideline 6 – Management of change
Guideline 7 – Acquisition of goods and services
Guideline 8 – Training, education and awareness
Guideline 9 – Information management
Guideline 10 – Communication
Guideline 11 – Contingency
Guideline 12 – Community relations
Guideline 13 – Accidents and incidents analysis
Guideline 14 – Product stewardship
Guideline 15 – Assessment and continuous improvement

2.9 BRASKEM

Braskem is the Americas' top thermoplastic resins producer. With 36 industrial plants spread across Brazil, United States and Germany, the company produces over 16 million tons of thermoplastic resins and other petrochemicals per year. Braskem is the world's leading biopolymers producer with its 200,000 tons Green PE plant that produces polyethylene from sugarcane-based ethanol.

Braskem in order to standardize procedures across all its units, consolidated its practices and strategies in the areas of EHS, integrating them into a program entitled "SEMPRE" Excellence in HSE. "SEMPRE" is supported on sixteen (16) Strategic Elements (*EE's – Elementos Estratégicos*) which are a set of practices related to the care of Quality, Health, Safety and Environment and principles within the HSE Integrated Management System comprising:

EE 01. Commitment and Leadership
EE 02. Governance and Organizational Synergy
EE 03. HSE Identification and Risk Analysis, Aspects and Impacts
EE 04. Standards and Procedures
EE 05. Products and Processes Information
EE 06. Behavioral and Cultural Development
EE 07. Competencies and Skills Management
EE 08. Management of Contractors
EE 09. Communication with Stakeholders
EE 10. Projects and Developments
EE 11. Facilities and Equipment Integrity
EE 12. Aspects and Impacts of Product Management
EE 13. Management of Change
EE 14. Crisis and Emergency Management
EE 15. Lessons Learned
EE 16. HSE Verification

3. Comparison of Requirements

A comparison among the elements/guidelines of 09 entities mentioned above is presented below. In order to enhance the understanding, the requirements of process safety management systems will be compared with the 20 elements defined by CCPS.

Tables 1 to 4 show comparisons divided in 04 groups defined by CCPS:

- Commit to process safety
- Understand hazard & risk
- Manage risk
- Learn from experience

Table 1 – Group I – Commit to Process Safety

GROUP I – COMMIT TO PROCESS SAFETY								
CCPS	**DNV (ISRS8)**	**PETROBRAS (Guideline)**	**ANP (Practice Management)**	**CETESB (Element)**	**FEPAM (Element)**	**OSHA (Requirement)**	**API (Element)**	**BRASKEM (EE)**
Process Safety Culture	Process 1 Leadership Process 2 Planning and Administration Process 5 Compliance assurance Process 8 Communication and Promotion Process 14 Risk Monitoring.	D1 Leadership and Accountability D5 Operation and Maintenance D8 Training, Education and awareness D9 Information Management D10 Communication D12 Community Relations D13 Accidents and Incidents Analysis D15 Assessment and continuous improvement	Nº 1: Safety Culture, Commitment and Management Responsibility Nº 2: Involvement of Personnel	Law Nº 13.577 - July, 2009 / Article 12	1. Allocation of Responsibilities	Employee Involvement	1. Leadership & Commitment 2. Safety and Environmental Information	1. Leadership and Commitment 2. Governance and Organizational Synergy 6. Development of Behavior and Culture
Compliance with Standards	Process 5 Compliance assurance	D2 Planning and Administration	Nº 6: Performance Monitoring and Continuous Improvement	Law Nº 13.577 - July, 2009 / Articles 26, 28	2. Information on process safety	Compliance Audits	2. Safety and Environmental Information	1. Leadership and Commitment 2. Governance and Organizational Synergy
Process Safety Competency	Process 5 Compliance assurance Process 7 Training and Competence Process 8 Communication and Promotion	D1 Leadership and Accountability D3 Risk Evaluation and Management D5 Operation and Maintenance D6 Management of change D7 Acquisition of goods and services D8 Training, Education and Awareness D9 Information Management D10 Communication D15 Assessment and continuous improvement	Nº3:Qualification, Training and Personnel Performance Nº 8: Information and Documentation Management Nº 4: Work Environment and Human Factors	4. Identification of hazards 9.1.2. Review of process risks.	5.Training	Process Safety Information Employee Training	7. Training	7. Competencies and Skills Management

Note: Do not add page numbers. Do not refer to page numbers when referencing different portions of the paper

GROUP I – COMMIT TO PROCESS SAFETY								
CCPS	**DNV (ISRS8)**	**PETROBRAS (Guideline)**	**ANP (Practice Management)**	**CETESB (Element)**	**FEPAM (Element)**	**OSHA (Requirement)**	**API (Element)**	**BRASKEM (EE)**
Workforce Involvement	Process 8 Communication and Promotion	D10 Communication	Nº 2: Involvement of Personnel	9.1.1 Process safety information 9.1.6 Training of human resources	5. Training 6. Management of change	Employee Involvement.	7. Training 8 Assurance of Quality and Mechanical Integrity of Critical Equipment	1. Leadership and Commitment 2. Governance and Organizational Synergy 6. Development of Behavior and Culture
Stakeholder Engagement	Process 1 Leadership Process 15 Results and Review	D12 Community Relations	Nº 8: Information and Documentation Management	8.1 Social risk	5. Training 6. Management of change	Contractors	3.2.2 Hazards Analysis for Mobile Offshore Units 6.4 Contractor Selection 8. Assurance of Quality and Mechanical Integrity of Critical Equipment	2. Governance and Organizational Synergy 9. Communication with Stakeholders

The first group (Commit to Process Safety) reflects the transparency and the commitment of high level management in matters related to process safety.

Note that all entities demonstrate through elements/guidelines that commitment. This group shows the sustainability of the implementation of the process safety management, once require the involvement of leadership, workforce involvement as well as stakeholder engagement.

Table 2 – Group II – Understand Hazard & Risk

GROUP II – UNDERSTAND HAZARDS & RISK								
CCPS	**DNV (ISRS8)**	**PETROBRAS (Guideline)**	**ANP (Practice Management)**	**CETESB (Element)**	**FEPAM (Element)**	**OSHA (Requirement)**	**API (Element)**	**BRASKEM (EE)**
Process Safety Information	Process 3 Risk Evaluation; Process 9 Risk Control Process 8 Communication and Promotion	D3 Risk Evaluation and Management D5 Operation and Maintenance	Nº 17: Safe Work Safety and Proced. Of Control for Special Activities	9.1.2. Review of process risks	2.Information about process safety	Process Hazard Analysis Nonroutine Work Authorizations Operating Procedures	3. Hazards analysis 6. Safe Work Practices	5. Products and Processes Information
Hazard Identification & Risk Analysis	Process 3 Risk Evaluation Process 9 Risk Control Process 14 Risk Monitoring	D3 Risk Evaluation and Management;	Nº 12: Identification and Risk Analysis	4. Identification of hazards	3.Risk Analysis	Process Hazard Analysis Nonroutine Work Authorizations Operating Procedures	3. Hazards analysis 6. Safe Work Practices	3. HSE Identification and Risk Analysis, Aspects and Impacts

The second group (Understand Hazard & Risk) reflects the importance of understanding the existent hazards and risk assessment. A comprehensive risk analysis shall be conducted by specialist and formal procedures need to be written.

Similar to Group I, all entities have elements to cover these issues. Particularly, the element "Hazard Identification & Risk Analysis" has specific guideline in all entities, which reinforce the importance of hazard identification. To better control and management of risk, it's necessary firstly to understand/know the hazards in the process plant.

In most of entities, the risk assessment process has a period for revalidation. Normally every 5 years is the frequency of revalidation. In case of changes, a risk assessment is required, to ensure that all risks continue under control (acceptable level).

Table 3 – Group III – Manage Risk

GROUP III – MANAGE RISK								
CCPS	**DNV (ISRS8)**	**PETROBRAS (Guideline)**	**ANP (Management Practice)**	**CETESB (Element)**	**FEPAM (Element)**	**OSHA (Requirement)**	**API / SMES (Element)**	**BRASKEM (EE)**
Operating Procedures	Process 5 Compliance assurance Process 7 Training and Competence Process 8 Communication and Promotion	D1 Leadership and Accountability D3 Risk Evaluation and Management D5 Operation and Maintenance D7 Acquisition of goods and services D8 Training, Education and awareness D9 Information Management D10 Communication D15 Assessment and continuous improvement	Nº 3 Qualification, Training and Personnel Performance Nº 4 Work Environment and Human Factors Nº 8 Information and Documentation Management	4. Identification of hazards 9.1.2. Review of process risks	5.Training	Process Safety Information Employee Training	7. Training	04 Standards and Procedures 05 Information on Products and Processes
Safe Work Practices	Process 3 Risk Evaluation Process 9 Risk Control	D3 Risk Evaluation and Management D5 Operation and Maintenance	Nº 12 Identification and Risk Analysis Nº 17 Safe Work Safety and Proced. of Control for Special Activities	4. Identification of hazards 9.1.2. Review of process risks.	3.Risk Analysis	Process Hazard Analysis Nonroutine Work Authorizations	3. Hazards analysis 6. Safe Work Practices	04 Standards and Procedures
Asset Integrity & Reliability	Process 7 Training and Competence Process 8 Communication and Promotion	D1 Leadership and Accountability D3 Risk Evaluation and Management D5 Operation and Maintenance D8 Training, Education and awareness D10 Communication	Nº 15 Operational Procedures Nº 10 Project, Construction, Installation and Deactivation	4. Identification of hazards; 9.1.2. Review of process risks. 9.1.3 Management of change 9.1.7 Incident Investigation	4.Operating Procedure	Operating Procedures; Pre-Startup Safety Review.	5. Operating procedures	11 Facilities and Equipment Integrity
Contractor Management	Process 10 Asset Management	D5 Operation and Maintenance; D6 Management of change; D7 Acquisition of goods and services D8 Training, Education and awareness	Nº 5 Selection, Control and 3rd Party Management Nº 11 Critical Elements of Operational Safety Nº 13 Mechanical Integrity	9.1.4 Maint. and ensuring the integrity of critical systems 9.1.6 Training of human resources; 9.2 Risk Management Program	7. Assurance integrity of critical equipment and protective systems	Mechanical Integrity of Equipment	8. Assurance of quality and mechanical integrity of critical	08 Managing of Contractors

GROUP III – MANAGE RISK								
CCPS	**DNV (ISRS8)**	**PETROBRAS (Guideline)**	**ANP (Management Practice)**	**CETESB (Element)**	**FEPAM (Element)**	**OSHA (Requirement)**	**API / SMES (Element)**	**BRASKEM (EE)**
Training	Process 11 Contractor Management and Purchasing	D8 Training, Education and awareness D7 Acquisition of goods and services	Nº 3 Qualification, Training and Personnel Performance Nº 6 Performance Monitoring and Continuous Improvement	9.1 Risk Management Program I	4.Operating Procedure	Contractors	1.Leadership & Commitment 6. Safe Work Practices	04 Standards and Procedures 05 Information on Products and Processes 7. Competencies and Skills Management
Management of Change	Process 4 Human Resources.	D6 Management of change	Nº 16 Management of change	9.1.3 Modifications Management	6. Modifications Management	Managing Change	4. Management of change	13 Management of Change
Operational Readiness	Process 12 Emergency Preparedness	D11 Contingency D 13 Accidents and incidents analysis	Nº 14 Planning and Management of Major Emergencies	9.1.8 Emergency Action Plan	8.Emergency Planning	Emergency Preparedness	10. Emergency response and control	14 Crisis Management and Emergencies
Conduct of Operations	Process 2 Planning and Administration Process 7 Training and Competence Process 14 Risk Monitoring	D1 Leadership and Accountability D3 Risk Evaluation and Management D5 Operation and Maintenance D10 Communication	Nº 11 Critical Elements of Operational Safety Nº 13 Mechanical Integrity Nº 17 Safe Work Safety and Proced. of Control for Special Activities	9.1.8 Emergency Action Plan	4. Operating Procedure. 5. Training	Written Procedures Quality Assurance Nonroutine Work Authorizations	1.Leadership & Commitment 6. Safe Work Practices	04 Standards and Procedures 05 Information on Products and Processes
Emergency Management	Process12 Emergency Preparedness	D11 Contingency D 13 Accidents and incidents analysis	Nº 14 Planning and Management of Major Emergencies	9.1.8 Emergency Action Plan	8.Emergency Planning	Emergency Preparedness	10. Emergency response and control	14 Crisis Management and Emergencies

The Group III (Manage Risk) is the biggest group (9 elements). This group represents the accountability, i.e., defines the responsibility for manage the risk identified.

In this group is worth to highlight the element "Management of Change". Definitely, this element is very important in terms of process safety management. Most of the major process accidents occurred has a connection with a fail in this element.

Another relevant point is the element related to "Emergency Management". A comprehensive Emergency Response Plan shall be implemented to be reduced the consequence of serious accidents.

Table 4 – Group IV – Learn from Experience

GROUP IV – LEARN FROM EXPERIENCE								
CCPS	**DNV (ISRS8)**	**PETROBRAS (Guideline)**	**ANP (Management Practice)**	**CETESB (Element)**	**FEPAM (Element)**	**OSHA (Requirement)**	**API / SMES (Element)**	**BRASKEM (EE)**
Incident Investigation	Process 13 Learning from Events	D 13 Accidents and incidents analysis D11 Contingency	Nº 9 Investigation of Incidents	9.1.7 Incident Investigation	--	Incident Investigation	11. Investigation of incidents	15 Learning with deviations and Losses
Measurement & Metrics	Process 15 Results and Review	D1 Leadership and Accountability D9 Information Management D15 Assessment and continuous improvement	Nº 6 Performance Monitoring and Continuous Improvement	9.1.9 Audits	--	Compliance Audits	12. Audit of safety and environmental manag. program	16 HSE Verification
Auditing	Process 14 Risk Monitoring Process 15 Results and Review	D1 Leadership and Accountability D15 Assessment and continuous improvement	Nº 7 Audits	9.1.9 Audits	5.3 Requirements for obtaining the Operation license (LO)	Compliance Audits	12. Audit of safety and environmental manag.program	16 HSE Verification
Management Review & Continuous Improvement	Process 14 Risk Monitoring Process 15 Results and Review	D1 Leadership and Accountability D15 Assessment and continuous improvement	Nº 6 Performance Monitoring and Continuous Improvement	9.1.9 Audits	--	Compliance Audits	12. Audit of safety and environmental manag. program	16 HSE Verification

The fourth and the last group (Learn from Experience) represents the closing of PDCA cycle. This group reflects the commitment to implement and sustain the process safety management.

In this group there is a key element related to "Measurement & Metrics" - KPI – Key Performance Indicators. The most of entities reinforce the importance of this element. However there is a lack of definition of international KPIs focused in process safety. In order to create a common standard and promote sharing information related to process safety, some KPIs shall be defined and adopted by all entities.

4 Conclusion

Analyzing all references mentioned on this paper, there is similarity in almost all process safety management systems. The amount of elements/guidelines can be different from one system to another, however the content is practically the same.

In CCPS reference, there isn't one specific requirement related to process safety applied to new projects. However some others references, for instance ANP, PETROBRAS, BRASKEM, this issue (new project) is explicit. The authors consider highly relevant include in the process safety management the assessment in the design phase, i.e. several aspects can be identified and solved in the project phase, becoming the operational phase more profitable and safer.

Another factor worthy of note is the element focused on management of change. Definitely this element, present in all process safety systems, is actually very relevant for robust and complete process safety system.

In order to create common standard and promotes sharing information related to process safety management, some reactive and proactive KPIs – Key Performance Indicators shall be defined and adopted by all entities. The adoption of these indicators facilitates the process of continuous improvement, provides grants and support decision making by senior management of companies.

5. References

[1] CCPS. "Guidelines for Risk based process safety", American institute of Chemical Engineers, New York, NY. 2007.

[2] OSHA Occupational Safety and Health Administration, United States Department of Labor, https://www.osha.gov/Publications/osha3133.html (accessed Sept.20,2013)

[3] API. (American Petroleum Institute) Process Safety Performance Indicators for the Refining and Petrochemical Industries. ANSI/API RECOMMENDED PRACTICE 754 FIRST EDITION, APRIL 2010.

[4] ANP - Agência Nacional do Petróleo, Gás Natural e Biocombustíveis www.anp.gov.br (accessed Oct.10,2013)

[5] CETESB – Companhia Ambiental do Estado de São Paulo, www.cetesb.sp.gov.br (accessed Sept. 08,2013)

[6] FEPAM-RS - Fundação Estadual de Proteção Ambiental Henrique Luiz Roessler do Estado do RS, http://www.fepam.rs.gov.br/central/formularios/arq/manual_risco.pdf (accessed Sept.05,2013)

[7] International System Rating System (ISRS8). DNV Det Norske Veritas .2009.

[8] PETROBRAS – Petróleo Brasileiro S.A. - www.petrobras.com.br/pt/meio-ambiente-e-sociedade/preservando-meio-ambiente/downloads/pdf/diretrizes_seguranca.pdf. (accessed Oct.09,2013)

[9] BRASKEM . "Manual de Saúde, Segurança e Meio Ambiente", Bahia, BA. 2012.

Note: Do not add page numbers. Do not refer to page numbers when referencing different portions of the paper

Safety Leadership and Implementation- Independent of Culture

Kumar (Chris) Israni , Sarah Acton and Tianherng (Joshua) Yang
ERM
840 West Sam Houston Parkway North, Suite 600
Houston, TX 77024
sarah.acton@erm.com

Prepared for Presentation at
American Institute of Chemical Engineers
2014 Spring Meeting
10th Global Congress on Process Safety
New Orleans, LA
March 30 – April 2, 2014

UNPUBLISHED

Safety Leadership and Implementation- Independent of Culture

Kumar (Chris) Israni , Sarah Acton and Tianherng (Joshua) Yang
ERM
840 West Sam Houston Parkway North,
Suite 600
Houston, TX 77024
Sarah.acton@erm.com

Keywords: safety, culture, leadership, diverse, behavior, workforce,

Abstract

Multiculturalism has become abundant in today's workforce. When industries are diverse and multicultural in nature, innovation thrives and encourages the newest industry practices and wisdoms. Production almost always benefits from this type of environment; however, cultural differences within a workforce can initially complicate the process of creating and implementing a common safety culture. What may seem strange in one person's culture may be normal in another person's. A few of these potential conflicting differences are:

- Contractor companies may have safety rules that are not in alignment with those of the managing company.
- Mandatory safety rules could be considered suggestions or guidelines in some cultures.
- Certain groups may encourage risky behavior because not doing so is seen as socially undesirable.
- Symbols and colors on warning signs might have multiple meanings depending on individual cultures.
- Selected cultures may follow very stringent, non-flexible hierarchical structures that are not very open to change

Several leadership concepts preach and practice the idea of workers *owning* safety as common practice. However, in order for that to happen, and in order to best develop safety practices in the workforce, the respective management needs to first develop an environment based on the understanding of the different rationales and expectations of a multi-cultural workforce. Management can encourage the workers to "own" safe behaviors and empower employees to take action when they witness unsafe behaviors. Educating employees about off duty activities that can lead to unsafe acts at work is also valuable in the management of risk in the workplace. This paper will discuss how safety leadership can consider cultural differences when implementing a safety plan for a diverse workforce.

1. Introduction

The word "culture" can have different meanings depending upon the context of the conversation. The basic definition, according Mirriam-Webster [1], is:

- "the beliefs, customs, arts, etc., of a particular society, group, place, or time
- a particular society that has its own beliefs, ways of life, art, etc.
- a way of thinking, behaving, or working that exists in a place or organization (such as a business)"

The workforce within a company can exhibit all three definitions as each person has their own culture based on their own past experiences. However, culture is more than just one's nationality or one's religious beliefs. In another sense, culture can be defined as simply, "the way things are." That sounds simpler than what it means though because obviously, the way things are in the North America is very different from the way things are in Asia. The way things are in Africa is very different from the way things are in Europe and South America.

Culture can range from "my mother taught me to always throw a pinch of salt over my shoulder for good luck" to "I learned how to do it this way in college" to "we get a flu shot every year." These differences exist both on small and large scales which all affect the way a person acts on a day to day basis, and consequently in the workplace. The primary thing that keeps our cultures intact and consistent is that they are all founded on the same common belief system; we all *believe* that our cultures are the right ways of doing things; therefore we do not question them. The mother believes in a ritual's benefit, the student believes his teacher is correct, the patient believes the medicine will prevent disease. When interacting with people of the same culture, the established belief system is generally unproblematic; however when people from different cultures start mixing into the equation, everyone needs to take a step back and consider all of the different cultures that are present.

Similar to this, the word "safety" can have different meanings depending upon the context of the conversation. The basic definition, according Mirriam-Webster [2], is:

- freedom from harm or danger : the state of being safe
- the state of not being dangerous or harmful
- a place that is free from harm or danger : a safe place

The workforce within a company can exhibit various approaches to the concept of safety, whether it is willingly embracing this concept or adopting it with resistance. Irrespective of the nature of this acceptance, every company wants its workforce to come to work all intact and also to be able to return to their families all intact. All the while, the integrity of the environment and the wellbeing of the dwellers around the company premises should also be given the respect and credit due.

This article will help show that despite a workforce's differences in cultures, management can implement a unilateral safety management system which establishes a common belief system that everyone will naturally become a part of.

2. Differences in Culture

2.1 Global Problem

Many of today's oil and gas companies have become globally developed and have assets in nearly all parts of the world. Due to such an established global presence, a company's workforce is often multi-national and multi-cultural in nature. The cultural differences within such a diverse workforce should be considered during the development of a company's mission statement and overarching policies and procedures to ensure that the company runs safely and smoothly.

Cultural diversity opens the door for productivity and efficiency, especially when individuals share their knowledge and practices from different cultures and past experiences. Management can assess methodologies and take the best practices from different regions to decrease downtime and improve workflow.

Sometimes, however, cultural differences can clash and create barriers to moving forward which in turn decreases productivity and efficiency. Differences in culture can be a challenge especially while creating and implementing common safety values amongst a global workforce. A couple conflicting differences are:

- Between same kind of workers from different national or religious cultures
- Between different career types who come together to work on the same project

2.1.1 Clash between different national cultures

When blending safety cultures, it is important to understand foundations for beliefs. While many foundations for beliefs are passed down from ancestors and family traditions, some foundations for beliefs can be established by the policies of a previous employer; for example, subcontracting companies with safety rules that are not in alignment with those of the managing company. Symbols and colors on warning signs might have multiple meanings depending on individual cultures between companies.

Rules regarding safety could be misunderstood as suggestions rather than mandatory policies in some cultures. In many of these situations, the social norm is often to break the rules. Risky behavior is commonly encouraged in social settings because it is seen as macho or otherwise socially desirable. Other cultures may follow very stringent, non-flexible hierarchical structures that have been followed for generations and are not very open to change.

2.1.2 Clash between work ethic or job climates

Communication issues can exist, in addition to differing expectations of respect or acceptance of opinions between different types of workers who must collaborate on the same project. A young college graduate, for example, my expect a level of appreciation or respect due to his education that his coworkers feel may not be earned yet. Certain items may be called one thing in the office when drawing up drafts, but operations may refer to the item in completely different terms. All of these items should be strongly taken into consideration when dealing with a multicultural workplace.

2.2 Testimonies of Workplace Cultural Differences

Below are three testimonies of how work culture can differ depending on various factors. The difference can occur between different industries, different countries, or different languages.

2.2.1 Kumar's Example 1

Observation – "One of my earlier assignments in the late 90's was at a chemical manufacturing unit, in the western region of India. As for some background information on this facility, the processes and equipment used were not abreast of the technological advances with times. There was a lack of automation at this facility which meant that people had to handle all of the process materials. Understanding these limitations, the corporation would regularly provide feedback on potential hazards and the required mitigation such as facemasks, safety shoes, gloves and flame resistant clothing (FRC). On the daily operator rounds in the plant, I would observe that the majority of the workforce would not use facemasks while handling powders, or adorn the appropriate FRC. Their reasons were that it is too hot to use both the face masks and FRC. Dust collectors were provided, but they were deemed to be very noisy so they remained shut. Hearing protection was provided but was not used, as the workforce considered them as a nuisance. These folks repeatedly made visits to the in-house health clinic, complaining of congestion and discomfort in the ears however management had accepted that as a tolerable behavior."

Conclusion – The dangers of handling powders and noisy equipment exist in various chemical processes. These dangers are independent of location or region therefore the level of danger that exists is independent of location or region. But a major hindrance is the people's perception of safety. In this observation, it was clear that the concept of safety was not well embraced by the workforce, nor was it insisted upon by the management. Based on the author's several years of experience in the region, he had noticed similar behaviors towards safety of various workforce teams.

2.2.2 Kumar's Example 2

Observation – "While facilitating a Hazard and Operability Analysis (HAZOP) at an oil & gas site in Alberta, Canada, one of the items within the scope of this workshop was a team walk down. During the walk down, we were reviewing various release scenarios that led to pressure safety valve (PSV) lifting. In one specific case of a toxic release, it was noted that the PSV was not relieving to a safe location. This was noted only during the walk down, because during the HAZOP workshop, the team mentioned that the venting was at a safe location. One of the team members, who was also a new recruit to the facility and of the same culture as Kumar, mentioned that that was ok, and that in the worst case scenario we may lose a life, although it was very unlikely since the unit was out in the wilderness".

Conclusion – As a facilitator, Kumar had to explain to him the importance of human life and why the venting location had to be altered. Having the same cultural background as this team member made it easier for Kumar to explain the differences in perception in the region where the team member had just come from, in comparison the region of the plant site.

These conversations can sometimes be sensitive, but they need to occur as the end goal is always to protect human lives and the environment.

2.2.3 Sarah's Example

Observation – "As a Human Factors professional, I am dedicated to making the work environment safer and more fitted to the individual user. I focus on productivity and ease of use, but also on policy and procedures to make a person's job less complicated and as "fool-proof" as possible. When working on a railroad project, my colleagues and I discovered an interesting situation that seemed to be present amongst many companies. The policy book, or rule book, was quite thick and extensive. They were a combination of federal regulations and interpretations by the company as well as other company rules for work procedures. The problem, other than the extensive amount of rules to follow, was that the workers complained that they could not physically do their jobs if they followed all the rules. The management's expectation was that the workers had to break the rules in order to finish the jobs. However, if someone got hurt while breaking a rule, they did not qualify for workman's compensation and had to take leave without pay."

Conclusion – The carryover issues to other jobs will be that "rules" are meant, or expected, to be broken. Rules in these types of industries are almost always in place for safety reasons. When a person is used to working in one culture and moves into another, his habits may not change immediately which could result in accidents. In the first situation, someone's colleagues may be helping or watching out for him while he is breaking the rules, ready to assist if an accident occurs. When he moves to a new environment, his colleagues will not be expecting him to break rules and therefore are not prepared for the accident to occur.

The overarching issue in this scenario was improper prioritization. The management should have reworked the rules and regulations and found ways to change the working environment where people are not expected to break rules in order to successfully complete their jobs.

2.2.4 Tianherng's Example

Observation – "While working in Malaysia, I discovered a different work ethic and rule set than I worked with in the US. I was a relatively young employee there on transfer for a short period of time within the same company. I worked several weeks without much interaction from the others in my group. I wanted to be social and interact, but everyone kept to themselves. I finally discovered that unknowingly, I was sitting in a Sr. level desk and everyone was keeping distance from me out of respect.

Conclusion– The key factor here is to maintain an understanding of regional customs in the workplace. This concept of respect, essentially also transmits further to the fact that in certain regions, in spite of noticing a huge safety faux pas amongst peers and supervisors, employees/observers are hesitant to bring up these to the attention of the appropriate authorities. Reasons for such behaviors are not only due to fear of any backlash, but also due to fear of violating cultural respect limits.

2.3 *Creating a Common Work Culture*

2.3.1 The need for constructive criticism

Various corporations have rolled out behavior based safety programs, where the focus is aimed at human mannerisms, noted through observations. The takeaway message from such programs is to emphasize on following safe practices, and then to communicate this information in a positive, non-intimidating and educative manner [6].

While this practice is more visible in fieldwork and while its impacts have positive ramifications to Health, Safety and Environment (HSE) programs, the same concept is successful in the process safety management realm as well.

2.3.2 How management can help cross the cultural boundaries

The role of management is very crucial to the success of the company's safety management system. Figure 1 illustrates this in what is called the Bradley curve [3].

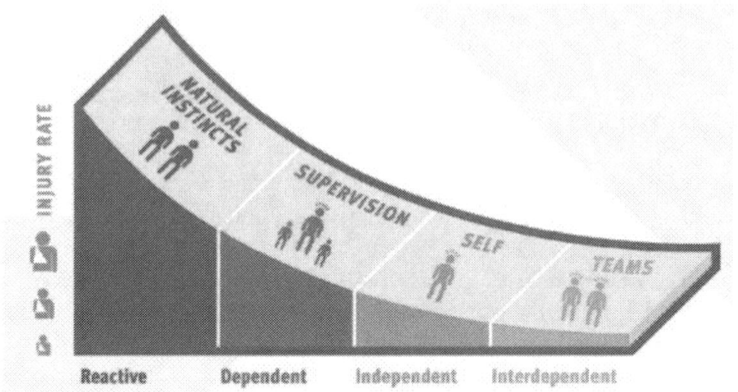

Figure 1 Bradley Curve

The right side of this curve indicates the independent and interdependent zones, which lead towards reducing the injury rate. It should be noted that these zones are highly individual based and will be effective when individuals adopt this safety understanding. One of the major factors to bringing this about would be to embrace safety culture universally and independent of individual's regional culture. According to a publication about safety leadership, the company's common safety culture needs to be "a reporting culture, a just culture, a flexible culture, and a learning culture" [4].

While steering towards this common safety culture, it is pertinent for organizations to have well defined indicators for safety culture [5]. This can be accomplished through adhering to several concepts:

- Communication – Actively sharing and reminding about safety in the workplace. Developing a culture means sharing information in effective ways that help the employees take ownership of the culture. Adequate communication leads to have a just culture as well in that employees are aware of new
- Empowerment– The "Stop Work Authority" is becoming a popular program amongst safety critical industries. Some companies have slogans around it such as "when in doubt, shut it down." The basic premise is that the company wants to convey to the employee that they have the authority and sometimes the responsibility to stop work without fear of repercussion.
- Lessons Learned– Lessons learned is a way of reporting and spreading the knowledge to others who may be asked to do the same job. This concept has been made into regulation in certain parts of the Oil and Gas world, such as control room management operations. The benefits of Lessons Learned is that the next person performing a similar task to the one being shared can learn from previous attempts and adjust behaviors accordingly
- Automation– Making the right decision in high stress situations can be a difficult challenge. Part of the safety culture's philosophy should be to improve decision

making capabilities through automation of equipment. Sometimes the addition of signage and periodic refresher training can help the workers

- "Owning" Safety– Many employees do not practice safe actions at home because there are no work rules or regulations guiding them to do so. However, instilling employees with a commitment of safety both in and outside of work is the ultimate success in a safety culture. If employees can make the connection to be safe because they have families and friends depending on them, it will be a more powerful motivator than obedience to the company's rules. If each person "owns" his or her actions and has a personal commitment to be safe then they will carry the attitude to non-work related activities as well.

3. Conclusion

From the information and testimonies that this report has provided, it is evident that one of the biggest challenges in the workplace is dealing with human interaction. Especially when diversity is taken into account, both in terms of culture/origin and experience levels and perceptions, communicating safe habits can sometimes be more challenging that it sounds. While a diverse workforce is essential to productivity and innovation, it may also lead to multiple points of view within an industry, especially when it comes to safety issues – occupational or process.

Creating a common safety culture that "sticks" and becomes socially established takes time and effort. It will not happen overnight, but it is achievable with persistence. Leaders should motivate and inspire their workforce by committing to a continual campaign for safety education and reminders rather than one time training [4].

The respective managements need to take into account cultural differences and work towards understanding the different rationales and expectations of a multi-cultural workforce in order to best develop safety practices. Management can consider cultural differences when implementing a safety plan for a multi-national workforce. Management should also encourage to "own" safe behaviors and empowering employees to take action when they see unsafe behaviors.

4. References

[1] "culture." *Merriam-Webster.com.* 2014. http://www.merriam-webster.com (January 17, 2011).

[2] "safety." *Merriam-Webster.com.* 2014.. http://www.merriam-webster.com (January 17, 2011).

[3] Dupont Bradley Curve Infographic. Dupont.com. 2014. http://www.dupont.com/products-and-services/consulting-services-process-technologies/articles/bradley-curve-infographic.html (January 17, 2014)

[4] Ruchlin HS, Dubbs NL, Callahan MA. The role of leadership in instilling a culture of patient safety: lessons from the literature. *Journal of Healthcare Management* 2004 Jan-Feb;49(1):47-58; discussion 58-9

[5] Research Report 367, "A review of safety culture and safety climate literature for the development of the safety culture inspection toolkit", prepared by Human Engineering for the Health and Safety Executive 2005.

[6] Geller, E. Scott, "Working Safe: How to help People actively care for Health and safety", 2nd Edition, CRC Press LLC.

Additional references not cited:

[7] Reason J. Managing the risks of organizational accidents. Aldershot, Ashgate; 1997.

[8] Center for Chemical Process Safety (CCPS), "Guidelines for Chemical Process Quantitative Risk Analysis" 2nd Edition, Hoboken, New Jersey: John Wiley and Sons, 1999

[9] N. Hyatt, Dyadem, "Guidelines for Process Hazards Analysis, Hazards Identification & Risk Analysis". Boca Raton, Florida, USA: CRC Press LLC.

[10] Senthil, Shan, "Cross Cultural Management", World Focus, www.asse.org 2012

The Legal Requirements of the PSM Standard: Finding Your Way

Michael Taylor, Esq.
Jackson Lewis P.C.
10701 Parkridge Boulevard, Suite 300
Reston, Virginia 20191
michael.taylor@jacksonlewis.com
703-483-8320

Prepared for Presentation at
American Institute of Chemical Engineers
2014 Spring Meeting
10th Global Congress on Process Safety
New Orleans, Louisiana
March 30 – April 3, 2014

UNPUBLISHED

The Legal Requirements of the PSM Standard: Finding Your Way

Michael Taylor, Esq.
Jackson Lewis LLP
10701 Parkridge Boulevard, Suite 300
Reston, Virginia 20191
michael.taylor@jacksonlewis.com
703-483-8320

Keywords: OSHA, Process Safety Management, OSHRC

Abstract

Over the last several years, the federal Occupational Safety and Health Administration issued record-breaking citations and penalties to the process safety management community. This trend is likely to continue as a result of the follow-up inspections regarding the Refinery National Emphasis Program and the on-going inspections regarding the Chemical National Emphasis Program. As a result of these types of inspections, OSHA often issues citations for failing to comply with best practices rather than actual compliance with the Process Safety Management standard, 29 C.F.R. § 1910.119. This typically happens when an employer considers a best practice as compliance with the PSM standard, and OSHA believes that there is a deficiency regarding the documentation or implementation of that best practice. Accordingly, Process Safety Engineers should know the distinction between best practices and compliance with the PSM standard. This, in turn, will help Process Safety Engineers develop written programs and procedures that are in compliance with the PSM standard. As a result, the many risks associated with OSHA citations and penalties alleging a violation of the PSM standard will be greatly minimized.

1. Introduction

This paper identifies and analyzes some of the risks associated with OSHA citations and penalties alleging a violation of the Process Safety Management standard, 29 C.F.R. § 1910.119. Those risks involve: (1) administrative liability; (2) civil liability; (3) criminal liability; (4) abatement costs; (5) follow-up inspections; (6) negative publicity; (7) Severe Violator Enforcement Program; (8) union organizing and collective bargaining; and (9) obtaining contract work. This paper also discusses the distinction between best practices and compliance with the PSM standard.

2. OSHA Citations – The Risks

There are several risks associated with OSHA citations and penalties. First, under the Occupational Safety and Health Act of 1970 ("OSH Act"), 29 C.F.R. § 651 *et seq.*, a citation item may be characterized as serious, willful, repeat, or failure to abate. A serious violation has a maximum penalty of $7000, a willful or repeat violation has a maximum penalty of $70,000, and a failure-to-abate violation has a maximum penalty of $7,000 per day. Over the last several years, OSHA issued record-breaking citations and penalties to the process safety management community as evident in the following chart:

Table 1. Top PSM Enforcement Cases Based on Total Issued Penalty

	Company Name	Inspection #	Total Issued Penalty
1	BP Products North America	311962674 308314640	$81,340,000
2	BP Products North America	308314640 308314988	$21,361,500
3	IMC Fertilizer/Angus Chemical	107607863 107607871	$11,550,000
7	CITGO Petroleum	110416880	$8,155,000
11	Phillip 66/Fish Engineering	106612443 107365751	$6,395,200
12	Hercules, Inc.	108662420 100490705	$6,328,000
13	Arcadian	102281292 102281128	$5,085,000
19	Arco Chemical Co.	110318540	$3,481,300
23	BP North America Inc. and BP Husky Refining LLC's Refinery	311611081	$3,042,000
24	Shell Oil Chemical Co.	103342093	$3,017,000
25	Union Carbide	110398310	$2,803,500

This trend is likely to continue as a result of the follow-up inspections regarding the Refinery National Emphasis Program and the inspections regarding the Chemical National Emphasis Program.

Second, in general, state workers' compensation exclusivity provisions prohibit employees from filing a civil cause of action against their employers. *See Larsen's Workers' Compensation Law § 100.01.* These provisions do not prohibit employees from filing a cause of action against another employer that may have caused their injuries. In this regard, approximately thirty two states authorize the use of OSHA standards as evidence of negligence at trial, and approximately fourteen states authorize the use of OSHA standards as negligence per se. In a state that authorizes the use of OSHA standards as evidence of negligence, a plaintiff may argue, for example, that the PSM standard is evidence that the employer had a duty of care. The plaintiff may also argue that noncompliance with the terms of the PSM standard is evidence that the employer breached its duty of care. The employer can still present evidence, however, that it acted reasonably and therefore did not breach its duty of care. In a state that authorizes the use of OSHA standards as negligence per se, a plaintiff may argue that the PSM standard *proves* that the employer had a duty of care. The plaintiff may also argue that noncompliance with the terms of the PSM standard *proves* that the employer breached its duty of care. The employer is prohibited from presenting evidence, however, that it acted reasonably and therefore did not breach its duty of care. Experienced plaintiff counsel will use the OSHA inspection (i.e., employee statements, documents, and photographs) and citations in the civil case.

Third, section 17(e) of the OSH Act states that an employer who willfully violates the law and that violation caused the death to an employee shall be guilty of a misdemeanor with a maximum of six months in prison. Moreover, under the Sentencing Reform Act of 1984, 18 U.S.C. § 3551 *et seq.*, an individual may receive a maximum criminal penalty of $250,000, and the organization may receive a maximum criminal penalty of $500,000. The United States Department of Justice handles the criminal prosecution. There may be State criminal prosecution as well. In the last few years, there has been an increase of federal criminal prosecutions.

Fourth, abatement may be extremely expensive and have minimal or no safety or health benefit, if at all. This could include the initial and on-going capital costs needed in order to implement the abatement.

Fifth, OSHA may conduct a follow-up inspection. In general, employers who tend to "rollover" after receiving significant citations and notifications of proposed penalties often receive more follow-up inspections (and therefore more citations and notifications of proposed penalties) than employers who "politely push back" after receiving the same. It is a myth that if an employer accepts the citations and pay the penalties as proposed, OSHA will not come back to the worksite for a follow-up inspection. OSHA is currently conducting follow-up inspections from the Refinery NEP.

Sixth, OSHA has been issuing negative press releases with very harsh language shortly after issuing citations and proposed penalties to an employer. OSHA calls such press releases "regulation by shaming" and asserts that press releases can be "very effective" in doing the right thing. In essence, the employer is presumed guilty prior to being able to defend itself during litigation.

A negative press release can harm an employer in many different ways. A negative press release may, for example, damage an employer's relationships with current customers and employees. A negative press release may also damage an employer's ability to obtain relationships with potential customers. A negative press release may also damage an employer's reputation within the local community. A negative press release can also impact civil and criminal litigation. Simply put, a negative press release can create an impression that the worksite is a very dangerous place to work. By filing a notice of contest, an employer can defend itself by achieving a favorable settlement agreement or winning at trial. In either case, the employer may want to consider issuing its own press release to minimize or neutralize the impact of OSHA's press release.

Seventh, OSHA may place an employer in the Severe Violator Enforcement Program ("SVEP"). On June 18, 2010, OSHA launched the SVEP. Certain criteria must be met before an employer can be put in the SVEP. For example, three or more willful, repeat, or failure-to-abate violations of the PSM standard would put an employer in the SVEP. Once an employer is in the SVEP, OSHA may conduct nationwide inspections of the employer at the same time or over a certain period of time. This can pose a significant risk to an employer and cause major business interruptions. An employer can get out of the SVEP by having the qualifying criteria withdrawn or re-characterized through a settlement agreement or after a trial.

Eighth, a union may use citations as leverage in the organizing and collective bargaining process. Many unions use citations and penalties as leverage in the organizing process in an attempt to persuade the employees for the need of a union. Many unions also use citations and penalties as leverage in the collective bargaining process in attempt to force an employer to agree to develop and implement other things related to safety or health in the workplace.

Finally, citations may impact the ability of the employer to gain private or government contract work. The government or other employers may evaluate the OSHA citation history for contractors prior to accepting their bid for work. This is becoming more prevalent in the private sector.

3. Best Practices and Compliance – The Distinction

OSHA often issues citations to employers for failing to comply with best practices rather than actual compliance with the Process Safety Management standard, 29 C.F.R. § 1910.119. This typically happens when an employer considers a best practice as compliance with the PSM standard, and OSHA believes that there is a deficiency regarding the documentation or implementation of that best practice. Accordingly, Process Safety Engineers should know the distinction between best practices and compliance with the PSM standard. This, in turn, will help Process Safety Engineers develop written programs and procedures that are in compliance with the PSM standard. As a result, the many risks identified above will be minimized.

When determining how to achieve compliance with the Process Safety Management standard, 29 C.F.R. § 1910.119, Process Safety Engineers should be generally familiar with the legal framework regarding the interpretation of the PSM standard. Indeed, the courts have often found that OSHA's interpretation is contrary to the plain language of a particular standard. The courts have also found that OSHA's interpretation of an ambiguous standard is not reasonable.

In a recent case, the federal Occupational Safety and Health Review Commission, the agency in charge of adjudicating workplace safety and health disputes between OSHA and the regulated community, summarized the law regarding when OSHA is entitled to deference regarding the interpretation of its standards. In *Cargill Meat Solutions Corp.*, 2012 OSAHRC LEXIS 29 (No. 10-1156, 2012), the administrative law judge concluded that OSHA is entitled to deference only if the interpretation is reasonable. *Id.* at 34 (citing *Martin v. OSHRC*, 499 U.S. 144, 158 (1991)). The ALJ explained:

> [T]he Commission evaluates whether the interpretation 'sensibly conforms to the purpose and wording of the regulation,' taking into account 'whether the Secretary has consistently applied the interpretation embodied in the citations, the adequacy of notice to regulated **[*35]** parties, and the quality of the Secretary's elaboration of pertinent policy considerations.

Id. at 34-35 (citation omitted).

When determining whether the interpretation "sensibly conforms to the purpose and wording of the regulation," the ALJ examined whether OSHA's position is contrary to the plain language of the standard. *Id.* at 35-36. Regarding whether OSHA "has consistently applied the interpretation embodied in the citations," the ALJ analyzed interpretation letters and directives as well as the OSHA's position in other adjudicative matters. *Id.* at 36-40. When examining the "adequacy of notice to the regulated parties," the ALJ reviewed the language of the standard, legislative history of the Final Rule, and letters of interpretation or directives that may have been issued. *Id.* 40-41. Finally, regarding whether OSHA has elaborated any pertinent policy considerations that support its interpretation of the standard, the ALJ reviewed whether OSHA has elaborated the policy considerations to support its interpretation of the cited standard. *Id.* at 41.

Given that the PSM standard is "performance-oriented," many provisions within the fourteen elements are inherently ambiguous. As a result, Process Safety Engineers may want to analyze prior enforcement actions, interpretation letters, compliance directives, and the legislative history of the PSM standard in order to confirm the requirements of the PSM standard. Depending on the issue, this may necessitate the assistance of counsel, particularly someone who has experience with the PSM standard.

In addition to being generally familiar with the legal framework regarding the interpretation of the PSM standard, Process Safety Engineers should also be familiar with some best practices that have been equated with compliance with the PSM standard. Here are some examples:

Section 1910.119(c)(3), which applies to employee participation in process safety management, states that employers "shall provide to employees and their representatives access to process hazard analyses, and to all other information required to be developed under this standard." The cited standard does not indicate whether these documents need to be available to employees electronically or hard-copy. This may be considered a best practice. Due to the performance-oriented nature of the standard, an employer may make the documents available either electronically or hard-copy.

Section 1910.119(d)(3)(i)(B), which addresses process safety information, states that "information pertaining to the equipment in the process shall include: Piping and instrument diagrams (P&IDs)." The cited standard does not state that P&ID(s) shall be current and accurate at all times. This may be considered a best practice. Due to the performance-oriented nature of the cited standard, an employer may use red-lined copies of the P&ID(s) for a reasonable period of time until the changes can be finalized.

Section 1910.119(f)(1)(i)(D), which addresses operating procedures, states that employers "shall develop and implement a procedures for emergency shutdown including conditions under which emergency shutdown is required, and the assignment of shutdown responsibility to qualified operators to ensure that emergency shutdown is executed in a safe and timely manner." The cited standard does not state that the emergency shutdown procedure shall indicate who has the authority to make the decision to shut down the covered process. Nor does the standard state that the emergency shutdown procedure shall include the names of the employees who have shutdown responsibility. This may be considered a best practice. Due to the performance-oriented nature of the cited standard, an employer may use job titles in the procedure to indicate who must perform the specific steps to shut down the covered process.

Section 1910.119(f)(1)(ii) states that operating procedures "shall contain operating limits, consequences of deviation, and steps to correct." The cited standard does not state that the information must be contained in the actual procedures themselves. This may be considered a best practice. Due to the performance-oriented nature of the cited standard, an employer may have the information contained in a separate document.

Section 1910.119(f)(4), which address safe work practices, states that employers "shall develop and implement safe work practices to provide for the control of hazards during operations . . . ." The cited standard does not state that the work practices must be in writing. This may be considered a best practice. Due to the performance-oriented nature of the cited standard, an employer does not have to have written work practices for the control of hazards during operations.

Section 1910.119(g)(1)(i), which address operator training, states that employers "shall train employees in an overview of the process and in the operating procedures." The cited standard does not state that employers are required to train employees regarding what constitutes process safety management. This may be considered a best practice. Due to the performance-oriented nature of the cited standard, an employer simply has to provide an overview of the covered process in which the employee is being assigned along with the operating procedures associated with that particular covered process.

Section 1910.119(g)(3) states that employers "shall ascertain that each employee involved in operating a process has received and understood the training . . . ." The cited standard does not state that employers must administer a written test to employees to ensure that they understood the training. This may be considered a best practice. Due to the performance-oriented nature of the cited standard, an employer may choose any particular method for determining whether employees understood their training. This may be a computer-based test, field verification, or a verbal quiz.

Section 1910.119(i)(1) states that employers "shall perform a pre-startup safety review for new facilities and for modified facilities when the modification is significant enough to require a change in the process safety information." The cited standard does not state that the PSSR must be in writing. Nor does the cited standard state that a checklist is required. This may be considered a best practice. Due to the performance-oriented nature of the cited standard, employers can establish their own method for ensuring that the safety requirements are performed.

Section 1910.119(j)(5), which addresses mechanical integrity, states that employers "shall correct deficiencies in equipment that are outside the acceptable limits (defined by the process safety information in paragraph (d) of this section) before further use or in a safe and timely manner when necessary means are taken to assure safe operation." The cited standard does not state that all deficiencies in equipment must be corrected before further use. Rather, the cited standard allows employers to provide a temporary solution until a permanent solution is implemented.

Section 1910.119(l)(3), which addresses management of change, states that employees "involved in operating a process and maintenance and contract employees whose job tasks will be affected by a change in the process shall be informed of, and trained in, the change prior to startup of the process or affected part of the process. The cited standard does not require that the information and training be in writing. This may be considered a best practice. Due to the performance-oriented nature of the cited standard, employers can select the manner in which employees receive the information and are trained.

Section 1910.119(l)(5) states that if a "change covered by this paragraph results in a change in the operating procedures or practices required by paragraph (f) of this section, such procedures or practices shall be updated accordingly. The cited standard does not state that the operating procedures shall be updated prior to implementation of the change. This may be considered a best practice. Due to the performance-oriented nature of the cited standard, employers can rely on red-lined versions of operating procedures until the procedures can be updated. There is no timeframe set forth in the cited standard for updating the operating procedures.

Section 1910.119(n) states that "the emergency action plan shall include procedures for handling small releases." The cited standard does not state that the emergency action plan must contain a definition of what constitutes a small release. A general definition of what constitutes a small release may be considered a best practice. A precise definition is virtually difficult to accomplish considering that chemicals pose different threats at different quantities.

Section 1910.119(o)(4), which addresses compliance audits, states that employers "shall promptly determine and document an appropriate response to each of the findings of the compliance audit, and document that the deficiencies have been corrected." The cited standard does not require employers to explain why certain findings were not accepted or rejected. Due to the performance-oriented nature of the cited standard, employers can simply reject or accept findings.

With these examples of best practice and compliance-related distinctions, along with the legal framework noted above, Process Safety Engineers will be equipped with tools necessary to help differentiate best practices from compliance with the PSM standard. This, in turn, will help Process Safety Engineers develop written programs and procedures that are in compliance with the PSM standard and minimize the risk of receiving OSHA citations for failing to develop or implement best practices.

4. Conclusion

OSHA continues to issues record-breaking citations and penalties to the process safety management community. OSHA often issues citations for failing to comply with best practices rather than actual compliance with the Process Safety Management standard, 29 C.F.R. § 1910.119. This typically happens when an employer considers a best practice as compliance with the PSM standard, and OSHA believes that there is a deficiency regarding the documentation or implementation of that best practice. Accordingly, Process Safety Engineers should know the distinction between best practices and compliance with the PSM standard. This, in turn, will help Process Safety Engineers develop written programs and procedures that are in compliance with the PSM standard. As a result, the many risks associated with OSHA citations and penalties alleging a violation of the PSM standard will be greatly minimized.

353790 A Recipe to Excel: Beyond Technical Competency

Abstract:

The presentation will highlight the fundamental soft skill sets needed to be successful in managing diverse work forces in an organization. The skill sets are invariant of the type of organizational environment. This could span R&D driven organizations to manufacturing & production to engineering, procurement and construction driven organizations and to other service industries.

To be successful, it is taken for granted for any technically driven project or organization, that to carry out his/her assigned responsibility, the manager needs to have technical skills and understanding. The factor that differentiates a truly successful manager, however, is the manager's extraordinary strength in soft skills, sometimes referred to as people skills. The necessity of these skills is becoming more important and pertinent than ever before. The reason behind this special requirement is the diversity of groups involved in any work environment. Such diversity is due to rapid globalization. Challenges associated with diversity, when managed properly, bring strength rather than weakness.

In this presentation the importance of effective communication will be emphasized. Effective communication is the most difficult skill to acquire and practice but the most rewarding when a manager/leader acquires and proficiently applies it. Through effective communication the leader brings team spirit and trust. Trust in turn brings success.

Management Careers in Consulting and Engineering

R. Benson Pair, Jr., PE, BCEE

Chief Technology Engineer - Environmental

KBR

KBR

Career Opportunities in Consulting and Engineering

Consulting companies and Engineering firms offer engineers a wide range of career opportunities in management

KBR

Management Careers in Consulting and Engineering

- Technical and Technology
- Engineering
- Project Management
- Engineering Management
- Project Controls
- Offices
- Business Units
- Corporate
- Sales
- IT/IM
- Procurement
- Construction

KBR

Consulting

- De-centralized, serving regional markets
- Smaller offices and smaller projects compared to EPC firms
- Historically has been engineering only, but that is changing
- May be focused on one market sector: Industrial or Government (Federal/State/Local)

KBR

Engineering, Procurement, Construction

- Fewer, larger offices
- Larger projects
  - Contract value
  - Schedule
  - Task force can be >100
- Often serves Industry and Government

KBR

Technical Skills

- Engineering Disciplines
- Accounting
- Cost and Scheduling
- Contracts
- HSE

KBR

People Skills

- Leadership
- Communication: Listening and talking
- Negotiating
- Respect for others
- Patience
- Humility in yourself / Pride in others
- Sense of Humor
- Delegation
- Mentoring

KBR

Career Development - The "Y" Approach

- Recognizes and values the technical specialist
- Shows transition from generalist to technology specific roles
- Identifies three broad bands of skill levels
- Continues to recognize and value the technical leaders who combine technology and management skills
- Shows transfer routes to Project Engineering / Project Management and Sales Coordinator / Sales

KBR

The "Y" Model for Career Development

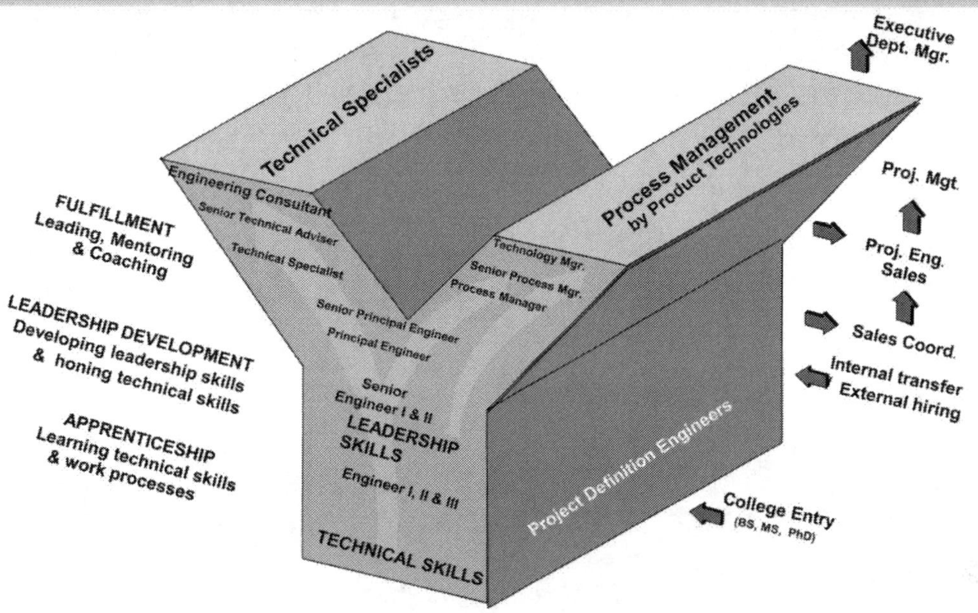

KBR

Management Is…

… a set of well-known processes, like planning, budgeting, structuring jobs, staffing jobs, measuring performance and problem-solving, which help an organization to predictably do what it knows how to do well. Management helps you to produce products and services as you have promised, of consistent quality, on budget day after day week after week.

KBR

...(Still) not Leadership

Leadership is associated with taking an organization into the future, finding opportunities that are coming at it faster and faster and successfully exploiting those opportunities. Leadership is about vision, about people buying in and about empowerment and most of all about producing useful change.

Source: "Management Is (Still) Not Leadership", John Kotter, Konosuke Matsushita Professor of Leadership, Harvard University, *Harvard Business Review*, January 9, 2013

KBR

Management and Leadership

Managers

- administer
- maintain
- focus on systems and structures
- rely on control

Leaders

- innovate
- develop
- focus on people
- inspire trust

Source: "The Wall Street Journal Guide to Management," Alan Murray

KBR

Honesty

"Your business and its employees are a reflection of yourself and if you make honest and ethical behavior a key value, your team will follow suit."

Source: "Top 10 Qualities that make a Great Leader", Tanya Prive, Forbes, 12/19/2012

KBR

Engineering Ethics

Engineer's Creed

As a professional engineer, I dedicate my professional knowledge and skill to the advancement and betterment of human welfare

I pledge:

- To give the utmost of performance;
- To participate in none but honest enterprise;
- To live and work according to the laws of man and the highest standards of professional conduct;
- To place service before profit, the honor and standing of the profession before personal advantage, and the public welfare above all other considerations.

In humility and with need for Divine Guidance, I make this pledge.

Adopted by the NSPE, June 1954

KBR

353824 Management Careers in an Industrial Setting

Abstract:

Management careers in an industrial setting offer a broad range of opportunities and challenges.

It certainly is important to gain a clear understanding of your personal goals and objectives before embarking on a career in management.

What are your motivations? Do you have the skills and aptitude? Are you willing the do the things that need done?

Learning and development are essential. Time commitments are significant.

Rewards can be amazing when you see people & operations develop beyond expectations.

The key becomes tapping into the capability of the team and help move the organization forward.

353793 Managing Through Extreme Change

Abstract:

The Department of Energy Weapons Complex, a nuclear chemical enterprise employs hundreds of thousands of scientists and engineers. Through its birth during the Manhattan project in the 1940's to the end of the Cold War, the DOE Weapons Complex was a rapidly expanding enterprise that provided career growth and service towards a national mission. After the Cold War ended, the weapons complex has been slowly decreasing in size through a series of layoffs, furloughs, reductions in benefits and contract changes. While the DOE Complex was originally built by world class industrial companies, the current contract practices and potential legal liabilities have changed the field of companies that now operate the complex for the Department of Energy. These new types of companies formed as limited liability corporations, are typically partnerships of engineering and construction firms, aerospace companies or not for profit research companies. Managing businesses in a declining industry and opportunities within the nuclear chemical industry will be presented.

353905 You Can Have it All, But Not at the Same Time

Abstract:

Women (and men) in science and engineering have struggled with the Career Life Balance. With the rise of dual income couples and more female breadwinners, the working world has a goal of becoming better at balancing work and family life for both men and women. As managers women may face additional challenges not experienced by our male counterparts. Challenges occur in all sectors of the workplace: private industry, academia and government. Numerous articles and books have been written on the subject. Recently the Huffington Post featured and article titled "Do Women Make Better Bosses?".

Managing Volunteers Is A Little Like Herding Cats

1

Joseph J. Cramer, PhD, PE

Director Emeritus, Program Development

Careers in Management Session

AIChE Spring Meeting

April 1, 2014

New Orleans, LA

How did I select this title?

2

- Clearly chosen to attract interest
- Analogy has some validity
- Both cats and volunteer Ch.E.s are:
  - Intelligent
  - Fiercely Independent
  - Can scratch when irritated
- Strict military style of management won't work
- More than 60 million volunteers in USA
- Volunteers (and their management) are very important

Background Information

3

- Have been managing volunteers for >30 years – almost entirely involving AIChE
- Three degrees in Chemical Engineering
- Didn't like life in a lab or plant
- Didn't like just doing calculations all day
- Gravitated towards project management and later department management
- Enjoyed interaction with people, like dealing with crisis situation but also liked doing planning to avoid the crisis in the first place
- Slowly transitioned from managing 'direct reports' to 'volunteers'

Types of Authority

4

- Line Authority (Traditional)
- Staff (Expertise) Authority
- Coercive/Reward Authority (penalties and rewards)
- Functional Authority (specific tasks for designated time)
- Referent Authority (based on charisma, charm, personality)

Developing Your Style of Managing Volunteers

5

- Expertise (spoke the language, been there/done that)
- Rewards (important to volunteers even if they won't admit it)
- Granting Functional Authority
- Referent Authority (personality based-honesty, credibility, trust, knowledge)
- Motivating (nurturing enthusiasm)

Volunteers vs Employees

6

- Even disinterested workers can be motivated by compensation
- Disinterested volunteers disappear (or worse, can be a negative force)
- Poor workers can be fired (not always easy)
- Poor volunteers are difficult to fire
- Must word harder to really engage volunteers
- Must match tasks with skills - carefully

Effective Volunteers

7

- Must understand the direction
- Must buy into it (engagement)
- Must have an interest (enthusiasm)
- Must have ability to comply or execute
- Good communications must be maintained
  - Clear
  - Concise
  - Consistent
  - Good listening skills

Useful Skills for Volunteer Manager

8

- Reasonable technical proficiency of knowledge
- Good Communications skills
- Enjoy interacting with people
- Goal oriented but
  - Process sensitive
  - Good at building and maintaining relationships

The Benefits

9

- The organization can accomplish so much more
- The volunteer learns about 'managing'
- Volunteers acquire management experience more quickly
- The experience is transferrable
- The skills are good for a lifetime
- Relationships can last for life

10

Thank you for your attention.

Any questions?

Comments?

Anecdotes?

Keys to Avoid Shortcomings in Your MOC System

Revonda P. Tew
Process Improvement Institute, Inc.
1321 Waterside Lane
Knoxville, TN 37922, USA
rtew@piii.com

Tim Waugh
Process Improvement Institute, Inc.
twaugh@piii.com

Matthew LaPlante and Dan Pastirik
Irving Oil Refining, G.P.
340 Loch Lomond Road
Saint John, New Brunswick, Canada E2L4H6
Matthew.LaPlante@irvingoil.com
Dan.Pastirik@irvingoil.com

Prepared for Presentation at
American Institute of Chemical Engineers
2014 Spring Meeting
10th Global Congress on Process Safety
New Orleans, LA
March 30 – April 2, 2014

UNPUBLISHED

Keys to Avoid Shortcomings in Your MOC System

Revonda P. Tew
Process Improvement Institute, Inc.
1321 Waterside Lane
Knoxville, TN 37922, USA
rtew@piii.com

Tim Waugh
Process Improvement Institute, Inc.
twaugh@piii.com

Matthew LaPlante and Dan Pastirik
Irving Oil Refining, G.P.
340 Loch Lomond Road
Saint John, New Brunswick, Canada E2L4H6
Matthew.LaPlante@irvingoil.com
Dan.Pastirik@irvingoil.com

Keywords: MOC, Management of Change, Process Safety Information, eMOC, KPIs, Key Performance Indicators, MOC Workflow

Abstract

When thinking about MOCs on a case by case basis it is easy to forget the overall role of a change management system. Think of the MOC system as an umbrella protecting the plant's current process safety design. The MOC system helps to validate the adequacy of the proposed process changes based on the plant's current design. This means the MOC system needs to ensure that any impacts of the change to other PSM elements' implementation are identified and addressed. This requirement makes an MOC system inherently complex, requiring extra care to design the MOC workflow for ease of use and efficiency across the organization. If key components are not addressed, an MOC system can quickly become a dog's breakfast. The MOC system must address many aspects of change, specific circumstances and nuances that if unchecked can weaken an organization's perspectives on MOC and override the discipline to carry out day to day change management. Therefore, an MOC system's workflow must be designed to identify and address how the proposed change can affect implementation of other PSM elements and be constantly monitored and assessed. For sustainability and accountability, associated key performance indicators must also be established, reported and communicated.

1. Introduction

This paper discusses how to ensure the MOC system addresses a change's effect on implementation of other process safety management systems and discusses common MOC workflow challenges that can undermine implementing an efficient, compliant and effective MOC system. This will be illustrated based on the recent eMOC system rollout at Irving Oil, New Brunswick, Canada.

MOC is often considered the most important element of a Process Safety Management program, because it can prevent the undermining of safeguards provided by the other elements. Poorly managed changes can degrade the quality of Process Safety Information, introduce new risks in plants which have already undergone a Process Hazard Analysis, and / or reduce the mechanical integrity of equipment, to mention a few.

An effective MOC process must facilitate the review, approval, and documentation of many complex changes at a time. This can become an administrative nightmare leading to MOC process failure, without a robust tracking system. An electronic tracking system has the functionality to manage a large number of MOCs at one time. That is why Irving Oil implemented an electronic MOC (eMOC) system in 2011.

Also, examples of common MOC system gaps discovered throughout audits across small and large sized facilities and a variety of chemical processes will be presented, such as updating Process Safety Information (PSI) and other documents, managing temporary changes, quality of risk reviews and the MOC process work flow.

How to ensure the MOC system addresses a change's effect on implementation of other elements of PSM will be accomplished by:

- Presenting the relationship between Process Safety Information (PSI) and MOC and how together these elements influence the implementation of all process safety elements
- Discussing Irving Oil's eMOC system
- Discussing how simple changes can become unexpectedly complex and require revising many documents that support day-to-day implementation of many PSM elements.

How common MOC workflow challenges can undermine implementing an efficient, compliant and effective MOC system will be illustrated. Training and utilization of eMOC systems have resulted in a better understanding of MOC by organizations as a whole, and generally leads to the recognition of the need for many more MOC's.

2. Relationship Between Process Safety Information and Management of Change

2.1 MOC Protects PSI

Process Safety Information is an element of PSM which documents the process safety design. It is the foundation for process safety. Ninety-nine percent of changes are changes to PSI (others being organizational changes, some facility changes and stand-alone procedural changes). Management of change is the umbrella, protecting the process safety design **(Figure 1)**. Not only is current PSI needed to evaluate the technical basis and evaluate the risk of the change, the PSI that is changing due to the process change must be identified and must be updated to reflect the change.

Relationship Between MOC, PSI and Other PSM Elements

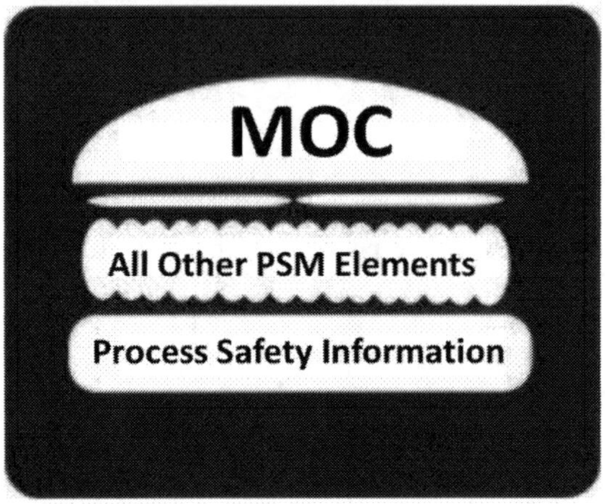

Figure 1 – MOC protects the process safety design (PSI), which is the foundation on which the implementation of all PSM elements is based.

2.2 MOC Protects Other PSM Elements

When a process change is proposed and evaluated, the quality and thoroughness of the MOC review extends beyond the change itself. The implementation of a change will require identifying other elements of PSM that must be changed/updated. For example, if a new instrument is added to a process:

- Operating procedures must be updated with instructions for operating, monitoring, and troubleshooting the instrument, and should have a description of its process safety function
- Mechanical Integrity program must be updated. The instrument data sheets must be filed properly, inspection, testing and preventive maintenance program established and entered into the Computerized Maintenance Management System (CMMS) and spare parts ordered
- Training guides must be updated to include the instrumentation, safe operating limits, consequences of deviations, and troubleshooting guidelines
- All related design and specifications of the instrument must be filed properly in the specific PSI documents such as loop sheets, P&IDs, etc.

2.3 Day-to-Day Use of PSI

On a day-to -day basis plant staff relies on the PSI documents to make decisions concerning process changes. Therefore, if documents are not updated to reflect changes as they occur, eventually the process safety design documents will be out of date and future process change decisions will be made on outdated information. PSI is the basis for implementing all elements of PSM. Some examples where PSI is used:

- SOP content
- Training materials for operators
- Emergency response plans
- Training for contractors and mechanics
- Establishing ITPM for equipment and instrumentation
- Conducting Process Hazards Analyses (PHAs)
- Conducting incident investigations
- Evaluating MOCs

2.4 MOC Procedural Step to Protect PSI

To keep PSI updated and accurate, a critical step in the MOC process is to identify all the documents associated with elements of PSM that require updating, to reflect the process change with respect to the PSI. Many companies' MOC systems lack a mechanism to help the MOC owner identify all documents that contain PSI to be updated. **Figure 2, "Documentation Checklist,"** is how Irving Oil identifies all the documents related to PSM implementation that may need updating to reflect a process change. However, identifying the documents is only the first step. Not creating an action plan, and tracking these actions to update PSI and other PSM documents to completion, are common MOC system shortcomings which will be discussed below, along with other MOC workflow challenges.

3. MOC Workflow Opportunities

MOC process workflows have some variations and commonalities across companies. Irving's workflow has 8 major steps **(Figure 3):**

1) Initiate the change
2) Initial Approval and assign a change owner
3) Change-specific Form and Change Type Selection
4) Assign Functional Reviewers
5) Functional Review Completion
6) Develop Action Plan
7) Final Approval
8) Complete Change Request

Our client has found opportunities for incremental improvement at several steps of the process, which will lead to a more efficient and accurate representation of all the changes through the overall MOC implementation. Each is discussed along with solutions being addressed.

IRVING

Documentation Checklist

NOTE: This list is NOT all-inclusive; consult your plant/function library to determine if any other records and/or documentation will need to be updated as a result of this change request.

Processing
- ☐ Operating Manuals
- ☐ Training Material
- ☐ Standard Operating Procedure(SOP)
- ☐ Standard Job Procedure (SJP)
- ☐ Asset Operating Procedure (AOP)
- ☐ Operating Best Practice (OBP)
- ☐ Operating Instruction (OPI)
- ☐ Department Policies
- ☐ Quality Manual
- ☐ Training records
- ☐ Process Description
- ☐ Product Specifications
- ☐ Refinery Energy Model
- ☐ Refinery Utility Capacity Report
- ☐ Loss of Utility Emergency Procedures
- ☐ Shift Log / Turnover Log
- ☐ Shift Schedule for Plant
- ☐ Shift Personnel List

Piping
- ☐ Steam & Trap Docs
- ☐ Pipe Specs
- ☐ Pipe rack drawings
- ☐ LX drawings
- ☐ Piping isometrics

Organizational
- ☐ Corporate and Refining Phone List
- ☐ Organizational Charts
- ☐ Outlook Distribution lists
- ☐ On-Call Schedule
- ☐ Vacation Schedule
- ☐ Cintellate Hierarchy
- ☐ Oracle Hierarchy

Technical-PTS
- ☐ Process Flow Diagrams (PFDs)
- ☐ Piping and Instrument Diagrams (P&ID)
- ☐ Safe Upper and Lower Limits
- ☐ Maximum Intended Inventories
- ☐ Evaluation of the Consequence of Deviations
- ☐ Material and Energy Balances
- ☐ Description of the process materials and process chemistry
- ☐ Hazard Effects of mixing two chemicals
- ☐ Design basis
- ☐ Control Narratives

Health, Safety & Environment
- ☐ Evacuation Routes and Strategies
- ☐ Emergency Response Manual (ER)
- ☐ Health & Safety Instructions (HSI)
- ☐ Process Safety Instructions (PSMI)
- ☐ Environmental Instruction (EMSI)
- ☐ Training database
- ☐ OH Manual
- ☐ C. of A. records
- ☐ Waste Characterization
- ☐ Disposal Permit
- ☐ Process Hazard Analysis Report
- ☐ MSDS

Civil
- ☐ Equipment Plot Plans
- ☐ Underground drawings

Instrumentation
- ☐ SIS Instrument Check Procedures
- ☐ JX,PX drawings
- ☐ Loop diagrams
- ☐ Instrument Datasheets
- ☐ AMS Files

Reliability
- ☐ Materials of Construction Drawings and Datasheets
- ☐ Design Codes and Standards
- ☐ Equipment Datasheets
- ☐ Vessel Drawings
- ☐ Fire water system drawings
- ☐ Underground Drawings
- ☐ Lubrication records
- ☐ Vibration Database Records
- ☐ Spare Parts
- ☐ PD-18
- ☐ Equipment specification sheets
- ☐ CMMS records
- ☐ Vendor information
- ☐ Asset Criticality Assessment (RELI 138)
- ☐ Design Calculations
- ☐ EWO/TWO
- ☐ Reliability Instruction (RELI)

Maintenance
- ☐ Repair Procedures
- ☐ Open Issue List
- ☐ Basic Equipment Maintenance Plan
- ☐ Materials Mgt. Instruction (MMI)
- ☐ Maintenance Instructions (MNTCI)
- ☐ Repair Procedures

Document Control/Management
- ☐ All *IOR Instructions, Policies & Standards*

Documentum:
- ☐ Bill of Material
- ☐ Breaker Testing Records
- ☐ Calculations
- ☐ Canadian Registration Number (CRN)
- ☐ Certificate of Compliance
- ☐ Data Sheets
- ☐ Drawing (Engineering Drawing, Isometric Loop, P&ID, PFD, Plot Plan, Sketch)
- ☐ Drawing (Above Grade Piping, Absorber, Boiler, Building, Column, Compressor, Concrete, Contractor, Cooling Tower, Design Data, Drum, Electrical, Exchanger, Fin Fan Coil, Foundation, Furnace, Grading, Insulation, Misc. Mechanical Equipment, Paving, Piping Isometric, Pump & Driver, Reactor, Stripper, Structural Steel, Tank, Vessel, Underground Piping
- ☐ eAM Equipment Data
- ☐ Emergency Procedures
- ☐ Engineering Equipment Summary
- ☐ Engineering Work Order (EWO)
- ☐ Feed Package
- ☐ Forms
- ☐ Inspection Record
- ☐ Inspection Test Plan
- ☐ Install Op Maintenance Manuals
- ☐ Instructions
- ☐ Instrument Index
- ☐ Isometric
- ☐ Line List
- ☐ Manufacturers Data Reports
- ☐ Material Testing Report (MTR)
- ☐ Parts List
- ☐ Performance Curves
- ☐ Principles & Philosophies
- ☐ Process Design Data
- ☐ Project PEP, PPM

FIGURE 2 – Documentation Checklist used to help identify all documents impacted by a process change

3.1 Step 1 – Initiate the Change

In step 1, the change originator is responsible for initiating a request for a change, and follows through to submission of the change request to the initial approver. The MOC procedure contains a list of process/physical change and Replacement in Kind (RIK) examples. This reference helps the change originator correctly recognize a process change. With approximately 500 MOCs generated every quarter (typical for refineries the size of Irving Oil's), the refinery staff identifies process changes and recognizes the need to manage change.

3.1.1 Challenges

There is a need for a formal monthly audit mechanism to monitor how many uncontrolled changes are occurring, such as auditing work orders, operation's log books, etc. This performance indicator, when used properly, can help personnel to better recognize change. Also, communicating the root causes of incidents related to MOC system weaknesses, such as failure to properly evaluate and approve process changes, helps to educate employees and change the culture.

The protocol for MOC evaluation has been fortified with creation of a role profile and the assignment of a position (MOC Coordinator). Part of the role profile involves ongoing assessment of the MOC system and further development of performance indicators.

One of the most challenging MOC issues is overlooking changes in positions or organizational structure. Much accountability can fall through the cracks if the MOC system is not utilized for organizational changes. Not only for leadership roles, but if the roles and responsibilities are not examined for individuals throughout the organization, the implementation of PSM elements can be affected. For example, Irving's role profiles do not always reflect PSM responsibilities. An operator may be familiar with the Incident Reporting software on his shift and is designated as responsible for entering incidents. If he is moved to another unit, retires, or resigns, it is not always identified that this role needs to be filled or that another employee needs to be trained. Consider the hypothetical example of a person retiring who is the "Warden" for a department and floor in the event of an Emergency. This personnel change was not identified through MOC and the Warden role was not reassigned. If an emergency occurred and the role of the Warden not identified, confusion could arise (i.e. headcount, muster location, etc.). This can be addressed through the rollout of PSM "Duties and Accounabilities" for each role profile.

Because some individuals do not understand what constitutes a true "emergency" change, some changes are classified as emergency changes to allow expeditious approval. This increases the chance that these changes are not evaluated properly or timely, possibly resulting in intolerable process safety risk. It is difficult for staff to stop a change that is already in the process of being executed. There is a KPI for the number of emergency changes. While there is no predicted number of emergency changes a facility should experience, reporting the trend of this KPI can alert the MOC coordinator to investigate the emergency MOCs and intervene if the system is being misused.

3.2 Step 2 – Initial Approval

Step 2 is carried out by the initial approver. The initial approver reviews the justification for the proposed change. If the initial approver does not understand the existing risk of making the change, or if he knows the risk is high, he can require that a formal assessment be performed. A document is available in the eMOC system to aid in determining if a formal assessment should be performed. This ensures that the proper information is supplied to the Functional Reviewers and Final Approvers. The initial approver needs to be aware of their responsibilities as they are "Approving to Develop," not approving to implement the change. The initial approver assigns the Change Owner per the MOC procedure. The initial approver completes the initial risk assessment and determines if a formal risk assessment is required (note: in the United States, OSHA 1910.119 requires safety and health risk reviews for all process changes).

3.2.1 Challenges

An Initial Approver may sometimes overlook the Formal Assessment Requirements checklist which is used to determine if a formal assessment is required. Additionally, if the change constitutes the formal assessment being technical (HAZOP / LOPA, or FMEA, for example) an operation, such as a refinery, may have limited resources available to conduct these MOC formal risk assessments. Many have been trained and are participative, but only a select number of individuals may be active in facilitating formal assessments. Sometimes, consultants are hired to facilitate the formal risk assessments. The long-term objective is to develop and deliver an in-house PHA training package for technical members of the area teams, to lead the analyses for MOCs in their own areas.

3.3 Step 3 – Form and Change Type Selection

The change owner selects the specific change types and selects and completes the Change Specific Forms. If a formal assessment was requested, it is completed at this step before proceeding to the functional review (this is not clear on our chart **Figure 3**).

3.3.1 Challenges

The change owner may not select the correct (or only one of multiple) specific change type(s), which makes it difficult to determine the required functional reviewers. Typical change types include equipment changes, chemical changes, process control changes, facility changes, material management or logistics changes, analytical changes, tool changes, work process and personnel changes, and changes to standards, policies and administrative controls. The assigned change owner must be aware of the proposed change's scope, otherwise the proper functional reviewers may not be selected and the risk associated with the change may not be fully identified and evaluated. One of the most stringent responsibilities of the functional reviewers is to "crosscheck" with the subject matter experts to determine if a formal risk assessment is warranted. As mentioned under Step 2, there are many resources prepared to participate in the risk reviews, but limited resources are sufficiently trainted to facilitate or conduct risk reviews. Additionally, for some types of changes, specific forms are required such as the Temporary Facility Form or the DCS Change Forms. If these forms are not selected and completed, it could result in critical information missing for Functional Reviewers to perform their assessment.

Also, if the specific change type is incorrect, then accurately trending the specific types of changes implemented is impossible. While there is no set expected percentages for types of changes, PSM managers often trend this data to possibly alert staff of procedural problems, mechanical integrity issues, etc.

3.4 Step 4 – Assign Functional Reviewers

The change owner selects reviewers based on the specific change type and a Functional Review Matrix which identifies the subject-matter experts (SME) who have the responsibility to review the change based on the scope and type of change.

3.4.1 Observation

If Step 3 is not effectively carried out (not selecting the specific change type), the change owners may inadvertently select the incorrect required functional reviewers. To manage this issue the MOC coordinator monitors compliance with the MOC procedure and communicates through the Health Safety and Environmental (HSE) meeting to Refinery Leadership Team (RLT).

Another issue is that certain personnel are the SMEs for the majority of the MOCs. This work overload can cause insufficient review of important MOC issues,, and also create an MOC backlog.

3.5 Step 5 – Functional Review Completion

The reviewers have the responsibility to review the proposed change, identify potential risks, provide input to the mitigation plan, complete the review form and attach supporting documentation.

3.5.1 Challenges

Sometimes reviewers complete the functional review form's checklist, which is a tool to help guide the reviewer through the review process. It may be not mandatory, and there may be no minimum required documentation of what the reviewer considered in their review and pertinent information of which the functional reviewer is aware may not be communicated to the change owner. If the reviewers determine the change will present an intolerable risk, often the supporting information, such as the risk consequence and likelihood, are omitted. This is a very important checkpoint. There is a field to check whether the risk is tolerable or not, but this does not necessarily record the consequence nor the probability of risk, which leaves the change owner in a quandry. Some functional reviewers may be unsure of their roles and believe they are approving the change, not reviewing the risk. A mandatory checklist is being implemented which should help clarify what was considered, improve the review process and provide critical information needed for final approval. There is the risk that the functional reviewer who is the subject matter expert can overlook a potential hazard ("silo effect"). The change owner should be aware of this and consider all functional reviews for assessing this risk.

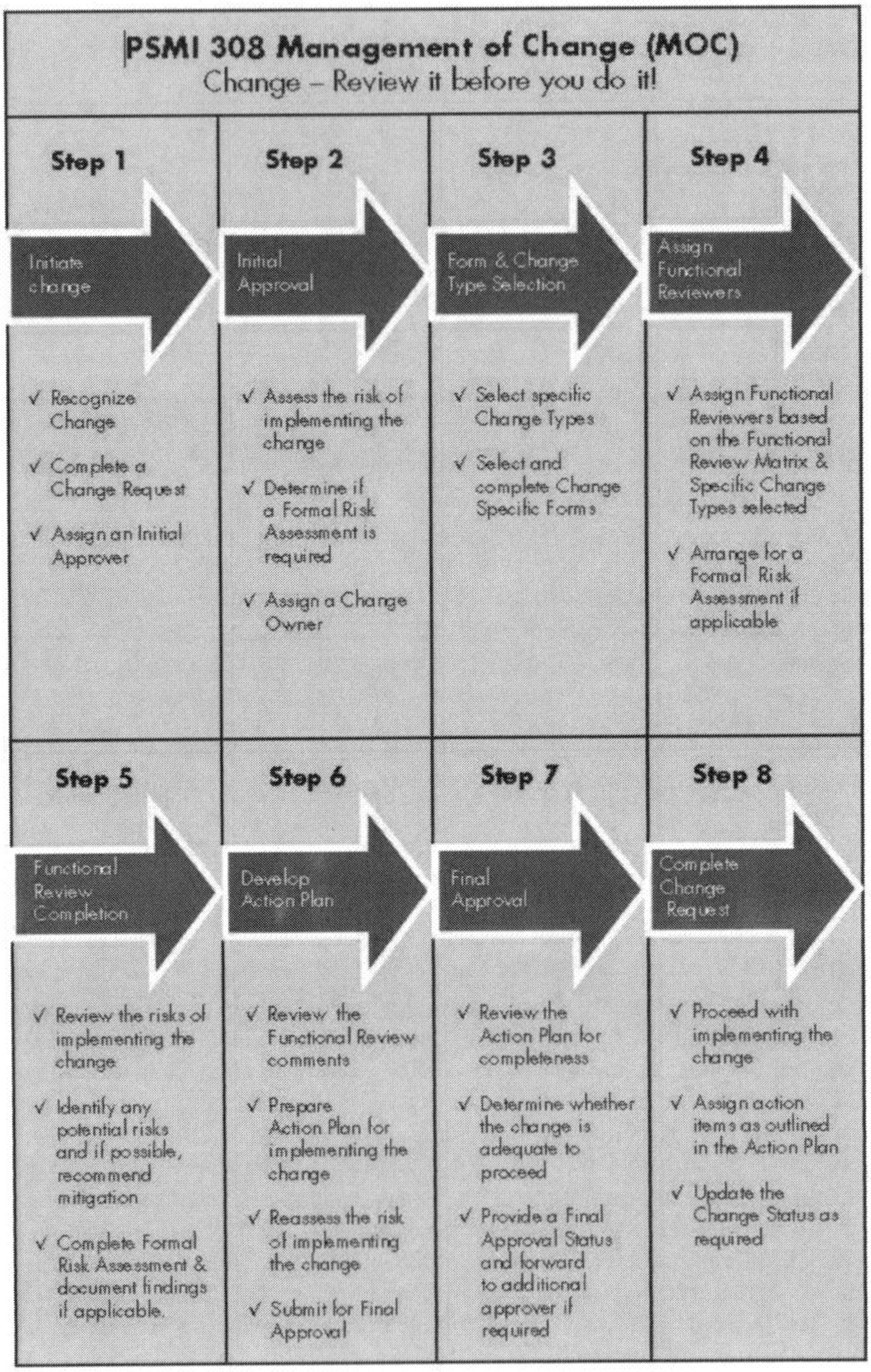

Figure 3 – Irving's MOC Process Workflow

3.6 Step 6 – Develop Action Plan

The MOC owner reviews the Function Reviewers' comments, prepares the action plan, reassesses the risk and submits for final approval. The change owner should now be aware of the requirements (identified by functional reviewers) to address in order to implement the change safely. The action plan is developed using functional review information along with the document/training checklist.

3.6.1 Observations

This part of the process can be very challenging. Change owners must document their action plan, which includes all steps required to completely implement the change. Issues can arise if they are not addressing all the concerns raised by functional reviewers, and are not reassessing the risk of implementing the change correctly. The change owner could consequently not select the appropriate final approver. There is no hold point for the creation of the action plan, and change owners could overlook this part of the process, which can result in the identified risk remaining unmitigated and critical PSM documents required for day-to-day implementation of PSM elements not being updated **(refer back to Figures 1 and 2)**.

3.7 Step 7 – Final Approval

Final approvers review the action plan for completeness, determine if the change is adequate to proceed, and provide final approval status.

3.7.1 Challenges

If uncertain of the requirement to review the MOC for quality, final approvers can run the risk of approving changes with risks inherent to any of the previous steps. Changes can potentially be approved even when the change owners fail to document adequate mitigation for the identified intolerable risk associated with the implementation (discussed in Steps 3 and 4). MOCs may be cancelled or closed because the process change was implemented before MOC approval, especially in the case of Emergency MOC's. These changes still need to be reviewed for adequacy and approved to ensure hazards have been identified and addressed and to ensure all PSI and other documents/systems used daily for PSM implementation are updated to reflect the change.

3.8 Step 8 – Complete Change Request

The change owner proceeds with the implementation, assigns action plan items, and updates the Change Status as required. This step also consists of:

1. Proceeding with installation/PSSR
2. Completing documentation updates
3. Ensuring all assigned action items are completed
4. Proper closure of the MOC in a timely manner which includes replacing temporary changes with a new MOC or returning the change to its original state.

3.8.1 Challenges

When a change is perceived as implemented successfully, change owners can quickly “close the book” and move on to other work assignments. They need to consistently assign actions to the individuals responsible for completing the various tasks required for change implementation and follow-up the actions. There must be tracking to ensure the tasks are completed, and to indicate when the change has been completely implemented and processed for closure. When action plans are developed, the timely completion of action items needs to occur consistently.

Installation and PSSR – Some change owners could be unaware of the PSSR requirement or choose not to execute one, thinking the MOC covered the initial start-up requirements. The PSSR is a vital part of the PSM cycle to help ensure MOC requirements have been met. The PSSR helps ensure changes are implemented as described in the MOC, the action plan is developed and items that must be addressed before start-up are flagged.

Not completing action items/not properly closing out the MOC – As mentioned above, change owners sometimes fail to assign actions for each aspect of the change i.e., drawings and procedures. For those action items assigned to staff, the items may not be followed-up on. This again results from the change owner’s expectation that the change has been successfully implemented and not recognizing the importance of getting the ball from the “red zone” to the “end zone”. Some of the resulting oversignts may include improper updating of Process Safety Information, insufficient information (or incorrect information) used to conduct a PHA, outdated or incorrect operating procedures and other document updates associated with PSM implementation. The PSSR process, if conducted properly, should catch those omissions related to action plan development and implementation. When the MOC coordinator reviews the implemented MOCs and notes missing PSSRs when required, the change owner and area managers should be notified and a PSSR conducted. This is the best way to quickly flag any installation omissions and action plan deficiencies that could increase the risk.

Timely MOC Closure – There is no maximum time limit for closing permanent MOCs. The eMOC system is great at tracking action items, if the action items are assigned, which is not the case for some of the MOCs. The number of emergency and temporary changes is reported as a KPI however, the number of days late is not published as a KPI.

4. Key Aspects of an Effective MOC Process

As mentioned above, steps can be taken to fortify the eMOC system and workflow. The integrity of an MOC system can be compromised, and it is prudent to address these anomalies in a timely fashion, project management style. To avoid these MOC oversights, the opportunities listed below should be considered:

- Electronic MOC - challenges with software
- MOC Training
- Additional Key Performance Indicators
- Miscellaneous MOC challenges

4.1 MOC Software

The MOC program is managed through an electronic EHS software system. The system has proven reliable with regards to '"uptime." However during times when the system is unavailable, a paper-copy version of the MOC process is available. The MOC coordinator and Systems administrator manage MOCs during those times on behalf of the organization. The software is easy to use and is very customizable despite software speed issues and occasional bugs. The electronic system is open to the organization (with permissions), so that all changes, regardless of status, can be viewed through a number of search operations. One of the greatest advantages of the eMOC system is tracking the status of MOCs and implementation action items. The MOC system is easier to execute and manage when staff knows where the MOC is in the workflow versus the former paper trail. The new MOC coordinator is responsible for monitoring the system, both from a process execution and approval point of view. Monitoring the action item execution is especially important. Below are some things to consider when selecting and implementing MOC software:

- **Make sure the mandatory workflow steps and completion of mandatory forms are not skipped**, by designing the electronic MOC system so that users cannot advance in the eMOC workflow process until the hold points or permissives have been met. The lagging solution is to rely only on the approvers and PSM staff to catch omissions and require corrections.
- **MOC software also needs mandatory fields for selecting the specific change types**. This will allow auditing of the change requests to ensure the proper functional reviews are being completed.
- **Some MOC software allows the assignment of Functional Reviewers before the Formal Risk Assessment is completed**. The software workflow is being changed so that, when a Formal Risk Assessment is required, assignment of the Functional Reviewers will not be enabled until the assessment is complete and associated documentation is attached to the change request.
- **Functional Reviewers currently have an electronic "considerations checklist" to remind them of specific items to consider** when reviewing the proposed change for risks. These checklists should be mandatory. PSM staff should be able to audit what each reviewer considered. Mandatory completion of the Functional Review Form improves the review process and provides critical information needed for final approval.
- **MOC software should also require Functional Reviewers to indicate the potential consequence and likelihood of the risk** when the risk is identified as intolerable. This information is required to enable the Change Owner to develop actions to mitigate the risk to an acceptable level. A new section can be added to the Functional Review form insisting that the Functional Reviewer document the potential consequence and likelihood of the risk associated with the proposed change. The highest risk identified by the Functional Reviewers will then be transferred to the Action Plan form at Step 6 of the process.

4.2 Training Opportunities

The initial training for the new eMOC process was extensive, and all employees were made aware of the need to manage changes, and provided an overview of the MOC process and electronic system. It needs to be followed by role-specific training for the Change Owners, Functional Reviewers, and Final Approvers. Listed below are some things to consider when developing training for an MOC process:

- **Functional Reviewers should be advised not to review any change request which does not have sufficient information to identify risks.** In these situations, the Functional Reviewers should contact the Change Owner, and request the completion of a formal assessment. Also, the organization should train additional hazard review facilitators in each process area, to conduct the required formal assessments (risk reviews).
- **Change Owners must document an action plan at Step 6**, outlining the tasks required to properly implement the change. This can be addressed through the role-specific training.
- **The requirement for a Pre-startup Safety Review must be recognized by the Change Owners.** The PSSR Coordinator should be tasked with reviewing all MOCs as they are approved, to determine if a PSSR is required as part of the change implementation, and then contact the Change Owner to initiate the PSSR.
- **Final Approvers should be reviewing the Functional Reviews, and the Action Plan, prior to approving the change.** This can be addressed through role-specific training.
- **Change Owners must assign actions for tasks to be completed** as part of an approved MOC. This in turn provides an indication as to when the change is completely implemented, and prompts the Change Owner to complete / close the MOC. Again, this can be addressed through role-specific training.
- **All personnel should have role-specific PSM training**. This training can provide individuals with the knowledge they require to properly complete their part of the process. Their MOC accountabilities can be spelled out in a "PSM Duties and Accountabilities Template", which can be part of the "awareness" training.

The MOC coordinator's role is to conduct training sessions, personal training, and continuously look for ways of improving and streamlining the process. With respect to rolling out an MOC program, management support is needed to ensure all levels in the organization are educated on their roles and responsibilities, and those of others.

4.3 Key Performance Indicators

KPIs communicate issues with MOC and are therefore the first step in the improvement process. Currently Irving tracks three MOC performance indicators:

- # of temporary MOCs
- # of emergency MOCs

- # of expired MOCs – those temporary and emergency MOCs that have exceed their time limit. Temporary MOC's must be returned to their original state within the time period approved for the specific temporary MOC. Emergency MOCs are rare in occasion and are supposed to either be completed to satisfy all normal MOC requirements or returned to the original state within 24 hours.

Irving is currently developing these additional KPIs:

- The number of MOCs approved with an intolerable risk identified
- The number of MOCs approved without an action plan
- The number of MOCs closed without actions being assigned. More KPIs will be developed to monitor the system changes proposed previously

Other possible KPIs include:

- The number of permanent MOCs that are late for closure expressed as a number and percentage of overall MOCs
- The number of permanent MOCs that are late for closure reported as less than 30 days late, greater than 60 days, greater than 90 days
- The number of temporary MOCs that are late for closure reported as less than 30 days late, greater than 60 days, greater than 90 days
- The number of incident root causes related to inadequate MOCs
- "Look Ahead" metrics for temporary MOCs that will expire in 14 days, 30 days, 60 days
- Establish KPI to show specific steps in MOC workflow that were not performed properly
- Number of unreviewed/unapproved implemented changes

4.4 Other MOC Considerations

- **Require that when unapproved changes are discovered**, staff ensure an MOC is initiated to properly review, approve and document the implemented change and any corrections to the installed design are identified and corrected
- **Train and coach staff on PSSRs** as PSSRs are integral to ensuring the MOC system is properly followed
- **Establish time limit for closing MOCs** to help ensure documents are updated as soon as possible
- **If MOCs are being cancelled or closed** because the process change was implemented before MOC approval, then these changes still need to be reviewed for adequacy and approved to ensure hazards have been identified and addressed and to ensure all PSI and other documents/systems used daily for PSM implementation are updated to reflect the change

5. Common MOC Opportunities Observed During Audits

These are examples of findings observed while auditing other facilities (not Irving):

- Not conducting a safety review for procedural changes – All process changes require an evaluation of the safety and health impacts. Changes to procedures should be reviewed to ensure all pertinent information is included and all parts of the procedure are updated. A good technique is to use a checklist composed of questions to help evaluate the procedural change and use of What-If or Procedural HAZOP technique where skipping steps and performing the steps incorrectly are evaluated to ensure that safeguards are in place and adequate to mitigate the consequences and prevent the causes of the procedure being performed incorrectly.
- When procedures are required to be updated as a result of operating limit changes, equipment additions/deletions, piping configurations (adding/deleting valves/adding instruments), etc., often the redline SOP is not part of the MOC scope and therefore, not reviewed and approved as part of the scope. Often the MOC owner may flag that an SOP needs updating however, the SOP is redlined and updated after the MOC is approved, either through a separate MOC or only through a document change notice. The problem is procedural changes are changes that must meet the minimal requirements for MOCs (OSHA and industry standards), which includes evaluating the safety and health impacts of the procedural change. Many MOC systems do not require a safety review for procedural changes or if the SOP change is handled separately from the main change, there is opportunity to overlook a safety issue since the overall change is evaluated separately.
- Procedures for jumpers (bypassing a safety system) and other temporary changes may have special forms and systems to manage the change and may not be tracked. Often individual production areas manage jumpers through a paper system due to their intended short duration. There must be oversight through auditing and reporting to ensure jumpers are consistently managed properly since most jumpers are bypasses of safety systems and therefore the risk is greatly increased when these safety systems are unavailable.
- MOC of programming changes or removing equipment from service are types of changes that once done, the change is implemented. This can lead to changes being implemented prior to affected personnel knowing about the change. Special PSSRs for these changes are necessary to ensure training occurs and staff on-shift when the change is made know exactly when the change is implemented.
- Not updating PSI for temporary changes – Temporary changes can last for a short duration or for up to 6 months or a year. If PSI is not updated for temporary changes, process change evaluations completed while the temporary change is in effect, may not consider the temporary change in the process.

- Not having a system to check out P&IDs – for large facilities with project and process engineering groups this can help ensure that proposed design changes are not based on P&IDs that will change in the near future as other MOCs or projects that are in progress are implemented.

6. Conclusion

To avoid making a dog's breakfast out of your MOC system the electronic or paper based system must require certain tasks in the workflow. Development and delivery of initial and ongoing staff training are critical to ensure responsibilities and accountability are understood and that personnel have the skills and understand the requirements to execute their responsibilities to effectively carry out the MOC tasks. Constant monitoring by PSM staff and reporting implementation throughout the organization via KPIs are keys to achieving and sustaining an effective MOC system.

Mechanical Integrity 101 For Process Safety Professionals – How Safety is impacted

Robert Smith
ABB Consulting
3700 Sam Houston Parkway S.
Houston TX 77042
Robert.Smith@US.ABB.com

Prepared for Presentation at
American Institute of Chemical Engineers
2014 Spring Meeting
10th Global Congress on Process Safety
New Orleans, LA
March 30 – April 2, 2014

UNPUBLISHED

Keywords: Mechanical Integrity, Process Safety Management, Safety, Plant Operations, Maintenance & Reliability

Abstract

This paper prepares you to evaluate your existing practices versus recognized codes and standards and then to efficiently improve or develop and implement your mechanical integrity (MI) program. Topics also include linkages of MI to Process safety and reliability, What a “Good” MI Program looks like, and How to measure success.

This paper is intended for the PSM Manager or Coordinators and is also useful for Plant managers or anyone seeking an overview of Mechanical Integrity programs and the linkages to Process Safety Management.

The paper outlines the elements of a Mechanical Integrity Program, typical best practice elements with background information on what is required for each element of a program as well as discussing inspection, linkages to Process Safety Information and Quality Assurance.

1. Overview of Mechanical Integrity

Mechanical Integrity isn’t a project, it’s a process that lives and takes new shape and direction when appropriate. Mechanical Integrity requires discipline, and it requires that everyone be involved starting with company and plant leadership.

Mechanical Integrity is intertwined with Process Safety Management as well as Reliability & Maintenance. While there are individuals and teams that specialize in each specific function it is important to acknowledge how they inter-relate. It is difficult, if not impossible, to have one area excel if they all are not equally as strong.

As it relates to OSHA Citations, Process Safety Information is the number one cited area and Mechanical Integrity is number 2. [1]

Figure 1.

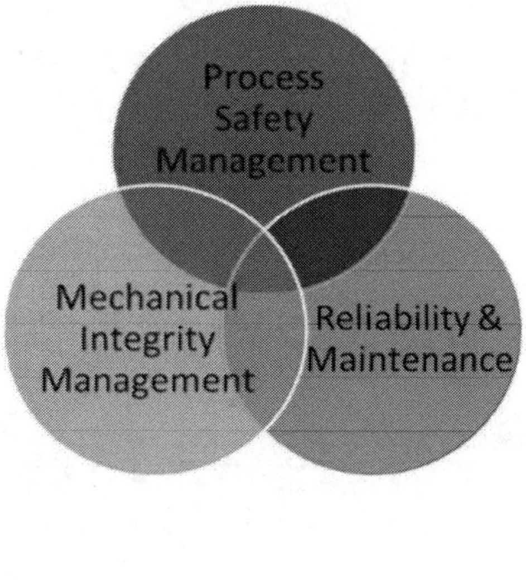

2. A Look at the standard

First let's review the standard from 29 CFR 1910.119. The OSHA standard in summary states:

- Operations must establish and implement written procedures to maintain the ongoing integrity of process equipment such as tanks, vessels, piping systems, pumps, valves, controls, interlocks, etc.
- Must correct deficiencies.
- Train employees and contractors involved in maintaining the integrity of equipment.
- Perform periodic inspection and testing, following "recognized and generally accepted good engineering practices,"(RAGAGEP) and document that inspections have been done.
- Correct equipment deficiencies outside acceptable limits before further use or in a safe and timely manner.
- Develop a quality assurance program to ensure that:
    - Equipment for new plants is suitable for use in the process, and is properly installed; and
    - All maintenance materials, spare parts and equipment are suitable for intended use.

3. The Total Mechanical Integrity Program

A total mechanical integrity program consists of the lifecycle depicted in Figure 1. Within this process there are a number of avenues where the program can attain excellence or dwindle to sub-standard.

Figure 2.

Let's examine the process, the first two steps Design and Construction are typically handled by an Engineering and Procurement Contractor (EPC). Operation and Maintenance typically falls to the Operating location, while inspection and modification typically are a mix of the operating plant and outside contractors and consultants. Typically the two biggest areas where issues arise out of and take a program from great to average or worse are in the areas of Operation & Maintenance and Inspection and modification.

Design and Construction

Rarely it seems the Safety Professional is consulted on the Design and construction of an asset or asset modification. Right or wrong, at the end of the day the safety professional has to deal with the outcome and has to ensure that all the relevant data is not only received from the EPC, but is also filed and retained as it relates to the Process Safety Information (PSI) element of Process Safety Management (PSM)[3]. Some of this information needed is:

- Block Flow Diagram
- Design Basis Information
- Equipment involved in the system
- Piping & Instrumentation Diagrams (P&IDs)
- Start-up and Operating Procedures
- Process Critical Equipment
- Process Flow Diagrams (PFDs)
- Process Hazards Analysis
- Safety Critical Equipment
- Vendor Data

Operation and Maintenance

Operation and Maintenance can be an interesting area to look at as there are often issues. Process safety is obviously dependent on these two areas often run by two different departments. Simply put, Operations runs the plant and Maintenance fixes it. Simple enough in concept, but factor in what each department is measured on and this can become an issue. Operations is measured on making the plant run and producing the maximum amount of material, whatever that is, in the shortest amount of time.

Maintenance is given the task of creating and maintaining a reliable plant environment. This can create conflict in the organization as downtime for repair costs not only the labor and material of repair, but downtime in some amount.

The conflict is in striking a balance between Operations, Maintenance and last but not least Safety and Process Safety Management.

Process Safety professionals are often caught in the middle on this. Everyone wants the plant to run and be profitable and safe; however, these objectives can easily become at odds and are highly dependent on the overall culture of the organization.

Inspection and Modification

It is critical that good inspection protocols are adopted, followed and documented. The process holds true in either scenario where internal resources are used or external contractors are utilized. A best practice is for the operating company to have a documented inspection program.

Modification of equipment or systems needs to be planned, documented and controlled. The use of a documented Management Of Change (MOC) system is critical to the success of any system or equipment modification. Management of Change programs are the third most cited areas as it relates to OSHA PSM audits [1]

Elements of a Mechanical Integrity Program

Often times engineers are hired and told "make the plant more reliable". If it were only that simple. Mechanical Integrity isn't a project, it's a process that lives and takes new shape and direction when appropriate. Mechanical Integrity requires discipline, and it requires that everyone in the plant be involved. [2]

Leadership Commitment

Let's consider the foundation of a well-engineered reliability initiative. We have to understand where we are starting from, so a bit of up front legwork is required:

- Do you have senior leadership championing this cause?
- Do they really understand what their role is within the Mechanical Integrity program?
- Do they know that they are the "chief roadblock-removers" and "the directors of assimilation"?

A successful program requires that management be the head cheerleaders for the cause and instills the expectation of everyone working toward the common goal of improving reliability. Through leadership's direction and guidance, the processes will start to take hold and the plant's culture will start to evolve. As roadblocks arise, it is leadership's responsibility, built on the foundation of training and process, to remove them so the goal of reliability improvement can proceed.

Accurate List of Equipment

The first step of an effective mechanical integrity program is to compile and categorize a list of process equipment and instrumentation. This list would include:

- pressure vessels
- storage tanks
- process piping
- pressure relief and vent systems
- fire protection system components
- emergency shutdown system
- interlocks
- pumps

Critical Equipment & Inspection

Following a general list, a list of Critical equipment should be complied. This critical list should be separated into both Process Critical and Safety Critical Equipment. While not covered in this paper, typically a good Process Hazard Analysis (PHA) will outline this equipment and related instrumentation.

Once this equipment is identified the equipment should be risk ranked and all the associated documentation should be kept for it specifically, this will include:

- Vendor Data
- Design Basis
- Maintenance records

Once the list is established it is very important to understand the Mean Time Between Failure (MTBF) and Mean Time To Repair (MTTR) data accuracy and know if there is an up-to-date Bill of Materials (BOM). MTBF and MTTR combined with the criticality of the equipment will set the foundation for the inspection frequency.

If the plant does not have all of this data there are several sources in addition to the plants Computerized Maintenance Management System (CMMS). Some of these include:

- National Board Inspection Code
- American Society for Testing and Material (ASTM)

- American Petroleum Institute (API)
- National Fire Protection Association (NFPA)
- American National Standards Institute (ANSI)
- American Society of Mechanical Engineers (ASME)

The applicable codes and standards such as the above provide information to help establish an effective testing and inspection frequency, as well as appropriate methodologies for inspection. Inspection in this context refers to both external and internal inspection - - covering areas from general condition to corrosion rates. While the Process Safety professional does not need to be an expert in inspection, it is helpful to understand the elements of a good program.

Each plant operation needs to develop procedures to ensure that tests and inspections are conducted properly and that consistency is maintained even where different employees may be involved. Appropriate training should be provided to maintenance personnel to ensure that they understand the preventive maintenance program procedures, safe practices, and the proper use and application of special equipment or unique tools that may be required. This training is part of the overall training program called for in the standard.

Quality Assurance Program

Quality Assurance (QA) is a way of preventing mistakes or defects in manufactured products and avoiding problems when operating a plant environment. A quality assurance system is needed to help ensure that a Mechanical Integrity Program starts off right and stays sustainable. The quality assurance program is an essential part of the mechanical integrity program and will help to maintain the primary and secondary lines of defense that have been designed into the process to prevent or mitigate unwanted chemical releases. "As built" drawings, together with certifications of coded vessels and other equipment, and materials of construction need to be verified and retained in the quality assurance documentation and as part of a quality Process Safety Information (PSI) program. Equipment installation jobs need to be properly inspected in the field for use of proper materials and procedures and to assure that qualified craftsmen are used to do the job. The use of appropriate gaskets, packing, bolts, valves, lubricants and welding rods need to be verified in the field. Also, procedures for installation of safety devices need to be verified, such as the torque on the bolts on ruptured disc installations, uniform torque on flange bolts, proper installation of pump seals, etc. Any changes in equipment that may become necessary will need to go through the MOC process.[4]

4. Conclusion

Summary: Mechanical Integrity

This paper has looked at how Mechanical Integrity is not just an operations and maintenance function, but has deep ties into safety and Process Safety Management of every facility. Much like Process Safety the best programs are a collaborative effort between departments and senior management. Below are some points to consider to create a successful mechanical integrity program:

1. Preventive maintenance programs should be established, funded and staffed to sufficient levels to avoid the need for "breakdown" maintenance.

2. Turnarounds need to be held often enough to avoid "breakdown" maintenance. Units need to stay down until scheduled repairs are completed.

3. All maintenance work should be performed by trained and experienced crafts workers.

4. Requirements for, and documentation of, contractor training should be equivalent to that of regular employees.

5. All maintenance work must be done using proper equipment, installation procedures, safety devices and according to applicable codes and standards.

6. It is critical to actually implement the written mechanical integrity program. A good MI program requires preventive maintenance be performed within 3 months and detailed records of maintenance and inspections be kept for the life of the equipment.

7. If your workplace frequently experiences "breakdown maintenance", your mechanical integrity program is not working. [4]

Keep in mind that while Mechanical Integrity is "owned" typically by some combination of Maintenance and operations, from an OSHA perspective the person that own Process Safety Management and it's compliance is the person that is accountable for a solid, documented MI program.

Appendix C from OSHA 25 CFR 1910.119 References)

5. References (examples given to show format)

[1] AFPM 2013 National Occupational & Process Safety Conference; May 14, 2013, The Woodlands, TX, OSHA update Presentation

[2]Appendix C to §1910.119 -- Compliance Guidelines and Recommendations for Process Safety Management (https://www.osha.gov/pls/oshaweb/owadisp.show_document?p_table=STANDARDS&p_id=9763

[3]Process Safety Management Guidelines for Compliance, OSHA 3133 1994, https://www.osha.gov/Publications/osha3133.html

[4]Process Safety Management of Highly Hazardous Chemicals -- Compliance Guidelines and Enforcement Procedures. 9/13/1994, https://www.osha.gov/pls/oshaweb/owadisp.show_document?p_table=DIRECTIVES&p_id=1558

Additional references not cited:

[5] CCPS. Guidelines for Risk Based Process Safety, Center for Chemical Process Safety. New York, NY. 2007

Enabling Performance Management – Producing vs. Collecting Metrics Data

Alfonsius Ariawan
DuPont Sustainable Solutions
1919 Minnesota Court, Mississauga, Ontario
Alfonsius.B.Ariawan@dupont.com

Prepared for Presentation at
American Institute of Chemical Engineers
2014 Spring Meeting
10th Global Congress on Process Safety
New Orleans, LA
March 30 – April 2, 2014

UNPUBLISHED

Enabling Performance Management – Producing vs. Collecting Metrics Data

Alfonsius Ariawan
DuPont Sustainable Solutions
1919 Minnesota Court, Mississauga, Ontario
Alfonsius.B.Ariawan@dupont.com

Keywords: Performance Management, Metrics, Data Management, Dashboard

Abstract

The topic of process safety metric selection has been widely discussed. Yet, some organizations still feel inadequate with their process safety performance monitoring. Is the data collection process as important as metric selection in ensuring the effectiveness of performance management? The concept of producing vs. collecting data is a shift in paradigm that could further increase the effectiveness of the overall performance management process. Currently, non-paper-based systems are available to facilitate this. This presentation will discuss the characteristics of a robust system and the benefits of having such systems in an organization.

1. Introduction

We cannot improve what we do not measure. That does not, however, mean that we can improve just because we measure. To be useful, what we choose to measure needs to point us to where improvements need to be made and what actions need to be taken. These measurements should help assure us of the safe operations of our units when they are safe, and alert us to take action when it is necessary to do so. Not having the right measures in place may give an organization a false sense of safety and even worse, drive complacency.

Before the Texas City incident in 2005, BP safety performance looked impressive. Fatality trends had been falling and injury rates had decreased from 1.4 per 200,000 working hours in 1999 to 0.53 per 200,000 working hours in 2004[1]. Unfortunately, this statistic had masked some critical issues that ultimately resulted in one of the deadliest industrial incidents in the U.S. The Baker report released in 2007 points to a number of causes – one of which was the reliance on metrics that are unsuitable as process safety performance indicators[2]:

"BP has emphasized personal safety in recent years and has achieved significant improvement in personal safety performance, but BP did not emphasize process safety. BP mistakenly interpreted improving personal injury rates as an indication of acceptable process safety performance...

BP's reliance on this data, combined with an inadequate process safety understanding, created a false sense of confidence that BP was properly addressing process safety risks."

Unfortunately, BP had another incident in 2010. The investigation of the Macando event points to more or less the same issue. Chemical Safety Board (CSB) investigator Cheryl MacKenzie said[3]:

"The emphasis on personal injury and lost work-time data obscures the bigger picture: that companies need to develop indicators that give them realistic information about their potential for catastrophic accidents. How safety is measured and managed is at the very core of accident prevention. If companies are not measuring safety performance effectively and using those data to continuously improve, they will likely be left in the dark about their safety risks."

Following these incidents, there was a rise in the number of references on the subject of process safety metrics. In late 2007, the Center for Chemical Process Safety (CCPS) issued a guideline for the development and use of Leading and Lagging Process Safety metrics[4]. In response to the CSB recommendation to develop a guidance document on PSM metrics following the BP Texas City catastrophe, the American Petroleum Institute (API) team published the API Recommended Practice 754 in 2010[5]. More recently, the European Chemical Industry Council (CEFIC) initiated some work on the topic of Performance Indicator[6]. Despite all these resources, however, some organizations still feel inadequate with their Process Safety performance monitoring. Could it be that this feeling of inadequacy has more to do with how metrics are collected and used? Is the data collection process as important as metric selection in ensuring the effectiveness of performance management?

This paper considers the elements of a performance management program and discusses the common challenges that an organization often faces during its implementation. It proposes characteristics of a robust program and discusses how a non-paper-based data management system can overcome the common barriers surrounding metrics data collection. The right system will allow organizations to produce metrics data, rather than needing to collect them – a powerful shift in paradigm.

2. Elements of a Performance Management Program

A performance management program should include the following four elements. It is important for all of these four elements to be executed well in order to have an effective program.

2.1 A Selection of Metrics

Careful thought should be given to the choices of metrics to monitor. The collection of metrics should contain both leading and lagging indicators, tiered appropriately, such as per the API 754 recommendations[5]. Data should be collectable at a reasonable frequency, and able to provide signals and directions for specific actions to be taken. The selected metrics need to be sensitive to relevant changes. This is particularly important with higher tiered metrics, so effective trouble shooting can take place before it is too late. The type of metrics data can also make a difference. Data of continuous type (e.g. cycle time for completing incident investigation, number of days between an incident and when key learnings are shared, number of days between when an action item is completed vs. when it is due) is better for trending, allowing for some extrapolation to be made. By having such data, an organization may get an earlier warning of potential issues and can proactively address the situation as warranted. In contrast, discrete metrics (e.g. number of overdue action items), while still meaningful, provide a view of what already happened – that is, a number of action items are already overdue, before the organization can react to them. Of course, many times, the choices of metrics are limited by the practicality of data collection.

2.2 Data Collection Process

The data collection process is probably the most difficult element of a performance management program to implement. Data needs to be collected in a coordinated manner, at some established frequency. This is not an easy task. This can be especially challenging in an organization that is made up of multiple entities or businesses. Generally, data is collected by each entity and is then summarized, reported and rolled up into a corporate database or dashboard for review. Leadership plays a key role in ensuring that the entire organization understands the value of the process. Leadership also needs to set the appropriate expectation for timely compliance.

The data collection process can also be resource intensive. Collection, consolidation and reporting of data take considerable effort and can leave personnel feeling frustrated. Organizations may feel that they have chosen the wrong set of metrics when, in fact, it is the supporting process they are struggling with. This is where having a non-paper-based data management system in place is helpful. We will explore this topic further below.

2.3 Monitoring and Review Process

A key element of the performance management program is the monitoring and review of the metrics data. Unless regularly reviewed, metrics data becomes a meaningless output of a reporting process. A review process needs to be established at some frequency and is typically conducted by the leadership team. To facilitate the process, metrics need to be summarized and presented by way of a dashboard. The format and complexity of the dashboard depends on the reviewing audience. At a corporate level, the dashboard tends to be simpler, allowing the audience to get a good understanding of the status of the organization quickly, without the need for much analysis. The dashboard should be sufficiently sensitive to changes in the data. Typically, dashboards at this level are presented graphically or in tabular format. At site level, however, a dashboard can be quite detailed. In addition to enabling the assessment of a site's safety performance, a site dashboard may also be used to diagnose operational discipline or other operational issues, which may affect safety performance if not addressed early (tier 4).

It is not uncommon to have more frequent reviews of metrics data by site. A visual board can be put up in a control room or maintenance shop, for example, to assist with this. At typical DuPont sites, a weekly site leadership meeting includes a review of the PSM data listed on the board[6]. A visual board is a great communication tool and having it in place helps with the overall site engagement. It also allows for a more continuous monitoring of performance.

2.4 Action Initiation and Tracking Process

The test of a good performance management program is the program's ability to provide confidence on the status of process safety in the organization. This confidence needs to be translated to something actionable; otherwise, the organization will risk having the program turn into nothing but an elaborate reporting process. The leadership team needs to use the metrics data to guide them on the appropriate actions to take. For example, a safety stand down may need to be initiated following a disappointing safety performance, or the leadership team may decide to conduct an organization-wide Safety Perception Survey to gauge the organization's safety culture. At the site level, the data may show that personnel turnover is contributing to some of the observed trends, in which case, the site may want to put a special effort to strengthen its MOC-Personnel management and evaluate its learning and development program. On the flip side, an organization should also be able to confidently use the data as the basis for rewarding and recognizing individual units or sites, or the organization as a whole, following a stellar performance. This is an effective and objective way to motivate an organization.

Sometimes, it is not clear as to what causes certain trends and what actions are best taken. In such cases, further data analysis may be required. This is when having additional data is helpful. With a paper-based performance management program, collecting such data will be challenging. On the other hand, having a non-paper-based system will allow an organization to query additional data with more ease.

As part of the performance management program, it is also important to include an action tracking process. This is to ensure that corrective and preventative actions are executed before pre-defined due dates. Performance against this requirement may also be included in the collection of metrics being tracked.

3. Common Challenges in Implementing a Performance Management Program

An organization looking to implement a performance management program is often faced with several common challenges. In many cases, these challenges stem from the failure to execute one or more of the above elements. In addition to these, there are other functional and cultural issues that an organization often needs to overcome.

3.1 Functional Issues

3.1.1 Unclear Definitions

Many organizations fail to provide clear metric definitions. Definitions are typically available, but they do not adequately address the practical nuances that may be encountered during the data collection process. Ultimately, the organization needs to be confident in the data that it is monitoring and using for the basis of actions and rewards. Not having clear definitions may cause data to be unreliable and inconsistent. It may provide a false sense of safety, or conversely, an unnecessary sense of paranoia, and the consequences could be serious. The personnel tasked to collect the data may also feel frustrated and confused, and may begin questioning the value of the process.

3.1.2 Practical Difficulty in Capturing Data

While the intent is noble, some data is just not practical to capture and track. Metrics need to be selected only if data is collectible. This issue is typically encountered when metrics are selected without review by the organizational units or sites that are ultimately responsible for data collection. It is important that sites understand what needs to be collected and confirm that they are able to do so. Sites are to be encouraged to discuss and share suggestions. If a metric is deemed not practical to collect, the organization should drop it from the list and find an alternative.

3.1.3 Inconsistent Formatting/Reporting

When an organization is made up of multiple units or sites, consolidation of data may become a challenge. Usually, issues are encountered in the form of inconsistent data or report format, untimely submission of data or incomplete reporting. Having a standard template will help with format consistency, but only to a certain extent. Typically, this issue resolves itself after some period of time as the organization gets used to the process.

3.1.4 Intensive Resource Requirement

The coordination of data collection and the consolidation, formatting and rolling up of information take considerable effort. Many organizations are resource constrained, and so personnel are usually assigned to these tasks in addition to their day-to-day job. Workload usually spikes near the end of the month and many struggle with this. This often translates to late reporting and an overall non-robust program.

3.2 Cultural Issues

3.2.1 Reluctance to Share and be Compared

During the initial phases of performance management program implementation, there can be a general reluctance to share data for fear of comparison with other sites. Organizations with multiple sites typically encounter this issue – especially those with a less mature safety culture. Competition is healthy, but it is really the spirit of collective improvement through sharing that leadership should try to foster.

3.2.2 Fear of Repercussion

If a site does not understand the intent of the program, it may be reluctant to share performance data for fear of being punished. The site may point to the unclear definition of metrics as a reason for not submitting data, or use the ambiguity in the definition to submit a more favorable interpretation of the metrics. Leadership needs to ensure that expectations are clear and that all personnel understand the purpose of performance management as a program to drive the organization towards continuous improvement.

3.2.3 Questioning of Value

With the extra workload associated with data collection, particularly nearing the reporting period, personnel may feel frustrated and start to question the value of the process. With no visibility of the review process and the subsequent actions taken, personnel may feel that they are collecting data for the sake of reporting. It is important that any output of the review be communicated back to the organization at an appropriate frequency, such as through regular safety or town hall meetings.

4. Characteristics of a Robust Program

It is clear that to have an effective performance management program, the four elements described above need to be functioning properly with the highlighted issues addressed. A robust program, however, will need to be more than just functional. It is suggested that a robust performance management program also exhibits the following additional characteristics:

- Less dependency on manual intervention: Manual handling and sharing of data can be both a source of frustration, delay and error. A program that is less dependent on human resources would be more robust. Installing a non-paper-based (IT) data management system to assist with the data collection and sharing process can help reduce this dependency.

- More continuous vs. intermittent data collection: For example, an organization should know the number of overdue audit recommendations at any given time rather than just once a month. In order to achieve this, data collection needs to be event- rather than time-triggered. Having a more continuous data collection process will also regulate personnel workload.

- Ability to produce additional data for analysis: Sometimes the need arises for additional data to be produced and analyzed, due to unexpected metric trends. This is possible when data is collected beyond just what is reported for performance management review. With a paper-based data management system, this may not be practical. But, with an IT-based system, queries can be set up to enable further analysis or diagnosis of metric trends, in order to pinpoint specific issues to be addressed.

- Linked to a centralized corrective and preventative action tracking system: Any action resulting from the review of performance management data will need to be captured and tracked for progress. A robust system will link the performance management program to a centralized corrective and preventative action database to facilitate tracking and ensure completion of actions.

These additional characteristics of a robust program are difficult to achieve with a paper-based data management system. Fortunately, there currently exist information technologies that can assist organizations with non-paper-based data management. This topic is explored further below.

5. Non-Paper-Based Data Management System

Many organizations are now turning to IT tools to help them with data collection, consolidation and presentation for their performance management program. This is primarily driven by resource constraint. Some organizations build their own data collection templates and dashboards using Microsoft Excel or a similar program, and some use commercial software. Research by the Aberdeen Group highlighted that best-in-class companies tend to invest in an integrated safety system that connects with plant automation data directly[7]. Such a system allows for real-time performance monitoring and provides visibility to plant operation for diagnostic purposes. Having the right non-paper-based IT data management system in place can certainly help with the robustness of a performance management program.

For example, a data management system that works by managing process safety workflows can make performance management more effective and robust. The system works by collecting information as events occur, such as when the need to initiate a change arises, an emergency drill or PHA is conducted, or when a personnel transfer occurs. A user will create the appropriate record into the system indicating that an event has occurred, or is about to occur. The system will then guide the organization through the workflow as defined and configured into the system. It will trigger appropriate notifications, requests for necessary approvals, or requests for other additional information as necessary, until all requirements to close the record are completed. It essentially functions to electronically guide the organization through its process safety standard.

The idea of collecting data as it is available or required in the workflow is a change from how a performance management program is traditionally done. With this, data collection is now part of the management of the specific PSM element. No longer are personnel feeling that they are collecting data for the sake of reporting. The use of the software will provide them with an understanding and clarity on the relevance of the data being collected in the overall workflow. Users will get an appreciation that the data being collected is consistent with the requirements of the organization's PSM standard. As a bonus, the system will also force certain operational discipline. Overdue items are highlighted immediately and reminders automatically triggered. Short-cuts or incomplete information on certain forms can be configured so that they are not possible when using the software.

Collecting data as part of workflow management using an IT system will also allow the organization easier access to more information. Information that is not typically reported for performance management purposes is now available – electronically. For example, the leadership team can now easily find out if the number of near misses is disproportionally higher with overtime workers, in which case a review the organization's fatigue management maybe warranted. This allows for a more effective diagnosis of trends, facilitated by data queries or built-in search capabilities.

Having data available on demand for management review and diagnosis is powerful. Data no longer needs to be collected specifically for the purpose of performance management review at some regular frequency. Instead, data can now be produced anytime. The system will reflect all changes real-time into the pre-built dashboards. The idea of being able to produce and report data as needed vs. having to collect data reflects a shift from how performance management is traditionally implemented. With this, functional and cultural issues highlighted above can be alleviated, if not eliminated altogether.

So, what kind of data management system is most effective? DuPont represents the 14 PSM elements in the spokes and wheel model, as shown in Figure 1. An ideal data management system should be able to manage all 14 PSM elements in an integrated manner to get the full benefit of the system. For example, an incident that triggers a record into the incident management module should be able to be linked to the resulting MOC record, if one of the recommendations from the investigation is to execute a process change. Subsequently, the MOC record should be able to be linked to a Process Hazards Analysis (PHA) record, which may be triggered by the request for the change, and so on. All action items should also be managed centrally, so as to enable users to have a complete view of all their obligations. This is certainly more preferable than having multiple systems managing the various PSM elements, which will require resources to download and consolidate data from the multiple source points. In fact, having multiple systems will defeat the benefit of having an IT-based data management system. Also, because workflows may change over time, the chosen system needs to be configurable so as to give the organization sufficient flexibility to accommodate future changes.

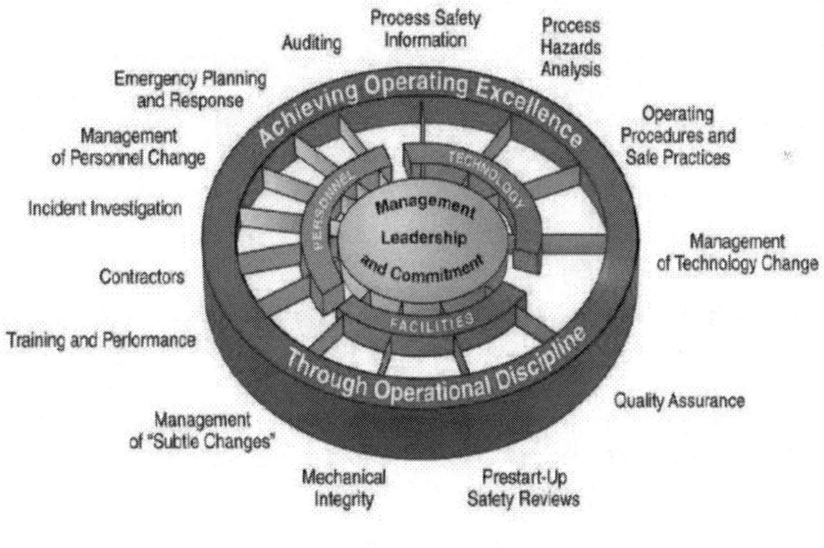

Figure 1. DuPont PSM Model

But the system is just a tool to facilitate the overall objective. Supporting business processes, operational discipline, leadership support and the right organizational culture are still important and needed for an effective performance management program. Also, the system will work only as well as the workflow that is configured into it. An effective system will have a practical, yet rigorous element management workflow configured into it.

6. Conclusion

Having a robust performance management system is critical for the safe operation of an organization. The choice of metrics is important, but equally important is the implementation of an effective data collection and rigorous review process, and the subsequent means to initiate and track actions. Many organizations struggle with these elements. They are usually faced with some common functional and cultural issues when implementing such system. Having a non-paper-based data management system can help. It allows for data to be collected as part of workflow management and be produced on demand. The concept of producing vs. collecting data for performance management is a paradigm shift that may help organizations with their journey of continuous improvement.

7. References

[1] "HSE Charting Tool," http://www.bp.com/en/global/corporate/sustainability/data-and-mapping-tools/hse-charting-tool.html#/health-and-safety/RIF-Group (assessed December 16, 2013)

[2] "The Report of the BP U.S. Refineries Independent Safety Review Pannel," http://www.bp.com/liveassets/bp_internet/globalbp/globalbp_uk_english/SP/STAGING/local_assets/assets/pdfs/Baker_panel_report.pdf (assessed December 16, 2013)

[3] "CSB Investigation: At the Time of 2010 Gulf Blowout, Transocean, BP, Industry Associations, and Government Offshore Regulators Had Not Effectively Learned Critical Lessons from 2005 BP Refinery Explosion in Implementing Safety Performance Indicators," http://www.csb.gov/csb-investigation-at-the-time-of-2010-gulf-blowout-transocean-bp-industry-associations-and-government-offshore-regulators-had-not-effectively-learned-critical-lessons-from-2005-bp-refinery-explosion-in-implementing-safety-performance-indicators/ (assessed December 16, 2013)

[4] Center for Chemical Process Safety: "Process Safety Leading and Lagging Indicators", American Institute of Chemical Engineers (2011)

[5] ANSI/API 754: "Process Safety Performance Indicators for the Refining and Petrochemical Industries", American Petroleum Institute (2010)

[6] "Guidance on Process Safety Performance Indicators," http://www.cefic.org/Documents/IndustrySupport/RC%20tools%20for%20SMEs/Document%20Tool%20Box/Guidance%20on%20Process%20Safety%20Performance%20Indicators.pdf (assessed December 16, 2013)

[7] Nuris Ismail: "Integrated Safety Systems: Ensuring Safety and Operational Performance", Aberdeen Group (2011)

Additional reference not cited:

[8] UK Health and Safety Executive: "Step by Step Guide to Developing process safety Performance indicators", HSG 254 (2006)

352260 Monetising Gas Hydrates: New Mode for Natural Gas Storage and Transportation Against LNG Storage and Transportation

Abstract:

Gas Hydrates are considered in the petroleum industry e.g. natural gas transportation through pipelines, during natural gas production from the crude oil wells. But this paper focuses on converting the nuisance of gas hydrates to an attractive means of transportation. New Technology development has been focusing on using gas hydrates as a way of converting gas to solids.

The Natural Gas Hydrate (NGH) is a clatherate compound in icy state and stabilised around -20 degree Celsius under atmosphere. This paper summarises about the cost effectiveness of the NGH value chain against the LNG (Liquefied Natural Gas) value chain and also in this paper entire processes and operations to create NGH pellets are designed and discussed. The calculations show that this process has a high initial plant cost but on a long term this process of NGH production is cost effective compared to LNG liquefaction process. Moreover, it is easy to handle the NGH at -20 to -30 degree Celsius compared to LNG at around -160 degree Celsius.

Thus, this unique method of transporting the natural gas in bulk solid form leads to reduction of the total investment cost for the entire value chain. Moreover, NGH is expected to be a plausible solution in monetising small to medium gas fields which are considered to be difficult to be developed economically using LNG.

352647 Towards the Development of a Control-Relevant Model of the Hydraulic Fracturing Process to Investigate Various Control Strategies

Abstract:

Hydraulic fracturing or 'fracking' is an increasingly important technique for developing energy resources across both the U.S. and the world, in which high pressure fluid is injected into the reservoir to crack the rock formation and provide highly conductive channels for oil and gas transport back to the wellbore. During this process, the underground environment is characterized by complex fluid dynamics, geo-mechanics, and multiphase flow phenomena. Surface equipment exerts control over the process from between hundreds of feet to several miles from the moving front of the fracture(s) below ground. The fluid pumped underground is a slurry consisting of water, proppant (often a special type of sand) to keep the created fractures open after the process finishes, and chemical additives such as cross-linkers, acids, and surfactants to maintain fluid viscosity, reduce corrosion, and perform other duties. During the fracturing process, pumping rate and fluid properties (viscosity and the concentrations of proppant and other additives) can be varied at the surface. The main objectives are to keep the fracture growth within a predefined region of the reservoir open by distributing the proppant evenly throughout the fracture so that oil and gas can flow after the fracturing fluids have leaked into the formation. Complications can include excessive growth of the fracture external to the target region that results in the proppant being spread so thinly that it is ineffective to keep the fracture open. Conversely, if a high concentration of proppant is added to the fracturing fluid mix too early in the process the condition called 'screen-out' can occur wherein the proppant enters the advancing fracture front and bridges the tip preventing further fracture growth and causing rapid pressure buildup in the system. A related issue is proppant settling, that is a large sand bank or dune forms near the entrance of the fracture at the wellbore resulting in only a small open region around the wellbore after the fluid leaks off. These three potential proppant-related challenges may have a large negative impact on oil well production.

Applying process control techniques to regulate the hydraulic fracturing process, via setting optimal pumping schedules, is a challenging problem not only due to the remoteness of the control action, but also due to the limited feedback information available because of the sensing system associated with the bottom-hole situation. To better understand these challenges, a mathematical model of the physics of the fracturing system is developed to determine the effect of different pumping schedules on the final fracture conductivity, the most widely used calculated performance metric. The physics-based model incorporates equations governing fracture propagation, fluid dynamics, proppant transport, and bank formation. The resulting system of nonlinear, partial differential equations is solved numerically using a moving grid method to track the fracture growth and particle tracking to predict the proppant movement and settling. Additionally, the 'shut-in' process, during which pumping is stopped and the rock formation walls are allowed to close as the fluid leaks off is a part of this model development. The final geometry of the channel available to oil and gas flow is predicted. Using this model, improved pumping schedules can be developed and tested to achieve near-optimal fracture conductivity.

352284 Shale Gas as a Potential Source of Unconventional Reserve for Crude Oil

Abstract:

With a decrease in the conventional reserves available for the recovery of oil, the alternative fossil fuel extraction industry is on a growth. As the development of unconventional oil and gas resources has been progressed, the heterogeneity and complexity of shale gas and tight oil reservoirs have become apparent. In the U.S. primarily shale gas and tight oil is expected to continue growing dramatically in coming decades. According to the U.S. Energy Information Administration, an assessment of 137 shale formations are identified in 41 countries other than those identified in U.S. At the local level, though the environmental impacts of shale gas and tight oil development may turn out to be significant.

Shale gas and tight oil are unconventional sources of petroleum that are found within the reservoirs with a very low permeability. The oil and gas contained within these reservoir rocks typically will not flow to the wellbore at the economic rates without assistance from technologically advanced drilling and completion processes. Usually, horizontal drilling coupled with multi stage fracturing is used to access these difficult to produce reservoirs. Unconventional natural gas and oil deposits are difficult to characterize overall, but in general are often lower in resource concentration, more dispersed over large area, and require well stimulation or some other extraction or conversion technologies.

Compliance: The Necessary Evil

Brian D. Rains
DuPont Sustainable Solutions
Brian.D.Rains@dupont.com

Prepared for Presentation at
American Institute of Chemical Engineers
2014 Spring Meeting
10th Global Congress on Process Safety
New Orleans, LA
March 30 – April 2, 2014

UNPUBLISHED

Compliance: The Necessary Evil

Brian D. Rains
DuPont Sustainable Solutions
Brian.D.Rains@dupont.com

Keywords: Compliance

Abstract

Compliance is not optional. Where process safety regulations are in place and enforced, failure to comply puts facilities and company reputations at risk. It could even land company leaders in jail. We could debate the actual role and need for regulations, however, that will not make them go away. The real question is, does compliance help a company (or facility) achieve improved process safety performance? Can compliance efforts interfere in the pursuit of zero process safety incidents by contributing to a complacent PSM culture? This presentation will detail DuPont's historical experience managing the seemingly competing goals of regulatory compliance and PSM excellence and make recommendations on how to fight against the Compliance Complacency Syndrome.

In 1967, psychiatrists examined the medical records of more than 5,000 medical patients as a way to determine whether stressful events might cause illnesses. Patients were asked to tally a list of 43 life events based on a relative score. Their results were published as what is commonly known today as the Holmes and Rahe Stress Scale. These "Top 10 Life Events" are listed here, plus their relative "Life Change Units" or stress units, as published in Wikipedia.

Life Event	Life Change Units
Death of a spouse	**100**
Divorce	**73**
Marital separation	**65**
Imprisonment	**63**
Death of a close family member	**63**
Personal injury or illness	**53**
Marriage	**50**
Dismissal from work	**47**
Marital reconciliation	**45**
Retirement	**45**

No. 8 on this list, right below marriage but well below Divorce (at No. 2) is 'Dismissal from Work.' Although I was not technically dismissed from my work, nor did I experience a divorce, in 2004 I did experience significant stress when I was part of a business unit divestiture. Overnight, I went from working for a company that I had been with for nearly 23 years to a new company. In many ways it felt like a dismissal and/or a divorce.

As I have analyzed the component parts of the stress associated with this change, differences in corporate culture were the main culprits.

The different cultures were understandable. My former company was large, with millions of shares being traded publicly every day. The Board of Directors consisted of all external, non-employee directors. The CEO was rotated every five to 10 years. My new company was even larger, but ownership was concentrated within a few family members, who also served as the Board of Directors and as the company CEO and executive vice president. They had held these positions for decades. Decision-making, ownership, and culture originated with the very few, most senior leaders.

Not surprisingly, nowhere were the two cultures between these companies more different than in their approach to compliance.

In my former company, rarely was the word (regulatory) compliance mentioned. We had strong corporate standards and expectations. We pursued a "Goal of Zero" with passion. There was a strong accountability culture in place. Without being explicated stated, regulatory compliance was something that just "happened" by following the corporate rules. I had even been introduced to the concept of assuming (or choosing) "regulatory compliance risk" as a possible business alternative.

In my new company, "10,000% Compliance" was the corporate mantra. This was communicated by the new company's leaders even before the official sale was completed. "Ten Thousand Percent Compliance" simply meant that 100% of the people were to comply 100% of the time. And they meant it. There was no margin for error. Needless to say, this added significant employee stress as we adjusted to this new, stark reality.

Compliance began with regulatory compliance. Tremendous effort was placed on fully understanding all applicable regulatory requirements. Training was conducted. Audits were performed. Systems were instituted to track new and emerging regulatory requirements. But it didn't stop there. Compliance with corporate standards was also a 10,000% expectation, as was compliance with local site procedures.

Individuals who could not, or would not, live up to the expectations associated with 10,000% Compliance looked for work elsewhere. Living through this abrupt transition in corporate culture, with particular emphasis on compliance led me to the observations that are summarized in the title for this paper: Regulatory Compliance; A Necessary Evil.

I firmly believe that both of these approaches to compliance have both merit and pitfalls. There are strengths and weaknesses of both. I will present these observations and personal conclusions in the body of this paper. I will label them as if they were extremes, which they are not. For clarity, I will call one culture "10,000% Compliance" and the second "Goal Zero."

Pros and Cons: 10,000% Compliance

Regulatory compliance is not optional. It must be a given. In a very real sense, license to operate potentially dangerous facilities is granted to companies from governments that are obligated to provide for and protect the welfare of its citizenry. Laws are enacted and enforced by governments in the discharge of these obligations and duties.

The history of Process Safety Regulations has clearly been in direct response to catastrophic incidents. The Seveso regulations in the European Community were some of the first Process Safety Regulations enacted by any governmental body. The regulation's common name comes from the small town, Seveso, in Italy where a significant process safety incident occurred in 1976. A few of the details are found below.

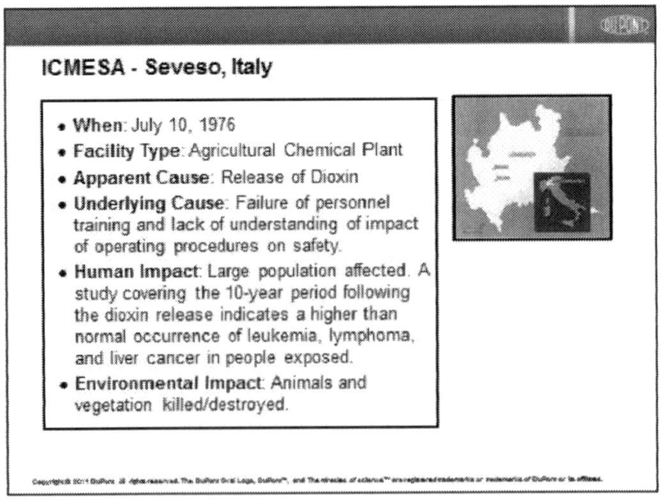

In the United States, a similar pattern exists. In fact, in its September 1992 'Compliance Guidelines' directive regarding compliance with 29 CFR 1910.119 regulations, better known as the PSM Regulations, OSHA stated that:

"In recent years, a number of catastrophic accidents in the chemical industry have drawn attention to the safety of processes involving highly hazardous chemicals. OSHA has determined that employees have been and continue to be exposed in their workplaces to the hazards of releases of highly hazardous chemicals which may be toxic, reactive, flammable, or explosive.

The requirements of the PSM standard are intended to eliminate or mitigate the consequences of such releases. The standard emphasizes the application of management controls when addressing the risks associated with handling or working near hazardous chemicals."

A similar pattern of regulatory reactivity and response has evolved in the Upstream Oil & Gas compliance environment following the April 2010 Deepwater Horizon incident in the Gulf of Mexico.

Indeed, regulatory compliance is not optional.

Regulatory compliance brings with it many benefits. Most benefits have to do with image and perception in the "court of public opinion," including:

- Building stronger credibility with regulators and government representatives.
- Establishing a stronger basis for trust with local communities; preserves a facility's "right to operate."
- Helping to avoid fines and helps prevent other court action and the potential negative financial impact.
- Helping to prevent reputational loss.

The primary challenge for organizations that adopt a compliance (only) orientation is that regulatory compliance will not necessarily ensure or even improve the organization's process safety performance. Complacency or an attitude of "compliance is good enough" can set in when compliance is the only management message. I will use a simple example to illustrate.

I believe all responsible operators of hazardous materials would agree that we are all still on a learning curve with respect to PSM excellence. Since 2007, the DuPont Corporate PSM Standard has been amended three times to reflect new knowledge gained from our own incidents, plus those from industry in general. Excellence in PSM requires a continual learning culture to drive continuous improvement. But yet the Seveso Regulations have gone through only one significant re-write in its history. And the U.S. OSHA regulations have not been formally amended a single time in the 22 years since they were issued. If an organization is relying on the regulations to keep pace with the ever changing and improving field and science of process safety management, I expect that their results will be inferior vs. their peers and will very likely be unacceptable.

I don't believe even the regulators believe that total compliance (only) with the existing regulations will yield zero (or maybe even top quartile) process safety incidents performance.

A 10,000% Compliance Culture does have a distinct impact on the overall management-to- employee dialogue and relationship. As intended, 10,000% leaves little room for interpretation as to what is expected of employees and their behavior. This can be a positive from an operational discipline perspective and can produce very consistent and reliable performance. However, the opposite is also possible when such a rigid standard causes an increase in unsatisfactory performance discussions up to and including terminations. An organization can quickly become demoralized when such activity highlights negative performance and the loss of experience and esprit de corps.

My '10,000% Compliance' observations are summarized in the table below:

10,000 % Compliance; STRENGTHS	**10,000 % Compliance; CHALLENGES**
#1: Builds stronger credibility with regulators and government representatives	#1: Can create a sense of complacency that "compliance is good enough"
#2: Establishes a stronger basis for trust with local communities; preserves a facility's "right to operate"	#2: Can place undue emphasis on compliance as the means to achieve excellent process safety performance
#3: Avoids fines and helps prevent other court action and the potential negative financial impact	#3: Can be demoralizing when such a strict "black and white" interpretation of individual performance is fully implemented
#4: Helps prevent reputation loss in the public's eye	
#5: Establishes clear employee compliance expectations	

<u>Pros and Cons: Goal Zero</u>

Most organizations and companies express their commitment to Safety and Process Safety Management excellence using the word "zero" or something similar, as seen below using examples from DuPont, Chevron, ExxonMobil, Dow Chemical plus others.

- ***Zero Harm to People or the Environment***
- ***No accidents, injuries or harm to the environment***
- ***Goal Zero***
- ***Zero is Attainable***
- ***Zero Fatalities***
- ***Zero Incidents Period (or ZIP)***
- ***Nobody Gets Hurt***

Needless to say, these statements of aspiration and direction are easier to articulate and communicate than they are to realize and achieve. But I believe we must applaud these organizations for making these statements during a period when catastrophic incidents continue to occur. In other words, these statements are made when, in many cases, how they are to be realized is not fully understood.

This speaks to the most important positive aspect to the Goal Zero Culture. Goal Zero drives the right behaviors:

- It drives a learning culture that is always attempting to learn and not repeat mistakes.
- It drives innovation and creativity in the pursuit of new ways to improve and reduce incidents.
- It drives greater involvement and engagement among all levels in the organization. Who benefits when incident performance improves? Everyone!
- It drives greater line management accountability for improving results at all levels.

One could also argue that even beyond the business benefits that Goal Zero may yield; it is simply the right thing to do. Injury to individuals and/or the environment is increasingly intolerable across the globe … and it should be. No organization has the right to inflict such damage.

There are some potential challenges to a Goal Zero pursuit, however. The first is that with such lofty aspirations and objectives, the organization may lose sight of how important regulatory compliance is. A certain type of arrogance and malaise can set in. Individuals can erroneously conclude that their work is on "a higher plane" than mere compliance. They may choose to not dedicate the necessary time and effort to stay abreast of changing regulatory requirement or the effort to be in full compliance.

A second potential challenge, and often the more difficult, is the issue of stakeholder expectations management. When organizations publicly declare their intentions to pursue "Goal Zero" objectives, sometimes the associated time table for that pursuit gets lost or drowned out. It is only human nature that people assume that if Goal Zero is possible, then, 'Why not now?' Why should I have to wait? What can be done to accelerate progress? And then if an incident does occur, trying to put the event in context of the long term objective can be very difficult.

A very current example of this is the debate underway in California about the potential benefit of the Safety Case Regime. An incident did occur at a company facility that embraces the Goal Zero ideal. Perfection, however, has not yet been achieved. I believe that with all good intentions, the community and governmental organizations are demanding improvement. But changing the status quo does not necessarily yield improvement. The company is spending considerable resources managing the expectations of its many stakeholders, both external and internal. Is this inappropriate? No, but it surely has the potential to divert attention away from efforts to identify and eliminate the root causes of the incident itself.

My 'Goal Zero' observations are summarized in the table below:

Goal Zero; STRENGTHS	**Goal Zero; CHALLENGES**
#1: Drives greater line management accountability for improving results	#1: Can lead to arrogance and organization indifference for regulatory requirements
#2: Has the potential to achieve process safety performance that exceeds what a compliance only approach may achieve	#2: Can create an impression in smaller jurisdictions where regulatory support may be limited, that a complete, local understanding of the law is not essential
#3: Creates a culture of greater creativity, involvement and employee ownership in the pursuit of excellence	#3: Managing expectations is a much bigger challenge as the public, governments and even employees assume that Goal Zero statements mean it is achievable NOW!
#4: Encourages a learning culture in support of continuous improvement	

Conclusion

Regulatory compliance is not optional, regardless of the country or jurisdiction. Too much of a business's long term future is at risk for any organization to demonstrate disdain for the law and its compliance. But the limitations of adopting a PSM performance objective based only on regulatory compliance must be recognized and rejected. Compliance will not lead companies interested in long term viability to a level of process safety performance that all of its stakeholders will expect and accept. A balanced approach that embraces both a regulatory compliance imperative as minimum essential but drives beyond and further towards zero process safety incidents is the only truly sustainable approach to Process Safety Management.

Engaging Senior Management in Process Safety: A Case History

David Kamrath
Air Liquide USA LLC
2700 Post Oak Blvd.
Houston, Texas, USA
david.kamrath@airliquide.com

Prepared for Presentation at
American Institute of Chemical Engineers
2014 Spring Meeting
10th Global Congress on Process Safety
New Orleans, LA
March 30 – April 2, 2014

UNPUBLISHED

Engaging Senior Management in Process Safety: A Case History

David Kamrath
Air Liquide USA LLC
2700 Post Oak Blvd.
Houston, Texas, USA
david.kamrath@airliquide.com

Keywords: Process Safety, Video, Case Study

Abstract

Traditional PowerPoint style presentations are good for transferring information but were not found to be an effective medium for engaging senior managers in process safety. Slide shows, videos and case studies of real process safety events proved to be much more effective. This paper describes how these methods were used to draw senior managers into actual process safety events, helping them to realize that the consequences could easily have been much worse and that the events were not "Acts of God" but could have been prevented through more effective and rigorous process safety management. Guidance is provided concerning how to create an effective video or case study and then deliver it in a way that achieves the results desired.

Good process safety management will improve plant reliability as well as safety. The paper shows how the root and contributing causes of reliability events can typically be traced to ineffective implementation of the same management system elements that are required for safe operations. The financial impact of improved reliability through the more disciplined approach to operations achieved by applying the elements of process safety management can then be readily demonstrated.

Finally, auditing is an element of every process safety management system but audit findings are often numerous and range from recommendations for improving a generally compliant work process to identifying serious non-compliances that need urgent attention. The paper describes how the results of audits can be processed to identify systemic issues and significant risks needing senior management attention.

1. Introduction

The leaders of a company create and nurture their company's culture through the policies they establish, the actions they take and the behaviors they reward or punish. Our desire as process safety professionals is that our leaders communicate, not only overtly in policies and procedures, but also subtly in the day to day priorities they set, comments they make and actions they take, that process safety is a core company value. But to get such a commitment from senior managers requires that they believe deeply that, first, their organization is vulnerable to catastrophic process events that could profoundly impact the company's employees, facilities and license to operate and, second, the most effective way to prevent these events is through the implementation of a strong process safety management program.

In many companies, especially when the concepts of process safety are new to the organization, the need for a strong, clearly visible and special commitment to process safety is not obvious to senior managers. They may already be devoting considerable resources to managing personal safety, can see steadily improving accident frequency rates and are satisfied that their operations are safe. In today's business environment where results are measured quarter to quarter, it is difficult to accord high priority and devote significant additional resources to preparing for process events that may or may not occur far in the future.

So what can we do, as process safety professionals, to overcome these barriers? How can we convince our senior managers that:

- Process safety risks are real and their potential impact is too significant to ignore.
- A process safety management system is needed to mitigate these risks. Existing occupational safety practices will not do the job.
- The methods employed to make processes safe will also improve operational performance and reduce cost.

The third bullet is important. A business exists to make a profit and, if it can't accomplish that goal, it will not be in business for long. The significant effort required for managing all the elements affecting process safety is much easier to justify when it pays for itself through improved plant performance.

2. Early Efforts at Air Liquide

In Air Liquide USA, our initial approach to bringing senior management on board the process safety effort was to present the facts in well crafted PowerPoint presentations. We emphasized the potentially very great cost of a major process safety incident, described how the management system could prevent such an event and defined the culture that management needed to build within the organization to make process safety management effective. We borrowed heavily from the "Business Case for Process Safety" pamphlet and the "Lessons from the Columbia Disaster" available from CCPS. We thought we presented the information in a compelling way but it was quickly apparent that we were getting little traction. We weren't making the sale.

What was wrong? According to our senior managers, much of our talk sounded like an insurance sales presentation and they weren't convinced they needed more insurance. However, we had included a few slides showing some Air Liquide process safety events that occurred around the world. They remembered those slides and the short stories we told about each one. That had grabbed their attention.

3. Making Process Safety Personal: The Power of Stories

PowerPoint presentations may be state of the art technology. But bullet points, graphs and data rarely engage people and move them to action. Stories, on the other hand, directly involve people by carrying them into a situation. They leapfrog the technology and bring the listener to the core of the human experience. They provide an emotional connection to others that moves people to take action. They simulate an event and allow people to learn from it without having to experience it themselves. Finally, the impact of a story can extend well beyond the original telling when listeners repeat the story to others.

The source material for process safety stories is the process safety incidents and accidents that have impacted the company directly or are relevant to the company's business. These are most often events from within the company but can also be external events that, either because of the technology or the causes contributing to the event, are directly relevant.

In Air Liquide, we used three methods to tell our stories:

- Slide Show
- Video
- Case Study

3.1 Incident Slide Show

Many of our North American senior managers were only aware of process safety related events that occurred within their businesses and so viewed them as extremely rare. But Air Liquide is a global company and, if you assemble all the events that occurred in all entities of the Air Liquide Group or that were relevant to the Group, the number no longer looks insignificant. So, instead of presenting principles and processes, we decided to tell the stories of some of the process related accidents, incidents and events that had occurred over the last 15 years:

- What happened?
- What was the impact on people?
- What was the impact on the plant and equipment?
- What was the impact on our customers?
- What caused the event?
- What management system elements could have prevented it?
- ***How worse could it have been?***

We gathered photographs that dramatically showed the damage resulting from about a dozen incidents covering all the processes Air Liquide operates. Some of these resulted in severe injuries or fatalities, some caused catastrophic damage to facilities and some resulted in no injuries with damage limited to local equipment but with the potential for a much more catastrophic outcome readily apparent. We assembled these photographs into a slide pack and presented it to all levels of management at every opportunity, quickly running through the answers to the questions above for each of the events. We were concise, answering the questions in no more than 1 to 2 minutes per slide. We asked two questions in the follow up discussion:

- Are you operating any plants of the types shown in the photos?
- For those you are operating, do you have similar risks and, if so, how have you mitigated them?

Feedback on this approach was encouraging. People personally identified with the incident stories. The risks and consequences in these events were real to them, not just abstract concepts.

3.2 Video of a Catastrophic Event

This initial success encouraged us to go one step further and make a video of a recent incident in one of our North American plants. No one was injured in the incident, but much of the plant was destroyed. The damage was significant enough to get the attention of everyone on the management team.

The video followed a similar structure to the photo presentation:

- What happened? What was the extent of the actual damage? How could it have been worse?
- What initiated the incident and how did it progress?
- Why did it happen? What were the causes that contributed to the incident?
- What must our managers and employees do if such incidents are to be prevented in the future?

It is critically important to emotionally engage the audience immediately in the opening scene of a video. Fortunately, we had video footage taken by an Air Liquide employee showing the event as it was occurring. While this was running, we used the device of a simulated phone call to emergency responders to describe what had happened. The audience was drawn into the situation. They could see the impact of the event and hear the urgency in the caller's voice. It was not hard for them to imagine how much worse the event could have been.

But what was it like for those trying to control the plant when the incident occurred? In an on-screen interview, the operator on duty described what happened from his perspective, the actions he took and how he was unable to control the situation. It was readily apparent in the interview that the event had affected him personally.

With the viewers emotionally engaged, we could transition to communicating information about the incident – how it started and progressed, what were the causes, how could it have been prevented. Following the US Chemical Safety Board video model, we used animation to recreate the event so viewers could see it unfolding as the narrator described each step in the process. The narrator then demonstrated how, if the elements of the management system had been properly applied, the event would have never occurred.

The video closed with a statement by a Senior Manager of the Air Liquide Group directly addressing the fact that, while an incident such as this is rare and many people have not experienced one, no one should fall into the trap of thinking that, because such an incident has not occurred in his plant, his operations are safe. He then strongly endorsed the management system as the method to prevent such incidents and requested that no one allow himself to be complacent in its application.

The 15 minute video was first presented at a meeting of the senior managers of all the Air Liquide businesses in North America. Most important was the discussion that followed. How much worse could the incident have been? Can we afford to have more such incidents? Are they really preventable? Is the process safety management system the most effective way to prevent them? Once people start thinking about these issues and actively discussing solutions, they were engaged in the process.

The video format had an additional benefit. We were able to easily communicate the story well beyond the original audience just by distributing it as a DVD and posting it on the company intranet.

3.3 Case Study: Discovering the Impact of Senior Management Policies and Decisions

Senior Management engagement is necessary but not sufficient. They also must understand how their policies, decisions, statements and actions can have a direct impact on process safety in their plants. As process safety professionals, we could have prepared a PowerPoint presentation telling them what they needed to know. But how effective would that have been? Business schools have found that managers at all levels, but especially senior managers, learn best if they feel they are discovering the knowledge for themselves. They help managers discover the most effective approaches to solving complex business problems through the case study method. We decided to take the same approach.

A good case study, like a video, tells a story in a way that engages the audience. But, unlike a video, it does not provide a causal analysis or recommend solutions. That is left to the case study participants. The case facilitator draws participants into the analysis of the case scenario and the subsequent development of solutions through the questions he asks.

There are a number of criteria that we used for selecting a process safety case for senior management:

1. Senior management policies and decisions must be clearly identified as contributing causes of the incident.
2. The event should be external to the company so none of the participants will react defensively.
3. Although external, the event must still be relevant and timely.

4. The causes of the incident should be well documented in the open literature by respected authorities.

The 2005 BP Texas City refinery explosion that resulted in the loss of 15 lives best met these criteria. This incident was investigated by two US government sponsored organizations, the Baker Panel and the US Chemical Safety Board, and the reports they published are readily available public documents. The reports cite a number of senior management policies and decisions as contributing to the accident. Texas City is just south of Houston, where Air Liquide's US headquarters are located, and a number of our managers have previously been employed by Amoco or BP, some in the Texas City refinery, making the event both timely and relevant to our management team.

The case was prepared in-house from information provided in the Baker Panel and US Chemical Safety Board reports. It opened with a brief description of the accident:

- What happened?
- What was the extent of fatalities, injuries and property damage?
- What was the impact on others working in the refinery and living in the surrounding community?
- What was going on in the refinery when it occurred?

The next section briefly described the operation of the raffinate splitter, the unit that lost containment, initiating the explosion and fire. After that, the actions and decisions made by those involved in the raffinate splitter start-up that led up to the event were described. Finally, the safety history and culture of the refinery as determined by the two investigations was presented.

We ran the case study at a meeting of our North American Senior Management Team. The case, which was a 5 page document, was provided to participants as pre-reading. The meeting attendees were divided into a number of small groups so that all would have an opportunity to participate. A facilitator was assigned to guide discussion for each group. The facilitators were provided with a Case Leader Guide and trained on how to facilitate the case before the meeting.

The facilitators started with a review of the scenario and then used the following questions to guide discussion:

1. What decisions and actions (or inactions) before and during start-up led to this accident?
2. What might have led those involved to behave the way they did? Didn't they care about safety?
3. What was the safety culture in the plant?
4. What policies, decisions and actions taken by BP's senior management could have encouraged such a culture to develop?
5. Which of these issues have you seen within Air Liquide?
6. What can you, as senior managers, do about them?

The facilitators summarized the responses to each question as bullet points on a flip chart. Figure 1 is the page from the Case Leader's Guide to help facilitators guide discussion and record responses for the first question above.

Figure 1
Case Leader's Guide: What decisions and actions (or inactions) before and during start-up led to this accident?

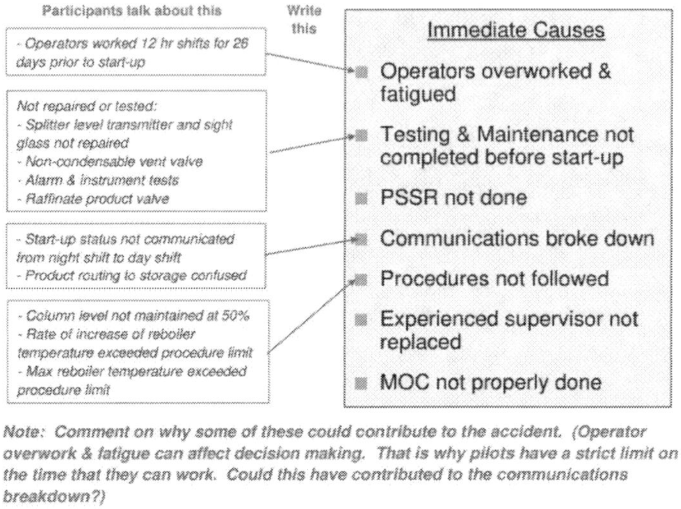

Of course, the answers to questions 5 and 6 were most important to our company since they brought the issues back to Air Liquide. Those answers were consolidated on PowerPoint slides and presented to the entire management team following completion of the case. The purpose was to develop a consensus view on the issues within our company and the actions senior managers needed to take to address them.

By the end of the exercise, there was much deeper understanding among our senior managers that the values they, as senior managers, express, the policies they implement and the decisions they make can have a profound impact, either positively or negatively, on the safety of their operations. They more clearly understood their role in the process and what they needed to do to have a positive affect.

4. Demonstrating Real Returns Today

Good process safety management is not just an insurance policy. The operational rigor required to keep a process safe (making sure operators are well trained, insisting that procedures are followed, assessing and mitigating risks associated with changes, assuring the integrity of assets, etc.) is also necessary for achieving reliable and efficient plant performance. When Air Liquide was first considering development of a process safety management system, I was invited to visit a plant that already had a strong system in place. I was very impressed with the system but I was concerned that maintaining the level of operational rigor it required must be very costly. I asked the plant manager how he could afford to do it and he replied that he could not afford not to. He then presented charts to me showing the improvement in plant reliability since the introduction of the management system and the resulting savings achieved. They dwarfed the cost of implementation.

We have had similar, though less dramatic, results in Air Liquide. Reductions in the number of reliability incidents from 2010 through 2012 are shown in Figure 2. The annual cost of reliability incidents over that time period fell by over 30%. These improvements, while admittedly over a relatively short period of time, have caught the attention of senior management. They are clear evidence that there is a near term financial return achievable through the operational excellence that process safety management imposes.

Figure 2

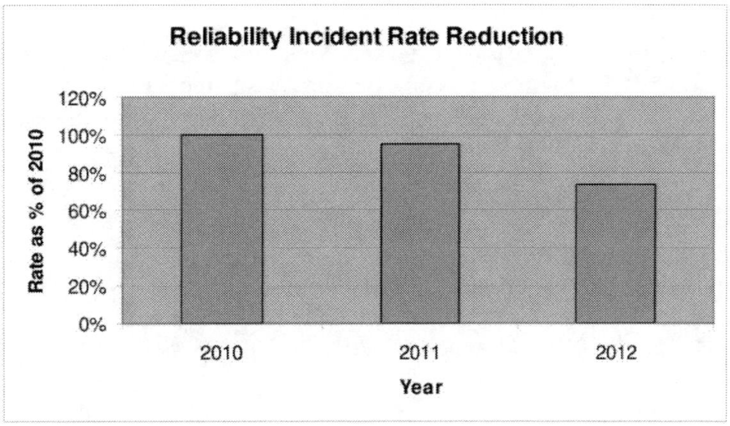

5. Reinforcing the Message

While videos and case studies can have a significant impact, the key messages they impart will tend to be lost over time if they are not reinforced. Everyone, including senior managers, needs to be occasionally reminded why their commitment to process safety is important. During the early stages of Air Liquide's process safety management effort, we found that our operations personnel did not always make the connection between process safety management principles and their day to day activities. Management system deficiencies were often not explored as possible causes of reliability and safety events. People were not yet thinking from a process safety perspective. By participating in the investigation and analysis of such events, those of us involved in process safety were able to help the teams recognize the management system issues that contributed to their occurrence. Discussions in the root cause analysis sessions and later in the presentations of those analyses to management reinforced the message that the rigorous implementation of the elements of process safety can prevent these undesired events from occurring.

Root cause analyses can identify management system deficiencies that directly contributed to a specific incident but do not shed much light on how common those deficiencies are. That is where the process safety audit program can help. Auditing is often viewed just as a process for testing whether activities are being conducted as required by the governing codes, regulations and corporate procedures. That is a necessary part of the process but it should only be the starting point. What management really needs to know is the level of risk these non-compliances create. What is the probability of the non-compliance resulting in an event and how severe could the consequences be? Are they limited to a single plant or are they systemic. Have they been identified as a root or contributing cause in any actual incidents? By addressing these questions in their reports, auditors can provide senior management with an independent view of significant residual risks in their operations that could be mitigated by more effective and rigorous implementation of elements of their process safety management system.

6. Conclusions

Senior Managers have more demands for their attention than they could possibly address. The skill in management at that level is to choose the ones that are most important and make sure they are properly handled. How can we ensure that process safety is on their priority list? At Air Liquide, we found that this could be best accomplished by connecting with senior managers on a personal basis through stories of real and relevant events and their actual or potential impact on people, the community and the organization. Slide shows that briefly presented a number of events demonstrated that, although rare, process safety events do happen and can carry very severe consequences. Videos can make the connection with process safety more personal by showing interviews with those who were involved in or affected by an incident. Case studies can be used to help senior managers understand how their policies and decisions can have a direct impact, positively or negatively, on process safety in their operations.

Senior managers, like anyone else, can be most easily engaged in a subject like process safety when they can be shown that it will help, rather than hinder them, in achieving their goals. As operations become more disciplined through the implementation of the elements of process safety management, improvements in plant performance will become apparent. Savings through better plant reliability are a near term benefit from good process safety management that can help to offset the cost of implementation.

Engaging Employees in Catastrophic Event Prevention

Greg Robinson
BST
417 Bryant Circle
Ojai, California 93023
bstusc@bstsolutions.com

Prepared for Presentation at
American Institute of Chemical Engineers
2014 Spring Meeting
10th Global Congress on Process Safety
New Orleans, LA
March 30 – April 2, 2014

UNPUBLISHED

Engaging Employees in Catastrophic Event Prevention

Greg Robinson
BST
417 Bryant Circle
Ojai, California 93023
bstusc@bstsolutions.com

Keywords: catastrophic event prevention, employee engagement, process safety, leadership, culture

Abstract

Implementing an effective process safety system goes beyond documented policies and procedures. It requires behavioral reliability in its execution. In a culture of commitment, employees have a personal value for safety that drives their performance. Safety isn't about just following the rules or completing a checklist; it's about going the extra mile for safety because it's the right thing to do.

Leadership actions have the greatest impact on creating a culture where employees feel accountability and ownership for process safety. To the extent that the culture helps employees feel comfortable raising issues, employees will be more willing to give the discretionary effort required to support all aspects of safety. The following essential areas for creating engagement are covered:

- Educate employees about the difference between process safety and personal safety.
- Help employees maintain a sense of vulnerability.
- Create a culture in which it's okay to "cry wolf."
- Communicate that process safety is important.
- Live up to your role as leaders.

1. Introduction

This paper looks at the concrete ways chemical organization leaders can engage employees in catastrophic event prevention and develop a culture of commitment—a workplace that supports and encourages dedication to the organization's values and creates an environment in which discretionary effort flourishes.

This paper explores what leaders need to examine to better understand what drives employee behavior and discusses things leaders need to consider in order to access workers' innate connection to safety, ensuring that the highest levels of process safety management are sustained across all levels of the organization.

There are five key areas leaders can focus on to get started:

Process vs. Personal Safety—Employees need to understand the differences between personal safety and process safety, including their role in raising process concerns to their leaders.

Maintain a Sense of Vulnerability—Employees learn to accept as normal the risk they live with every day. Maintaining a healthy sense of respect for workplace exposures is critical to creating a culture in which employees understand the importance of consistently following process management disciplines.

It's Okay to "Cry Wolf"—It is incumbent upon leaders to create a culture where employees are willing and able to speak up about process safety concerns so that steps can be taken to prevent a small problem from becoming a major event.

Leaders Need to Show They Care—Leaders need to create a working environment in which the contributions of employees to catastrophic event prevention are recognized and rewarded. Promoting these contributions signals to everyone in the organization the value leadership places of process safety execution.

2. Engaging employees in process safety

Engaging employees in process safety is an essential part of any system designed to prevent catastrophic events. Engagement breathes life into process safety and enables the particulars of an organization's objectives to be realized and sustained. Organizational culture plays a key function in building engagement. A culture that inspires behavioral reliability is especially critical in dispersed workforces where employees are required to act regularly without supervision.

Engagement begins when organizations establish their concern for the employee, ensuring that supervisors treat workers in an unbiased and fair manner and with dignity and respect. They also do this when they ensure that senior leadership and the organization demonstrate concern for employees' wellbeing. These things establish a good relationship with employees and foster a strong sense of reciprocity. In this environment employees tend to go above and beyond their job description because they feel it is the right thing to do for a peer, leadership, or the organization. This willingness to follow a procedure even when no one is watching or to step up and take on leadership in the moment is a sign of a very high-functioning organization.

Conversely, when employees feel no sense of connection to the organization or its leaders we cannot expect safety activities to flourish. Instead, the organization will likely meet resistance in trying to implement new programs or systems that require employee engagement.

3. Five ways to start engaging employees in catastrophic event prevention

Organizations that are susceptible to catastrophic events depend on a high level of compliance with rules and regulations to keep employees, the environment, and the community safe. To be sure, many organizations have proven to be highly effective at practicing and integrating the rules that make up a vital element of process safety management (PSM) systems. But in the face of serious concerns over the potential for catastrophic events, many leaders are looking for ways to engage employees in preventing catastrophic events, moving their workforce from a culture of compliance to one of commitment.

Engaging workers means involving them in process safety decisions. One way to do this is to ask questions of the people who are doing the work every day. Employees at the jobsite have a wealth of hands-on knowledge and are usually keyed into what needs to be done to ensure safety. By asking questions and encouraging workers to speak up, leaders not only engage employees in PSM, they also motivate them to take action when they see a risk or the potential for one.

Effective catastrophic event prevention also means using systems consistently and reliably. But how can industry leaders make sure their people are doing the right things at the right time and in the right way—even in the absence of supervision? What can leaders do to build a workforce that goes beyond simple compliance to one that understands and takes ownership of process safety? Recognizing the gap between good safety rules and their execution—and how to close it—is critical to achieving behavioral reliability in catastrophic event prevention.

There are five key areas leaders can focus on as they begin to create engagement:

3.1 Educate employees about the difference between process safety and personal safety

If this distinction is not understood, people may mistakenly believe that good performance in personal safety is the same as effective control of the potential for catastrophic events. While personal safety exposures are typically very apparent in the workplace, mounting process safety issues can be developing silently in the background. Employees have a role in raising process concerns just as they would alert management to obstacles in the workplace that impact their personal safety.

Even when good process safety systems are in place, it is incumbent upon the organization to ensure that the systems function as intended. Employees will not come forward with critical information if they believe that nothing will happen if they do, or, even worse, that such an effort on their part will be perceived as a nuisance. It is important for leaders to encourage individuals to voice their concerns and observations.

3.2 Help employees maintain a sense of vulnerability

Employees learn to accept as normal the risk they live with every day. Unfortunately their comfort level can serve to lower the rigor and urgency with which policies and procedures are implemented and maintenance items are addressed. This does not, however, change the risks associated with the work they are doing. Developing and maintaining a healthy sense of respect for workplace exposures is critical to creating a culture in which employees understand the importance of consistently following process management disciplines.

Leaders are in some ways more at risk of losing their sense of vulnerabilities than other employees. The sheer scope of their job means that they are continually monitoring and managing a wide range of threats to the wellbeing of the organization, and it is easy to lose a sense of urgency for process safety when the severity or frequency of accidents is low. But the absence of incidents does not indicate an absence of exposure. Nor does it mean that exposure levels are trending downward—in fact, the opposite may very well be the case.

Engaging employees requires leaders to maintain their own sense of vulnerability by continually gathering and encouraging the provision of information about the true state of workplace hazards, safety prevention mechanisms and practices, and the culture of the organization. Leaders maintain confidence by listening to the discussions about injury rates and trends. They listen to hear if there is more discussion about whether the classification is correct than about the event itself and the prevention plan. They look into whether there are systems that would discourage full disclosure.

3.3 Create a culture where it's okay to "cry wolf"

Too often employees feel that if they raise a concern that turns out to be nothing they will be punished or ridiculed. The path of least resistance is to assume that it is someone else's job to raise an issue or the observed deviation from normal is not a big deal. In a strong process safety culture the little boy who cried wolf would get the key to the city, assuming he brought facts and data to the table! Employees need to be knowledgeable enough to recognize when and how to intervene, and they need to be willing to do so. The challenge for leaders is to create a culture in which employees not only recognize when things start to go off track but are also willing and able to intervene to prevent a small problem from becoming a major event.

Engaging and motivating employees in process safety is about meaningful involvement. Workers have their own areas of expertise, namely how the work is executed on the floor and at the site and how the organization's systems and processes alter and influence the tasks being performed. Leveraging this knowledge—through honest inquiry and receptive dialogue—will gain leaders valuable insight at the same time it makes employees feel valued by the organization. It is important that leaders always remain open to workers' concerns and ensure that no issue is too small for their attention.

3.4 Communicate that process safety is important

One organization we worked with once had two small chemical releases in separate plants. At the next senior leadership team meeting, while the safety manager reviewed the incidents, the president wondered aloud why he had not been made aware of them at the time. The safety manager responded that the president must be doing something that makes people think these types of events are unimportant to him.

The leader's values, beliefs, commitments, interests, choices, and ways of doing things all send a message to workers about what the organization truly values and expects from its employees. The leader's actions and the way she communicates decisions help create and sustain the organization's commitment to catastrophic event prevention.

Leaders should look carefully at their behaviors and the words they say. Do they communicate the importance of process safety? Do their performance management systems and other business systems convey the importance of process safety? Determinations about staffing levels, supervisory development, promotions, budgets, or new projects all signal what is valued and what is important. Whether intended or not, a promotion is an implicit endorsement of a person's past leadership and actions with respect to safety. Decisions about staffing or organizational structure sometimes subtly—sometime profoundly—alter the interface between people and technology. The configuration and use of these and other business systems tell employees what is rewarded and what they are able to do.

3.5 Live up to your role as leaders

Leaders have both systemic and symbolic roles with respect to engaging employees in process safety. The systemic role is to ensure that the "consequence systems" in the organization recognize and support employees who raise process safety issues, no matter how seemingly small. The symbolic role is to promote the contributions that employees make to improving process safety and thereby demonstrating that management values their input.

How leaders reward workers and recognize their contributions to safety conveys a lot about what management values and lets people know just how serious the organization is about catastrophic event prevention. After all, what does it say about your commitment to process safety if you're promoting people with poor safety records and rewarding bonuses to individuals everyone knows is letting protocols slip.

No leader serious about reducing risks to injuries wants to be perceived as only paying lip service to safety. But no matter how noble your intentions, people are going to perceive your actions—and decisions—as the real indicator of what you think about safety. When employees believe you're not taking process safety seriously, they won't either.

That's why it's vital that leaders look closely at their reward and recognition systems to make sure they're really doing what they want them to. What you want them to do is support a culture that mitigates risk and encourages people to become active participants in the safety process. Rewards should go to folks who reduce *exposures* to risk rather than managers whose teams report the least injuries. Promotions should include an evaluation of the person's safety record. Without that your sending the signal that safety isn't really important to the organization and people who want to climb the ladder shouldn't worry about it.

Leaders also need to look at how employees perceive rewards. Are they valued and appreciated or seen as meaningless and mere tokenism? When workers think incentives are hollow, they will believe your commitment to process safety is equally empty. Make sure that your safety rewards and the ways people are recognized for their successes in process safety are on par with production and other performance rewards—or better.

4. Conclusion

Engaging employees in catastrophic event prevention requires a culture of commitment—a place where safety isn't about just following the rules or "checking the box," but a dedicated practice of going the extra mile for safety because it's the right thing to do. Leadership actions have the greatest impact on creating a culture in which employees feel accountability and ownership for the organization. To the extent that the culture helps employees feel safe and comfortable raising issues, employees will be more willing to give the discretionary effort required to support all aspects of safety, especially the prevention of catastrophic events.

5. References

[1] Erickson, J.A., "The Relationship Between Corporate Culture and Safety Performance," *Professional Safety*, May 1997

[2] Groover, D., "Taking Ownership of the Injury-Free Culture: 8 Questions Every Leader Needs to Ask & Answer," *Perspectives,* Fourth Quarter 2007

[3] Krause, T.R., *Leading with Safety*. Hoboken: Wiley Interscience, 2005

[4] Robinson, G., "Engaging Employees in Process Safety," *Safety+Health,* July 2013.

Who Are You and Why Does It Matter?

People are different in many ways. Many of these are purely physical such as height, weight, skin color, and hair color or style. Someone's ethnicity can also be pre-judged based on appearance factors. All people have subtle biases for or against these factors and can have an effect on how we relate to others, especially in a team situation. Many of these can be discussed openly as they are obvious as well as the normal biases we harbor.

There are other aspects of behavior unrelated to these physical characteristics which are much more critical when forming engineering teams and our behavior when part of such teams.

The first of these is what is called "social style", and it can be measured vary accurately by such assessment instruments such as the Myers Briggs (MBTI™) or 16Types™. These assessments measure preferences, I.e, what our normal behavioral preferences are. Anyone can behave differently or conform to a perceived needed behavior for some period of time. The larger this difference is, the more difficult it is to sustain this un-natural behavior. The social style preferences can be divided into four major areas:

Extroversion/introversion (enjoy socializing or prefer to be alone)

Sensing/iNtuitor (data analysis based solely on facts/data or "gut feel" and intuition)

Thinking/Feeling (making conclusions based solely on facts and data or concern for impact on people and softer implication)

Judging/Perceiving (desire for closure or desire to leave the door open to additional possibilities)

These four preferences are grouped together and given a summary classification such as ENTP or ISTJ. "N" is used to indicate the Intuitor preference since "I" has already been used in the "introversion" categorization. Many people are on the borderline between preferences and are flexible in their preferred behavior.

An additional important measurement is an individual's preference when it comes to problem solving style. Some people are strongly analogic in their approach to problems (use the tried and true, follow known procedures, etc.), while others prefer to get "out of the box" and try new things, make unrecognized (by others) connections, etc. This aspect of preference can be measured by the Kirton KAI™ or the Buffalo Creative Process Index (BCPI™). Obviously, problem solving teams, depending upon their charter, would benefit from different preferences in this assessment.

This workshop will review these assessments and use exercises to illustrate the differences and how to use these preferences pro-actively in team formation.

Assessing the Correct Risk

Jack Chosnek, PhD, PE
Knowledge*One*
PO Box 451620, Houston, TX 77245
jc@knowledge1.net

Prepared for Presentation at
American Institute of Chemical Engineers
2014 Spring Meeting
10th Global Congress on Process Safety
New Orleans, LA
March 30 – April 2, 2014

UNPUBLISHED

Assessing the Correct Risk

Jack Chosnek, PhD, PE
Knowledge*One*
PO Box 451620, Houston, TX 77245
jc@knowledge1.net

Keywords: PSM > Technical > Risk > LOPA

Abstract

During either hazard identification (HAZID) or process hazard analysis (PHA) a measure of risk has to be determined by the team. Except maybe in the cases of full Quantitative Risk Assessment (QRA), the risk determination is somehow subjective and dependent on the composition of the team doing the analysis. One of the problems is arriving at the right consequence. Although Consequence Analysis (CA) will provide the potential effects of a release, it doesn't provide a definitive answer (the ultimate consequence will depend on factors such as early or late ignition, presence of personnel, etc.) and this uncertainty can result in choosing the wrong risk. Complicating matters, the team has to decide whether to use a consequence with personnel impact, or one with economic, environmental, or perhaps company reputation, in arriving at the risk. The decision may rest between a low consequence-high probability personnel impact and a high (or very high) economic consequence with lower probability. The solution resides in applying Layer of Protection Analysis (LOPA) in a judicious manner. We can't apply LOPA to all the scenarios discovered by the PHA team as it would be extremely time consuming. Most teams will select high risk scenarios to apply LOPA. This could lead to the determination of the wrong risk because the team has already assigned a perceived probability to the consequence when they chose a consequence level. For example, the team decided that a personnel injury would not be half as severe as the resulting economic damage for the particular scenario. By selecting only the economic scenario for further analysis because the other *seems* to have a low risk, a potentially high risk could be ignored. Only by doing LOPA for the high-consequence-level scenarios rather than the high-risk scenarios the uncertainty of the risk can be reduced. The paper will expand on this concept and show that a more effective PHA, both from the risk and the time efficiency considerations, can be conducted.

1. Introduction

In conducting a Process Hazards Analysis (PHA) it is our desire to prioritize hazards in order to resolve the potentially most dangerous first and, if resources are available, work on the others. This prioritization not only makes sense, but it is absolutely necessary to eliminate or mitigate the high hazards given that resources are limited and not all the hazards can be worked on. Risk, a function of consequence and probability, is the most used criterion for prioritization and it is an appropriate one since we want to mitigate the high consequence events, but we will want to put our efforts on this endeavor only if these events are probable. That is, since we need to choose, we want to work only on scenarios or events that have high consequence and a high probability of occurring, or in other words, high risk. But what if we can't assess those risks correctly? Are we leaving behind potential events with hazardous consequences that may have a reasonable probability of happening?

In order to be sure we could conduct Quantitative Risk Assessment (QRA), but this process is very laborious and expensive and cannot be justified for regular PHAs. It is possible that after going through the process of zeroing on the correct risk we will find a risk that is inordinately high and has to be reduced. If we can focus on these risks then QRA may be the appropriate tool for their reduction or mitigation. On the other hand as we will see, we can be selective and apply judiciously a semi-quantitative analysis such as Layers of Protection Analysis (LOPA) ending with a better risk assessment without incurring the expense of QRA.

2. Erring on Assessing Risk

For most companies in the process industries assessing risk is almost a continuous practice brought about the realization that due to the intrinsic nature of the business incidents can happen with potential loss of life or injury, harm to the environmental, and/or grave economic loss. Thus, PHAs have become a way of life for these reasons, augmented by the existing regulatory requirements (e.g. the PSM Standard 29 CFR 1910.119, RMP, and others for specific industries), and the need to properly manage change in the face of complex operations. This is a positive development as identifying hazards and reducing risks obviously increases safety and wellbeing for everybody.

The increased activity in the process hazards analysis area though brings some dangers. Companies have not yet fully factored in their planning the necessary resources for conducting PHAs and resolving their recommendations. Pressures abound to minimize the time spent on PHAs and even more on coming up with fewer recommendations. The result is a tendency to underestimate risk either consciously or subconsciously, the latter being probably still the more prevalent mode. Even when the company is prepared to spend the time and money there may not be enough trained and experienced people to set up a knowledgeable team or there may not be a well-qualified and experienced facilitator available.

How is risk underestimated? Since risk is a function of consequence and probability, we can underestimate consequences or underestimate probability or even worse, underestimate both at the same time. Probability underestimation is enhanced by overestimating the effectiveness of safeguards. These will be discussed further. It is not uncommon, though, to be using a risk matrix that has ill-fitting consequence and probability levels. Although building a proper risk matrix is outside the scope of this paper, the influences of the matrix on the behavior of the team that leads to the wrong assessment will be discussed.

Even though underestimating risk can lead to operating under higher risks than thought, overestimating risk can also have negative effects as it can lead to significant overwork or on the other hand, paralysis, if all risks are indistinguishable high and it's impossible to know where to start to reduce risk.

The basis for analyzing hazards and correctly determining their risks is the discipline brought about by the PHA process. Whether the PHA is conducted using a HAZOP, a 'What If' or any other methodology, a consequence scenario needs to be developed first from a proposed deviation, the level or severity of the consequence determined next without consideration to safeguards, then followed by listing the safeguards that specifically prevent the scenario from occurring or that mitigate the consequence, and only then determining the probability of the event. Deviating from this procedure can contribute to risk assessment errors as will be seen.

3. Erring on Assessing Consequence

Even when formal consequence analysis (CA) is performed, only the possible ultimate effects of a process release are learned. The PHA team still needs to develop well a plausible scenario and arrive at the suitable and correct ultimate consequence. Most of the time it is a team member having direct or indirect experience with a similar scenario, or the facilitator, that will dictate the ultimate consequence. Since a process failure or deviation can result in many paths, it is important for the team to decide which path will be followed and which the most probable outcome is.

As opposed to QRA where an Event Tree can be designed to include all of the (reasonably) possible outcomes and assign a probability fraction to each branch of this tree, in a qualitative assessment a choice has to be made among the most probable outcomes. This is the where the team can go astray as the experience of the team members will weigh heavily on this decision.

We have two common types of errors in describing consequences: vagueness and undeveloped scenario.

3.1. Vague Consequences

In this case the consequence is not well defined either due to the company's requirement to categorize consequences (e.g., safety, environmental, economic, company reputation, etc.) and the mixing of many categories into one outcome, or lack of definition on what the outcome really is.

An example of the first type would read like this:
"...closure of valve XV-xxx would lead to loss of reflux with potential overpressure of the column T-xxx, column damage, loss of production, potential loss of containment, fire, explosion, and personnel injury."

When the team is trying to determine the level of the consequence, for example using definitions as in Table 1, it will be difficult to decide what is being rated: safety or economic loss. The difficulty lies in that one type of outcome may be more probable but less severe, while the other more severe but less probable, both estimations being dependent on the experience of the team members. If tower damage was the intended consequence the rating could have been 'Marginal' (level 2), while if there had been fire and explosion the resulting rating could have been 'Severe' (level 3) or even 'Catastrophic' (level 4).

The same example can be used to describe lack of definition. Let's suppose that the team decided that the outcome falls in the health and safety category. The outcome is still very vague because we don't know if we are rating the impact of a fire or an explosion on personnel, or on equipment reverting to an economic consequence. If the team has little concept of the radiation produced by a fire, or the impact of overpressures from an explosion, the rating of the consequence will have a very wide range.

Table 1. Consequence and Probability Definitions

Consequences	Probabilities
1. Low Safety– Light Injury/ illness Economic – Loss of <$100M	**P1. Improbable** Not expected to occur in 1,000 years.
2. Marginal Safety– Lost time injury/ illness Economic – $100M < Loss <$5MM	**P2. Rare** Not expected to occur in the life of the plant
3. Severe Safety– Significant injury/ disability Economic – $5MM < Loss <$50MM	**P3. Occasional** Likely to occur once or more times in the life of the plant but less than once in 10 years
4. Catastrophic Safety– One or more fatalities Economic – Loss of >$50 MM.	**P4. Probable** Likely to occur more than once in 10 years

3.2. Undeveloped Scenarios

An example of an undeveloped scenario would be half of the description of the previous example: "…closure of valve XV-xxx would lead to loss of reflux with potential pressurization of column T-xxx." It is implied here that there are no real consequences from this failure. The thinking may have been that the controls would correct the situation and nothing further would happen. The description could have gone a step further and still not accomplish the task: "…closure of valve XV-xxx would lead to loss of reflux with potential overpressure of the column T-xxx and loss of containment". We still don't know what the effect of loss of containment would be and any rating of this consequence at this stage would be untrustworthy to say the least.

The extreme of assuming that the outcome is just a pressure rise without pursuing the scenario further would only result from a very inexperienced team. This eventuality is possible but not the most probable. It would be hard to dismiss the consequences of a failed shut-off valve in a reflux line without some good explanation. If the consequence stood as just an operational upset, it wouldn't get rated or get a rating of 'Low'.

This scenario could easily have the complete range of outcomes, from 'Low' as indicated above, to a possible 'Marginal' or 'Severe' if the outcome was purely economic, or 'Severe' to 'Catastrophic' if the outcome was safety related.

4. Erring in Assessing Probabilities

If assessing the correct consequence is difficult, then assessing the correct probability is more difficult when doing it qualitatively. Since we typically look at failures that occur less often than the life of the process, it would be hard to find even an experienced worker that would have seen many failures of the same kind in his plant (if he has seen too many, it is time to leave his current employment – it is not a safe working place!). Thus we have to rely on experts that know the failure rates in their fields, or look for data.

So here is the first reason we underestimate probability: the experienced people in our plant may be very good at process design, or inspections, or chemistry, or environmental laws and calculations, or how things operate, but have not much experience on how, and how often, things fail. In addition, all have been trained to make things work no matter the difficulties, and are reluctant to accept the idea of processes or equipment failing or trained people making mistakes. Thus, they will tend to assign a low probability to most failure scenarios, unless having personally witnessed the failures or read about them. Even then, it will be in their own area of expertise, so the less multidisciplinary the team is the higher the chance that probability will be underestimated.

Another reason that leads us to underestimate probability is the overuse and misuse of safeguards. The more safeguards that we list, whether they are applicable or not, the more we will get the feeling that the probability of the scenario is low. The type of safeguards that are overused (they are considered much more effective than what they really are):

- Alarms
- Operator training
- Operating procedures
- Maintenance program
- Inspection program
- Lockout/tagout program (LOTO)
- Carseal program

In reality, as CCPS points out [1], training, operating procedures, maintenance, and inspections are already built in into the published failure rates. Having them doesn't further reduce the probability of the event; not having them would increase its probability. The only time that any of these items or programs would add to reducing the probability is if the program is specifically geared towards the scenario. For example, if loss of containment is caused by corrosion due to the presence of hydrogen sulfide and sometimes water, an enhanced inspection program of the parts of the process where this is possible would indeed reduce the probability of piping failure.

The LOTO program is a good safeguard if it is well established program and the safeguard is being applied to valve misalignment resulting from incorrect operations/maintenance interactions. It many times gets misapplied because the scenario where it is used doesn't involve a handover between maintenance and operations, but an operational error during normal operation. As far as the carseal program, it many times is given a higher value than it deserves, as many of these programs are not audited frequently enough.

Alarms is another safeguard that is often overestimated. The team needs to question whether the alarm will be effective in eliciting a response from the operator (will the alarm be silenced and forgotten?), whether the operator will have the time and wherewithal to identify the potential hazard and respond, and lastly, whether the operator has the time and means to stop the sequence of events that are leading to a potentially catastrophic event. Sometimes it just means closing a manual valve but that valve may be 20 feet up in the air, or it may be a very large valve, or if it involves opening a valve against high pressure, it may not be possible to accomplish it without mechanical help.

The existence of a risk matrix with risk levels that require certain actions can also influence the team's behavior and lead it to evaluate a lower probability in order to arrive at a lower risk. This results because the team believes that a given scenario shouldn't have such a high risk, or because management has provided guidance to minimize the number of recommendations which will depend on the risk level. Doing away with the risk matrix is definitely not the way to correct this.

On the other side of the coin, overestimating probability has its own problems as it distracts us from addressing the correct critical consequences.

5. Converging on the Correct Risk

Avoiding the pitfalls described above will lead to a better risk assessment but it may not be enough. When faced with potentially high consequences we should take a second look at their description and their category. If we have doubts on the final outcome, consequence analysis with the help of models may be desirable.

Let's consider the example of loss of reflux discussed above. We weren't certain if the consequence was level 3 or 4, and in assessing probability the team hesitated between level P2 and P3 (Table 1). Looking at the risk matrix shown in Table 2, we can see that if we have just one level of uncertainty each in selecting consequence and probability we could have risks ranging from 'A' (Critical) to 'C' (Moderate). If we selected a consequence of level 4, we could have risks of level 'A' or 'B' and if we selected a consequence of level 3, we could have risks 'B' or 'C'. If we had arrived at a risk 'C' we wouldn't be obligated to do anything to reduce risk and would continue to operate under a high risk. If we had arrived at a risk 'B' and the real risk was 'A', we could be operating under a very grave risk for a significant amount of time.

Table 2. Risk Matrix

Probability				
P4	C	B	A	A

P3	D	C	B	A
P2	D	C	C	B
P1	D	D	D	C
	1	2	3	4

Consequence

Table 3. Risk Definitions

A	Critical – risk has to be reduced before continuing operations
B	High – risk has to be reduced within a year
C	Moderate – risk should be reduced if cost effective
D	Low – risk is acceptable and no action is required

Therefore, if the same scenario can lead to two (or more) severe consequences, one should not be chosen over the other, but the two should be analyzed further. Given the uncertainty of assessing probability we could be missing some critical risks if we don't do this. In order not to disrupt the flow of the PHA, the worksheets could be marked for further analysis at the end of the PHA, maybe with the addition of more knowledgeable team members.

In effect, any scenario that leads to a severe consequence of the highest levels should be marked for further analysis. In the case of the example all consequences having a level 3 or 4 should be analyzed further.

The correct risk from these cases with high consequence will depend on how well we assess their probability. For this we can improve the process by using LOPA. With LOPA we can analyze each safeguard (layer of protection) and better determine if it is specific to the scenario (no generalized protections), independent of other safeguards (no double-dipping) and how robust it is by looking at its probability of failure on demand (pfd) [1].

Besides improving the descriptions of the consequences, we should also strive for better description of the safeguards as it will lead us to a better estimation of the probability and will ultimately improve the LOPA process [2].

Why not use LOPA for all the scenarios? This could be done but the time invested in hazards analysis could triple. The risk reduction that can be obtained from the low consequence events doesn't justify this extra effort.

6. Conclusions

We are aware of the uncertainties in assessing risk when performing PHAs. In order to arrive at a better determination of risk it is necessary to avoid pitfalls in assessing consequence and probability.

For estimating consequence we should:

- Conduct some Consequence Analysis for outcomes that result in toxic dispersions, fires or explosions to better understand their effects.
- Avoid vagueness in describing the consequence. The outcome should be clear enough to easily assess its severity.
- Develop all scenarios to their end in order to achieve a good understanding of the steps in the sequence of events and properly assess their probability.
- Separate consequences which fall into more than one category (safety, economic, etc.) for individual review.
- Consider separately consequences within the same category that could reach different probable levels, if those levels are high.

For estimating frequency we should:

- Avoid generalized safeguards and use only specific safeguards for the scenario.
- Apply a reasonable value to the robustness of the chosen safeguards.

Finally, select all the scenarios that have the top two levels of consequence (the worst consequences) and conduct a LOPA on them.

7. References

[1] Center for Chemical Process Safety (CCPS), *Layer of Protection Analysis, Simplified Process Risk Assessment*, AIChE, New York, NY 2001.

[2] Bayutt, P., "Conducting Process Hazards Analysis to Facilitate Layers of Protection Analysis", *Process Safety Progress*, Vol. 32, No. 5, p. 282-286, September 2012.

Additional references not cited:

[3] Van Scyoc, K., and Hopkins, R., "Anatomy of the Risk Matrix," *Proceedings of the Mary Kay O'Connor Process Safety Center International Symposium*, p. 232, Oct. 2012.

349748 Perspectives on Ragagep

Abstract:

Recognized and Generally Accepted Good Engineering Practice or RAGAGEP is a fundamental building block of OSHA's PSM standard. OSHA has announced that it intends to publish a Request for Information (RFI) regarding it's Process Safety Management (PSM) standard. This paper will analyze the responses submitted to OSHA's RFI questions and determine what industry positions and best practices are on RAGAGEP.

Note to abstract reviewers: This abstract is a place holder - this paper may need to be further refined after OSHA issues it's RFI. The paper will reflect the views of the authors and does not represent OSHA policy.

Relief Systems Design: Simplifying Assumptions Gone Wrong

Nicholas Cristea
Smith & Burgess, LLC
Address
Nicholas.Cristea@smithburgess.com

Jason White, PE
Smith & Burgess, LLC
Address
Jason.White@smithburgess.com

Prepared for Presentation at
American Institute of Chemical Engineers
2014 Spring Meeting
10th Global Congress on Process Safety
New Orleans, LA
March 30 – April 2, 2014

UNPUBLISHED

Relief Systems Design: Simplifying Assumptions Gone Wrong

Nicholas Cristea
Smith & Burgess, LLC
Address
Nicholas.Cristea@smithburgess.com

Jason White, PE
Smith & Burgess, LLC

Keywords: Relief systems, FCCU, Reactor Trains, PSVs

Abstract

The process of evaluating an entire facility's relief and flare systems design can be extremely time consuming and costly. In most cases, simplifying assumptions or generalities are utilized in order to make this process less painful. Unfortunately, doing this can result in even more costly or painful results if the team executing the work does not know when to forego these assumptions and generalities. This paper will look at some of the common areas where the generally reasonable engineering assumptions do not apply. The paper will delve into two case studies to give a more in depth account of what might be missed and how to properly account for the cases. In the first case, a Fluid Catalytic Cracker Unit (FCCU) will be examined to show how some of the simplifying assumptions could actually lead to costly mistakes; however, taking extra time during the initial analysis would have prevented the mistakes. In the second case, when looking at high pressure reactor train, the recycle gas compressor system needs to be analyzed in extensive detail. The usual simplifying assumption is to account for hydrogen going through each possible pressure safety valve (PSV) during a relief case instead of actually looking at how the hydrogen in the system may be relieved. These two cases are the first in a series of situations where it is important to put the time and energy into performing sound engineering analysis rather than applying simplifying assumptions.

1. Introduction

In order to perform a relief systems analysis in a timely manner, there are some assumptions that are made to simplify the analysis and improve consistency. These assumptions help by making the analysis be less time consuming and allow normal plant operations to not be disrupted, by providing enough guidance for multiple engineers to work in parallel and increase the percentage of the work that can be performed without requesting many specific details. However, making simplifying assumptions across the board may lead to larger relief loads being calculated or some scenarios being misidentified. This paper will look at two different cases of where simplifying assumptions may cause for errors and analysis and how to avoid these pitfalls. The first case will be look at the slurry pump around on the Main Fractionator of a Fluid Catalytic Cracker Unit (FCCU) and the second case will look at what happens during a global failure in a Cat Feed

Hydrotreater (CFHT). These two cases will show how the simplifying assumptions that are commonly used would result in errors in analysis.

2. Case 1 - Fluid Catalytic Cracker Unit (FCCU)

2.1 Do you have a Relief Scenario?

The first step when looking into a FCCU Main Fractionator, the engineer must ask the pertinent question of whether or not the air compressor for the reactor can continue or not? Once this question has been asked, then the detailed analysis can begin. There are two basic ways to operate the air blower to the reactor, which are either in constant pressure or constant volume mode. If the blower is running in constant volume, then the reactor effluent may still reach the reactor as the Main Fractionator pressures up. Credit may be taken by "backing off" the compressor curve to reduce the amount of reactor effluent entering the system. If the blower is running in a constant pressure mode, then as the Main Fractionator pressures up the change in pressure will decrease, resulting in a decrease in driving force for the vapor to enter the column would be lost. The worst event would be a backflow case into the regenerator, but this paper will not cover this scenario.

2.2 The Slurry Pump around

The most important pump around for the Main Fractionator is the slurry pump around. The slurry pump around performs many tasks, but the most important one is that it quenches the reactor effluent. [1] The bottom of the column is extremely hot, but it is actually cooler than the reactor effluent entering the column. This pump around quenches the heat entering the column is by mixing with the reactor effluent upon entry into the column.[2] This quench of the reactor effluent is essential to the operation in order to allow for the fractionation of the reaction products to occur in the column. Usually, the pump around will be split into two sections before returning to the column. See Figure 1 for schematic of the set-up. Depending on the mode of operation of the FCCU, the slurry that is sent above the reactor effluent will act as the quench; whereas, the slurry sent below the reactor effluent will help balance the heat absorbed by the slurry coming into contact with the reactor effluent. This configuration is designed to prevent coke formation at the bottom of the column. [1]

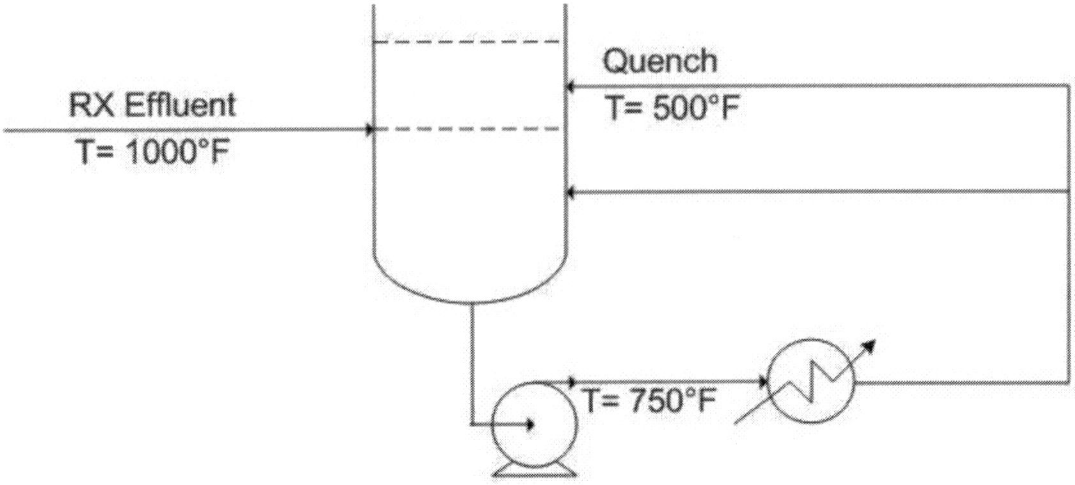

Figure 1 – The slurry pump around is very important as it acts as a quench for the reactor effluent. For a typical FCCU in operating for gasoline optimization, the reactor effluent is entering the column at ~1000 F.[5] The slurry is providing a quench volume to allow for fractionation.

The first scenario covered at will be the partial loss of the slurry pump around. When looking at the partial loss of the slurry pump around, the other pump arounds need to be looked at to see if they will continue. The pump arounds may be lost due to loss of the quench to the reactor effluent, resulting in a significant increase in vapor traffic up the column and subsequent drying of the trays. Without the quench, the lighter or potentially all the pump arounds may be lost, but it must be investigated further to see how the system may function. In all likelihood, the heavier the pump around the better the probability of that pump around continuing, which will not be known until the column is simulated. The assumption that the partial loss would not result in the other pump arounds being lost is not a good assumption. It helps to simplify the case, but it does not take into account what actually may be happening in the column. Upon simulating the scenario, it may be discovered that only one additional pump around is lost or maybe more. This more detailed analysis would lead to a smaller relief load because there would be more liquid traffic in the column, which would result in decreased vapor traffic leaving the top of the column. As stated in the above paragraph, the slurry pump around helps to aid in providing the needed quench to the entering reactor effluent, but if it is partially lost, then other pump arounds may be lost as well. The loss of additional pump arounds would result in a hotter and heavier relief stream, which would allow for a more accurate steady state representation of the relief composition.

The second scenario that will be covered is the total loss of the slurry pump around. When this occurs, there is a strong likelihood that the rest of column's pump arounds would be lost as well and the entire reactor effluent may need to be relieved, due to the subsequent drying of the trays preventing liquid draw to the pump arounds. In this event, it is likely the wet gas compressor will trip due to either the increased feed temperature or the increased molecular weight, which creates the need for the required relief rate to be the entire effluent stream.[3,4] The need to relieve the entire reactor effluent may result in a blocked outlet due to the likely compressor trip.

See Figure 2 for schematic of the set-up. The total loss of the slurry pump around would result in the loss of the main quench to the reactor effluent, which relatively quickly would result in the entire liquid traffic in the column to be vaporized. Without any of the slurry to contact the reactor effluent first, the liquid traffic in the column is not heavy enough to absorb the heat and remain a liquid. The pressure relief valves (PSVs) would not be able to handle this large relief load. There are many procedures in place to prevent this from happening. Most likely, the slurry pump around will be have two pumps with one being motor driven and the other turbine driven. This is to try to prevent a total loss of the slurry pump around.

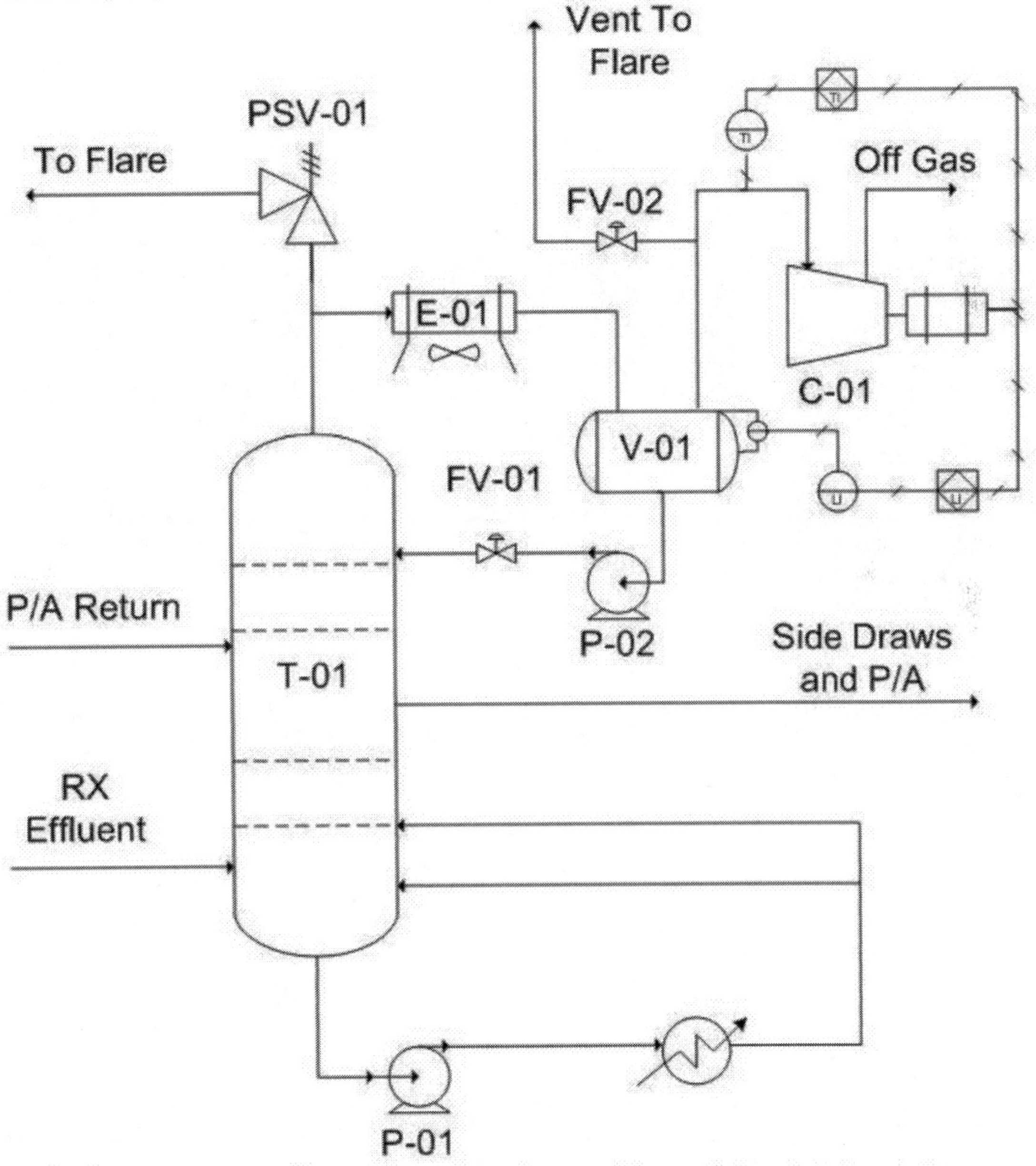

Figure 2 – The potential loss of the slurry pump around would result in the reactor effluent drying up the column preventing the other pump arounds from operating. The wet gas compressor would likely trip due to the high temperature or molecular weight resulting in a relieving scenario equivalent to the entire reactor effluent stream.

3. Case 2 - Cat Feed Hydrotreater Reactor Train (CFHT)

3.1 Individual PSV Analysis

In and around a hydrotreater reactor train are many locations where hydrogen may need to be relieved for certain scenarios. When looking at any high pressure reactor train, there are many interconnected scenarios that need to be evaluated, but those will be talked about later. The scenarios sizing individual pressure relief device installations will be based on an independent failure and not a global case. A global case can be described as one independent failure leading to multiple simultaneous and related events, or as covered here, simultaneous releases. The sizing basis of a PSV should account for the worst possible single initiating event for the protected system.[3] A common example of a case that is usually misidentified is the failure of the charge pumps drawing from the reactor feed surge drum. When the charge pump fails, high pressure hydrogen may flow backwards through the check valve and cause a two-phase relief as the feed to the drum is expected to continue. See Figure 3 for a schematic of the case. Failure of the check valve should be considered a latent failure and therefore not considered to be a double jeopardy event.[3] Some companies allow for credits to be taken for safety critical check valves, which puts responsibility on operations to ensure the check valves are operating properly, tested frequently, and therefore not expected to become stuck.

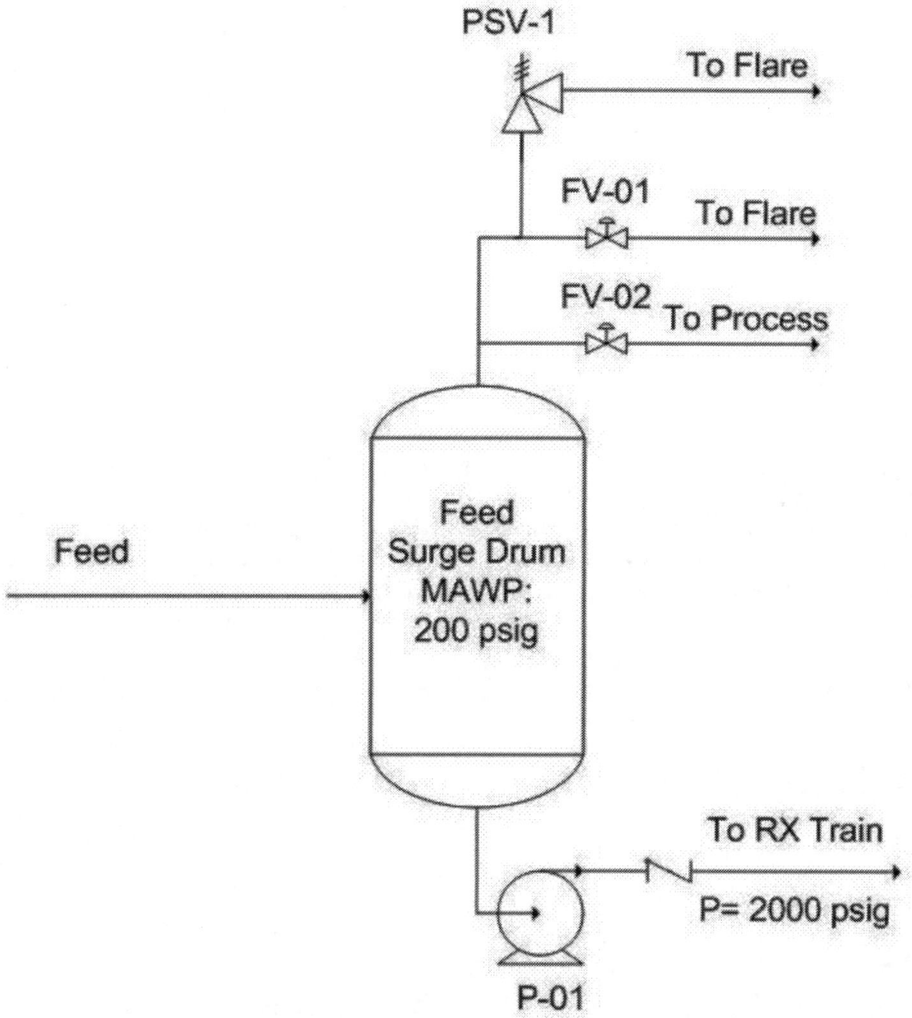

Figure 3 – In the event of the charge pump (P-01) failing, there is a possibility for hydrogen to backflow into the surge drum. The check valve may limit the flow to some extent, however the failure of this check valve is not considered to be a "double jeopardy" scenario. The resulting scenario is often the controlling scenario for the feed surge drum system and typically was not considered during the original design.

Another challenging system to perform scenario analysis is the compressor system for the Cat Feed Hydrotreater. There are several events that can result in overpressure relief, most of which result in hydrogen being relieved. The compressor could trip due to a high temperature reading from upstream cooling being lost, which may result in a blocked outlet and possible backflow around the suction KO drum feeding the compressor. This scenario would usually be the worst case for the suction drum relief device and would likely result in a large relief device. The backflow would most likely occur through the spillback line on the compressor rather than through the compressor as it is likely the compressor is a reciprocating type. The backflow may occur particularly when the system includes multiple compressors in parallel, resulting in high pressure hydrogen downstream of the system. The scenarios covered are for individual relief device sizing basis which requires the worst plausible case be protected. However, if this is the only step that is done, then the global analysis may over predict the required load being sent to the flare. The end result may be an undersized flare header or flare, but that may not be the case.

.

3.2 Global Analysis

In addition to the individual relief systems installation analysis, the performance of the pressure relief disposal system should also be analyzed. To perform this correctly global overpressure scenarios (total power failure, partial power failures, instrument air failure, etc.,) should be identified and quantified where required. The importance of understanding the global events cannot be understated, as individual relief installations must cover the worst case event for that system; it may not be physically possible for 2 such independent events to happen simultaneously. [6] Specifically for a hydrotreater reaction train, failure to consider these effects will result in an analysis where significantly more hydrogen is released than is physically possible. This type of oversight will almost certainly result in identified deficiencies in the disposal system that could result in costly modifications or operational changes. The problem stems from a detailed analysis not being performed on the actual hydrogen in the system and asking the question of what would happen given a specific scenario. See Figure 4 for a simplistic schematic of a reactor train.

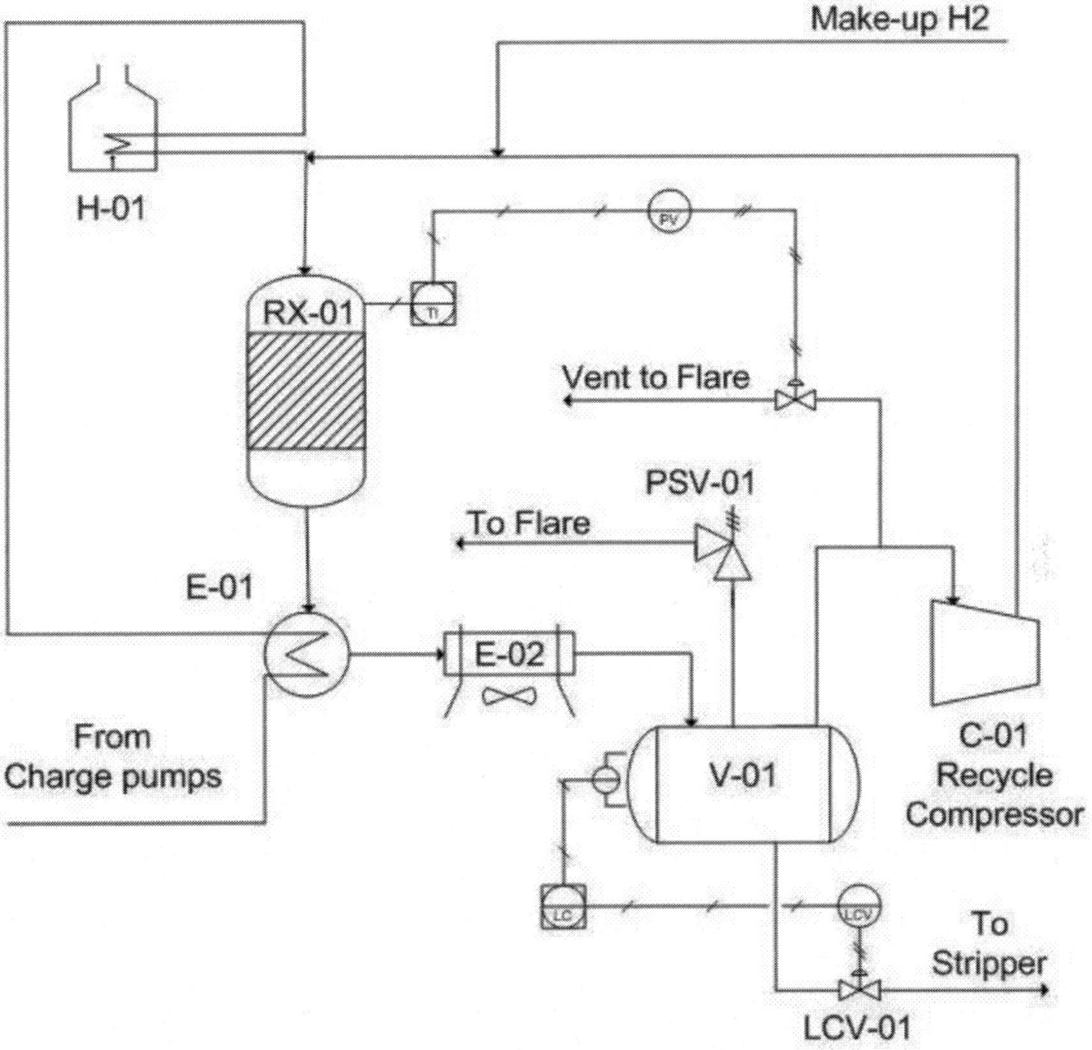

Figure 4 – In the event of a scenario where the compressor is tripped, there are 3 separate paths for the hydrogen to relieve: through the charge pumps as described earlier, through the PSV of V-01, and through the depressuring valve. Each of these discharge locations was individually sized for the worst case, which cannot happen simultaneously.

The hydrogen has many different ways to leave the reactor system during an overpressure event. The hydrogen may follow the usual flow path downstream or it could go backwards through many different pathways; however, until the whole system is evaluated as one system, there is no way to be sure. For example, there are usually depressuring valves located around the suction to the make-up compressor. By looking at the logic of the interlocks, it can be possible to decipher when the reactor loop would be depressured. Sometimes just the peak depressuring rate may be used as the hydrogen in the unit. The interesting thing to note for high pressure reactor loops is the possible presence of two different depressuring valves. One will be a low depressuring and the other will be a high depressuring. Again, which one opens will be based on what instrumentation has gotten a bad reading. If the depressuring rate is not used, it is imperative to not double count the hydrogen in the train. Due to the nature of hydrogen gas, a small amount will still take a large volume in the header and cause sizing problems.

Each of the potential partial power failures, and other global failures for that matter, should be evaluated by looking at the whole reactor loop to understand what is expected to happen during such an event. Which case should be considered for disposal system sizing, the depressuring or having different loads going through the respective PSVs within the system may be different for each global scenario. The reactor loop will more than likely be contributing one of the largest loads to the disposal system. The partial power failures around the different make-up compressors in parallel will be the hardest to evaluate because certain suction drums may be relieving the blocked outlet and backflow rate, while others may be functioning normally. It is important throughout this system to not double count hydrogen. See Table 1 below for a comparison from a conservative to a detailed analysis.

Unit	Conservative Load	Detailed Analysis
Hydrotreater	415,865 lb/hr	347,718 lb/hr

Table 1: Showing the comparison of performing a detailed to a conservative analysis for a Hydrotreater.

4. Conclusion

The two examples discussed in this paper identify specific areas where generalized assumptions and methodologies are not ideal when performing a complete relief systems design or analysis. The simplifying assumptions are extremely important to the analysis, but it is equally important to know when to deviate from them. In Case 1, the slurry pump around may be only partially lost, but it may still result in the column drying up. This would still lead to the reactor effluent needing to be relieved; however, the relief temperature would be lower. In Case 2, the flare analysis around the reactor train needs to have a detailed analysis to account for the hydrogen being sent to the flare. If all the individual loads and the depressuring rate are sent to the flare, then an existing flare is highly likely to be identified as undersized, while a new flare is likely to be overbuilt at considerable unnecessary cost. The analysis needs to be performed to see when the depressuring valves will open and what systems would be affected by the depressuring valves. The adequacy of the flare network depends of the detailed analysis. The use of simplifying assumptions may result in a larger flare network, which would likely cost millions to implement.

5. References

[1] Clark, Dan and Golden, Scott, "Improving FCCUs: bottoms system upgrades," *Digital Refining*, 3rd Quarter, 2003.

[2] White, Steve and Fulton, Scott, "Specifications-importance of getting them right," *Digital Refining*, 4th Quarter, 2003.

[3] ANSI/API Standard 521, "Pressure-relieving and Depressuring Systems," American Petroleum Institute, 6th ed., 2014.

[4] GPSA, Engineering Data Book: Volume I 10th Edition, Gas Processors Suppliers Association, Tulsa, OK. 1987.

[5] Meyers, Robert A., Handbook of Petroleum Refining Processes 3rd Edition, McGraw-Hill, New York, 2004.

[6] White, Jason and Spearow, Jason, "Heat Integration and Relief Systems Design," Process Safety Progress, Volume 32, Issue 3, 2013.

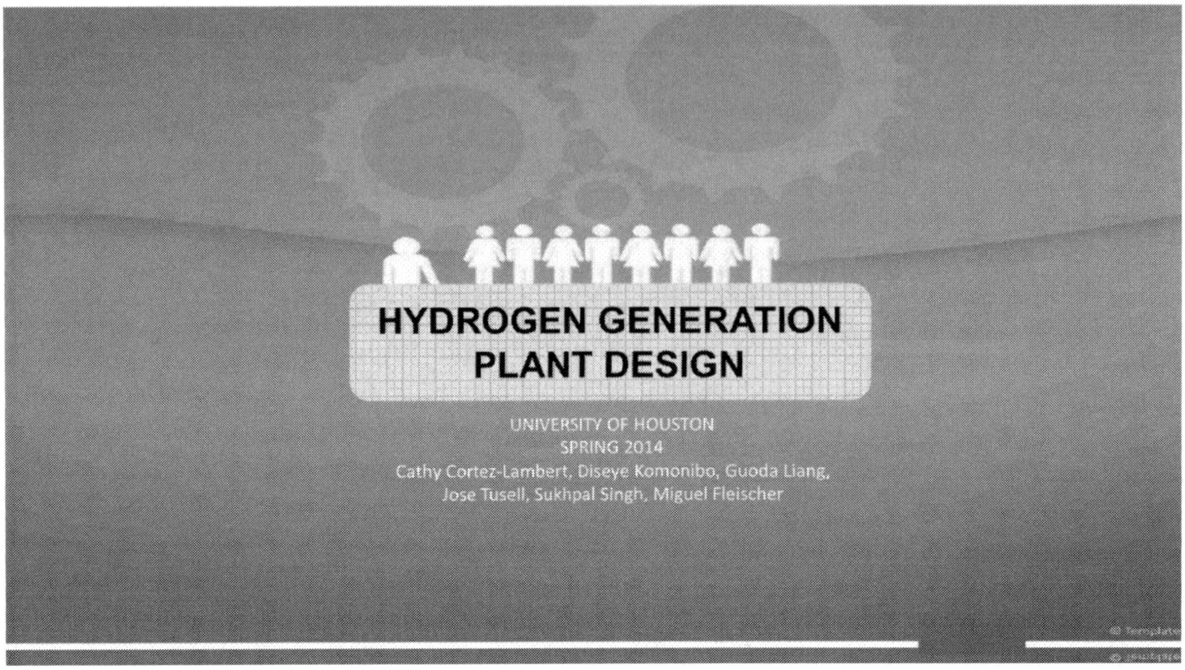

Outline

- Introduction
- Plant Specifications
- Health, Safety & Environmental
- Equipment Specifications
- Heat Integration
- Cost Analysis
- Recommendations
- Questions

Introduction

- **Chemical Engineering Design I at UH (fall)**

  - ➢ The objective is to design an entire plant given a feedstock and product specifications
  - ➢ Calculations are done without the use of a computer-aided design software package
  - ➢ Teams are selected by the professor (usually 4 or 5)
  - ➢ Industry representatives review the design twice in the semester to provide feedback
    - ➢ Fluor is the current sponsor
    - ➢ Previous sponsors: Dow, ExxonMobil, Fluor, Oxy, Sabic, Shell (2)

- **Chemical Engineering Design II at UH (spring)**

  - ➢ The exact same project is assigned to basically the same teams
  - ➢ Calculations are done with the use of Aspen to optimize the fall design
  - ➢ Each team is assigned a different location in the world
  - ➢ Profitability calculations are also done to compare the various locations
  - ➢ Industry representatives review the design twice in the semester to provide feedback

Introduction

Project

Design a Hydrogen Generation Plant based on Steam-Methane Reforming

Plant Specifications: Block Diagram

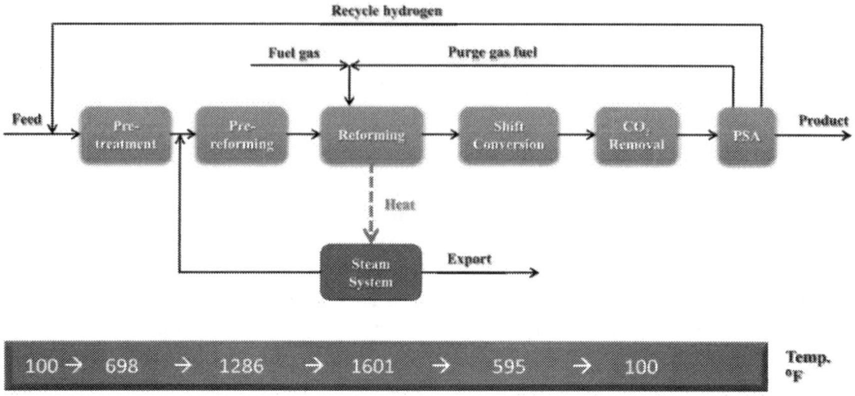

Plant Specifications: PFD Drawing

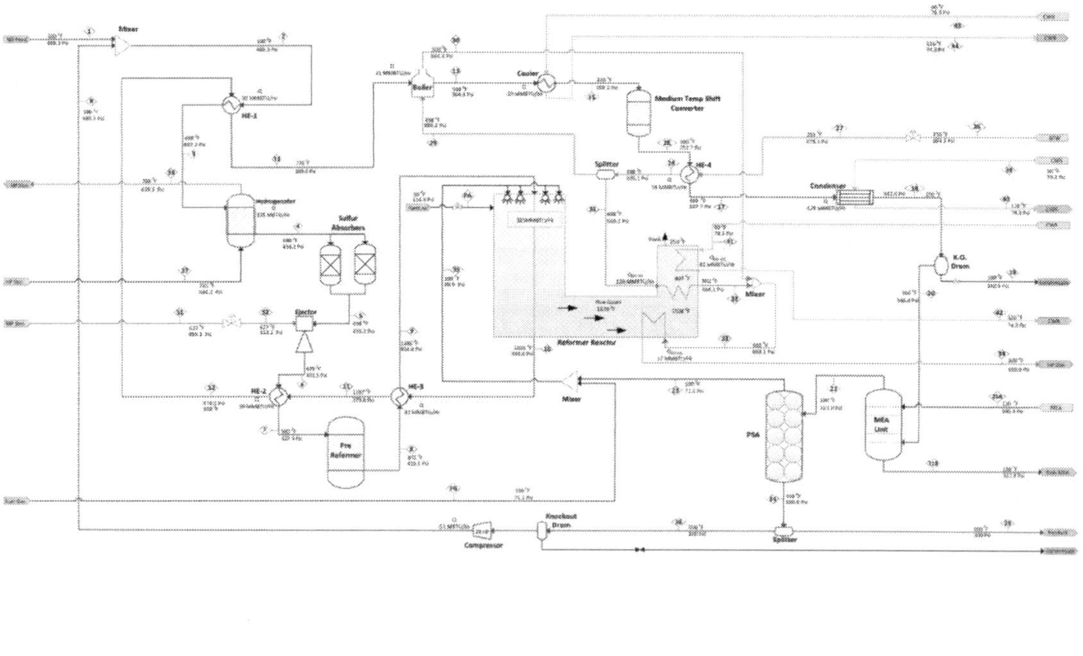

Health, Safety & Environmental

- **Environmental**
  - Emissions
    - Stack Gas
      - CO_2 Tax Permit
  - ZnS Containment
  - Cooling Towers
    - Monitor Emissions

- **Safety**
  - OSHA Standard Requirements
  - Process Hazard Analysis (PHAs)
  - Inspections & HAZOPs
  - SOPs and Training Enforcement
  - Strict PPE

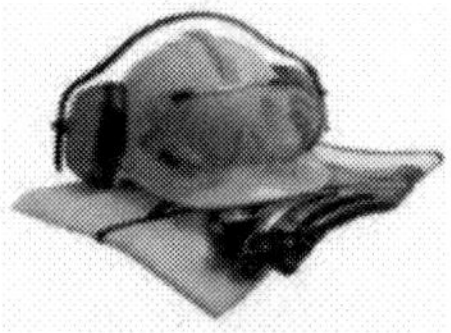

Hydrogenator

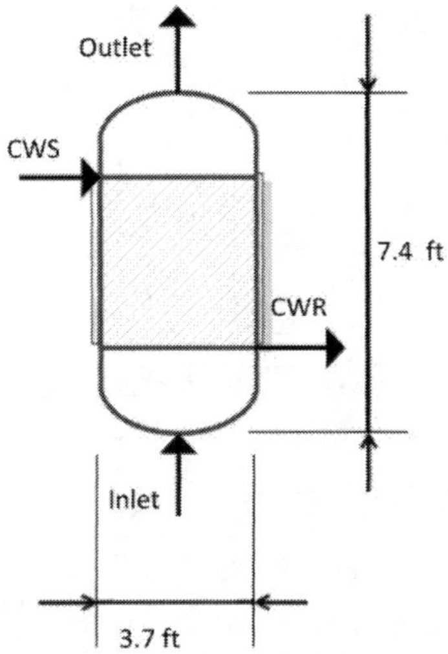

Operating condition: 708°F, 639psi
Residence time ---- 2.6 s
Over 99.5% **conversion**
Co-Mo/Al_2O_3 as catalyst
Reactor size---79.8 **cubic feet**
Constructure material: SS316l

Element	316L
Chromium	16.0 min.–18.0 max.
Nickel	10.0 min.–14.0 max.
Molybdenum	2.00 min.–3.00 max.
Carbon	0.030
Manganese	2.00
Phosphorous	0.045
Sulfur	0.03
Silicon	0.75
Nitrogen	0.1
Iron	Balance

Sulfur Absorber

Residence time ---- 1.8 s
Reactor size -- 57.2 cubic ft
Recycle period ---- 20 days
Two ZnO beds

HS&E:
SS 316l
Sealed storage
Containment tank
Personal protective equipment

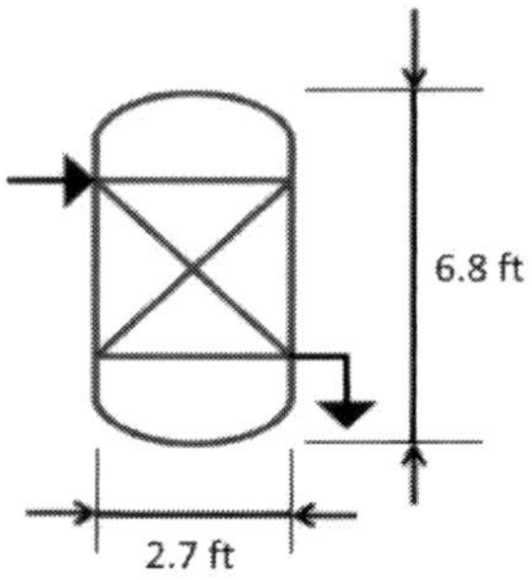

Pre-Reformer

- Capacity Increase
- Energy Savings
- Crack >C_1 Hydrocarbons
- Larger Operational Flexibility

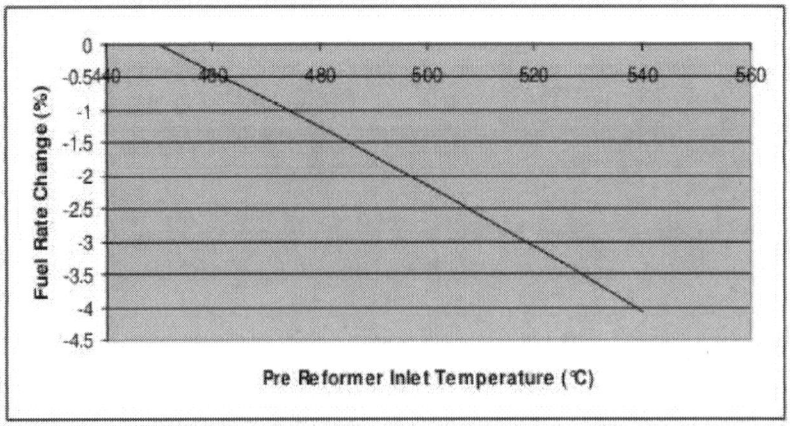

Pre-Reformer

Packed Bed ADIABATIC RX with ΔT=139°F

- $C_nH_m + nH_2O \rightarrow nCO + \left(n + \frac{m}{2}\right)H_2$
  - $C_2H_6 + 2H_2O \rightarrow 2CO + 5H_2$
  - $C_3H_8 + 3H_2O \rightarrow 3CO + 7H_2$
  - $C_4H_{10} + 4H_2O \rightarrow 4CO + 9H_2$

  X 100%

- $CH_4 + H_2O \rightarrow CO + 3H_2$
- $CO + H_2O \rightarrow CO_2 + H_2$

  X 7.5%

$286679\frac{lbm}{hr}$ @980°F, 427.4PSIA

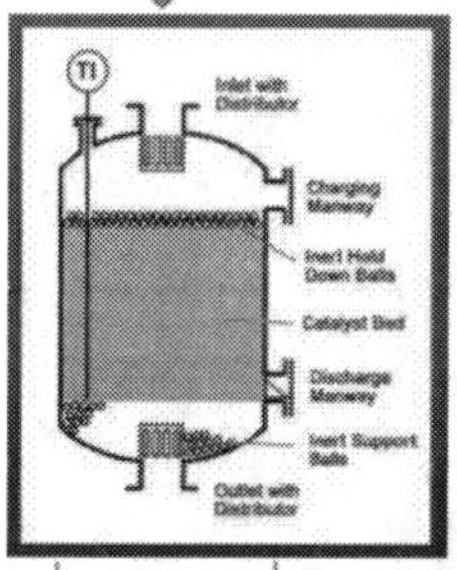

$\tau = 2\ sec$

15'2"

5'1"

MOC: SS 405

$286679\frac{lbm}{hr}$ @ 841°F, 419.5PSIA

Reformer

Reactions

$CH_4 + H_2O \rightarrow CO + 3H_2$

$CO + H_2O \rightarrow CO_2 + H_2$

X = 78.7%

Requires

ΔT=317°F

Q = 325MM$\frac{BTU}{HR}$

SIZE	
Heat Flux	≈ 17,000$\frac{BTU}{Hr\ Ft2}$
Volume	= 11,860 GAL
Tubes Diameter	= 4.5"
Tubes Length	= 43'0"
# of Tubes	= 334
MOC	= HP-MA

$286679\frac{lbm}{hr}$ **@1284°F**, 414 PSIA

- Excess 1.2% Air
- Plant Fuel
- PSA Purge Fuel

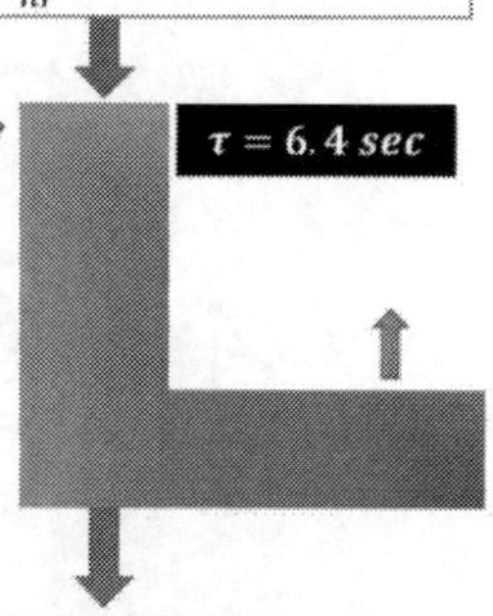

$\tau = 6.4\ sec$

1601°F, 384 PSIA

Medium Temperature Shift Converter

Adiabatic Operation

Conversion: 95%

Residence Time: 4.5 s

Reactor Size: 646 cubic feet

H/D Ratio: 2

Cost: $ 188,000

Material of Construction: Inconel® 600

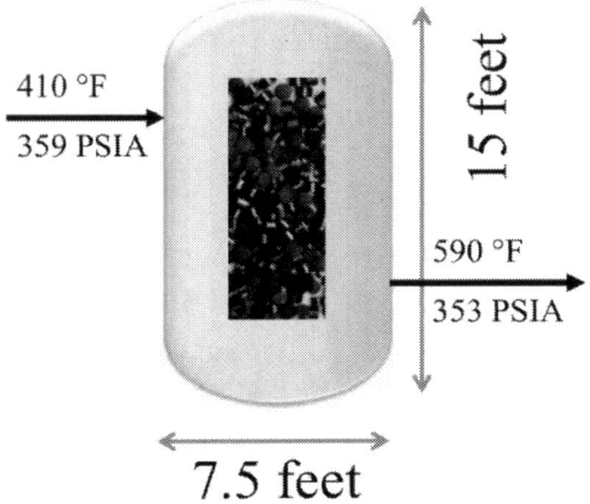

MEA Unit

Acid Gas Absorber			
Service:	CO2 Absorber	Tag:	MEA Absorber
Ave. Process T:	110 F	Average Process P:	323PSI
Process Stream:	Hydrocarbon	Absorbent:	MEA + Water
Column Diameter:	8.38 FT	Number of stages:	20
Column height:	54 FT	Tray spacing:	24"
MOC:	316 stainless steel	Loading Factor:	0.1 CO2/MEA
ΔH:	Adiabatic	Tray type:	Sieve trays
Cost		$1,018,019	

Pressure Swing Adsorber

- **Operating at** 100°F, 300 psia
- 5A zeolite, activated carbon
- Efficiency
  - 10% H_2 adsorbed
- 10 beds
  - 1 adsorbent bed per column
  - Cycle Time: 12 minutes
  - Column Volume: 954 ft^3
- Purge Tail Gas
  - ~ 99.9% CH_4 removal
- Cost
  - Carbon Steel
  - $332,615 per bed
  - 10 beds: $3,326,149

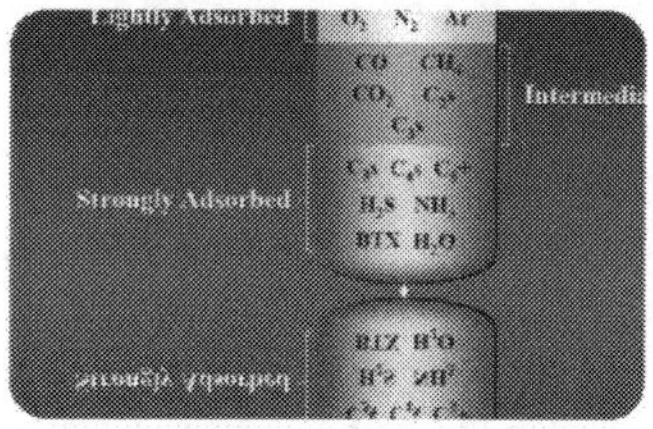

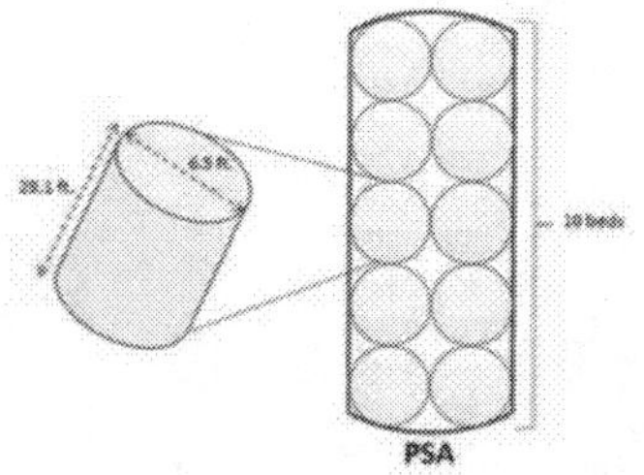

Heat Integration

Reformer Efficiency			AIR			Temperature of Flue Gases
Heat			20 % Excess Air & RH 68%			°F
Provided	560	MMBtu/Hr	N_2	281982	lbm/Hr	3434 Adiabatic Flame Temp.
CH_4 Cracking	325	MMBtu/Hr	H_2O	5817	lbm/Hr	1629 Convection Section
Saturated Steam	115	MMBtu/Hr	O_2	85617	lbm/Hr	1528 Convection Section
Superheated Steam	17	MMBtu/Hr	Total Air	373416	lbm/Hr	807 Convection Section
Stack Cooler	82	MMBtu/Hr	Total Fuel	37273.284	lbm/Hr	250 Atmosphere Stack
Total Heat Used	457	MMBtu/Hr				

Reformer Overall η	Furnace Tubes η
81.62%	58.00%

Total Steam Export

217,773 lbm/hr @ 600 °F, 658 PSI

Plant Heat Overall Efficiency

60%

Cost Analysis

Equipment name	Temp. range (°F)	Max P (psig)	MOC	Diameter (ft)	Volume (ft^3)	Heat Transfer Area (ft^2)	Power (BTU)	Quantity	Cost
HE1	[100-938]	474	SS416			1539		1	$ 235,686
HE2	[698-725]	418	SS405			4530		1	$ 388,390
HE3	[679-1197]	405	SS302			5179		1	$ 417,198
HE4	[841-1601]	660	CS			26641		1	$ 512,475
Reformer Tubes	[1286-3434]	400	HP-MA			16911		1	$ 1,276,463
Steam SH	[498-776]	648	SS302			530		1	$ 174,321
Boiler	[90-548]	654	CS			4486		1	$ 152,502
Steam Boiler	[251-590]	654	SS302			4972		1	$ 436,182
Cooler	[90-269]	350	CS			785		1	$ 66,254
Stack Cooler	[502-1629]	65	CS			4676		1	$ 130,916
Condenser	[498-1528]	333	CS			36319		1	$ 604,150
Hydrogenator	[90-807]	630	SS316l	3.70	79.8			1	$ 63,207
Pre-reformer	[410-590]	417	SS416	5.01	295.3			1	$ 173,958
MTSC	[841-980]	345	Inconel 600	7.44	646.0			1	$ 187,602
KO POT 1	[100-100]	319	CS	5.96	500.0			1	$ 199,688
KO POT 2	[100-100]	285	CS	3.02	54.3			1	$ 27,678
ZnO Absorber	[100-100]	443	SS316l	2.72	57.2			2	$ 228,916
PSA Unit	[698-698]	308	CS	6.46	954.2			10	$ 3,326,149
MEA Unit	[100-138]	326	SS416	8.60	998.5			1	$ 1,018,019
Compressor	[106-192]	475	CS				40	2	$ 68,539
Reformer Furnace	[1286-3434]	100	Various				220201	1	$ 15,618,714
Total								32	$ 25,307,008

Cost Analysis

Direct Costs	% of CBM	Amount
Equipment Purchases	1	$ 25,307,008
Equipment Installation	0.47	$ 11,894,294
Instrumentation and Controls	0.36	$ 9,110,523
Piping	0.68	$ 17,208,765.51
Electrical Systems	0.11	$ 2,783,771
Buildings	0.18	$ 4,555,261
Yard Improvements	0.1	$ 2,530,701
Service Facilities	0.7	$ 17,714,906
Indirect Costs		
Engineering and Supervision	0.33	$ 8,351,313
Construction Expenses	0.41	$ 10,375,873
Legal Expenses	0.04	$ 1,012,280
Contractor's Fee	0.22	$ 5,567,542
Fixed Capital Investment	4.6	$ 116,412,237
Working Capital	0.89	$ 22,523,237
Sub-Total Capital Investment	**5.49**	**$ 138,935,474**
OSBL	0.3	$ 41,680,642.35
Contingency	0.2	$ 27,787,094.90
Total Capital Investment		**$ 208,403,212**

Recommendations

- ➢Improve heat integration
- ➢Replace MTSC with low and high temp. shift converters
- ➢ Continuous improvement on process
- ➢Strive to achieve excellence in quality

Questions?

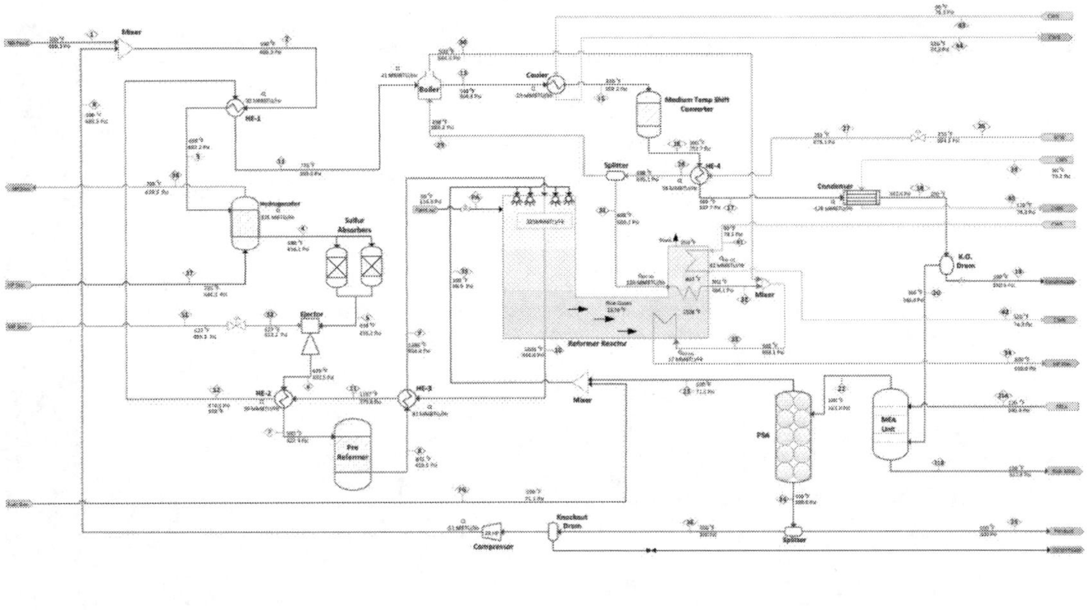

Ethylene Oxide Production

CENG: 4310 Final Design Project

Andrew Jones · Erica Diamond · Alex Tir · Morgan Weber

Tulane University

Goal

- Production Rate of 120 kTa of Ethylene Oxide
- Evaluate Process Design Options
  - Isothermal Reactor vs. Adiabatic Reactor
  - Oxygen vs. Air

$$C_2H_4 + 0.5\,O_2 \rightarrow C_2H_4O$$

$$C_2H_4 + 3\,O_2 \rightarrow 2CO_2 + 2H_2O$$

$$C_2H_4O + 2.5\,O_2 \rightarrow 2CO_2 + 2H_2O$$

Ethylene Oxide Production

- Isothermal Reactor
- Oxygen

- Annual Net Sales: $210 million
- Annual Net Cost: $137 million
- Annual Net Profit: $73 million

12 Year Cumulative Cash Flow:
$352 Million

Process Overview

- Process Description
  - Reactor Feed Prep
  - Reactor
  - Separation
  - Purification
- Safety
- Economics
- Recommendations
- Conclusion

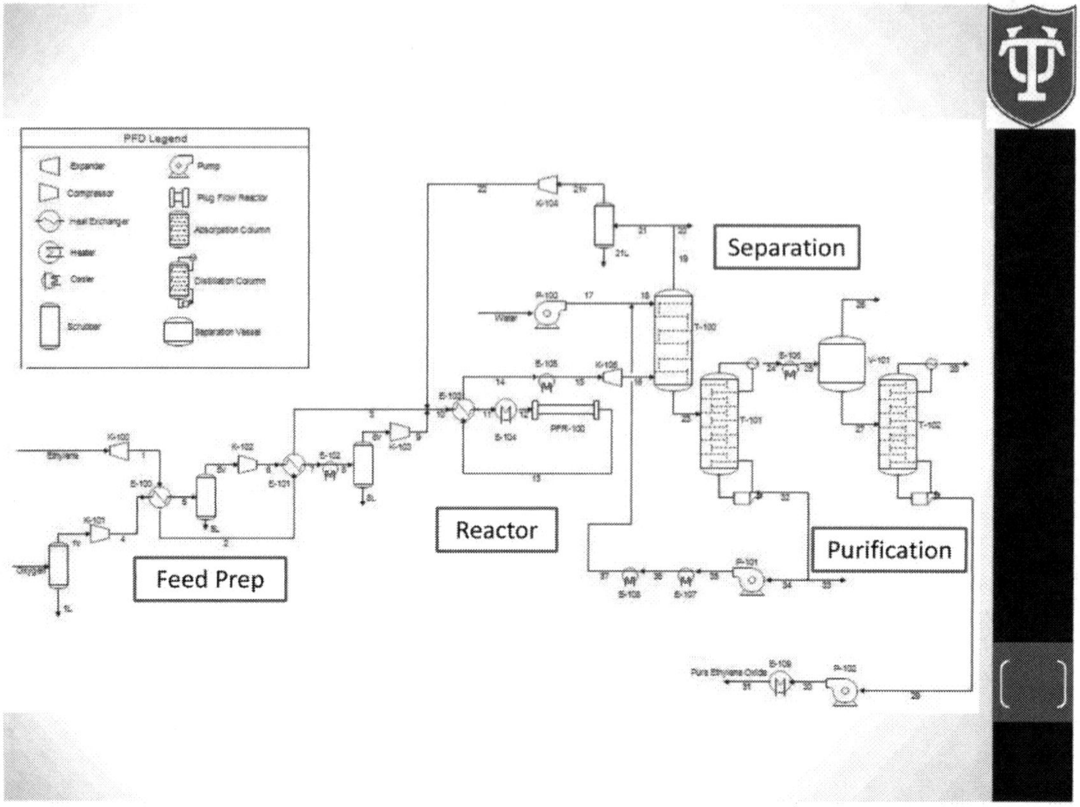
PFD Legend
Separation
Reactor
Feed Prep
Purification

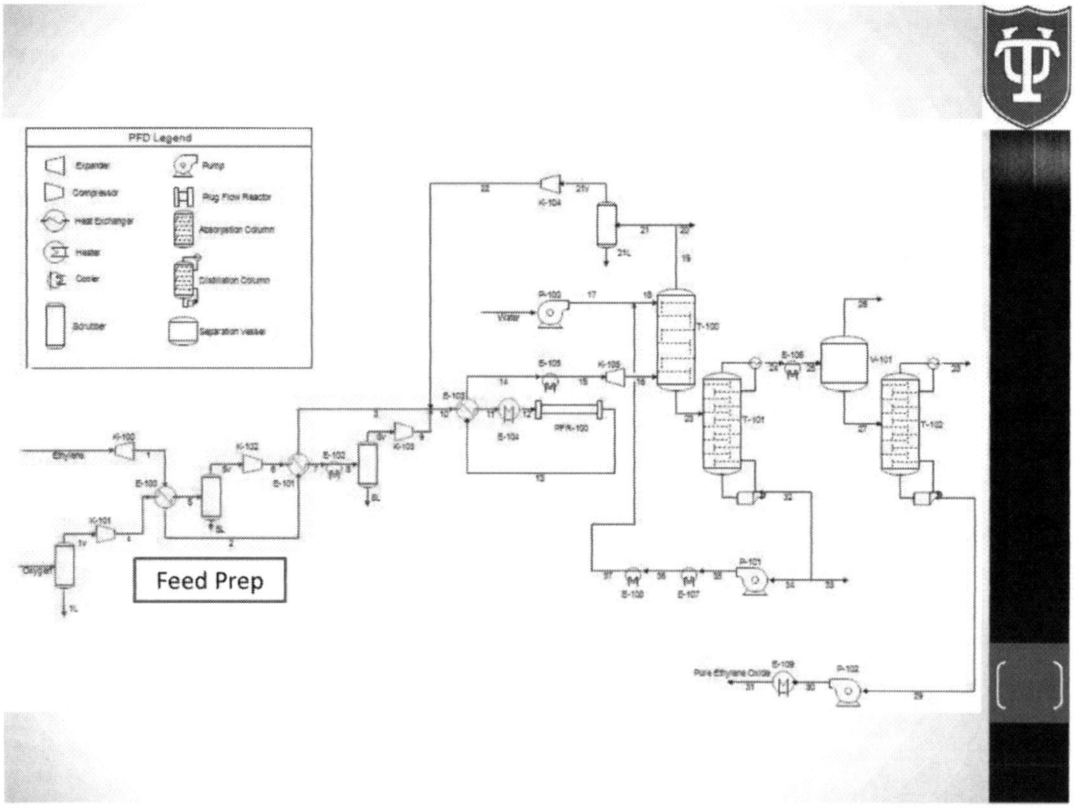
PFD Legend
Feed Prep

Reactor Feed Prep

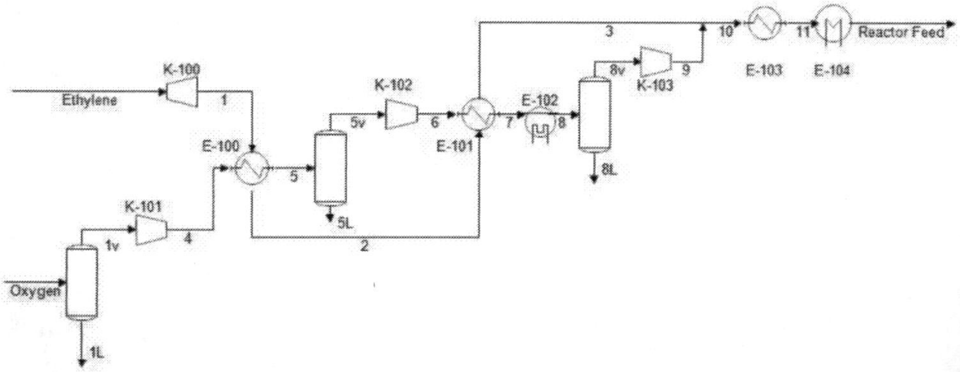

Reactor Feed Prep

Ethylene Depressurization

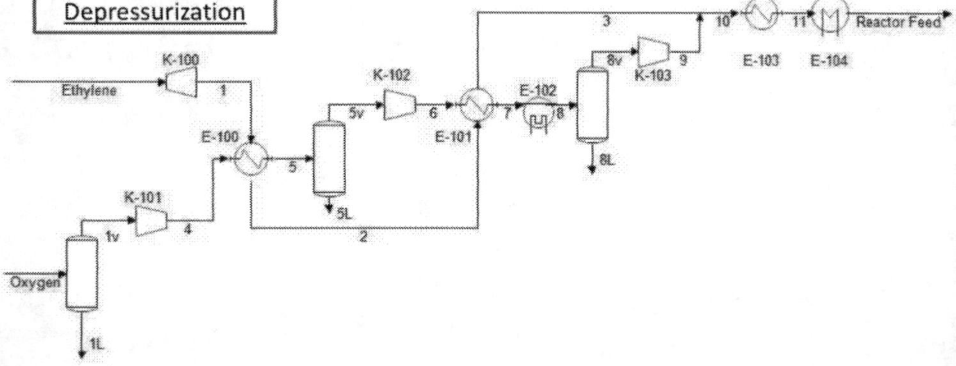

Reactor Feed Prep

Ethylene Depressurization

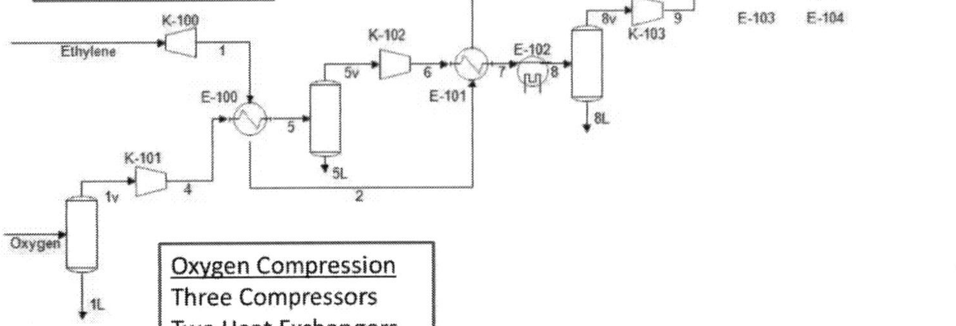

Oxygen Compression
Three Compressors
Two Heat Exchangers
Cooler

Reactor Feed Prep

Ethylene Depressurization

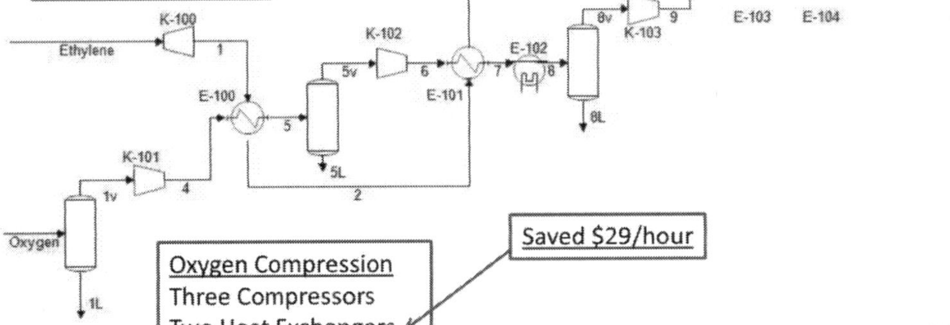

Saved $29/hour

Oxygen Compression
Three Compressors
Two Heat Exchangers
Cooler

Reactor Feed Prep

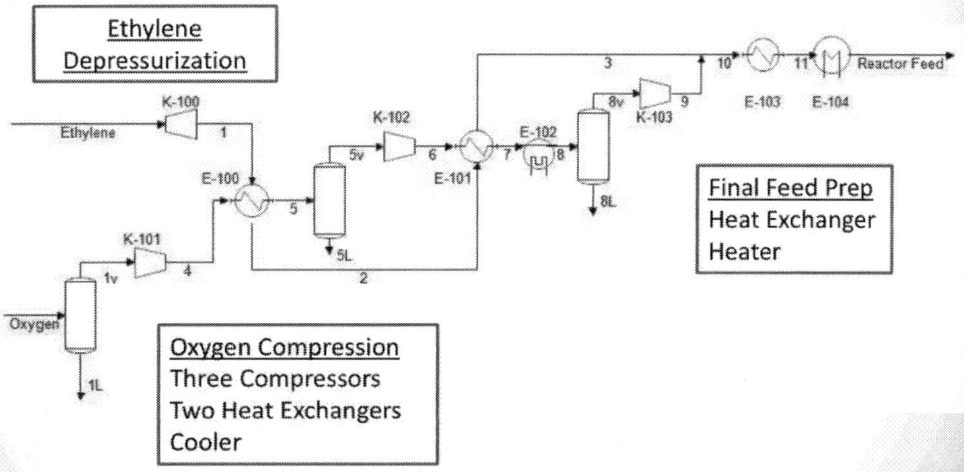

Oxygen vs. Air

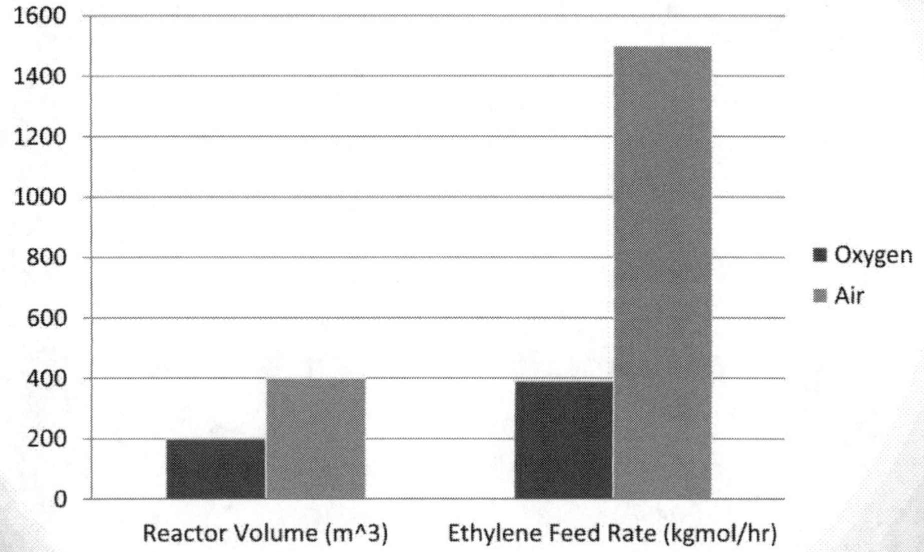

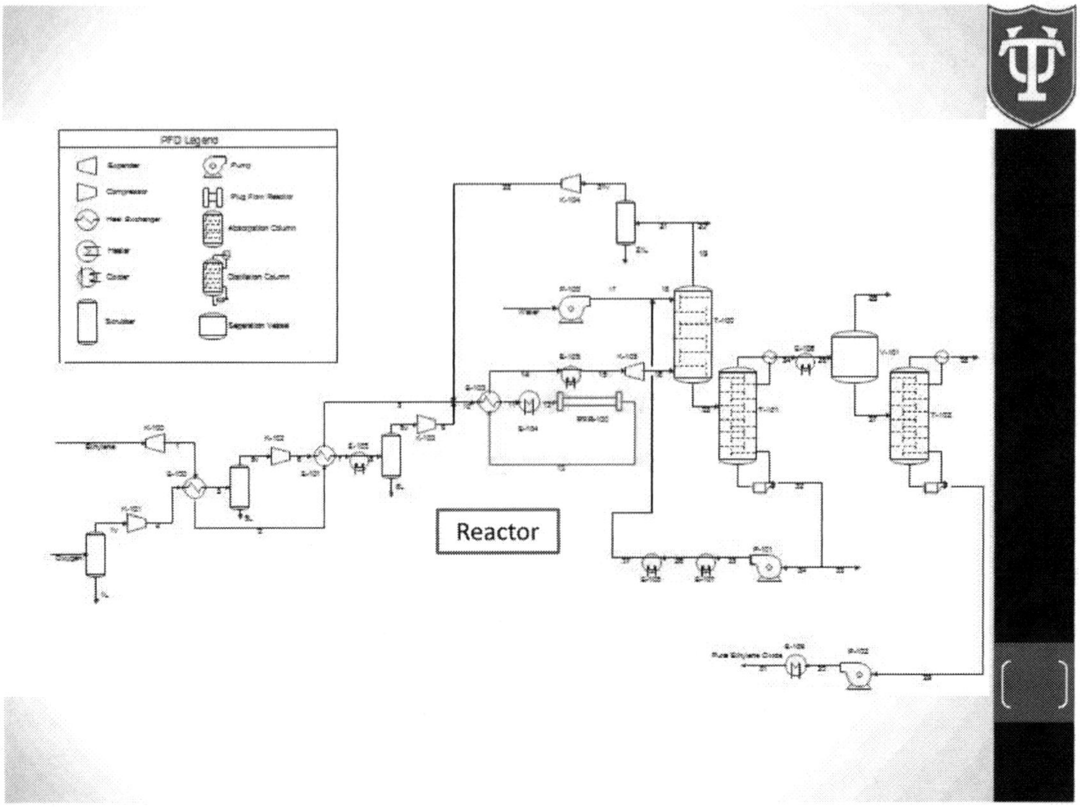
Reactor

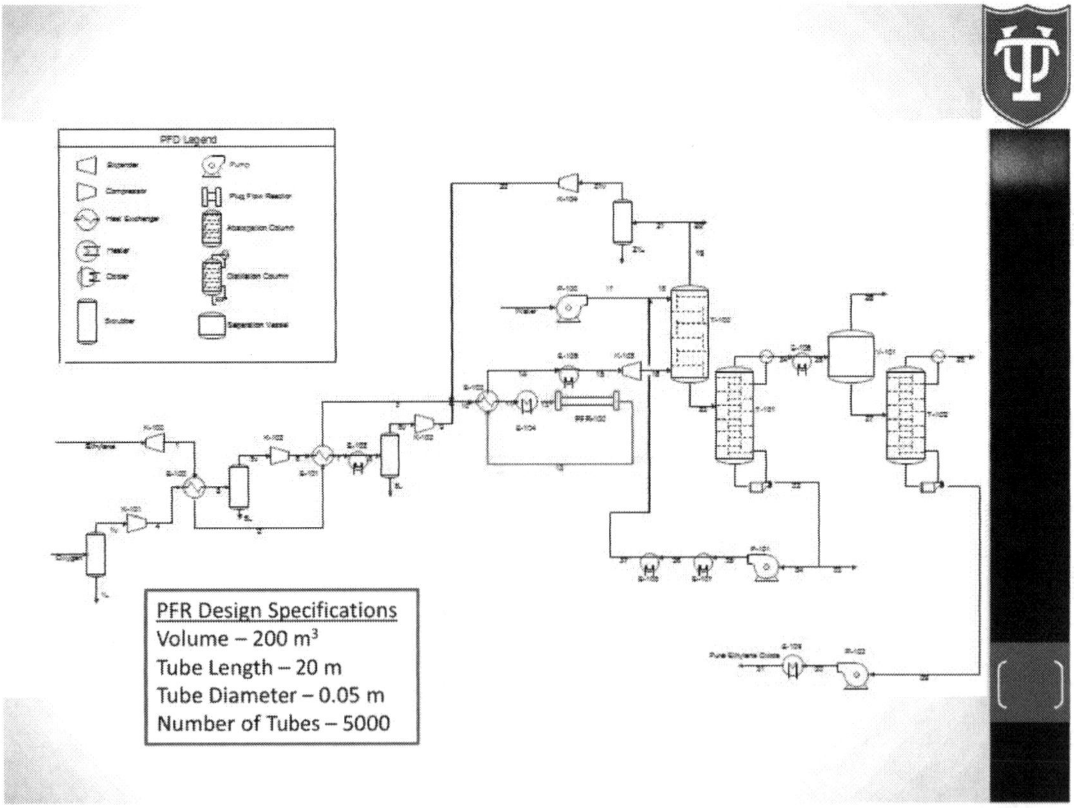
PFR Design Specifications
Volume – 200 m3
Tube Length – 20 m
Tube Diameter – 0.05 m
Number of Tubes – 5000

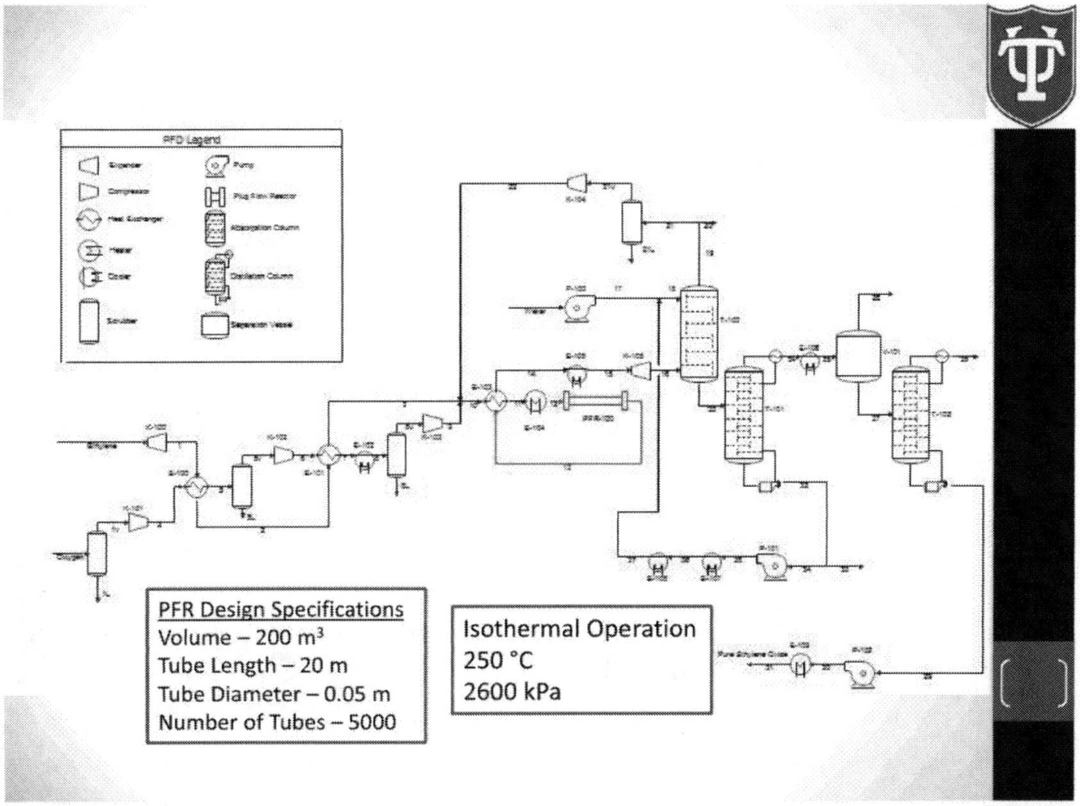

Case Study: Reactor T & P

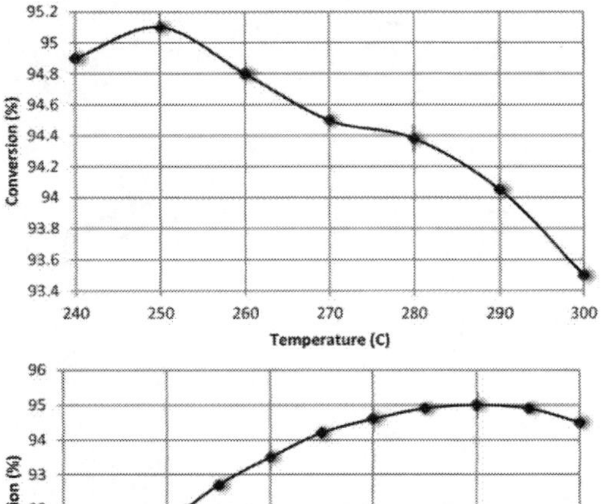

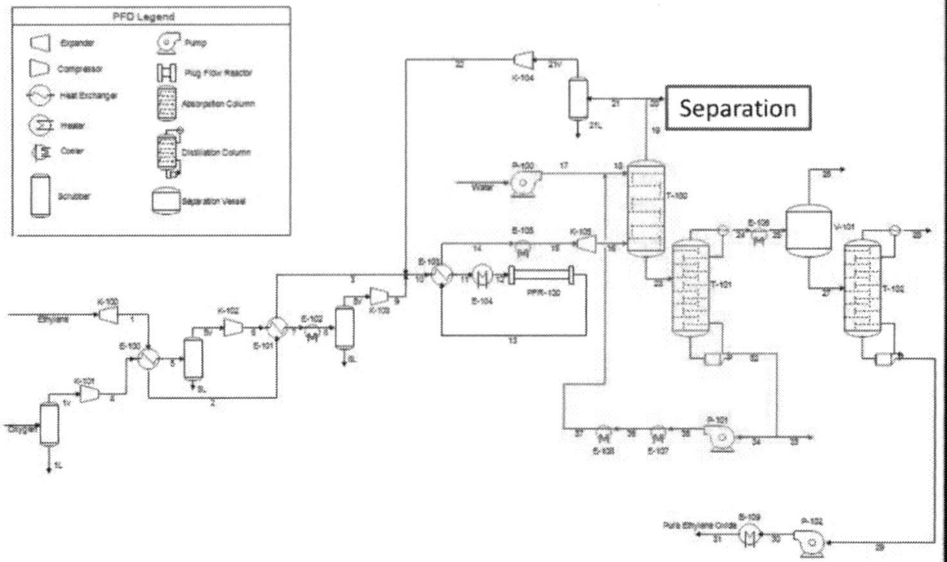
PFD Legend
Separation

Separation

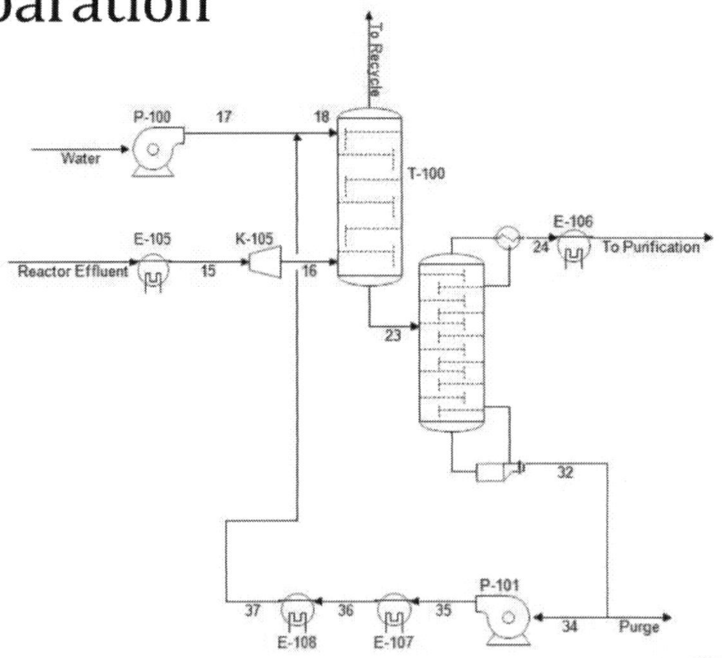
To Recycle
P-100
17
18
Water
T-100
E-105
K-105
E-106
Reactor Effluent
15
16
24
To Purification
23
32
P-101
37
36
35
34
Purge
E-108
E-107

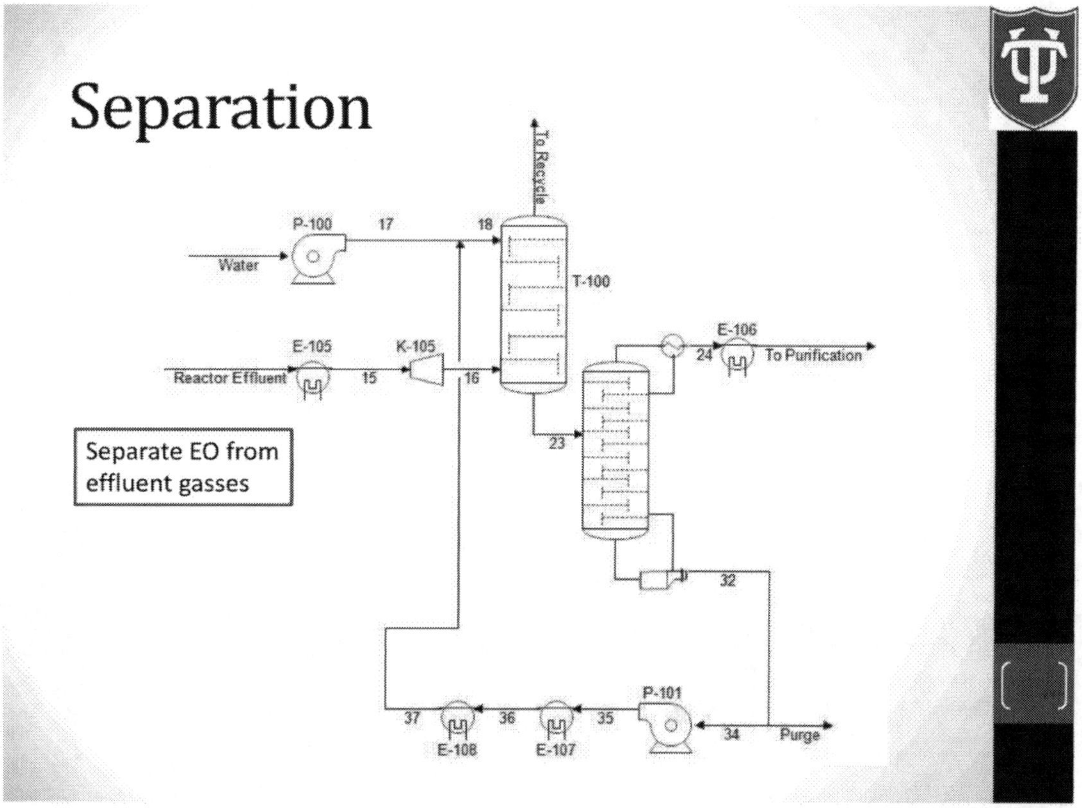
Separation
To Recycle
P-100
17
18
Water
T-100
E-105
K-105
E-106
24
To Purification
Reactor Effluent
15
16
Separate EO from effluent gasses
23
32
P-101
37
36
35
34
Purge
E-108
E-107

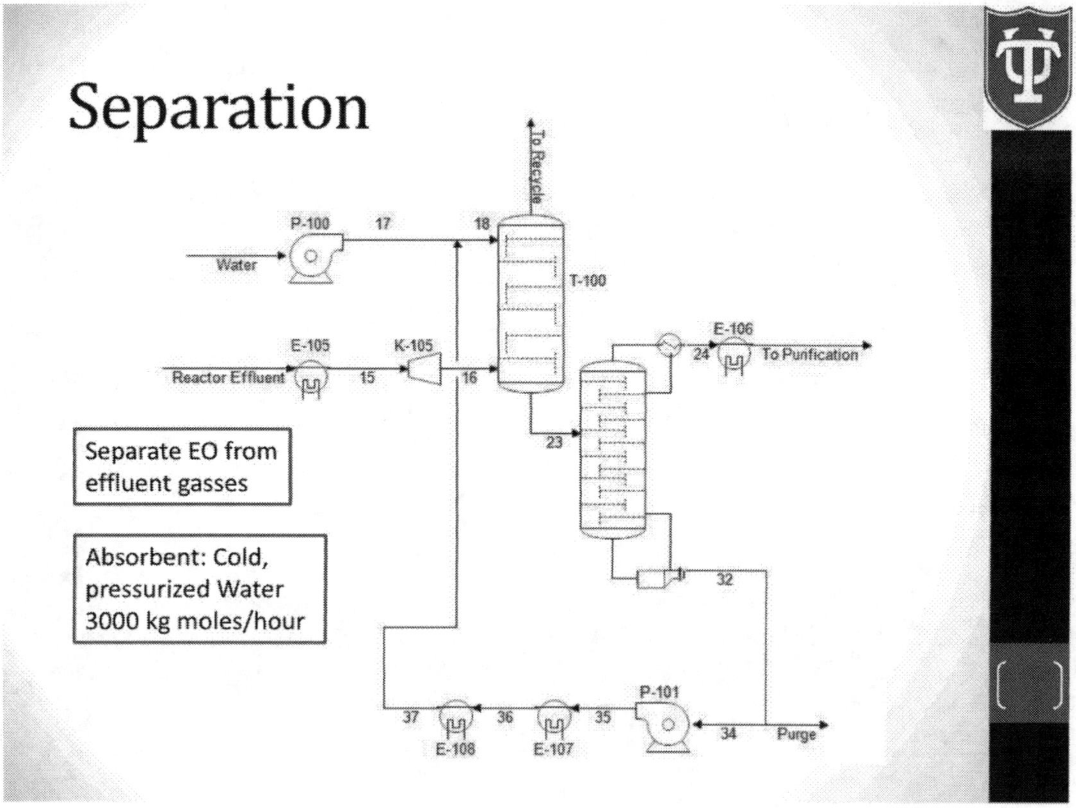
Separation
To Recycle
P-100
17
18
Water
T-100
E-105
K-105
E-106
24
To Purification
Reactor Effluent
15
16
Separate EO from effluent gasses
23
Absorbent: Cold, pressurized Water 3000 kg moles/hour
32
P-101
37
36
35
34
Purge
E-108
E-107

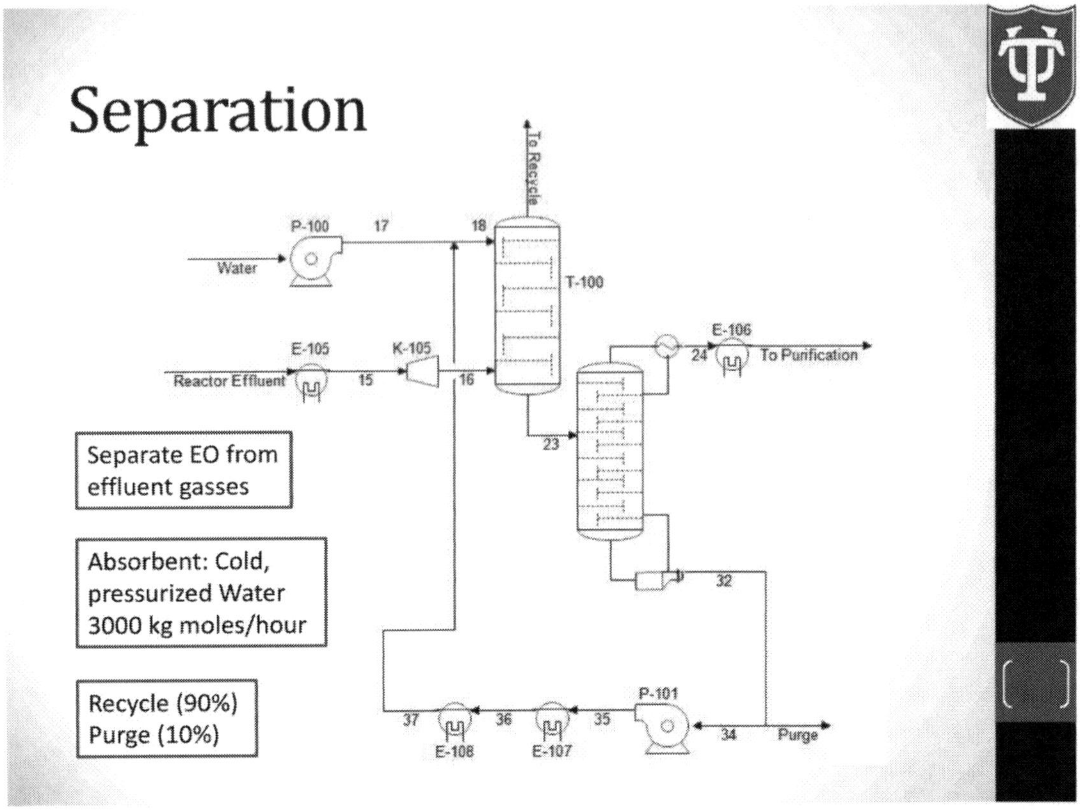
Separation
Water
P-100
17
18
To Recycle
T-100
Reactor Effluent
E-105
15
K-105
16
23
E-106
24
To Purification
32
P-101
34
Purge
35
E-107
36
E-108
37
Separate EO from effluent gasses
Absorbent: Cold, pressurized Water 3000 kg moles/hour
Recycle (90%)
Purge (10%)

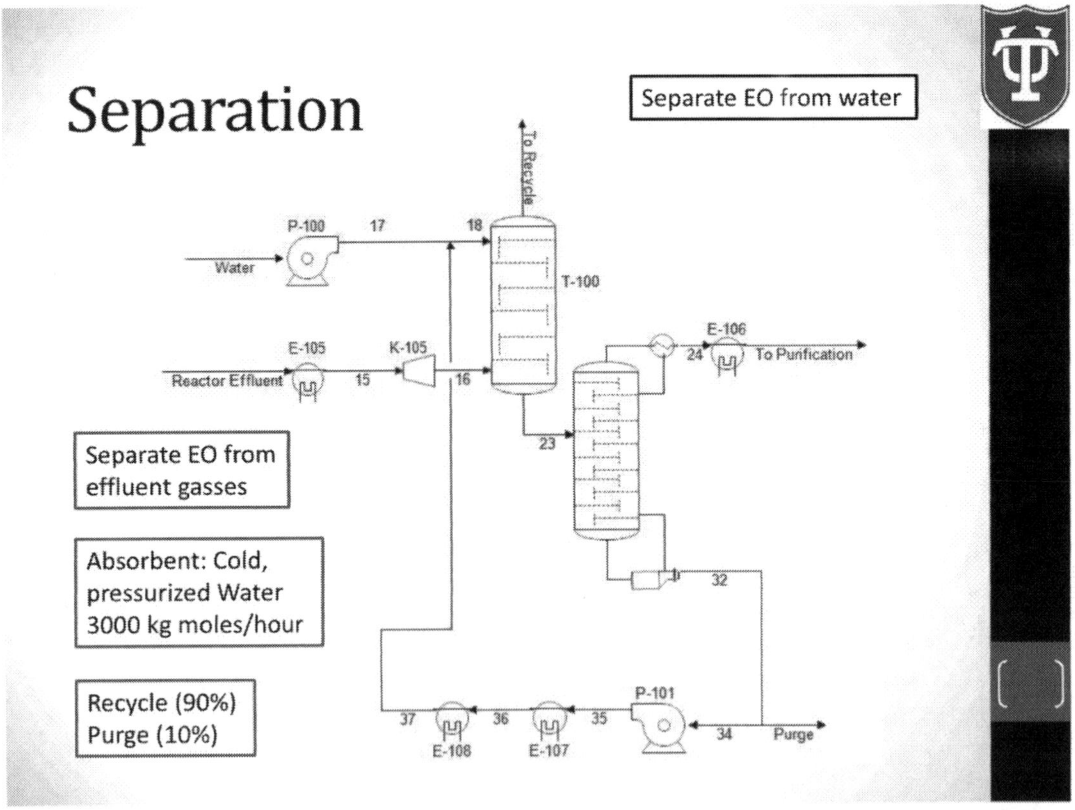
Separation
Separate EO from water
Water
P-100
17
18
To Recycle
T-100
Reactor Effluent
E-105
15
K-105
16
23
E-106
24
To Purification
32
P-101
34
Purge
35
E-107
36
E-108
37
Separate EO from effluent gasses
Absorbent: Cold, pressurized Water 3000 kg moles/hour
Recycle (90%)
Purge (10%)

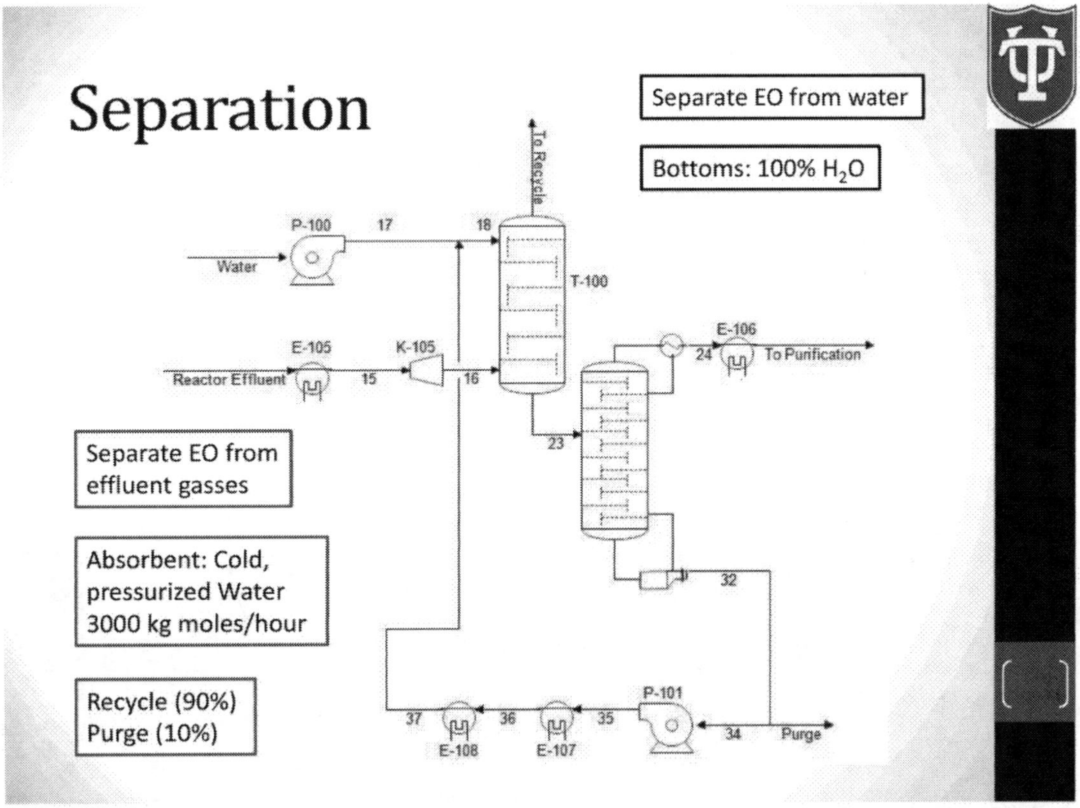
Separation
Separate EO from water
Bottoms: 100% H_2O
To Recycle
P-100
17
18
Water
T-100
E-105
K-105
E-106
24
To Purification
Reactor Effluent
15
16
23
Separate EO from effluent gasses
Absorbent: Cold, pressurized Water 3000 kg moles/hour
32
Recycle (90%)
Purge (10%)
P-101
37
36
35
34
Purge
E-108
E-107

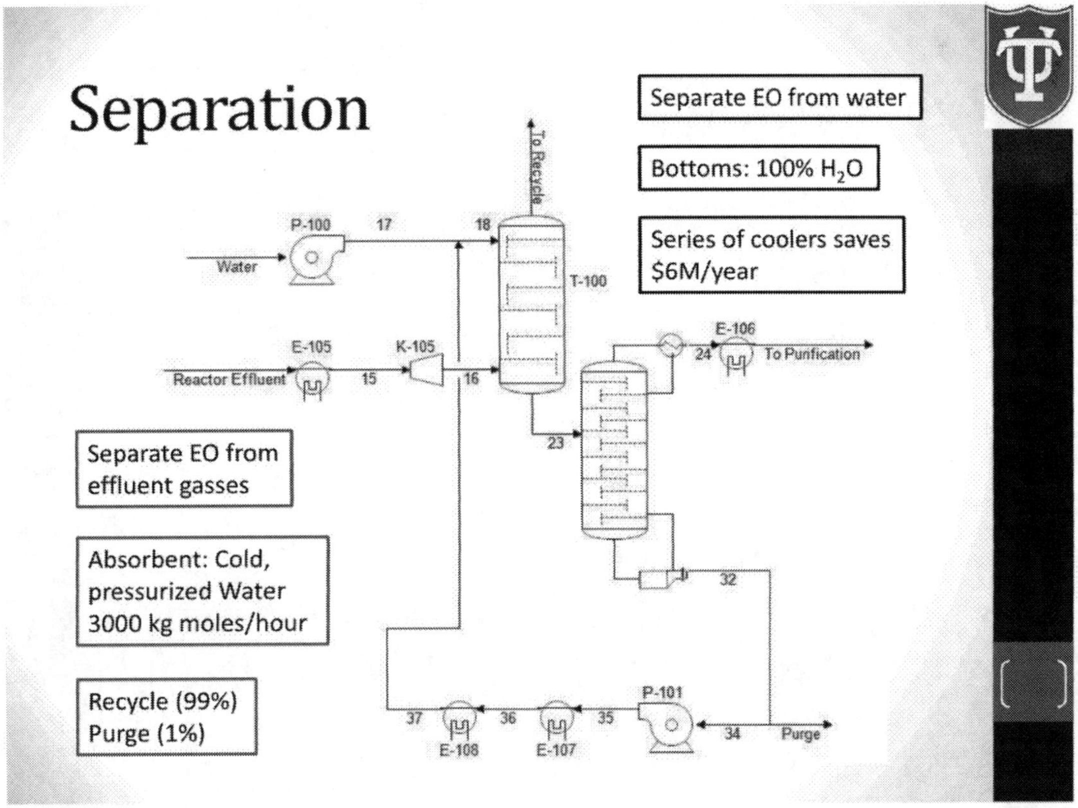
Separation
Separate EO from water
Bottoms: 100% H_2O
Series of coolers saves $6M/year
To Recycle
P-100
17
18
Water
T-100
E-105
K-105
E-106
24
To Purification
Reactor Effluent
15
16
23
Separate EO from effluent gasses
Absorbent: Cold, pressurized Water 3000 kg moles/hour
32
Recycle (99%)
Purge (1%)
P-101
37
36
35
34
Purge
E-108
E-107

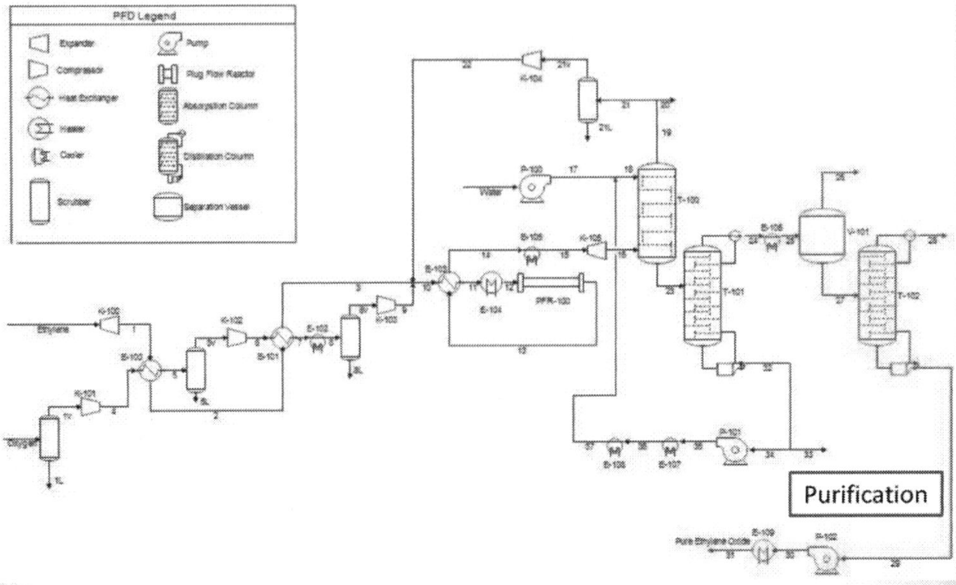

Purification

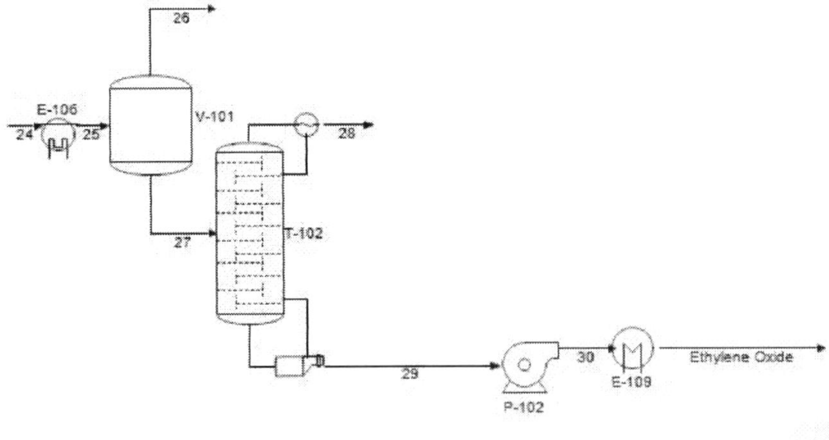

Purification

Most of the oxygen leaves in the separator

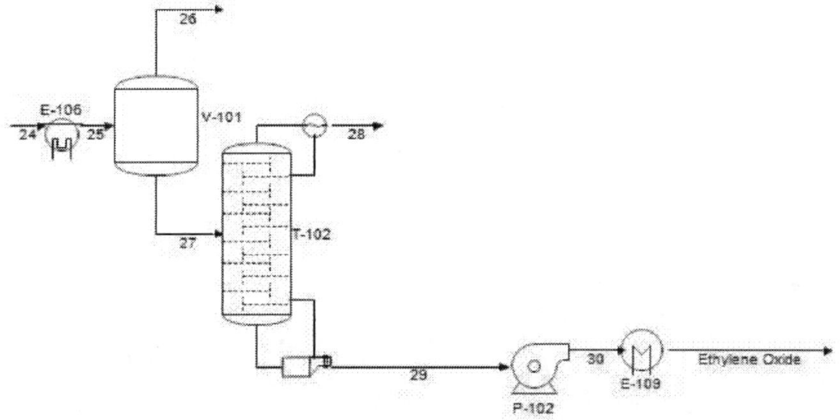

Purification

Most of the oxygen leaves in the separator

Carbon dioxide is mostly separated here

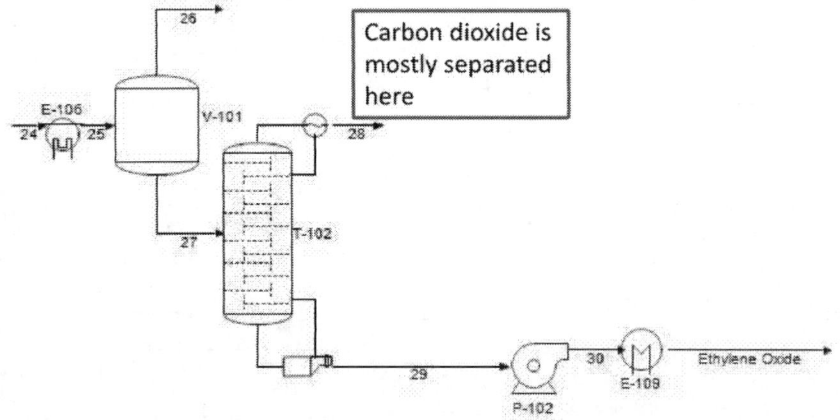

Purification

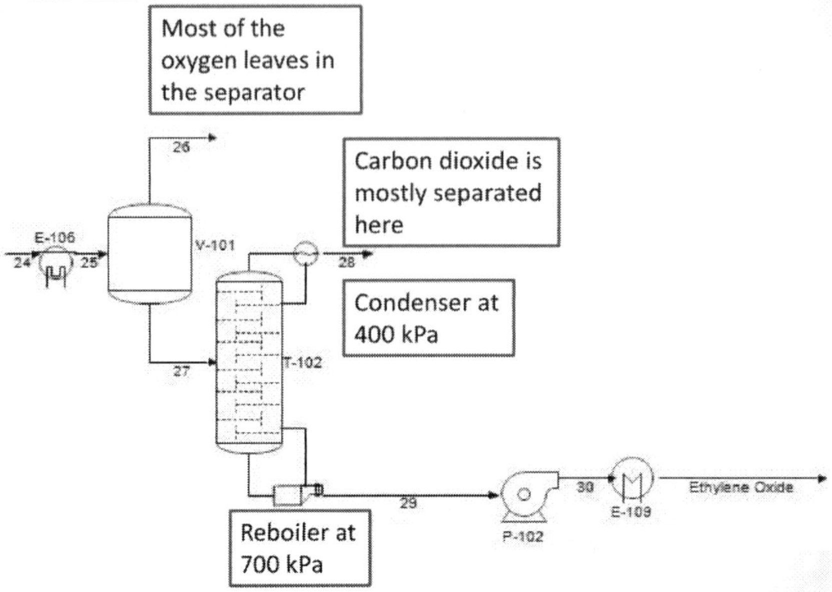

Purification

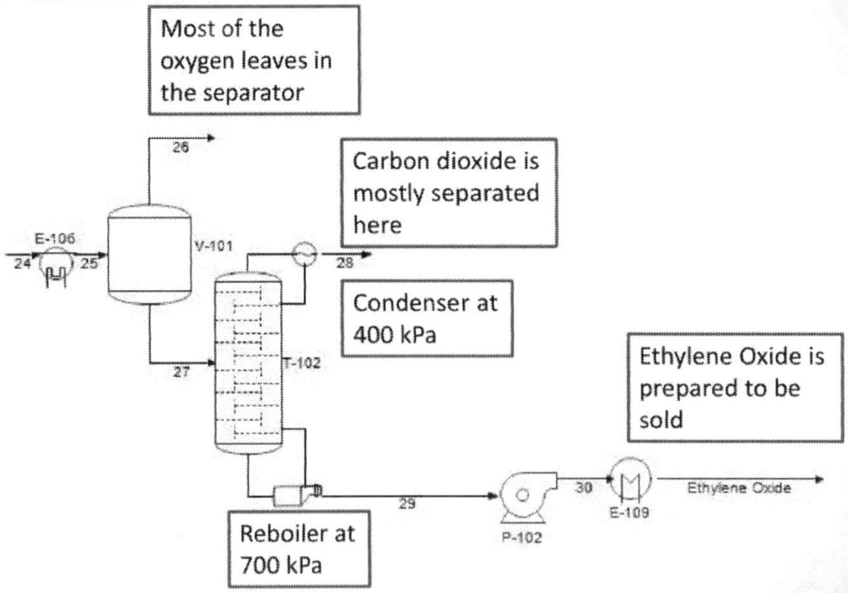

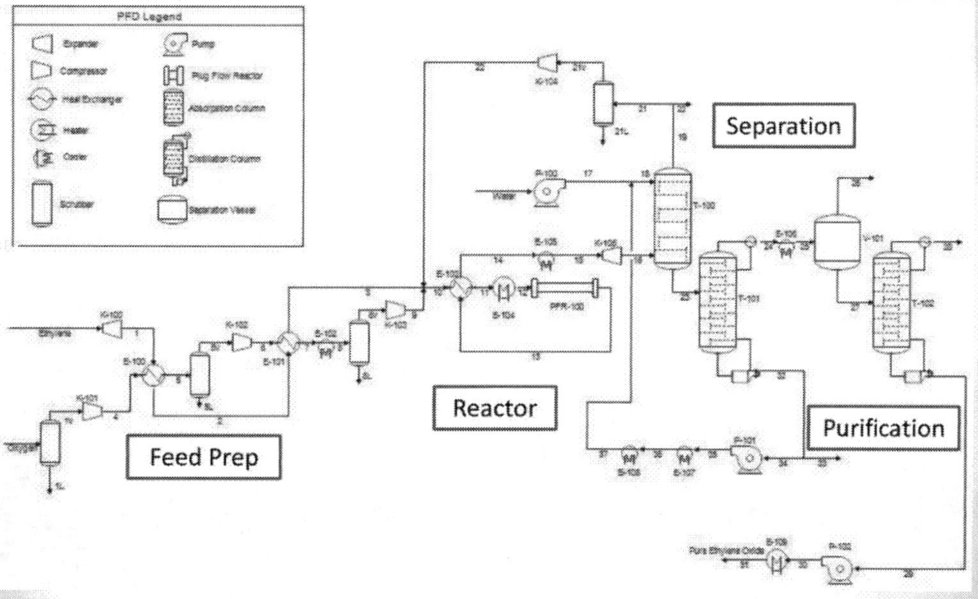

Safety

- Ethylene
  - Flammability limits
  - Safeguards
    - Oxygen shutoff system
    - Eliminate ignition sources
- Ethylene Oxide
  - Reactivity Hazards
  - Safeguards
    - Avoid reactive hotspots
    - Size relief systems for decomposition of ethylene oxide
- Pressure

Economic Summary

Total Capital Investment	$24,000,000

Economic Summary

Total Capital Investment Equipment and Construction Costs (10% overdesign) Working Capital Land	$24,000,000

Economic Summary

Total Capital Investment Equipment and Construction Costs (10% overdesign) Working Capital Land	$24,000,000
Yearly Product Revenue 10-Year Production	$210,000,000

Economic Summary

Total Capital Investment Equipment and Construction Costs (10% overdesign) Working Capital Land	$24,000,000
Yearly Product Revenue 10-Year Production	$210,000,000
Yearly Cost of Production	$137,000,000

Economic Summary

Total Capital Investment Equipment and Construction Costs (10% overdesign) Working Capital Land	$24,000,000
Yearly Product Revenue 10-Year Production	$210,000,000
Yearly Cost of Production Raw Materials Utilities Operation Labor Annual Capital Charge	$137,000,000

Economic Summary

Total Capital Investment Equipment and Construction Costs (10% overdesign) Working Capital Land	$24,000,000
Yearly Product Revenue 10-Year Production	$210,000,000
Yearly Cost of Production Raw Materials Utilities Operation Labor Annual Capital Charge	$137,000,000
Yearly Profit Margin	$73,000,000

Economic Summary

Total Capital Investment Equipment and Construction Costs (10% overdesign) Working Capital Land	$24,000,000
Yearly Product Revenue 10-Year Production	$210,000,000
Yearly Cost of Production Raw Materials Utilities Operation Labor Annual Capital Charge	$137,000,000
Yearly Profit Margin	$73,000,000
Cumulative Cash Flow (Present Value) Investment Payoff Inflation and Interest Rates	**$352,000,000**

Economic Summary

Total Capital Investment Equipment and Construction Costs (10% overdesign) Working Capital Land	$24,000,000
Yearly Product Revenue 10-Year Production	$210,000,000
Yearly Cost of Production Raw Materials Utilities Operation Labor Annual Capital Charge	$137,000,000
Yearly Profit Margin	$73,000,000
Cumulative Cash Flow (Present Value) Investment Payoff Inflation and Interest Rates	**$352,000,000**

Utilities and Optimization

Utility Type

Utilities and Optimization

Utility Type
Ambient Water
Refrigerated Water
Natural Gas
Electricity

Utilities and Optimization

Utility Type	Unit
Ambient Water	
Refrigerated Water	
Natural Gas	
Electricity	

Utilities and Optimization

Utility Type	Unit
Ambient Water	Coolers Reactor Condensers
Refrigerated Water	
Natural Gas	
Electricity	

Utilities and Optimization

Utility Type	Unit
Ambient Water	Coolers Reactor Condensers
Refrigerated Water	Coolers (below 25°C)
Natural Gas	
Electricity	

Utilities and Optimization

Utility Type	Unit
Ambient Water	Coolers Reactor Condensers
Refrigerated Water	Coolers (below 25°C)
Natural Gas	Heaters Reboilers
Electricity	

Utilities and Optimization

Utility Type	Unit
Ambient Water	Coolers Reactor Condensers
Refrigerated Water	Coolers (below 25°C)
Natural Gas	Heaters Reboilers
Electricity	Compressors Expanders

Utilities and Optimization

Utility Type	Unit	Optimization Method
Ambient Water	Coolers Reactor Condensers	
Refrigerated Water	Coolers (below 25°C)	
Natural Gas	Heaters Reboilers	
Electricity	Compressors Expanders	

Utilities and Optimization

Utility Type	Unit	Optimization Method
Ambient Water	Coolers Reactor Condensers	• Series of coolers • Controlled tops of separators • Heat exchangers
Refrigerated Water	Coolers (below 25°C)	
Natural Gas	Heaters Reboilers	
Electricity	Compressors Expanders	

Utilities and Optimization

Utility Type	Unit	Optimization Method
Ambient Water	Coolers Reactor Condensers	• Series of coolers • Controlled tops of separators • Heat exchangers
Refrigerated Water	Coolers (below 25°C)	• Ambient water first, then use refrigerated water
Natural Gas	Heaters Reboilers	
Electricity	Compressors Expanders	

Utilities and Optimization

Utility Type	Unit	Optimization Method
Ambient Water	Coolers Reactor Condensers	• Series of coolers • Controlled tops of separators • Heat exchangers
Refrigerated Water	Coolers (below 25°C)	• Ambient water first, then use refrigerated water
Natural Gas	Heaters Reboilers	• Controlled bottoms of separators
Electricity	Compressors Expanders	

Utilities and Optimization

Utility Type	Unit	Optimization Method
Ambient Water	Coolers Reactor Condensers	• Series of coolers • Controlled tops of separators • Heat exchangers
Refrigerated Water	Coolers (below 25°C)	• Ambient water first, then use refrigerated water
Natural Gas	Heaters Reboilers	• Controlled bottoms of separators
Electricity	Compressors Expanders	• Energy from expanders was recovered for use elsewhere

Sensitivity Analysis

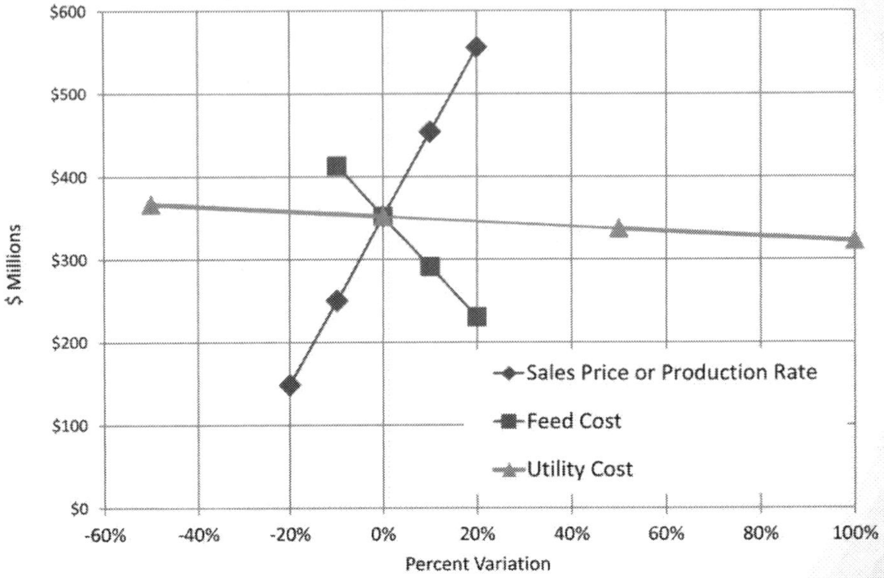

Optimization Comparisons

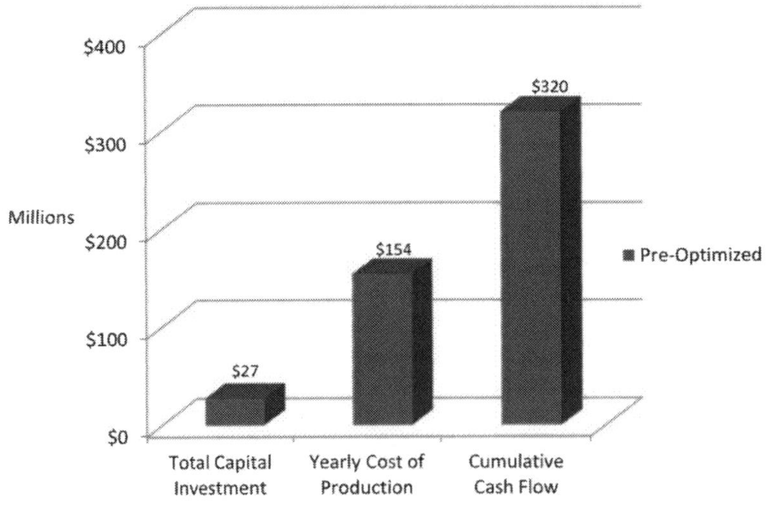

Optimization Comparisons

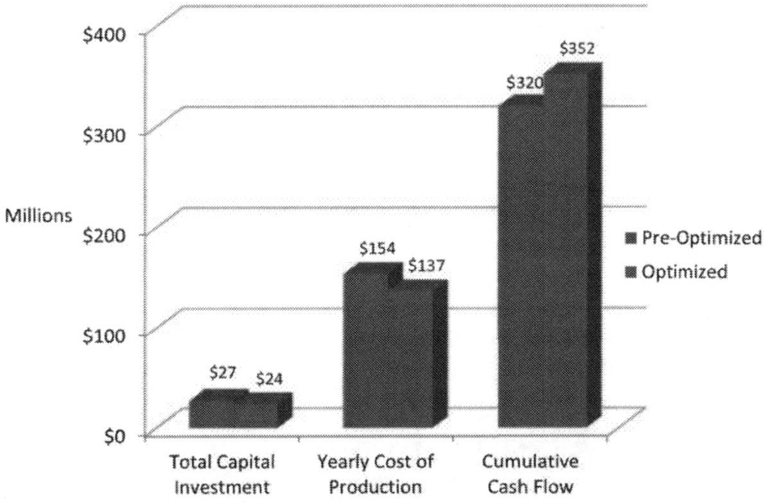

Recommendations

- Quality Check on lab data
  - Flammability envelope of ethylene in oxygen
  - Reanalyzing kinetics
- Ethylene Oxide market analysis→Higher production demands
- Valves vs. Expanders
- Higher Purity of Ethylene

Conclusion

- ✓Production Rate of 120 kTa Ethylene Oxide
- ✓Optimized Isothermal Reactor Operation
- ✓Pure Oxygen Feed

Conclusion

- ✓Production Rate of 120 kTa Ethylene Oxide
- ✓Optimized Isothermal Reactor Operation
- ✓Pure Oxygen Feed

- ✓Annual Revenue: $210 million
- ✓Annual Cost of Production: $137 million
- ✓Annual Profit: $73 million
- ✓"Worst-Case" Economic Studies

Conclusion

✓Production Rate of 120 kTa Ethylene Oxide
✓Optimized Isothermal Reactor Operation
✓Pure Oxygen Feed

✓Annual Revenue: $210 million
✓Annual Cost of Production: $137 million
✓Annual Profit: $73 million
✓"Worst-Case" Economic Studies

12-Year Cumulative Cash Flow: $352 Million

Acknowledgements

- Dr. David Wallace and Dr. Kory Blocker
- Department of Chemical and Biomolecular Engineering
- INEOS

Ethylene Oxide Production

CENG: 4310 Final Design Project

Erica Diamond · Andrew Jones · Alex Tir · Morgan Weber

352960 Economic Optimization and Hazard Analysis of a Hydrofluoric Acid Catalyzed Alkylation Process

Abstract:

A hydrofluoric acid catalyzed alkylation process was modeled using four series/parallel reactions starting with propylene and isobutane. A target of 5,000 barrels per day of 95% pure 2,2-dimethylpentane was specified. The product was also defined to contain less than 1% of heavies (1-nonene and 2,2-dimethyl-octane). The system was modeled using the process simulator ChemCAD and priced using cost equations from Turton, Bailie, Whiting and Shaeiwitz (2012). Both topological and parametric optimizations were considered. The recycle streams increased the complexity of the simulator significantly, but were required to accurately model the concentration build-up in the system. The impact of reactor residence time, separator recycle purity, and feed ratio on the economics will be presented. Process hazards will be identified and suggestions to mitigate hazards will be presented.

Shell Global Solutions

How to Be a Successful Manager in a Technical Organization

Jonathan H. Worstell
Principal Scientist

Shell Global Solutions (US) Inc
Westhollow Technology Center
Houston, Texas

Overview

Career Choice

Types of Industrial Organizations

Management per Industrial Organization

Problems Encountered by First-year Managers

Differences between Managing and Leading

Development of Leaders

Summary

Choices

Early Career Decision

Technical or Management

Lifetime Success

Matching personality with Career

- Look at yourself
  - Quiet / Loud
  - Retiring / Social
  - Introvert / Extrovert
  - Contemplative / Intuitive
- Ask spouse for their assessment
- Ask friends for their assessment
- Ask colleagues for their assessment

Moving Into and Up the Management Organization

Know Thyself

Know Organizational Style of Company

Linear

Matrix

Lean

Know Characteristics of Organizational Style

Task-oriented

People-oriented

Leader-oriented

Organizational Changes

1840 – 1985
Linear

1970 – 1995
Matrix

1995 – 200?
Lean

Shell Global Solutions 5

Linear Organization

Military Organization

First Large Organizations

Civil Organizations

Railroads First Such Organizations

Shell Global Solutions 6

Linear Organization

General

Brigadier General

Colonel

Lieutenant Colonel

Major

Captain

Lieutenant

Sergeant

Troops

Characteristics of Linear Organization

Positions well defined.

Authority = Responsibility.

Initiatives from Top to Bottom.

Organization acts / reacts quickly.

Manager makes decision, then announces it.

Individual acts / reacts quickly.

Individual executes Manager's decision.

Planning.

Done in detail and passed downward

Tasks.

Given to individuals;

Accomplished by individuals.

Comfort level.

Matrix Organization

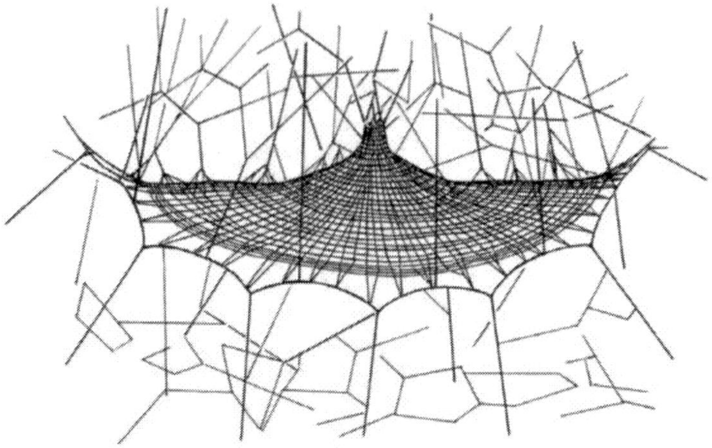

Shell Global Solutions 9

Characteristics of Matrix Organization

Position not well defined.

Authority << Responsibility.

Difficult to ascertain responsibility;

Easy to blame others;

Easy to be blamed.

Initiatives from within the "web".

Organization acts / reacts slowly.

Decisions made by committee.

Committee acts / reacts slowly.

Decision-making by consensus.

Planning.

Done by committees.

Tasks.

Given to committees;

Accomplished by committees.

Comfort level.

Shell Global Solutions 10

Lean Organizations

Assume you need 4,096 workers to make your chemical widgets.

How to organize your company?

2^0	1	4^0	1	16^0	1
2^1	2	4^1	4	16^1	16
2^2	4	4^2	16	16^2	256
2^3	8	4^3	64	16^3	4096
2^4	16	4^4	256		
2^5	32	4^5	1024		
2^6	64	4^6	4096		
2^7	128				
2^8	256				
2^9	512				
2^{10}	1024				
2^{11}	2048				
2^{12}	4096				
Total	8191		5461		4353

Shell Global Solutions 11

Characteristics of Lean Organization

Positions reasonably well defined.

Authority < Responsibility

Initiatives from Bottom to Top.

Organization acts / reacts rapidly.

Manager sets limits within which teams function;

Teams make decisions within limit space.

Team acts / reacts quickly.

Teams execute mission per their decisions.

Planning.

Mission oriented.

Tasks.

Given to Team Leader;

Accomplished by team.

Comfort level.

Shell Global Solutions 12

Upward Movement in Linear Organization

- Work safely.
- Demonstrate technical knowledge.
  - Be technically best among peers;
  - Be technical resource for group.
- Utilize manager's competencies.
- Gain interest of a Mentor.
  - Mentor must be someone "high" in the organization;
  - Mentor eliminates intermediate positions in the organization.
- Establish individual goals.
- Accomplish tasks ahead of schedule and under budget.
- Demonstrate "managerial" skills.
  - "Get 'er done --- Got 'er done";
  - People skills --- "what are those";
  - Communication skills --- "listen to me".

Upward Movement in Matrix Organization

- Work safely.
- Demonstrate "committee-think".
  - Emulate bureaucratic thinking;
  - Study bureaucracies;
  - Learn how to utilize other people's knowledge.
- Utilize committee's technical knowledge.
  - Know who is technically competent.
- Gain interest of a Mentor.
  - Mentor must be someone "nearer" center of the organization;
  - Mentor provides protection.
- Establish committee mission statement.
- Accomplish tasks without "hurting anybody's feelings".
- Demonstrate "managerial" skills.
  - People skills --- "keep everybody happy" --- "get along with everybody";
  - Communication skills --- sell ideas to committee.

Upward Movement in Lean Organization

- Work safely.
- Demonstrate technical knowledge.
  - Be technical resource for team;
- Utilize organization's technical knowledge.
  - Know who is technically competent within organization.
- Gain interest of a Mentor.
  - Mentor must be a peer in the organization;
  - Mentor alerts organization to potential management talent.
- Establish team goals.
- Accomplish tasks as a team.
- Demonstrate "managerial" skills
  - People skills --- lead team through team commitment;
  - Communication skills --- information conduit --- keep team fully informed.

Shell Global Solutions 15

Identity Transformation During First Year in Management

Previous Life	**Manager's Life**
Specialist	Generalist
Doer	Organizer
Meeting Presenter (Seller)	Meeting Leader (Buyer)
Independent	Interdependent
Individual Relationships	Networks
Task Responsible	People Responsible
Seeker Penitent	Forgiver

Shell Global Solutions 16

Psychological Adjustment During First Year in Management

Learning
What is a manager?

Developing
Interpersonal Judgment

Gaining
Self-knowledge

Coping
Stress and Emotions

Myth Busting --- The Reality of Managing

Managerial Parameter	Myth	Actuality
Operative Principle	Authority	Interdependency
Key Players	Subordinates	Network
Power Source	Formal Authority	Leadership Skills
Focus	Manage One-to-One Interactions	Manage Team
Goal Attainment	Control through Compliance	Commitment through Empowerment
Challenges	Coping with Complexity	Coping with Complexity and Change
Essential Competencies	Technical	Technical, Human, and Conceptual

Critical Issues During First Year in Management

Reconciling Initial Expectations with Reality

Handling Inordinate Number of Subordinate Conflicts
with Self
with Peers

Understanding and Meeting Boss' Ambitious Demands

Critical Issues --- Tabulated

	Manager's Function	Manager's Role	New Managers	Subordinates	Superior	Peers
Leadership	Agenda Setting	Boss	X		X	
		Technical Leader	X	X		
		Organizer		X		
		Business Person			X	
	Network Building	Supervisor	X			
		Administrator	X		X	
		Politician	X			
		Team Leader		X	X	
		People Manager		X		
		Liaison with Organization		X	X	
		Integrator with Organization			X	
		Negotiator with Organization				X

L. A. Hill, *Becoming a Manager: How New Managers Master the Challenges of Leadership*, Harvard Business School Press, Cambridge, MA, 2003.

Critical Growth During First Year in Management

Must evolve from

Boss-centered Management

to

Subordinate-centered Leadership

Leadership

Highly Studied

Poorly Understood

Management controls people

Leadership motivates people

Managing and Leading

Managers	Leaders
Develop Plan	Present Mission
Establish Budget	Set Direction
Organize Staff	Manage Context
Control Staff	Align People to Mission
Solve Problems	Motivate People
	Encourage Commitment
	Encourage Performance

Success in Management

All About

Leadership

How Does Industry Develop Leadership?

Leadership Development in Industry

Little instruction per Leadership
Internships ?

MBA programs
Little, if any, discussion about Leadership
Lack "people-skill" courses
Do not prepare individual for day-to-day management realities*

Stress short-term perspective

Impart a subliminal message
Do not have to know technology being managed
Do not have to know people being managed

* L. A. Hill, *Becoming a Manager: How New Managers Master the Challenges of Leadership*, Harvard Business School Press, Cambridge, MA, 2003.

Learning and Demonstrating Leadership Skills

For Lean Organizations

Be Safety Committee Chair
Be Diversity and Inclusiveness Chair
Be Active in United Way
Participate in Professional Organizations
Participate in Leadership of Professional Organizations
Participate in Leadership of Church Organizations
Coach Youth Sports Teams

Business Leadership

? – 1915
Entrepreneurial Capitalism
"Building market share for my company so it can survive."

1915 – 1985
Managerial Capitalism
"Building my empire and ensuring its survival."

1985 - ?
Egotistical Capitalism
"Obtaining the largest possible compensation for myself."

Training Committed Leaders

Is there an example for training leaders ?

Yes.

Leadership Development in Military

Command Level	Military Education	Years of Service	Rank	Course Duration	Civil Education
Tactics	Service Academies ROTC	Pre-Comm.	Sergeant Student	2 – 4 Years	
	Officer Basic Course	0 -1	Lt.	12 Weeks	
	Captain's Career Course	4 - 6	Capt.	12 Weeks	Apply for Graduate School (MS, PhD) Utilization Tour
	Combined Armed Services and Staff School	5 - 7	Capt. (Senior)	12 Weeks	
Operational	Command and General Staff College	10 - 14	Major	9 Months	Apply for Graduate School in Administration or International Relations
Strategic	Army War College	16 - 18	Lt. Col (Senior) Col.	1 Year	
	Joint Command Assignments	12 - 20	General	2 – 3 Years	

Learn from Other's Career Progression

Political Leaders	Military Leaders	Scientific Leaders	Business Leaders
Winston Churchill	George Marshall	Ernest Lawrence	Alfred Sloan
Teddy Roosevelt	Ike Eisenhower	J. Robert Oppenheimer	J. P. Morgan
Ronald Reagan	Hiram Rickover	Ernest Rutherford	Pierre du Pont
	Leslie Groves		Walter Chrysler

Shell Global Solutions 29

Summary

Learning about management left mainly to chance.

Entering management means:
forming a new professional identity;
forming a new personal identity.

Becoming a Leader
Am I implementing the business strategy?
What are my future plans for the team?
Am I treating team members ethically?
Am I getting along with my organizational peers?
Am I interacting with my boss effectively?

Agenda Setting
Network Building

Shell Global Solutions 30

AIChE Spring 2014

MANAGEMENT VS. TECHNICAL CAREER PATH

SABIC Innovative Plastics
Pankaj S. Gautam

CHEMISTRY THAT MATTERS™

GLOBAL DISCLAIMER

No. 1

THE CHOICE!

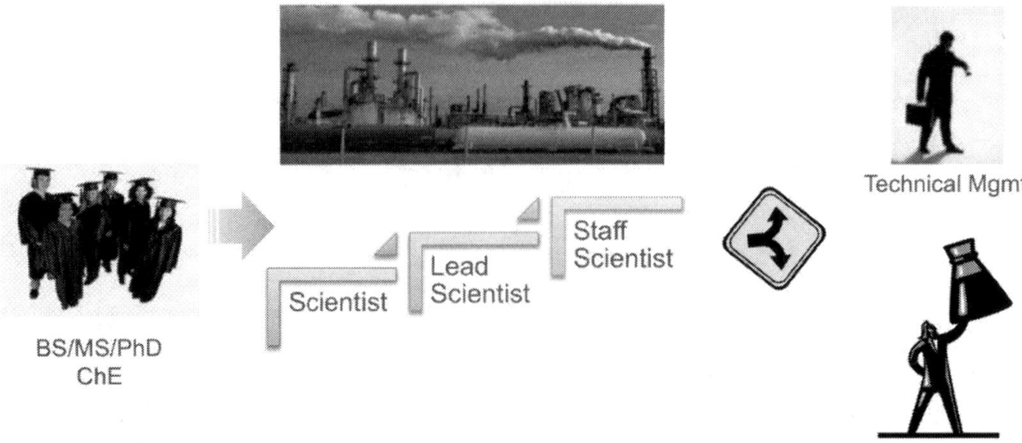

When You Come to a Fork in the Road, Take It!
- Yogi Berra

OUTLINE

- Elements of thought framework
- Career Roadmap
- Personal core competencies
- Gaps & development needs
- Lifestyle choices
- Role of luck
- Closing thoughts

سابك sabic

FRAMEWORK - SYSTEM VS. GOALS

Traditional Approach– Define your goals (SMART) and chalk out a detailed plan to attain the same.

Suggested Approach – Develop a *System** of incorporating behaviors/skills complementary in nature that add value and remain open to the possibilities

System Oriented Approach	*Goal Oriented Approach*
- Works well when future is uncertain **- Frees one up to take advantage of changing circumstances, Avoids "sunk cost " mindset** **- Plant multiple seeds and seek "market" feedback to select the right projects**	**- Works well when the path is "well trodden** **- Allows one to focus energies on execution to produce results**
- High personal energy needed **- Commitment to continuous "personal" value addition** **- Tolerance to failure**	**- "Sunk cost" mindset makes quitting difficult** **- Excessively risk averse behavior** **- Behavioral disorders when failures are encountered**

Consider a *System oriented approach* for career planning

* - Scott Adams, "How to fail at almost everything and still succeed,"

No. 4

سابك sabic

FRAMEWORK – WHAT IS "SYSTEM"

"System" is a set of behaviors and activities that incorporated as part of daily life that improve one's market value.

Industry landscape

Leading teams

Cross-business exposure

Finance & Strategy

Communic ation

System =	System 1*	System 2*
	• Automatic, quick, involuntary	• Needs deliberate effort

The chains of habit are too light to be felt until they are too heavy to be broken

System oriented approach converts desired behaviors from conscious effort (System 2) to *second nature (System 1)*

* - Daniel Kahneman, "Thinking fast and slow,"

No. 5

CAREER ROADMAP

Technical Career		**Management**	
Chief Scientist	Business unit level tech expertise, Competitive process/product knowledge, mentor	Program Mgr	Program deck management, Talent recognition & nurturing,
Research Fellow	Cross-business technology expertise, Lead strategic programs, Advise Mgr	Global Tech Mgr	Build program portfolio, Shape Tech strategy, Mgr talent nurturing, Coaching, Communication
Corporate Fellow	Multi-business tech expertise, Mission & Technology strategy, Advisor to VP/CTO	CTO/VP Tech	Mission & culture, Business strategy, Resource allocation, X-functional alignment to drive strategy, Portfolio mgmt, Communication

Use "Reference groups" to develop competencies for each career path

PERSONAL CORE COMPETENCIES

"Broadframing" – Daniel Kahneman

Technical
- Breadth
- Depth

Breadth refers to cross-business exposure while depth refers to expertise. One should strive for a "T" shaped distribution

People
- Trust
- Visionary

Trust = Character + Competence

Vision = Insight + Articulation

Network
- Internal champions
- External network

A social network of internal champions and mentors and an external network of peers, managers and experts critical to success

No. 7

PERSONAL CORE COMPETENCIES

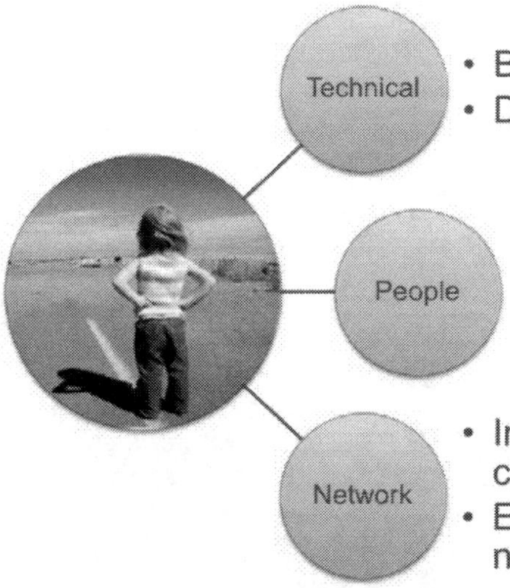

- Technical
  - Breadth
  - Depth
- People
  - Trust
  - Visionary
- Network
  - Internal champions
  - External network

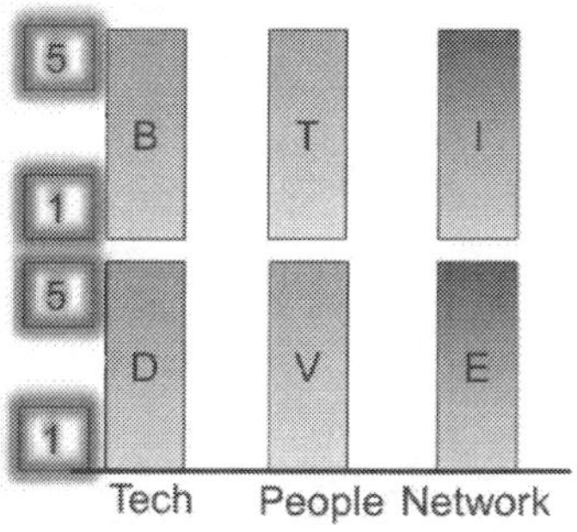

Understanding who you are must precede determining where you want to go

No. 8

GAP ANALYSIS & DEVELOPMENT NEEDS

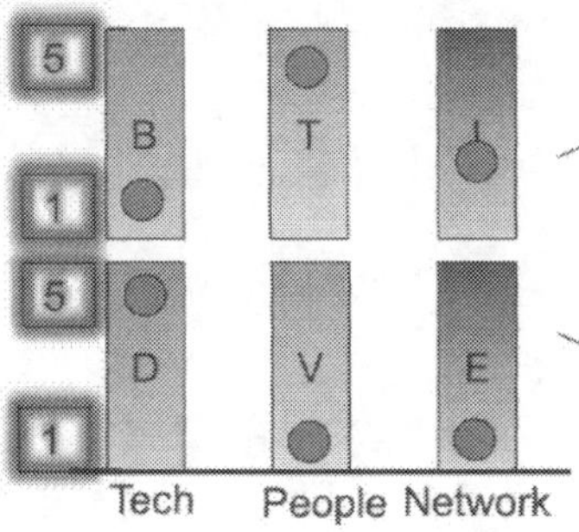

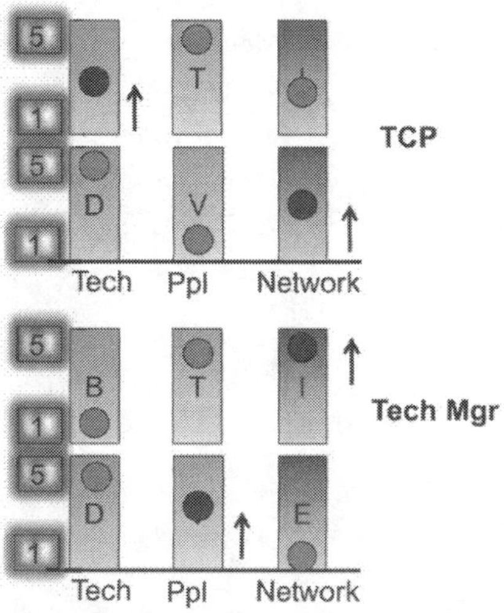

- Rule of thumb: Focus your energies in 2 areas and aim to improve by ~ 2 notches (Low ->Med or Med-> High)
- Compile a “Reference group” of people in each bucket (TCP, Mgr) and use their skill set to normalize needed skills for a career path

No. 9

GAP ANALYSIS & DEVELOPMENT NEEDS

Develop a personal "System" based on GAP analysis

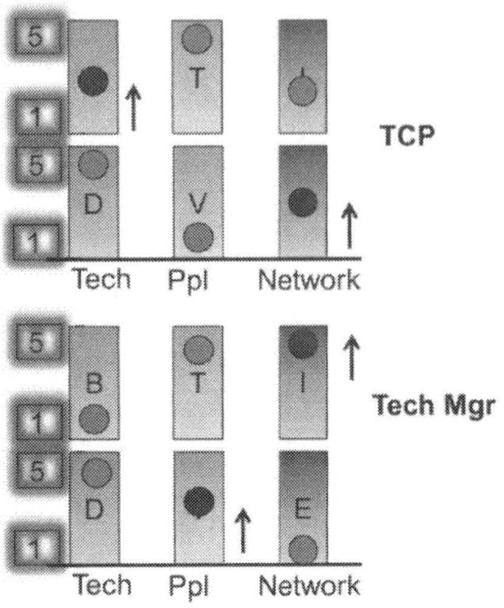

- Volunteer for projects outside your unit
- Bubble assignment or sabbatical in a different business unit
- Initiate collaboration with academia, consortiums, govt labs, and reach out to relevant industry groups
- Membership of Industry Conferences (AIChE)

- Improving visibility and finding internal champions or mentors
- Take courses in finance, business strategy and apply those to solve problems outside your normal job demands
- Lead teams and work on setting agenda and articulating vision

No. 10

LIFESTYLE CHOICES

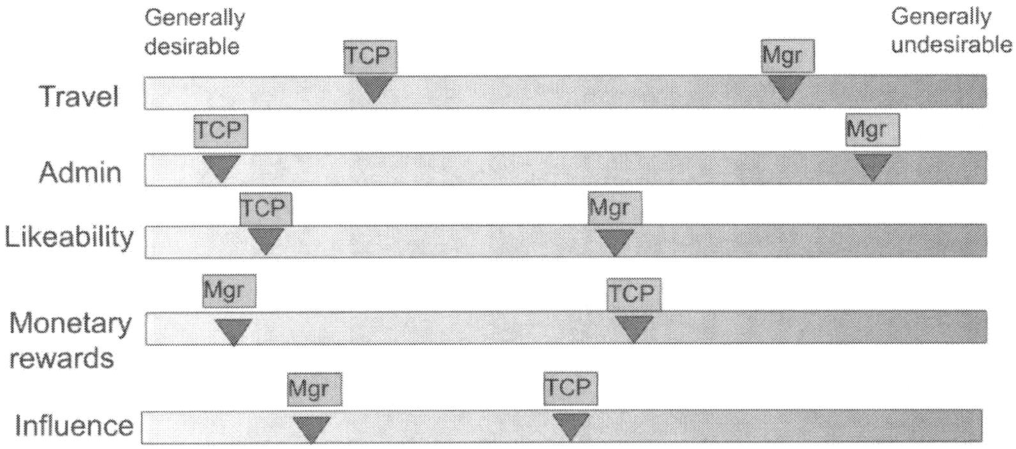

Choose lifestyle that is <u>sustainable</u> for you

No. 11

ROLE OF LUCK

Ecclesiastes 9:11 … *the **race is not** to the **swift**, **nor** the battle to the **strong**…but time and chance happeneth to **them** all.*

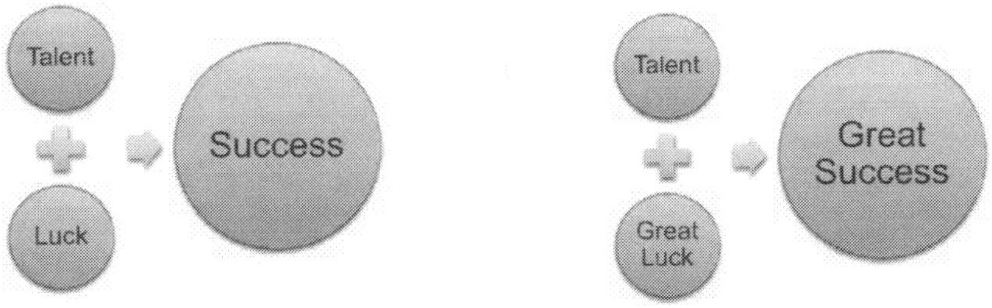

Often, luck plays a disproportionate role in great success stories

How can one get lucky ?

- Openness to accept assignments and jobs that appear to be apparently risky but offer chance of disproportionate success. Use your *System* to minimize downside risk
- Build your personal portfolio of activities and have 10-30% of your time devoted to long shot projects
- Often great success is found in activities that are *scalable*

Michael Mauboussin: "The Success Equation"

No. 12

CLOSING THOUGHTS

…if you know your enemies and know yourself, you can win a hundred battles
- Sun Tzu

- Understand your strengths and development needs in the context of chosen "Reference groups"
- Develop a *System* to increase your market value and use continuous feedback to allocate your energy to most promising project
- Strive hard but don't die trying. Don't let "sunk cost" mindset lead you to a dead end.
- Make a career choice that offers you a lifestyle that you can live with
- Build a personal portfolio of activities that allows you to participate in disproportionate success, i.e., get lucky

No. 13

353935 Managing Your Career---A Lifelong Journey

Abstract:

A successful and satisfying career in chemical engineering can best be achieved by continuously managing/updating your career goals and taking the necessary actions to achieve those goals. This topic is frequently discussed with a focus on the early stages of your career. Less focus is paid as you move to the middle and late stages of your career. Careful career management is just as important during the latter stages of your career.

In this presentation I will share my experiences and best practices from a career in process development, process engineering and leadership roles over 25 years. Ten best practices will be reviewed with personal examples and the impact they had on my career. These practices will be discussed in context of their application at early/mid/late stages of your career.

354003 Seeing and Taking Forks in the Road

Abstract:

Forks in the road are not obvious to someone driving at 100 mph and can be missed easily. The aggressive driver must therefore have heightened awareness of surroundings and signage to plan ahead and make appropriate navigational decisions. And so it is with a career. It is a fast journey with many forks in the road, forks that may take you to faster highways or dead ends, or forks that can avoid you staying on a highway that peters off into pasture. Awareness of your surroundings and of yourself is therefore key in making decisions on which fork to take. Self-awareness can be obtained through people, tools, visual and auditory acuteness, and feedback requests. Awareness of your surroundings can be obtained internally through office communications, budgetary decisions and the like, and externally through publications. Ultimately these signs must be internalized, and analyzed and then from this new knowledge new opportunities can be defined and envisioned. As with a road trip, these inputs and visions can help you create a roadmap for your career and ensure that once in a while you consider and sometimes take new forks in the road towards further career growth.

DISTILLATION FUNDAMENTALS

By

Mike Resetarits

Anand Vennavelli

AY Ogundeji

Fractionation Research Inc.
424 S. Squires Street, Suite 200
Stillwater, OK 74074

2 April 2014

Presented at YP Tutorials: Distillation and Absorption Equipment
AIChE Spring Meeting, New Orleans, Louisiana

Abstract

In the chemicals processing industry, distillation is the most commonly employed separation technique. Different chemicals have different boiling points. After a simple mixture flash, the vapor product contains more light compounds than does the liquid product. A series of flash vessels can be employed to effect purifications of the light and heavy compounds. Similarly, a series of equilibrium stages arranged vertically in a distillation column can effect simple and difficult separations. Column operating pressure is an important variable as is column pressure drop. A reboiler is required at the bottom of a column to boil the heavy liquid. A condenser is required at the top of a column to condense the light vapor and create a reflux stream that flows down the column and cleans the upflowing vapor. When the reflux flow rate is divided by the top (liquid) product rate, the resultant number is called the reflux ratio. Generally, reflux requirements are a function of the relative volatilities of the light and heavy components. Pumps, pipes, valves, meters, thermocouples and other instrumentation are all essential parts of a distillation unit. Batch, azeotropic and extractive distillations are also employed industrially.

Keywords: Distillation Separation Boiling Flashing

Acknowledgments

This tutorial paper is dedicated to the thousands of global engineers and technicians who design, build, operate and maintain distillation columns.

Introduction

When a liquid mixture is fully boiled, to dryness, the resultant vapor will have the exact same composition as the starting liquid. However, when a liquid mixture is only partially boiled, the lighter compounds (usually those with lower molecular weights) will leave the liquid first. The resultant liquid will be richer in heavier compounds (usually those with higher molecular weights). A partial separation has occurred, but the resultant vapor will not be pure and the resultant liquid will not be pure.

Now imagine a mixing chamber or a vertical column wherein a light liquid stream encounters a heavy vapor stream. Also imagine that the streams are not at compositional equilibrium. Some of the light components in the liquid stream might leave it and enter the vapor stream. Similarly, some of the heavy components in the vapor stream might leave it and enter the liquid stream. If, as the streams encounter each other, compositional equilibrium is achieved, it is said that an "equilibrium stage" or a "theoretical tray" of contacting has occurred. Such a compositional equilibrium will be achieved if the liquid and vapor phases contact each other intimately for a sufficient amount of time. When the equilibrium is attained, further contacting or mixing does not yield additional mass transfer.

Distillation begins with the boiling of a mixture. A distillation column can be viewed as a series of equilibrium stages arranged vertically. A liquid stream flows down from stage to stage and encounters a vapor stream that flows up from stage to stage. As compositional equilibria are attained in each of those stages, the downward-flowing liquid stream becomes heavier and more pure as the upward-flowing vapor stream becomes lighter and more pure. Distillation columns are capable of starting with a mixture and turning it into a very pure top product rich in light compounds, and a very pure bottom product rich in heavy compounds.

Discussion

Boiling Points

The science of distillation begins with the study of boiling points. The temperature at which a liquid boils depends upon the pressure that the top liquid surface is subjected to. When that pressure is atmospheric pressure, the boiling point is referred to as the "normal boiling point." **Table 1** gives the normal boiling points of several compounds of the "aromatics family" of chemicals. All of these compounds exist as liquids at room temperature and pressure. For distillation separations to occur, mixtures of these compounds need to be partially boiled. **Table 2** includes the normal boiling points of the main compounds that are found in a natural gas processing plant. All of these compounds exist as vapors at room temperature and pressure. For distillation separations to occur, mixtures of these compounds need to be partially condensed. Distillation always requires the presence of two fluids, a liquid and a vapor. In both of these tables, higher molecular weight compounds have higher boiling temperatures.

Bubble and Dew Points

The separation of benzene (B) from toluene (T) is an important industrial separation. It is usually effected via distillation. At atmospheric pressure, pure benzene boils at 80℃; pure toluene, at 111℃. A mixture of benzene and toluen e begins to boil across a wide range of temperatures; depending on the mixture composition. The lower curve of **Figure 1** gives the bubble point temperatures. Very unfortunately, taking any mixture of benzene and toluene, and raising its temperature to 80℃ or higher will not simply boil off all of the benzene, leaving pure toluene behind. When boiling most mixtures, the lighter component indeed boils first, but the resultant vapor includes very significant quantities of the heavier component.

Referring back to **Figure 1**, regarding B and T, the lower curve represents mixture bubble points. The upper curve represents mixture dew points. The area between the curve is the so-called "2-phase region."

Assume a cold 50/50 BT mixture (Point m_L in **Figure 1**) and assume that its temperature is being raised continuously. It will not begin to boil at 80℃; instead the bubble point will prove to be 92℃ (Point b_1 in **Figure 1**). The composition of the first bubble of vapor is represented by Point d_1. It is no surprise that d_1 is high in B content. As the 50/50 2-phase mixture is heated further, when the temperature reaches 95.5℃, rough ly half of the mixture will be vaporized. The vapor composition is represented by d_2; the liquid, b_2. The horizontal lines in **Figure 2** are called "tie lines." At this point, some separation of the two compounds has indeed been achieved, but the two phases are far from being pure. With additional heating of the mixture, the dew point (d_3) will be achieved at a temperature of 99℃. The v ery last droplet of liquid will have the composition represented by b_3.

If the mixture is heated all the way to 110℃, the resultant vapor (Point m_V on the graph) will have the same compositions as the initial liquid (m_L).

As the BT mixture was being heated, regardless of the temperature, there never came a point where the B or the T approached "purity" – not in the liquid phase and not in the vapor phase.

Flash Distillation

Figure 2 summarizes a simple flash distillation, again using the example of a 50/50 B/T liquid mixture. Imagine that the liquid mixture is heated to a temperature of 95.5℃ with a subsequent let-down to atmospheric pressure. **Figure 2** shows that a modest amount of separation has occurred. In fact, the after-flash vapor and liquid compositions are represented by Points b_2 and d_2 in **Figure 1.**

Figure 3 shows a series of flash distillations, where the resultant liquids are flashed again and again at ever-increasing temperatures. After 5 flashes, a T purity of 91% has been achieved. (For an industrial benzene column, the T purity target is closer to 99% than 90%.) Most importantly, **Figure 3** shows that a series of operations, in this case flashing, can be used to separate and purify chemical compounds that have different boiling points.

In **Figures 1**, **2** and **3**, boiling was always effected at atmospheric pressures, with boiling temperatures in the neighborhood of 80 to 110℃. Low pressure steam at 2.5 bara (36 psia) and 127℃ could be employed in the boilers (heat ex changers).

Vapor Pressures and Relative Volatilities

Generally, with a pure liquid, molecules are constantly escaping the top liquid surface and are becoming vapor. The pressure that those evaporated molecules exert is called the "vapor pressure" of the pure liquid. When a pure liquid is heated, the vapor pressure of the liquid increases. When the vapor pressure of the pure liquid equals atmospheric pressure, the "normal boiling point temperature" is achieved and evaporation becomes violent, i.e., the pure liquid "boils."

Different liquid compounds have different vapor pressures, and thus, different normal boiling points. Generally, smaller molecules have higher vapor pressures. When a liquid-vapor mixture is enclosed and heated, as the pressure of the vessel rises, the boiling point temperature also rises. If a vacuum is created above that mixture, the boiling temperature of the mixture becomes less than the normal (atmospheric pressure) boiling temperature. As an example, water can boil across a very broad range of temperatures. A very simple laboratory experiment can show how easily water can boil at room temperature. **Table 3** gives details regarding how pressure affects the boiling temperature of water.

Consider a mixture of small(1) and large(2) molecules. When that mixture is heated, and boiled, the smaller molecules will usually have higher vapor pressures and will escape the liquid phase first. The vapor phase composition will have more lighter compounds than the leftover liquid phase; a partial distillation separation will have occurred.

The ratio of the vapor pressures (VP) of different chemical compounds is often called the "relative volatility" and is often represented by "∝". The following simple equations apply:

$$\alpha_{12} = \frac{VP_1}{VP_2}$$

$$y_1 = \frac{\alpha_{12} x_1}{1 + (\alpha_{12} - 1)x_1}$$

$$x_1 + x_2 = 1$$

$$y_1 + y_2 = 1$$

where

x_1 = composition of Compound 1 in liquid phase

x_2 = composition of Compound 2 in liquid phase

y_1 = composition of Compound 1 in vapor phase

y_2 = composition of Compound 2 in vapor phase

For many chemical compounds, their vapor pressures are known across broad ranges of temperatures. These vapor pressures can be employed within the above equations to determine the separations that are achievable by boiling mixtures of compounds.

Figure 4 provides the vapor pressures of benzene and toluene across a broad range of temperatures. Notice how the graphed data converge at higher temperatures. At higher temperatures, the relative volatilities of binary compounds do tend to decrease, i.e., the separations become more difficult. **Figure 5** provides the vapor pressures of propylene and propane.

Figure 6 illustrates the ramifications of relative volatilities (essentially, vapor pressure ratios) on vapor-liquid equilibria. Notice that at α =1.1, the composition of component 1 in the vapor and liquid phases are just about equal. At α =3.0, the composition of the light component in the vapor phase is much higher than it is in the liquid phase.

Column Configuration and Ancillary Equipment

Figure 7 is a diagram of a typical distillation unit including the column and the heat exchangers. Column diameters can be as small as 2 inches or as large as 50 feet. Column heights can range from 12 inches to 350 feet. The inside of the column can be equipped with trays or packings – to mix the countercurrent liquid and vapor streams.

Figure 7 shows a feed stream entering the column about half-way up; this is not unusual. The feed stream can be a liquid, a flashing liquid or a vapor. For the sake of this discussion, it will be assumed that the feed stream is a liquid. The feed enters the column and joins the down-flowing reflux liquid stream. As the combined liquid stream flows downward, mass is exchanged with an upward flowing vapor. At the bottom of the column, most of the liquid stream is boiled in a "reboiler". The resultant vapor stream is called the "boil-up." In **Figure 7**, steam is being used to boil the liquid. A portion of the bottom liquid is not reboiled and is the final "bottoms product." All of the fluids that reach the bottom of the column are rich in heavy components, i.e., chemicals with high boiling points. The vapor leaving the reboiler is similarly heavy. As this heavy vapor flows upwards through the column, it encounters a lighter liquid. There is a compositional disequilibrium between the phases which causes light components to leave the liquid and heavy components to leave the vapor. In fact, as the vapor stream rises past the feed point, it becomes richer in lighter components. By the time that the vapor reaches the top of the column it is quite light. In **Figure 7**, the entire vapor stream enters a condenser which totally converts the vapor stream to a liquid. Some of this liquid is returned to the column as a "reflux" and some becomes the "overhead product." All of the fluids near the top of the column are quite light, including the liquid reflux stream. As this light reflux stream flows down the column it encounters a vapor stream that is relatively heavier. A compositional disequilibrium exists and heavy components in the vapor stream leave the vapor stream. Light components in the liquid stream leave the liquid stream. Some people regard the reflux stream as "clean'; they regard the reflux stream as cleaning the upflowing vapor stream.

In **Figure 7**, cooling water is employed as the heat sink medium in the condenser. Air (in a fin fan condenser) or refrigerants can also be used. Condensers are not actually positioned at the tops of columns. They are positioned about 30 feet above ground level. In fact, reflux drums are typically positioned underneath most condensers to collect condensates. Pumps are required to move the reflux liquids back to the tops of the columns. **Figure 7** depicts a typical distillation unit. There are many variations.

Reflux Requirements

"Reflux ratio" is defined as the reflux flow rate divided by the overhead product rate. Difficult separations, with low relative volatilities, require higher reflux ratios. Higher reflux ratios require more boil-up in the reboilers and higher steam rates and higher cooling water rates in the condensers and, overall, higher energy consumption rates. Additionally, higher reflux ratios lead to increased vapor and liquid traffics inside the column and the column diameter must be increased when the column is being designed. Using the benzene-toluene binary as an example, **Figure 8** shows that taller columns, with more theoretical stages, can lead to reduced reflux ratios. **Figure 8** also shows that as top and bottom purity targets are increased, so are the needs for additional column height and refluxing.

Atmospheric Distillation

Whenever possible, distillation columns are designed for, and operated at, near-atmospheric pressure. **Table 1** provided the atmospheric pressure boiling points of several compounds of the "aromatics family." When these particular chemicals are distilled, low-grade low-temperature steam can be employed in the reboilers. At the same time, cooling water or air can be used in the condensers to effect liquefactions. There are other occasions, and chemicals, where the associated distillation columns need to be run at high pressures or at vacuum conditions. Such columns are more expensive to build and operate. High pressure columns require many extra safety considerations.

Table 1 shows that the boiling points of benzene (B) and toluene (T) are quite different. Indeed, these compounds can be separated and purified via boiling – and distillation. Benzene columns are quite common, globally, and are operated at near-atmospheric pressures. **Table 1** includes other aromatic compounds which will be referred to later.

High Pressure Distillations

Distillation separations cannot occur with chemical compounds that are totally in the liquid state, or totally in the vapor state. Liquid chemicals need to be partially vaporized in reboilers. This is usually "easy" and can be effected using steam, hot water, etc. Vapor chemicals need to be partially or totally condensed in condensers and this often creates "difficulties." Cooling water and air are the most common heat sinks for hot chemicals. Cooling water and air are commonly assumed to be at 100°F = 38°C. The chemical compou nd vapors that they condense must therefore be at a higher temperature, like 130°F. As an alternative, a refrigerant could be employed as the heat sink fluid, but refrigeration is generally expensive and is generally avoided.

See **Table 2**, and in particular, the boiling points at atmospheric pressure, of pure methane and pure ethane. A natural gas plant demethanizer separates methane from ethane using distillation. If such a demethanizer is operated at atmospheric pressure, the methane-ethane mixture would need to be cooled to temperatures lower than -100°C to create a liquid phase. Instead, demethanizers are typically operated at pressures like 500 psia, to facilitate liquefaction. At pressures like 500 psia, the bubble and dew points of methane-ethane are high enough that cooling water and air can be used in the distillation column condenser.

Referring again to **Table 2**, this time noting the (atmospheric pressure) boiling temperatures of propylene and propane. The separation of propylene from propane is an important global process for polypropylene production. If a propylene-propane distillation column were run at atmospheric pressure, a (costly) refrigerant at about -50°C would be required as the heat sink in the overhead condenser.

Figure 5 provides vapor pressure data for these C_3 compounds. Most propylene-propane columns run at pressures like 300 psia. At such pressures, per **Figure 5**, these chemicals condense at temperatures like 130°F. Cooling water or air (at 100°F) can be used to effect these liquefactions in overhead condensers.

Vacuum Distillation

The polymer known as polystyrene is created using relative pure styrene monomer. Plants that produce styrene include distillation columns that separate ethylbenzene (EB) from styrene (S). Refer back to **Table 1** and note the atmospheric-pressure boiling points of pure ethylbenzene and pure styrene, 136 and 146°C, respectively. If an EB/S column were operated at atmospheric pressure, those chemicals would be subjected to such temperatures. Very unfortunately, such temperatures caused styrene monomers to combine repeatedly to form polymer molecules and those molecules solidify. Such atmospheric-pressure EB/S columns would plug up (foul) quickly. Instead, EB/S columns are actually operated at very deep vacuums, like 100 mmHga, to reduce the column temperatures and virtually eliminate the renegade polymerizations and the plugging.

There are other reasons why some distillation columns are operated at sub-atmospheric pressures and relatively low temperatures. At atmospheric pressures and higher temperatures, some molecules crack apart instead of boil. As an example, the second major distillation column in most refineries is typically run at vacuum pressures (and lower temperatures) to prevent large oil molecules from cracking.

Unpractical Distillations

Some distillation separations are quite difficult, but they are performed industrially nevertheless. Refer again to **Table 1**. The boiling points of EB and p-xylene are quite close, and yet, this distillation is performed industrially, but rarely. This is also true of the distillation of p- and m-xylene from o-xylene. Conversely, however, the boiling points of p-xylene and m-xylene are so close (0.8℃, actually) that this distillation is n ever performed industrially. As it turns out, the melting points of these two compounds are amazingly different (+13 and -47℃, respectively) and crystallization can be employed to separate them.

Closing

Thirty years ago, Dr. James Fair of the University of Texas at Austin, predicted that distillation would be a popular separation method for many decades. He was correct!

Figure 1: Benzene - Toluene Bubble and Dew Points; Pressure = 14.7 psia

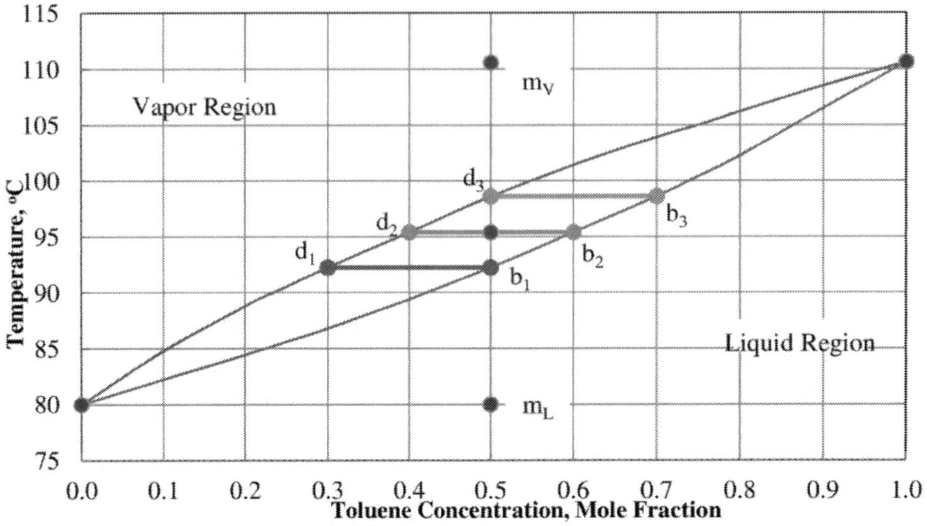

Figure 2: Flash Distillation

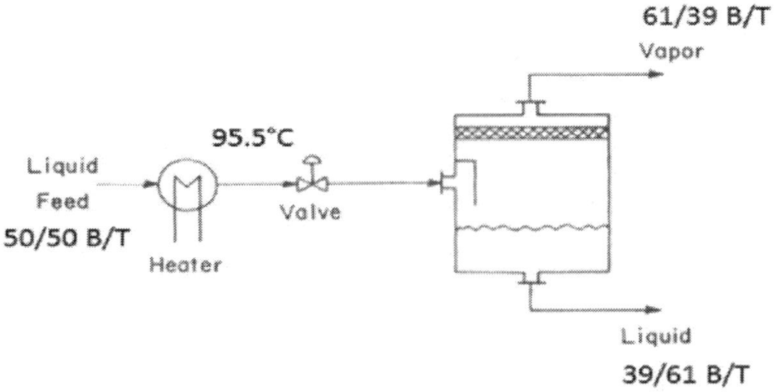

Figure 3: Successive Flash Distillations

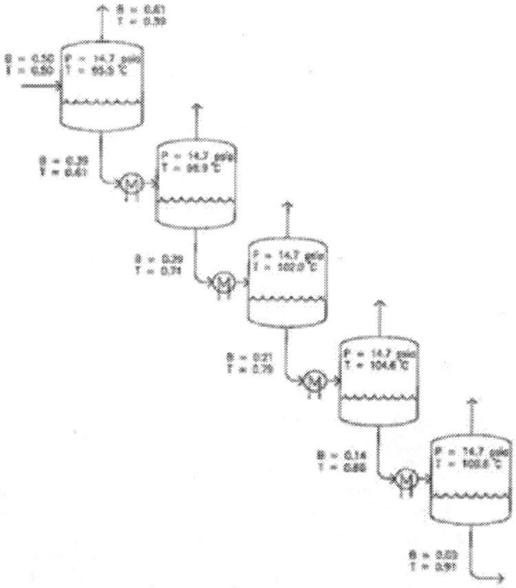

Figure 4: Vapor Pressures of Benzene and Toluene

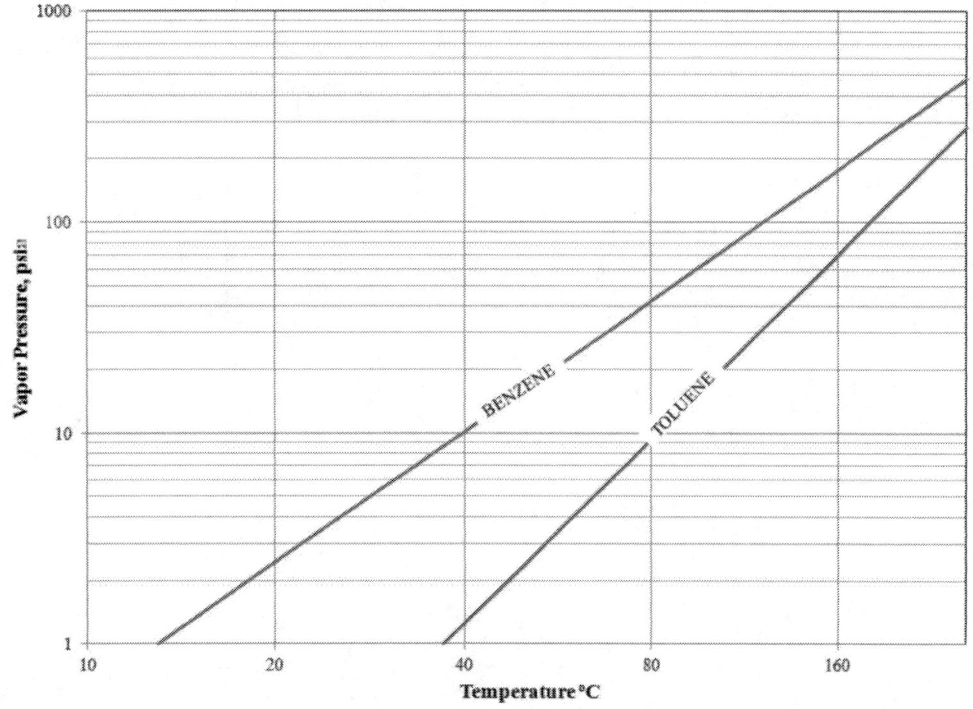

Figure 5

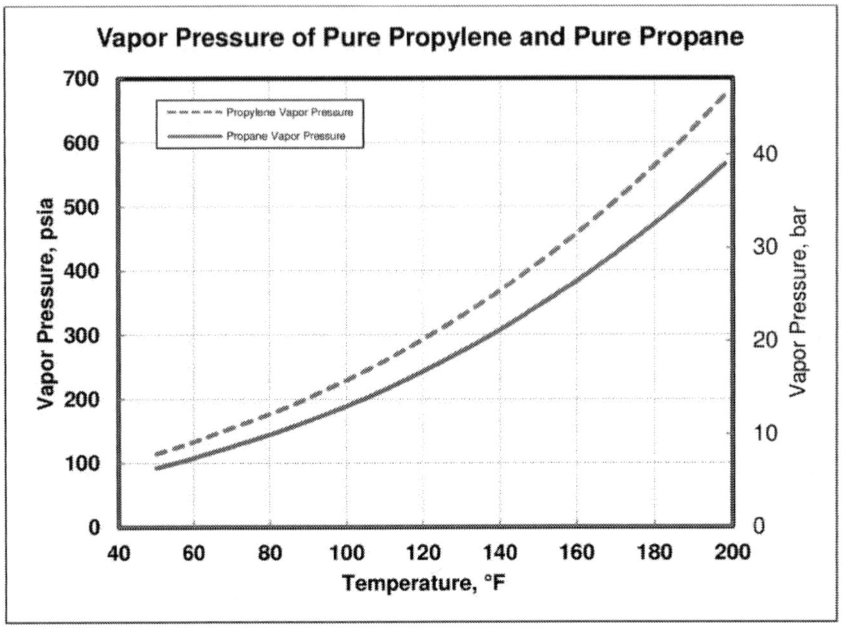

FIGURE 6 Vapor (y_1) versus liquid (x_1) concentration as a function of relative volatility.

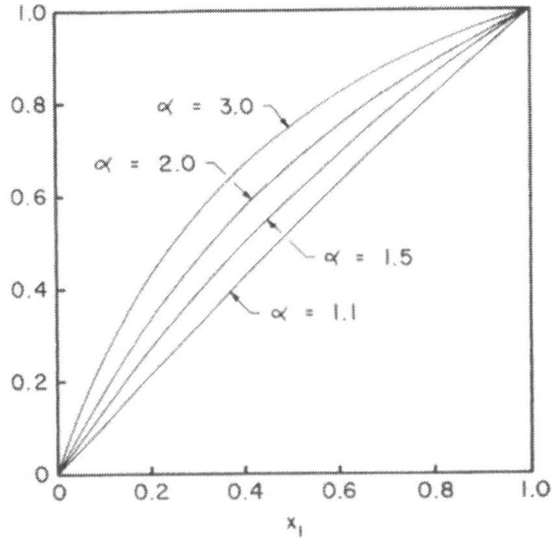

FIGURE 7 Schematic representation of a distillation column.

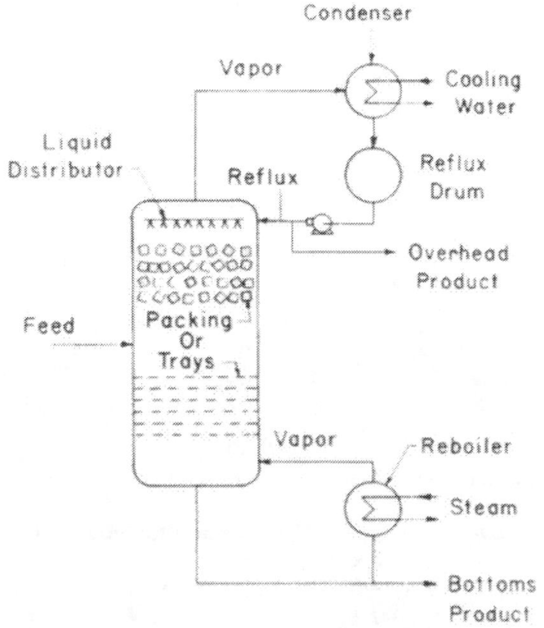

FIGURE 8 Theoretical stages versus reflux ratio (benzene-toluene at 1.0 bar). x_D, mole fraction benzene in overhead; x_B, mole fraction toluene in bottoms.

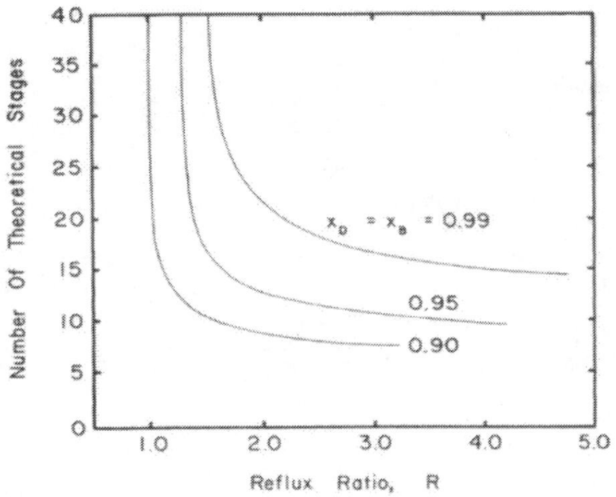

Table 1
Boiling Temperatures at Atmospheric Pressure

	Boiling Point, °C
Benzene (B)	80
Toluene (T)	111
Ethylbenzene (EB)	136
p-Xylene	138
m-Xylene	139
o-Xylene	144
Isopropylbenzene	153
Styrene (S)	146

Table 2
Boiling Temperatures at Atmospheric Pressure

	Boiling Point, °C
Methane	-161
Ethylene	-104
Ethane	-89
Propylene	-48
Propane	-42
Isobutane	-12
n-Butane	-1
n-Pentane	+36

Table 3
Water Boiling Temperatures

Pressure, psia	Temp, °C	Temp, °F
1.0	39	102
14.7	100	212
100	164	328
600	252	486

REFERENCES

1. Kister, H.Z., Distillation Design, McGraw-Hill, NY, NY, 1992.
2. Kister, H.Z., Distillation Operation, McGraw-Hill, NY, NY, 1990.
3. Lockett, M.J., Distillation Tray Fundamentals, Cambridge Univ. Press, NY, NY, 1986.
4. Strigle, R.F., Packed Tower Design and Applications – Random and Structured Packing, 2nd Edition, Gulf Publishing Co., Houston, TX, 1994.
5. Resetarits, M.R., Agnello, J., Lockett, M.J., and Kirkpatrick, H.L., "Retraying Increases C3 Splitter Column Capacity," Oil and Gas Journal (86), June 1988, pp. 54.
6. Resetarits, M.R. and Miller, R.J., "Increase Demethanizer Efficiency and Capacity," Hydrocarbon Processing, October 1991, pp. 79-81.
7. Resetarits, M.R., Navarre, J.L., Monkelbaan, D.R., Hangx, G.W.A. and van den Akker, R.M.A., "Trays Inhibit Foaming," Hydrocarbon Processing, March 1992, pp. 61-64.
8. Resetarits, M.R., Riter, J.B. and Weiland, R.H., "Distillation Technology: Cocurrent Flow Trays in Distillation Columns," Hydrocarbon Engineering, Vol. 7, No. 9, September 2002, pp. 85-88.
9. O'Connell, H.E., Plate Efficiency of Fractionating Columns and Absorbers, Trans. Am. Inst. Chem. Engrs., 42, 741, 1946.
10. Resetarits, M.R., Miller, R.J., Navarre, J.L., Linskeseder, M. and Reich-Rohrwig, P., "New Enhanced Capacity MD Tray Debottlenecks Deethanizer," IChemE Distillation and Absorption Conference, Birmingham, U.K., September 7, 1992.
11. Resetarits, M.R., Monkelbaan, D.R. and Miller, R.J., "Trays in Rocking Services," IChemE Distillation and Absorption Conference, Birmingham, U.K., September 7, 1992.
12. Resetarits, M.R., Urbanski, N.F., and Xu, Z.P., "Column Revamps," AIChE Spring Meeting, Chicago, IL, February 16, 1998.
13. Resetarits, M.R. and Pappademos, N., "Factors Influencing Vapor Crossflow Channeling," AIChE Annual Meeting, Reno, NV, November 8, 2001.
14. Resetarits, M.R. and Lockett, M.J., "Distillation," Encyclopedia of Physical Science and Technology, Vol. 4, 2001.
15. Resetarits, M.R. and Ogundeji, A.Y., "On Distillation Tray Weir Loadings," AIChE Spring Meeting, Tampa, FL, April 27-29, 2009.
16. Resetarits, M.R., Cai, T.J. and Whiteley, J.R., "On Distillation Tray Entrainment", AIChE Spring Meeting, Chicago, IL, March 2011.
17. Chambers, S., Cai, T.J., Vennavelli, A.N., Ogundeji, A.Y., Resetarits, M.R. and Redden, J.C., "Effect of Tray Spacing on Tray Performance," AIChE Spring Meeting, April 3, 2012.

Trays for Distillation and Absorption

By

AY Ogundeji
Mike Resetarits
Tony Cai

Fractionation Research Inc.
424 S. Squires Street, Suite 200
Stillwater, OK 74074

2 April 2014

Presented at YP Tutorials: Distillation and Absorption Equipment, AIChE Spring Meeting, New Orleans, Louisiana, April 2, 2014

UNPUBLISHED

Abstract

Distillation columns are essential to the chemical processing industries. They need to be properly designed and operated. Column internals, such as trays and packings, are essential to column performance. If column internals malfunction, the entire plant may suffer. Among column internal, trayed columns are widely used in distillation and absorption applications. This presentation will discuss the ranges of the liquid and vapor loads that a trayed column can handle. It will review the systems where trayed columns are particularly applicable. A brief history of trays and trayed columns will be presented. New generations of high performance trays, including augmented crossflow, counter current flow and cocurrent flow trays, will be discussed. The performance differences between trays and packings will be summarized. Column designs that provide high energy efficiencies will be briefly discussed. Extensive video footage of tray operations will be shown, from commercial size columns with various distillation systems ranging from deep vacuum to high pressure.

Keywords: Distillation Absorption Tray Flooding

OVERVIEW

Distillation is a process of separating mixtures into two or more products based on differences in their boiling points in a boiling mixture. The distillation process has been around for hundreds of years and, it is one of the most widely used processes in the industries.

Absorption is used to separate gas mixtures, remove impurities, or recover valuable chemicals. In absorption, there is a transfer of one or more species from the gas phase to a liquid absorbent. The species transferred to the liquid absorbent are referred to as solutes or absorbates.

Distillation and absorption employ trays and packings as column internals. This paper focuses only on tray internals. These column internals are used as phase contacting devices. There are different types of trays. The table below shows a rough comparison of the common tray types.

Type	**Sieve Tray**	**Valve Tray**	**Bubble-Cap Tray**	**Dual-Flow Tray**
Capacity	High	High	Moderate	Very high
Efficiency	High	High	Moderate	Lower than others
Pressure drop	Moderate	Moderate	High	Low to moderate
Turndown	About 2:1	About 3:1	Highest	Low
Cost	Low	About 20% higher than sieve	High. About 2-3 times more than sieve	Low

GENERAL

For distillation, a liquid phase flows countercurrently with a vapor phase, while for absorption, a liquid phase flows countercurrently with a gas phase. This is shown in **Figure 1**.

Figure 2 shows a drawing of a simple distillation column with trays. The figure shows that the column contains 20 trays. Feed and reflux streams are introduced into Tray 10 and Tray 20, respectively. Vapor is introduced into the bottom of the column via the "From Reboiler" nozzle.

The vapor rises and mixes with the liquid flowing down the column. The liquid flowing downward exits the column via the "To Reboiler" piping while the upward flowing vapor exits the column via the "To Condenser" nozzle.

Figure 3 shows the different tray configurations. Multi-pass trays are usually employed when liquid rates are too large for a 1-pass tray.

GENERATION OF TRAYS

There are different generations of trays. The generations include crossflow, augmented crossflow, countercurrent and cocurrent flow trays. The first generation is simple crossflow trays which include bubble-cap, sieve and valve trays, and which have the least capacities. The cocurrent flow trays (Shell Consep, Koch-Glitsch UltraFrac and UOP SimulFlow trays), which are the latest generation and have the highest capacities.

TRAY PARTS

Figure 4 shows a drawing of a 1-pass crossflow tray. It shows different parts of a tray. Liquid flows across each tray deck, over an outlet weir, and into a downcomer. The outlet weir purpose is to build a liquid head on the tray deck. A receiving pan collects and redirects the liquid onto the next deck through the downcomer clearance. The vertical distance between successive trays is called "spacing". The inner side of the downcomer is a vertical metal plate called the curtain. The column wall is also known as the shell.

Figure 5 shows a simplification of 1-pass crossflow tray hydraulics. Liquid enters the tray deck from the downcomer of the tray above by gravity. The liquid on the tray is aerated with vapor rising through the deck perforations to form a froth. The froth flows across the tray deck, over the outlet weir and into the downcomer where the vapor disengages. The distance between the top of the outlet weir and the top surface of the liquid flowing over the weir is called crest.

Figures 4 & 6 can be used to compare conventional and augmented crossflow trays, respectively. Augmented crossflow trays have truncated downcomers. That is, the downcomers have bottoms, and the bottoms are perforated. Augmented crossflow trays also have perforated decks below the bottoms of the downcomers as shown in Figure 6. Conventional crossflow trays do not have truncated downcomers. The higher open area of augmented crossflow trays yields higher capacities than conventional crossflow trays.

Figure 7 shows different types of tray decks. The choice of deck usually depends on the application. Vapor-liquid handling capacities and pressure drops of decks differ. For example, bubble-cap decks turn down better than sieve and valve decks. Bubble cap trays have higher pressure drops than sieve and valve trays. Moving valve decks have larger rangeabilities than simple sieve decks. Push valves reduce froth heights and increase capacities.

FLOODING

The performance of trayed columns is determined by many factors that include flooding, weeping and dumping. Flooding is excessive accumulation of liquid inside a trayed tower. When trays flood, there is sudden increase in column differential pressure and decrease in mass transfer efficiency. There are many different mechanisms that cause trays to flood. These mechanisms include choking, downcomer backup, spray entrainment and froth height flooding. More details about the mechanisms are as follows:

Choke Flooding

It occurs when the liquid flow rate increases to a point where the velocity of the biphase in the downcomer exceeds a certain limit. At this point, the biphase cannot even fit into the downcomer mouth and the biphase cannot be transported to the tray below. It happens when the downcomer top area (mouth) is too small to accommodate the aerated liquid above the outlet weir. A sketch is provided in **Figure 8**.

Downcomer Backup Flooding

It occurs because of high tray pressure drop, high liquid height on the tray and large downcomer clearance losses. When the backup of biphase in the downcomer exceeds the tray spacing (plus the outlet weir height), liquid accumulates on the tray above.

Spray Entrainment Flooding

It happens at low liquid rates when vapor velocity is increased to a point where the liquid on the tray is dispersed by the rising vapor into liquid droplets which are entrained to the tray above.

Froth Height Flooding

It occurs at high liquid and vapor rates. If the tray spacing is 24 inches or less, as vapor rate increases, a point is reached where the froth height equals the tray spacing. Liquid accumulates on the tray above.

Weeping

This phenomenon occurs at low vapor rates. At low vapor rates, when the pressure exerted by the vapor is insufficient to hold up the liquid on the deck, liquid starts to leak through the deck perforations. This liquid backmixing yields reduced mass transfer efficiency.

Dumping

It occurs at very low vapor rates. Excessive weeping will lead to dumping. With dumping, all the liquid on the deck will weep through the perforations. In this condition, no liquid flows over the outlet weir.

CLOSING

Roughly half of old and new columns contain trays; the other half, packing. Fundamental research regarding these contacting devices has been performed at Fractionation Research Inc. (FRI) since 1954.

REFERENCES

- H.Z. Kister, Distillation Design, McGraw-Hill, 1992.
- J.D. Seader, E.J. Henley and D.K. Roper, Separation Process Principles, John Wiley & Sons, 2011.
- M.J. Lockett, Distillation Tray Fundamentals, Cambridge University Press, N.Y., N.Y. 1986.

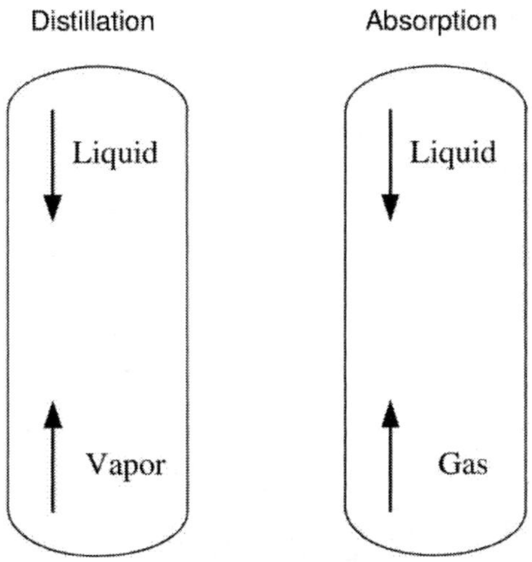

FIGURE 1.

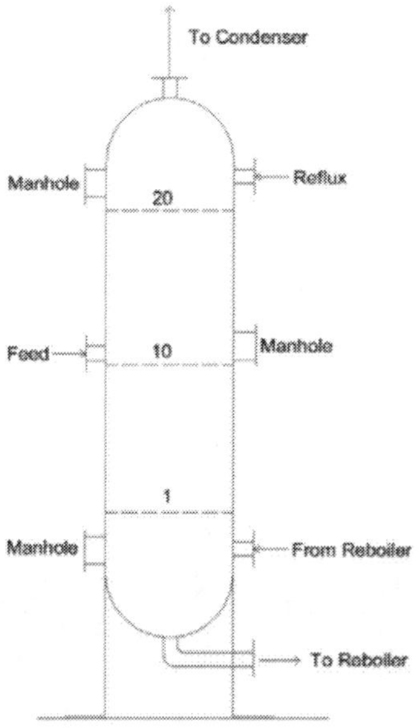

FIGURE 2. Simple Distillation Column with Trays

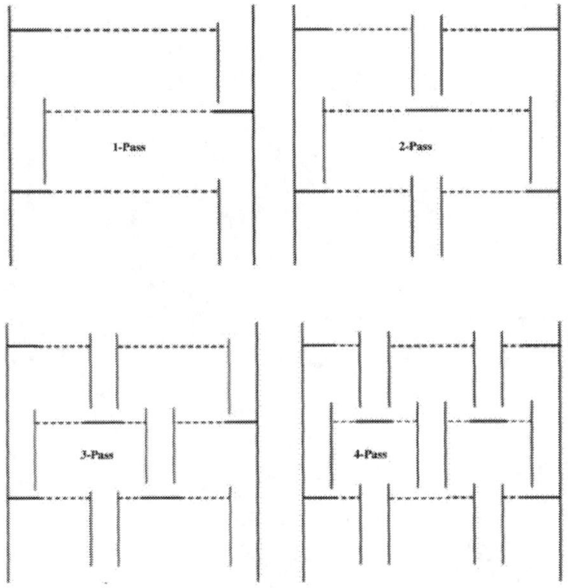

FIGURE 3. Crossflow Trays

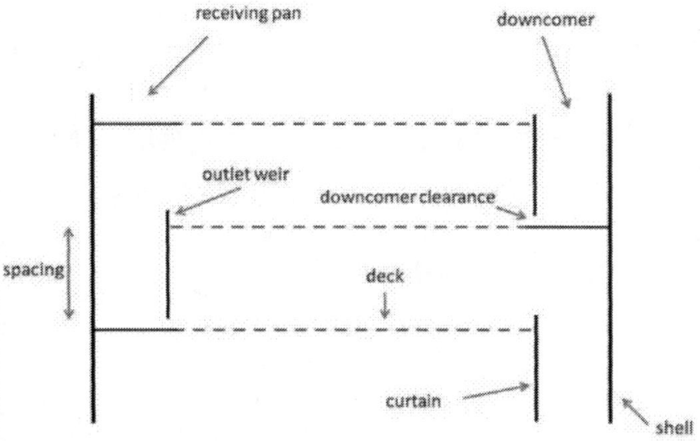

FIGURE 4. 1-Pass Crossflow Trays

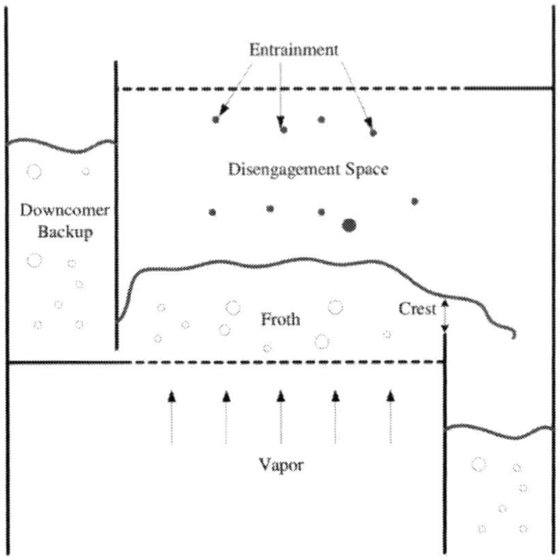

FIGURE 5. Flow Pattern on 1-Pass Crossflow Tray

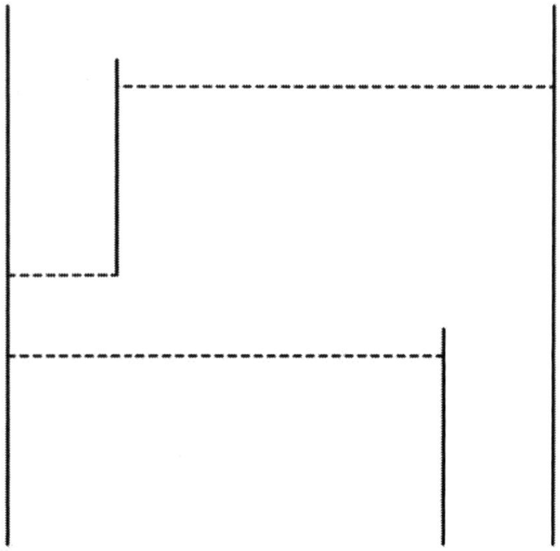

FIGURE 6. Augmented Crossflow Trays

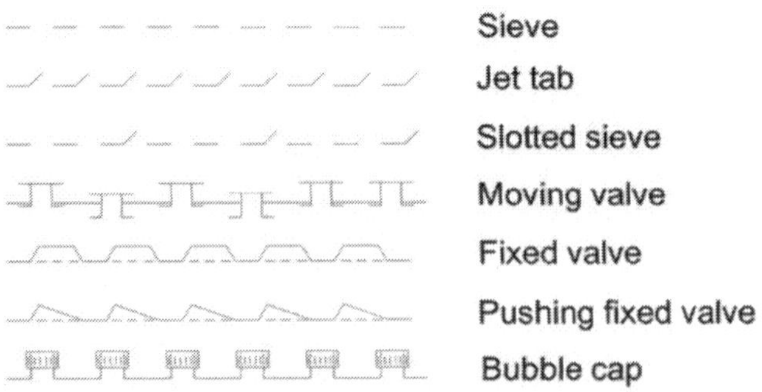

FIGURE 7. Tray Deck Types

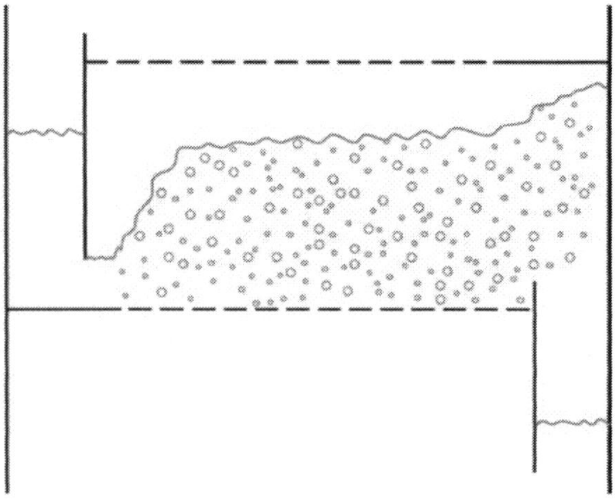

FIGURE 8. Choke Flooding

Structured Packings for Distillation and Absorption

By

Simon Chambers
Anand Vennavelli

Fractionation Research Inc.
424 S. Squires Street, Suite 200
Stillwater, OK 74074

2 April 2014

Presented at YP Tutorials: Distillation and Absorption Equipment, AIChE Spring Meeting, New Orleans, Louisiana, April, 2014

Abstract

Structured packings used in industrial practice can be broadly classified into three generations: wire gauze, standard, and high capacity corrugated sheet metal. Standard structured packings were a popular choice in the industry for many years. Newer high capacity structured packings, which comprise either "half-S" or "full-S" crimp angle shapes, can operate at much higher load points or maximum useful capacities (MUC's), and considerably lower pressure drops than standard structured packings.

Structured packings are favored in high vapor rate, low liquid rate distillation services and in absorption. They are particularly advantageous in vacuum columns, where low pressure drops are sometimes critical. For example, in Ethylbenzene-Styrene monomer columns, low pressure drops yield low liquid temperatures and, consequently, reduced renegade polymerization.

In this paper, structured packing physical characteristics, installation, and applications are discussed. In addition, basic performance characteristics, such as HETP, pressure drop, and capacity, and liquid distribution quality are reviewed. A 20-minute video was prepared by FRI to illustrate the basics of structured packings.

Keywords: Packings Distillation Absorption Low Pressure Drop

1 Overview

Structured packings are often the choice of internals for low-pressure distillation and absorption. Structured packings provide good efficiencies, low liquid hold-ups, and low pressure drops.

Structured packings are frequently used in:

- Chemicals and fine chemicals (e.g., Ethylbenzene/styrene distillation)
- Refining (e.g., vacuum and atmospheric crude oil distillation)
- Petrochemicals

Structured packings are best suited for vacuum to moderate pressure applications.

2 Generations of Structured Packings

Structured packings can be broadly classified into three generations – wire gauze, standard corrugated sheet metal, and high capacity corrugated sheet metal.

2.1 First Generation: Wire Gauze Structured Packings

First introduced in the late 1950's to early 1960's, these packings are most suitable for difficult separations and when there is a low pressure drop requirement. However, they are relatively expensive to fabricate. Examples include Sulzer Wire Gauze BX and CY, Koch Glitsch Goodloe, Montz A3-500, AMACS Woven Gauze. Figure 1 shows the photographs of some wire gauze packings.

2.2 Second Generation: Standard Structured Packings

Because of the high fabrication costs associated with the wire gauze packings, more cost-effective corrugated sheet metal structured packings were introduced in the late 1970's. These packings have since become the standard of the industry. Examples are shown in Figure 2.

Because these packings have a sharp change in angle between the layers, the vapor and the liquid have to change direction abruptly at the interface of the packing layers. Such a sudden change in the direction of the liquid and vapor flow can limit the performance of the packings.

2.3 Third Generation: High Capacity Structured Packings

The third generation, or high capacity structured packings, are very similar to the standard packings, except that they provide smooth transitions between layers for the liquid and vapor streams. Examples are shown in Figure 3.

3 Anatomy of a Corrugated Structured Packing

Structured packing comprises metal sheets that are textured, crimped, and often crimped – and assembled together as blocks.

3.1 Texture and Perforation

Texturing can be seen, for example, in the photograph of the Mellapak packing of Figure 2. Texturing promotes liquid film spreading while perforating minimizes vapor channeling and promotes vapor-liquid contacting. Texturing and perforating can increase the packing efficiency.

3.2 Blocks and Layers

Structured packings are usually built in layers of heights 200-300 mm, which are further segmented into multiple blocks. Figure 4 shows a full bottom layer of the packing with a few blocks of another layer placed on top of the bottom layer.

3.3 Crimps

The crimps are an important physical feature of a structured packing, and effectively determine the surface area and flow patterns. Referring again to Figure 2, the crimps of most of the structured packings are easily seen.

3.3.1 Crimp Angles

Based on the crimp angle from the vertical axis, structured packings are classified into X (30°), Y (45°), and Z (60°) packings. More-vertical crimps exhibit lesser hold-ups, lesser pressure drops, lower efficiencies, and higher capacities

3.3.2 Crimp Depth

Large crimp depths provide a low surface area, and conversely, small crimp depths provide a larger surface area. Obviously, to create more surface area, more metal is required - more metal means more cost, but also a higher efficiency. With more surface area, liquid hold-ups are higher, pressure drops are higher, and capacities are lower.

4 Installation of structured packings

First, the packing support grid is installed, and then, packing elements are lowered down by means of a belt or a rope to prevent damage to the packing. Metal guide plates or shoes are employed to facilitate installation of packing. The installer stands on a board to prevent damage to the packing sheets. As shown in Figure 4, each layer contains wall wiper bands that are bent out so they touch the column wall so that any liquid wall flow will be redirected back into the bulk of the packing. Each layer is usually rotated 90 degrees to minimize seam alignments of the packing blocks, and to prevent short-circuiting of vapor liquid traffic. Careful installation of the structured packing is vital for the column to work.

5 Performance Characteristics of Structured Packings

The key performance characteristics are HETP, Capacity or often, the Maximum Useful Capacity (MUC), and pressure drop.

5.1 Pressure Drop

Pressure drops of structured packings are lower than those of random packings, and much lower than those of trays. The pressure drop across a structured packing increases gradually with increasing liquid and vapor rates; the pressure drop increases sharply close to the flood point, as liquid begins to stack up in the bed.

5.2 HETP (Height Equivalent to a Theoretical Plate)

Structured packings use the concept of a "Height Equivalent to a Theoretical Plate", or HETP, to describe the performance of the packings. The HETP is defined as

$$\text{HETP} = \frac{\text{Packing Bed Depth}}{\text{Number of theoretical plates}} \tag{1}$$

The number of theoretical plates[1] required for a given separation is usually determined by a process simulation. An easy separation requires few theoretical plates; a difficult separation requires many theoretical plates.

For the same number of theoretical plates, a lower HETP results in a lesser packed bed depth, and consequently, a shorter column. Therefore, the lower the HETP, the more efficient the column is.

[1] Number of theoretical plates is also sometimes referred to as the number of theoretical stages, or the number of equilibrium stages.

For example, a highly efficient structured packing might achieve an HETP of about 300 mm (12 inches); a less efficient structured packing might achieve an HETP of 500 mm (20 inches).

5.3 Capacity or Maximum Useful Capacity (MUC)

The highest vapor rate at which the HETP of the column is good is called the MUC. A further increase in rate from the MUC will cause the HETP to rise rapidly, as the column progresses into a flooded condition.

5.4 Effect of packing specific surface area

Specific surface area, determined by the crimp characteristics, is an important factor affecting performance characteristics of structured packings.

High surface area structured packings exhibit high pressure drops and vice-versa. Typical pressure drop curves are shown in Figure 5.

High surface area structured packings exhibit low HETP's and vice-versa. Typical HETP curves are shown in Figure 6.

High surface area structured packings exhibit low MUC's and vice-versa. In Figure 6, the MUC point is indicated for both high surface area and low surface area packings. .

6 REFERENCES

Some of the best sources of information regarding structured packing are listed below:

Kister, H.Z., **Distillation Design**, McGraw-Hill, 1992

Kister, H.Z., **Distillation Operation**, McGraw-Hill, 1990

Perry's Chemical Engineer's Handbook, 8th edition, "Chapter 14 – Equipment for Distillation and Gas Absorption"

Strigle, R.F., **Packed Tower Design and Applications**, 2nd edition, Gulf Publishing Co., Houston, TX, 1987.

Vendor Resources:[2]

http://www.sulzer.com/en/Products-and-Services/Separation-Technology/Structured-Packings
http://www.koch-glitsch.com/masstransfer/pages/structured_packing.aspx
http://www.amacs.com/packing/structured/
http://amacs.com/wp-content/uploads/2012/09/AMACS-Montz-Brochure.pdf
http://raschig.com/-Structured-Packings

[2] The listing here is not comprehensive. The list of references is provided for the benefit of the reader. FRI does not endorse or recommend any particular vendor product over another, including those listed in this document.

7 FIGURES

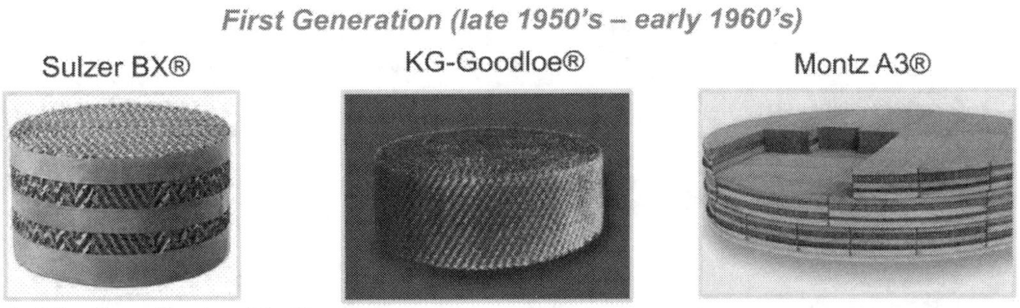

Figure 1: Wire Gauze Structured Packings

Figure 2: Standard Sheet Metal Structured Packings

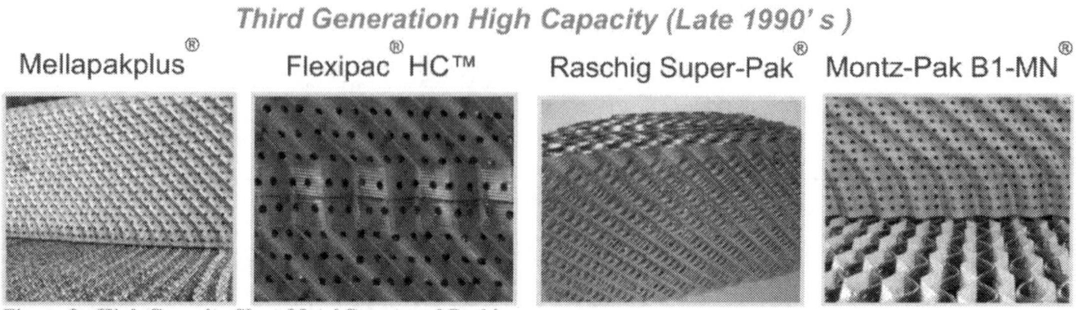

Figure 3: High Capacity Sheet Metal Structured Packings

Figure 4: Sulzer Mellapak 250.Y Packing - Packing Layer and Blocks

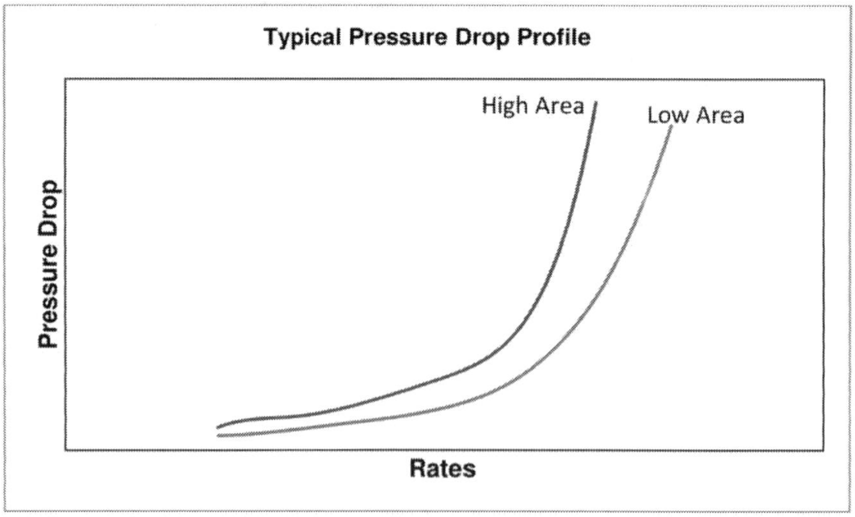

Figure 5: Typical Pressure Drop Curves

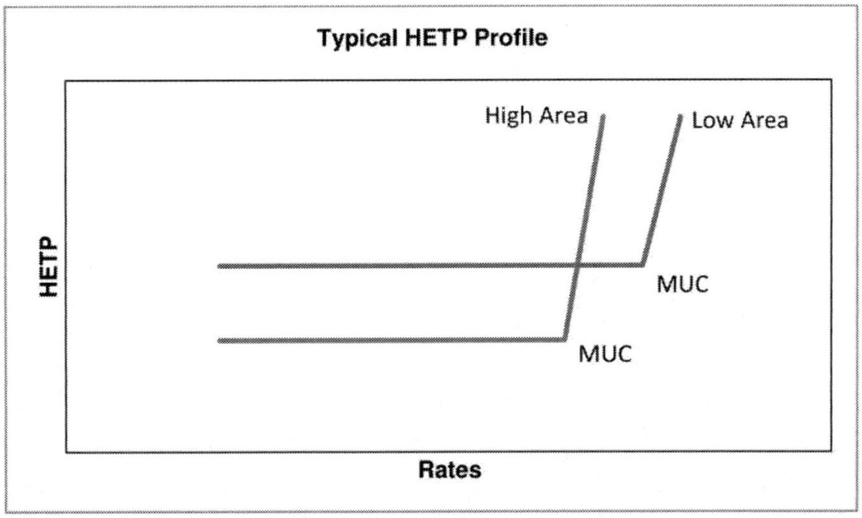

Figure 6: Typical HETP Curves

Liquid Extraction
Processes and Equipment

By

Mike Resetarits
AY Ogundeji
Tony Cai

Fractionation Research Inc.
424 S. Squires Street, Suite 200
Stillwater, OK 74074

Presented at YP Tutorials: Distillation and Absorption Equipment,
AIChE Spring Meeting, New Orleans, Louisiana, 2 April, 2014

UNPUBLISHED

Abstract

What is the best way to remove C_6, C_7 and C_8 compounds from a feed stream containing C_6, C_7 and C_8 compounds? How do you take a lubricating oil and remove certain compounds and improve its viscosity index (so that its viscosity does not change excessively with the oil's temperature)? What is propane deasphalting? The answers to all of these questions is "liquid extraction processes."

What technology was supported and funded by the actor Kevin Costner to help clean up the large 2010 Gulf of Mexico oil spill? What are Karr reciprocating plate columns, Oldshue-Rushton contactors and Podbielniak centrifuges? What the heck is a Graesser Raining Bucket? The answer to all of these questions is "liquid extraction equipment."

Imagine heavy and light fluids flowing vertically past each other with only a very small density difference. Imagine rotating equipment that works better without rotating. Liquid extraction is different – very different. There are very few experts.

Keywords: Extraction Aromatics Ternary Equilibria

Acknowledgment

The authors acknowledge the many contributions of Dr. Frank Seibert and Dr. Bruce Eldridge (both of the University of Texas at Austin) to the field of liquid-liquid extraction.

OVERVIEW

Some people say that chemical engineering is 50% Reactions and 50% Separations. In the field of Separations, distillation is the technique that is employed the most often, but liquid-liquid extraction, absorption, adsorption and membranes also play important roles.

Liquid-liquid extraction has many, and diverse, applications. Extraction devices can employ agitation to hasten mass transfer; there are many types of agitated devices. There are also many types of devices that do not employ agitation. Mass transfer calculations usually begin with liquid-liquid equilibrium (thermodynamic) data. Hydraulic calculations usually begin with force balances around droplets, and interfacial tension measurements.

GENERAL

As shown in **Figure 1**, when distillation is employed, a liquid phase and a vapor phase flow countercurrently. With extraction, a heavy liquid phase and a light liquid phase move countercurrently. With extraction, those two phases must be generally immiscible in each other. Chemical compounds in one phase, however, diffuse into the other phase. Very often, one liquid phase is aqueous; the other liquid phase is hydrocarbon. (Many people know that water and oil do not mix.) **Figure 2** shows a swarm of light liquid droplets (maybe "oil") rising through a heavier liquid (maybe "water"). At the top of the photograph, the light liquid droplets are seen to be coalescing.

Extraction columns usually have two feed streams. One is called the "Feed" and the other is called the "Solvent." The Feed stream of **Figure 3** is comprised of just two chemicals, A and B. The solvent stream is designated by "S". "A" is the "Solute" which ultimately transfers from the Feed stream to the solvent stream. After the Feed stream loses component A, the residual liquid, which is mostly component B, is often called the "Raffinate." (Generally, in the extraction world, the word "raffinate" refers to a low-value product stream.) In Figure 3, the product stream that is designated by "S + A" usually goes to a solvent recovery column that separates S from A. The resultant solvent stream is returned to the extractor. "A" is usually the final target product. The solvent recovery column is usually some form of distillation column, e.g., an extractive distillation column.

In trayed distillation and absorption columns, the vapor phase is dispersed as droplets in the downward-flowing liquid stream. As shown in Figure 3, extraction processes offer four different possibilities as follows:

1. Solvent = heavy phase; feed = light phase; heavy phase is dispersed
2. Solvent = heavy phase; feed = light phase; light phase is dispersed
3. Solvent = light phase; feed = heavy phase; heavy phase is dispersed
4. Solvent = light phase; feed = heavy phase; light phase is dispersed

Generally, the heavy phase is an aqueous phase and the light phase is a hydrocarbon phase. Most extraction experts believe that the more voluminous phase should be dispersed.

APPLICATIONS

A partial list of extraction applications is provided below:

- Pharmaceutical – production of penicillin
- Food – decaffeination of coffee
- Metals extraction
- Chemical and petrochemical
  - Aromatics recovery
  - Lube oil extraction
  - Acetic acid extraction
  - Desulfurization
- Liquid sweetening

BTX aromatics extraction is a particularly popular process, especially in refineries. A product stream from a catalytic reformer contains hexane, benzene, heptane, toluene, octane, xylenes and heavier aromatic and non-aromatic compounds. The aromatic (BTX) compounds are far more valuable than the non-aromatic compounds. The former cannot be separated easily from the latter via distillation columns.

Figure 4 shows a light liquid feed stream (C_6-C_8) entering the side of an extractor. A heavy liquid solvent stream enters the top of the extractor. There are at least two popular solvents – sulfolane and tetraethylene glycol. Both of these solvents have a strong tendency to absorb (or "extract") the benzene, toluene and xylenes, leaving behind the hexane, heptane and octane, i.e., the non-aromatic raffinate stream. Rich Solvent leaving the bottom of extractor goes to a (vapor-liquid) Solvent Recovery Column which separates the solvent from the BTX – the final product. That Recovery Column also generates a light reflux stream that is returned to the extractor. The BTX product stream goes to a series of distillation columns that separate and purify the benzene, toluene and xylenes.

Extraction can also be used to remove heavy aromatics compounds from a lubricating oil feed stream. The most popular solvents are furfural, phenol and NMP. The final product from the extractor is a lubricating oil that has an improved Viscosity Index, i.e., its viscosity is less dependent on temperature.

CONTACTORS

There are many types of liquid-liquid contacting devices. Some employ agitation, e.g., mixers, to hasten mass transfer. Others do not require or employ agitation. For those, in fact, agitation can be harmful, creating a totally-emulsified contactor wherein the light and heavy phases never separate out. The interface between any two liquids has an intermolecular "skin." The strength of that skin is called the "interfacial tension." Many extraction experts believe that liquid-liquid systems and interfacial tensions higher than 15 dynes/cm have a tendency towards large droplets and low interfacial areas for phase-to-phase mass transfer. Such systems require agitation to create more, smaller droplets. Liquid-liquid systems with interfacial tensions less than 15 dynes/cm have a tendency to easily produce many small droplets and a large interfacial mass transfer area. Agitation is not required with such systems; efficiencies are acceptable without it.

Generally, liquid phase diffusivities (diffusion coefficients) are much lower than vapor phase diffusivities. With a liquid-liquid extractor, diffusing solute compounds move slowly through one liquid phase, pass through the interface, and then move slowly through the second liquid phase. As a result of these slow mass transfers, extractor efficiencies are generally low. For example, a typical trayed extractor might have a tray efficiency of only 20%.

A partial list of agitated and non-agitated extraction devices is given below:

Agitated
- Mixer-settlers
- Podbielniak centrifuges
- Costner Centrifugal
- Rotating Disc Contactors
- Scheibel
- Karr reciprocating plate
- Kuhni

Non-agitated
- Packing
- Conventional trays
- Seal Deck and Rain Deck (UOP) trays
- Multiple Upcomer (UOP) trays

Figure 5 shows the UOP Multiple Upcomer (MU) tray. With this tray, the heavy phase is the dispersed phase. Each tray deck has many circular perforations which create a swarm of heavy phase droplets. As the droplets fall through the continuous light phase, mass is transferred. Those droplets coalesce on the tray below. The cycle is repeated again and again. As shown in Figure 5, the light liquid phase snakes its way up the column as the continuous phase.

There are many proprietary agitated extraction devices. Rotating disc contactors, such as the one shown in **Figure 6**, have often been used in Sulfolane aromatics extraction units. The Costner Centrifugal Extractor has been considered for off-shore oil-water separations. Podbielniak extractors are popular in pharmaceutical separations.

HYDRAULICS

For some liquid-liquid systems, at some temperatures, interfacial tensions can be measured easily with a ring tensiometer such as the one shown in **Figure 7**. Initially, the ring is lowered to the bottom of the containment cylinder. Then, the ring is pulled upwards, by a tin wire, until the interface between the two liquid phases is impacted. At that point, the interfacial "skin" resists any other upward ring movement. Eventually, the ring breaks through the skin with a force that is measured. That force is easily converted to an interfacial tension in dynes/cm. Thereafter, if the ring is pulled upwards through the light liquid phase, the liquid-vapor interface will soon be impacted. The force that will be required to pull the ring through that interface is a direct measure of the light liquid surface tension. For liquid-liquid systems at higher temperatures and pressures, interfacial tensions can be determined indirectly from the sizes of droplets that are created by sieve plates.

THERMODYNAMICS

Distillation calculations employ the concept of "equilibrium stages" to determine column heights. Vapor-liquid equilibrium laboratory data are employed to determine the number of equilibrium stages that are required to effect specified distillation separations.

Similarly, the extraction technology employs the concept of "equilibrium stages," and liquid-liquid equilibrium data, to determine staging requirements and contactor heights (sizes). Unlike distillation, extraction often employs ternary (triangular) diagrams to represent liquid-liquid equilibria. **Figure 8** provides a ternary diagram for the BTX example of Figure 4. The three apexes of the diagram represent pure components. Every position within the triangle represents a ternary composition. Every position under the curved umbrella represents a non-equilibrium mixture that sooner or later becomes two immiscible liquid phases that are in equilibrium with each other. The exact middle of the triangle represents a non-equilibrium mixture of 33% toluene and 33% heptane and 33% sulfolane. In a BTX extractor, Sulfolane is the solvent, toluene is the solute and heptane is the raffinate. Regarding this diagram, the feed stream is a toluene/heptane mixture.

In Figure 8, Points a, c and e represent light liquids that are high in heptane concentrations, low in toluene concentrations and almost absent of sulfolane. Points b, d and f represent heavy liquids that are high in sulfolane concentrations, high in toluene concentrations and almost absent of heptane. The almost-horizontal lines are called "tie lines." The points at the ends of each tie line represent light and heavy phases that are in equilibrium with each other. For example, Points a and b are in equilibrium with each other. The "Plait Point" represents the unique situation where the light and heavy liquid phases have the exact same compositions. Ternary diagram equilibrium data can be employed to determine the number of equilibrium phases that are required to effect target separations. Mass transfer models and computer programs exist for determining tray and packing efficiencies. More often than not, new extractors are designed using the efficiencies that were observed with similar extractors that are already operating.

REFERENCES

Some of the best sources of information regarding the science of liquid-liquid extraction are listed below:

- Treybal, R.E., **Liquid Extraction**, McGraw-Hill 1963.
- Lo, T.C., Baird, M.H.I., and C. Handson, **Handbook of Solvent Extraction**, John Wiley and Sons 1983.
- Laddha, G.S. and T.E. Degaleesan, **Transport Phenomena in Liquid Extraction**, McGraw-Hill 1978.
- Arlt, W., Macedo, M. Rasmussen, P., and J.M. Sorensen, **Liquid-Liquid Equilibrium Data Collection, vol. V, Parts 1-4,** DECHEMA Chemistry Data Series, Frankfurt 1997.
- **Perry's Chemical Engineer's Handbook**, 7th edition, "Liquid-Liquid Extraction Operations and Equipment", 1997. (revision in 2006)
- Walas, S.M., **Phase Equilibria in Chemical Engineering**, Butterworth Publishers 1985.
- Maddox, R.N., **Gas and Liquid Sweetening**, Campbell Petroleum Series, 1977.

FIGURES

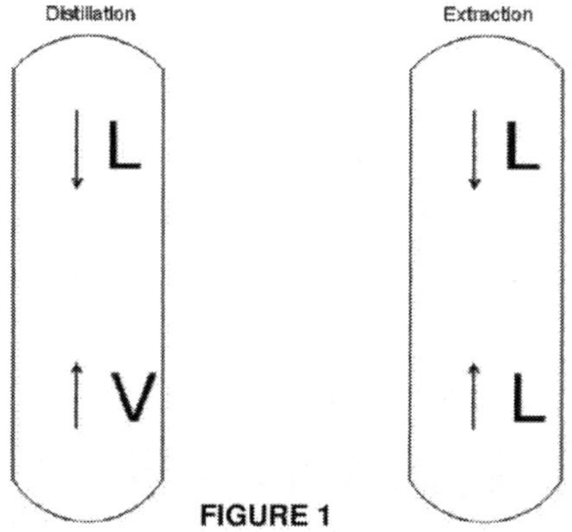

FIGURE 1

FIGURE 2

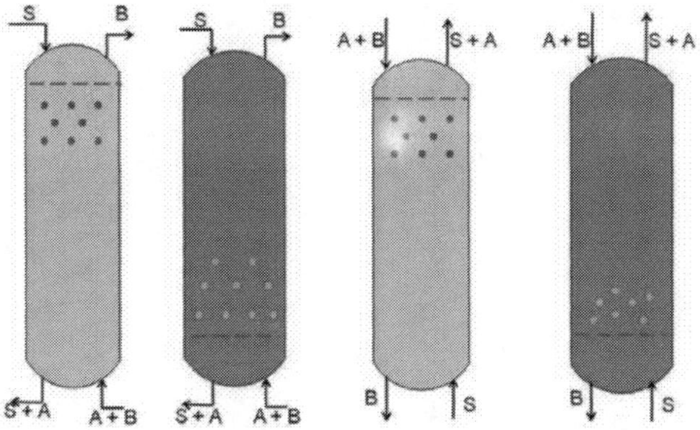

FIGURE 3. Extraction Options

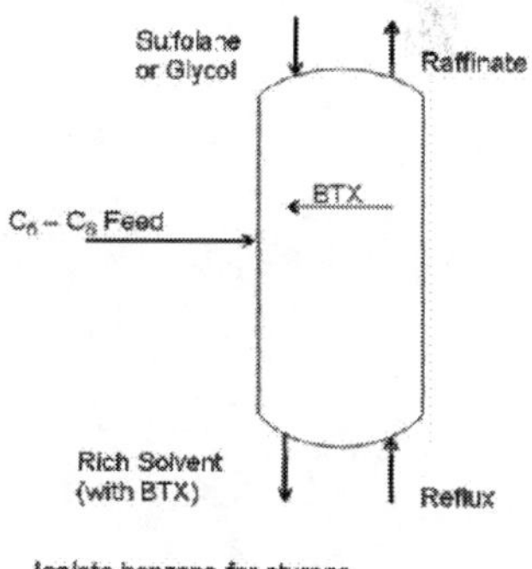

FIGURE 4. Aromatics Extraction

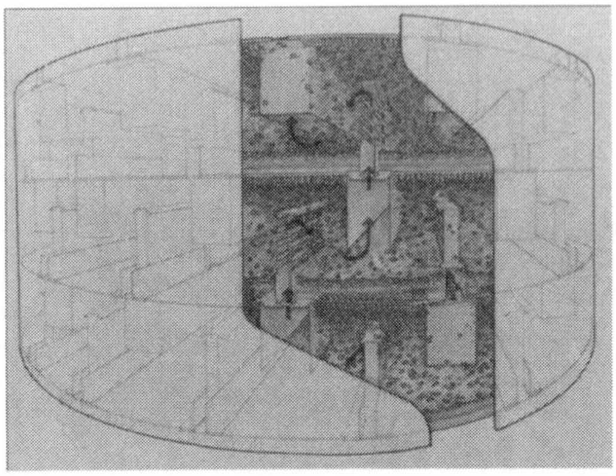

FIGURE 5. MU Tray Hydraulics

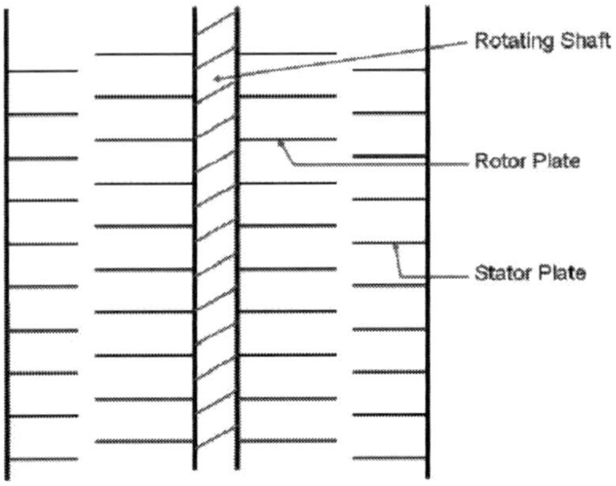

FIGURE 6. Rotating Disc Contactor

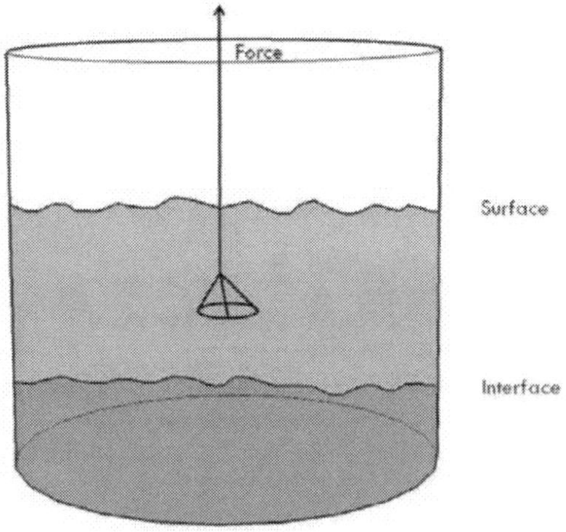

FIGURE 7. Ring Tensiometer

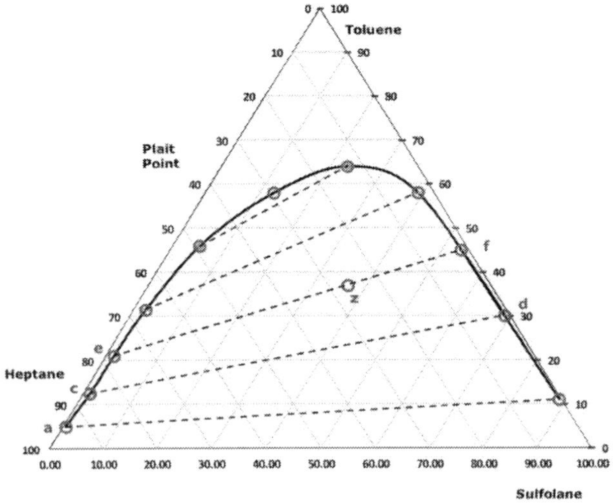

FIGURE 8. Heptane + toluene + sulfolane at 25°C, a type I system. [*Data taken from De Fre and Verhoeye,* J. Appl. Chem. Biotechnol., ***26,*** *pp. 1-19 (1976).*]

Acid Gas Absorption Fundamentals

By

Clint P. Aichele
James. R. Whiteley
Yash Tamhankar

School of Chemical Engineering
Oklahoma State University

Mike Resetarits

Fractionation Research Inc.
424 S. Squires Street, Suite 200
Stillwater, OK 74074

2 April 2014

Presented at YP Tutorials: Distillation and Absorption Equipment,
AIChE Spring Meeting, New Orleans, Louisiana, April 2014

ABSTRACT

Absorption is a standard unit operation that has been studied by chemical engineering students for decades. One of its primary uses is for the removal of acid gases, like CO_2 and H_2S, from natural gas streams. Historically, aqueous amines and hot carbonate solutions have been used to perform these absorptions. Many different amines, solution strengths, activators and passivators have been tested and employed industrially. Approximately half of the industrial towers are trayed; half, packed. Foaming adversely affects the hydraulic performance in these systems.

Acid gas absorption is receiving appreciable new attention as concerns rise regarding CO_2 in the atmosphere. In addition, the United States has recently experienced significant growth opportunities in shale gas production which require improvements in gas treating technologies. This paper reviews acid gas absorption and regeneration fundamentals. A PFD will be discussed at length, to illustrate important concepts. New absorbent solutions will be discussed, including ionic liquids. Flue gas applications will be discussed, including the difficulties associated with column pressure drops and very large column diameters. Rate-based modeling approaches will also be reviewed.

INTRODUCTION

Subsequent to the Industrial Revolution, global carbon dioxide emissions have increased from virtually zero to 9 billion total metric tons. These emissions originated from the natural gas, petroleum, coal and cement industries. Carbon dioxide, CO_2, emissions increase annually, as an increasing percentage of the global population consumes gas/oil/coal energy and drives cars. Carbon dioxide is a "greenhouse gas" and many people contend that such gases are contributing to the warming of the planet. As CO_2 emissions continue to increase, the technologies that reduce those emissions receive increasing attention from academia, industry and global governments.

For many years, and now, acid gas absorption units have been the primary means whereby CO_2 emissions have been reduced. Such units are – everywhere – and are relatively easy to understand. Several excellent books describe such units. [1,2,3]

DISCUSSION

Natural Gas Processing

When it first comes from the ground, natural gas contains many compounds, as shown in **Table 1**. Methane burns nicely in a furnace, but CO_2 does not burn at all. Therefore, the consumers who purchase natural gas do not want to receive, and pay for, CO_2. Typically, natural gas production companies must reduce the CO_2 content of natural gas down to 2% or less.

Raw natural gas sometimes contains hydrogen sulfide, H_2S, which is particularly-dangerous poison. Natural gas production companies must reduce the H_2S content of natural gas down to 4 ppmv or less.

Very recently hydraulic fracturing or "fracking" is yielding gases that are very high in ethane, propane and butane. These "natural gas liquids" or "NGL's" are very valuable in that they can easily be converted to ethylene and propylene. As fracking increases in the U.S., more raw natural gas streams are being created and all of those streams need the acid gases (CO_2 and H_2S) to be removed.

Flue Gases

"Flue gas" often refers to a combustion exhaust gas from the chimney of a power plant. Its composition depends on what is being burned, e.g., natural gas or coal. **Table 2** provides typical flue gas compositions. Globally, the CO_2 content of flue gases is receiving increasing attention, but the removal of that CO_2 is particularly challenging. Flue gases are at very low (near atmosphere) pressure and need to be compressed to flow (through absorbers). Many flues create huge volumes of such gases and those gases contain particulates and oxygen. Perhaps the largest challenge of all: What can be done with the CO_2 once we capture it? "Sequestration" is an umbrella word that includes all of those natural and unnatural places where CO_2 can be stored or utilized. For example, compressed CO_2 can be used for Enhanced Oil Recovery (EOR).

Solvents

Water can be used as a solvent for the acid gases in a natural gas stream, but water is very ineffective. The Benfield hot carbonate process employs aqueous potassium carbonate as the solvent. CO_2 and H_2S are removed from natural gas streams via the following overall chemical reactions:

$$K_2CO_3 + CO_2 + H_2O = 2KHCO_3$$

$$K_2CO_3 + H_2S = KHS + KHCO_3$$

Globally, for acid and gas removals, aqueous amine solutions are probably the most popular. Forty years ago, MEA and DEA were the workhorses. Then, MDEA became popular especially because of its ability to remove H_2S quickly and totally, without over-absorbing CO_2. Most recently, piperazine is being employed regularly. **Figure 1** is a partial list and classification of amines. When CO_2 is absorbed into an aqueous amine solution, many chemical reactions are involved. These chemical reactions are very exothermic. Amine solutions are very corrosive. Passivators are sometimes employed to reduce attacks on the metals of acid gas processing units.

Typical Unit

Figure 2 shows a typical amine absorption unit. Two processing columns are employed – an absorber and a solvent regenerator.

Referring to **Figure 2**, the Feed Gas enters the bottom of the absorber. Typical pressures of that gas range from 100 to 1200 psia. Absorber shells can be as thick as 3 inches. As the gas proceeds upwards through the column, it encounters a down-flowing liquid solvent stream. Lean liquid solvent enters the top of the absorber. This lean solvent stream must be low in acid gases (CO_2 and H_2S) or gases issuing from the lean solvent stream will contaminate the gas product stream. It is also important that the lean solvent stream be relatively cool to remove CO_2 to the lowest possible levels. Additionally, a too-warm lean solvent will flash off its acid gases too easily. The treated gas product stream that leaves the top of the absorber often goes to a train of distillation columns that separates and purifies the methane, ethane, propane, isobutane and n-butane.

Still referring to Figure 2, as the liquid solvent stream flows down through the absorber, the exothermic reaction associated with the chemical absorption causes the solvent to increase appreciably in temperature. Most of the temperature increase occurs near the bottom of the absorber. After the rich solvent leaves the bottom of the absorber, it often goes to a heat exchanger that increases the rich solvent temperature further. The hot rich solvent goes to the regenerator, where a massive flash occurs especially because the regenerator operates at near-atmospheric pressure. Some regenerators have two in-series flash chambers in their tops to allow additional flashing off of the acid gases. All of the vapors leaving the top of the regenerator (steam, CO_2 and H_2S) go to a condenser that liquefies the steam and returns the water to the top of the regenerator.

After flashing, the rich solvent flows all the way down the regenerator until it reaches a reboiler. That heat exchanger boils off a large portion of the water that is in the solvent. Very little amine is boiled off. The resultant steam flows upwards through the regenerator, and strips the acid gases out of the down-flowing solvent.

The lean solvent that leaves the bottom of the regenerator is very hot (depending on the reboiler temperature) and is low in pressure. A pump is required to increase the pressure of the lean solvent to a pressure somewhat higher than the absorber pressure.

A high-temperature lean solvent would be very ineffective at acid gas removal, especially due to vapor-liquid equilibrium (thermodynamic) considerations. The rich solvent can be used to decrease the temperature the lean solvent. The lean solvent temperature can also be reduced using a cooling water exchanger, as shown in Figure 2.

There are many variations of absorption unit processing schemes. Examples follow:

1. A semi-lean solution can be pulled from the side of the regenerator and fed to the side of the absorber
2. The bulk of the lean solution can be fed to the side of the absorber, with only a slip stream of the lean solution going to the cooling water exchanger, and the top of the absorber

Regarding the acid gases that leave the top of the regenerator, the H_2S can be converted to elemental sulfur in a Claus Unit. Historically, the CO_2 has been released into the atmosphere. It can, however, be compressed and utilized for Enhanced Oil Recovery (EOR).

Other Solvents

Aqueous amines and potassium carbonate solutions have historically been the most commonly employed absorption solvents. Some amine solutions contain a mixture of primary, secondary and/or tertiary amines. Some potassium carbonate solutions employ amines as "activators." Generally speaking, H_2S is absorbed quickly into amine solutions; CO_2 slowly. When an excess of CO_2 is absorbed by the lean solution, an excess of regeneration (and energy) is required. A perfect amine solution formulation would simultaneously exactly achieve the targeted CO_2 and H_2S product gas compositions.

Solvents other than aqueous amines and hot carbonate solutions have been employed and are being investigated

Physical solvents, ionic liquids, and enzymes have been explored as possible CO_2 capture solvents to replace aqueous amine absorption for flue gas applications[4]. These approaches have been investigated primarily for their ability to reduce the size and overall cost of acid gas absorption applications. Physical solvents require the physical solubility of the acid gas into the solvent; therefore, physical solvents are commonly used for high pressure applications such as synthesis gas scrubbing or natural gas absorption. Because physical solvent processes don't rely on chemical reactions to absorb the acid gases, the solvent regeneration energy is typically less than that associated with conventional aqueous amine absorption. Ionic liquids have received considerable attention for their potential ability to reduce the regeneration energy of the CO_2 removal process[5]. However, their large-scale use for CO_2 removal applications has not been achieved. Enzymes show promise as a way to enhance the reaction rate of CO_2 into solution, thereby reducing the size of the absorber. However, large scale demonstration of this technology at competitive costs remains to be seen. Despite these advances in solvent development, aqueous amine absorption remains the most competitive technology for CO_2 removal from flue gas, and it continues to be a technology of choice for natural gas treating and refinery scrubbing applications.

Column Internals

Historically, most hot carbonate columns have employed packings, especially random packings. For many years, most amine units have employed 20 trays in the absorbers and 20 trays in the regenerators. In recent years, however, packings have sometimes been employed in amine columns, especially when the lower pressure drops of packings have yielded processing benefits.

Foaming

Acid gas absorbers are known "foamers," i.e., the 2-phase mixtures on the trays or in the packed beds bubble excessively. These foams cause column capacities to be reduced. FRI has demonstrated that packings do not break foams, but the impact of the foaming is less with packings than it is with trays. On trays, foaming can be reduced, somewhat, via short froths, large deck holes and designing nearer the spray regime (as compared to the froth regime)[6] .

With time, aqueous amine solutions degrade and begin to carry particulates. It is generally held that these particulates cause foaming to increase. With operating untis, old amine solutions need to be filtered – and augmented regularly with fresh amine solution.

Equilibrium Stages and Process Simulations

With amine units and hot carbonate units, the absorber trays/packings and the regenerator trays/packings do not need to yield very many equilibrium stages (or "theoretical trays") to achieve the targeted absorption and regeneration separations.

McCabe-Thiele graphical constructions show huge areas between the operating and equilibrium curves. Step-offs between those curves show that a typical absorber only requires 1-3 equilibrium stages; likewise for a typical regenerator. At the very same time, in these mass transfer services, tray and packing efficiencies are low and difficult to calculate because of the chemical reactions and equilibria that are involved. Typical tray efficiencies are around 20%; some are as low as 5%. Because of these unusual circumstances and difficulties, rate-based theoretical calculations are increasingly preferred for unit simulations. For example, the ProTreat computer program of Optimized Gas Treating[7] is very versatile and can be used to calculate/predict column performances without using the concepts of equilibrium stages or tray/packing efficiencies.

Safety Considerations

Amine solutions are very corrosive. They can dissolve metal and burn skin. Acid gas absorbers are operated at very-high pressures, as high as 1200 psi. Pressure relief systems are, of course, required on such units.

In 1984, welding work was performed inside an amine absorber in Illinois. Following that work, Post-Weld Heat Treatment (PWHT) work was not performed. A weld failed due to stress corrosion cracking. The gas that escaped from the operating column ignited. The column ended up landing 1 kilometer away. Seventeen workers were killed. There was approximately $100 million in property damage. All pressure vessels must be properly designed and maintained.

Current Research

Acid gas absorption/regeneration is a very popular research area in 2014. At the November 2013 AIChE meeting in San Francisco, there were 22 sessions with "CO_2" or "carbon dioxide in their titles. Those sessions included 130 presentations! The presenters represented governments, academia and industry.

Many aspects of acid gas treating are now being studied globally. Two examples follow:

1. Dr. David Van Wagener (Ph.D. from the University of Texas with Dr. Gary Rochelle) focused on 2-stage flashing to reduce regeneration energy consumption[8].
2. Oklahoma State University is studying spray columns for the reduction of the capital cost associated with flue gas treating (see **Figure 3**)[9].

There are many other examples of on-going research work.

CLOSING

Acid gas removal from gas streams is not new. For many years, global gas plants have been removing CO_2 and H_2S from natural gas streams, especially using aqueous amines and hot carbonate solutions.

During recent years, acid gas removal has received appreciable fresh attention. Planet Earth is coming out of a relatively recent ice age, and so, it is generally held that Earth is warming. Many people believe that the warming is being exacerbated by human actions, especially the combustions associated with automobiles and power plants. Like the metabolism reaction, the combustion reaction yields CO_2 as a product. CO_2 is regarded as a "medium" strength" greenhouse gas. The reduction of global CO_2 emissions is correctly on many people's minds, these days, and in many scientists' laboratories.

REFERENCES

1. Maddox, R.N., Gas and Liquid Sweetening, 2nd printing, John M. Campbell, Norman, OK, 1977.
2. Humphrey, J.L. and Keller II, G.E., Separation Process Technology, McGraw-Hill, New York, NY, 1997.
3. Seader, J.D., Henly, E.J. and Roper, D.K., Separation Process Principles: Chemical and Biochemical Operations, 3rd Ed., John Wiley & Sons, Inc., Hoboken, NJ, 2011.
4. Figueroa, J.D., Fout, T., Plasynski, S., McIlvried, H., Srivastava, R.D., "Advances in CO_2 Capture Technology – The U.S. Department of Energy's Carbon Sequestration Program," International Journal of Greenhouse Gas Control, 2, 9-20, 2008.
5. Brennecke, J.D. and Gurkan, B.E., "Ionic Liquids for CO_2 Capture and Emission Reduction," The Journal of Physical Chemistry Letters, 1, 3459-3464, 2010.
6. Resetarits, M.R., Navarre, J.L., Monkelbaan, D.R., Hangx, G.W.A., and van den Akker, R.M.A., "Trays Inhibit Foaming", Hydrocarbon Processing, March 1992, p. 61-64.
7. Weiland, R.H. and Hatcher, N.A., Advanced Gas Treating: The Engineering Science, Optimized Gas Treating Inc., Houston, TX, 2011.
8. Van Wagener, D.H., "Stripper Modeling for CO_2 Removal by Monoethanolamine and Piperazine Solvents," PhD thesis, Univ. Texas Austin, 2011.
9. Tamhankar, Y., Whiteley, J.R., Resetarits, M., and Aichele, C.P., "Spray Droplet Characterization Inside a Glass Column Through Dense Wall Flow," Atomization and Sprays, 24, 115-128, 2014.

Table 1
Typical Natural Gas Concentrations

Gas	Feed	Product Specification
CH_4	70-95 %	-
C2-C4 (NGLs)	0-20 %	-
CO_2 (no heating value)	0-20 %	< 2 % for pipeline, < ~100 ppmv for cryogenic
H_2S (lethal)	0-5 %	< 4 ppmv
O_2	0-1 %	0.4 %
N_2	0-5 %	-
Ar, He, Ne, Xe	trace	-
H_2O	Equilibrium	4-7 lb/MMscf

Table 2
Typical Flue Gas Concentrations

Gas	Feed	Product Specification
CO_2	10-15 %	-
O_2	0-5 %	< 10 ppmv
N_2	70-75 %	-
H_2O	Saturated	< 1 %
SOx	~50 ppmv	-
NOx	~25 ppmv	-

Figure 1
Amine Chemistry

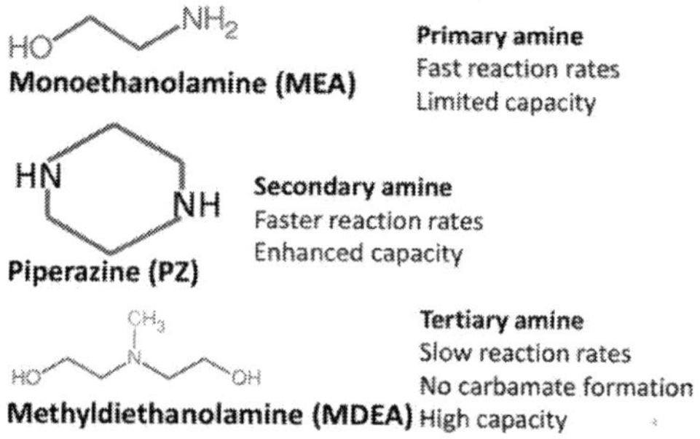

Figure 2
Amine Absorption

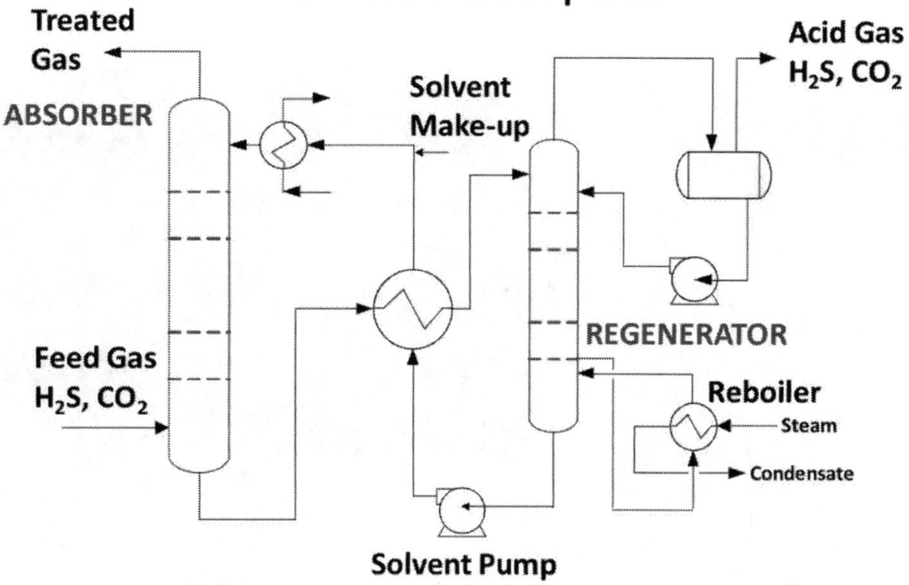

Figure 3
Oklahoma State University
Acid Gas Absorption Apparatus

350235 Radioisotope Applications for Obtaining Detailed Process

Abstract:

In all chemical plants and refineries there generally is not enough instrumentation in the right places in our processes. During the design phase, decisions were made between knowing more about the process, which means buying more instruments, and saving capital funds. When a problem appears with the operation of a vessel, there are always gaps in the knowledge of what is happening due to a lack of instruments in the places where knowledge is not routinely required. Process engineers and operations personnel need to know what is happening so they can take corrective action. When knowledge is missing, they often take actions that make matters worse.

For distillation and separation processes Chemical Engineers have historically relied upon plant process measurements such as flow rates, temperatures, and pressures and model or simulation results for data with which to solve troubleshooting or re-design problems. An array of on-line diagnostic services are available that can provide additional data offering real-time information on how pieces of process equipment are actually operating.

The most common application is Gamma Scanning. This test is primarily applied, but not necessarily limited to, distillation or separation columns. Gamma scans provide a density profile of the internal process of operating distillation columns and other process vessels. The density profile can be used to diagnose the hydraulic operating conditions of mass transfer devices such as damage to internals, flooding, degree of entrainment or weeping, liquid levels on trays and distributors, liquid distribution through packed beds, etc. The presenter will show case studies where scanning revealed vital process information that helped solve an operating problem, or helped make a revamp/re-design successful.

Additional derivatives from gamma scanning of detecting flow distribution patterns will be demonstrated. The CAT-Scan technique is a specialized horizontal gamma scan used to generate a topographic profile of the internal cross-sectional density of process equipment. This profile is useful for the detailed study of liquid flow distribution through packed columns as well as having applications beyond distillation columns. A hybrid application involving tracers and scanning on fixed bed catalytic reactors will also be discussed.

Another technique involving radioactive materials is the Tracer Test, in which a radioactive tracer is introduced into the process for investigating the process flows. External radiation detectors are placed at strategically chosen locations outside vessels and piping to monitor the tracer flow through the process. Either liquid, gas or a solid tracer can be utilized depending on process conditions and the fluid to be studied. This tracer technique can identify liquid or gas maldistribution in fixed or fluidized bed reactors. The residence time through a tank or reactor can be measured and compared with the calculated residence time. Observed differences can relate directly to the degree of mixing or channeling inside the reactor. These tracer techniques have also been used in identifying leakage of heat exchangers and valves.

AIChE Paper Number 150a

PHA Methodology and Training Practices Addressing Auto-Refrigeration Brittle Fracture Hazards – 25 Years Later

Craig R. Thompson
Consulting Engineer
Equistar Chemicals LP
A LyondellBasell company

Michael W. Korst
Principal Engineer
Equistar Chemicals LP
A LyondellBasell company

Prepared for Presentation at the 2014 Spring National Meeting
New Orleans, Louisiana, March 30th-April 3rd, 2014

PHA Methodology and Training Practices Addressing Auto-Refrigeration Brittle Fracture Hazards – 25 Years Later

Craig R. Thompson
Consulting Engineer
Equistar Chemicals LP
A LyondellBasell company

Michael W. Korst
Principal Engineer
Equistar Chemicals LP
A LyondellBasell company

ABSTRACT

Nearly 25 years ago, the Morris, IL Equistar Chemical ethylene plant experienced a brittle fracture failure of a heat exchanger. Subsequent to that incident, the company undertook a program to identify auto-refrigeration brittle fracture (ARBF) failure risks throughout the company's processes and to mitigate those hazards. The company's effort to prevent a repeat of this type of incident also includes a detailed ARBF awareness and response training program, as well as a "Lessons Learned" training program. This paper will present details of these efforts, summarize the focused PHA methodology utilized to identify ARBF scenarios and provide an overview of the continuing efforts to address ARBF risks.

NOMENCLATURE

Auto-refrigeration (AR): The unintentional and uncontrolled change in phase, from liquid to vapor, of a hydrocarbon that results in refrigeration. The resulting low temperatures for certain materials of construction can cause the equipment to become brittle.

Brittle Fracture (BF): Failure (catastrophic crack growth) of carbon steel or low alloy steel equipment which contains a flaw greater than a required critical flaw size when exposed to low temperatures and a stress above a minimum value. Brittle Fracture is a primary concern because failures progress in a break-before-leak fashion rather than the preferred leak-before-break fashion when in a ductile condition.

Independent Protection Layers (IPL): A device, safeguard or action that is intended to prevent or mitigate specific, hazardous events.

Maximum Allowable Working Pressure (MAWP): The maximum gauge pressure adjusted for liquid head for a component in its operating position at the design temperature, based on calculations using the current minimum thickness, exclusive of thickness required for future corrosion allowance and supplemental loads. This pressure value for vessels is normally calculated using ASME Boiler and Pressure Vessel Code, Section VIII, Div. 1 or 2.

Minimum Allowable Temperature (MAT): The minimum permissible lower metal temperature limit for a given material at a thickness based on its resistance to brittle fracture. It may be a single temperature at some rated condition or an envelope of allowable operating temperatures as a function of pressure. The MAT is derived from mechanical design information, materials specifications and/or materials data.

Minimum Design Metal Temperature (MDMT): The lowest acceptable operating temperature for an existing vessel based on material brittle fracture resistance at MAWP. The MDMT concept has been used for ASME Section VIII, Div. 1 since late 1987.

Susceptible Equipment: Equipment susceptible to brittle fracture failure when subject to auto-refrigeration. Susceptible equipment generally contains a LPG under normal operating conditions or has the potential to be exposed to LPG due to process upset (overflow of an upstream vessel).

INTRODUCTION

Brittle fracture failure of equipment and piping attributed to low temperature conditions has serious process safety consequences as evidenced by several industry incidents. One of the more significant auto-refrigeration brittle fracture (ARBF) events occurred in 1989 at the Equistar Chemical Morris Ethylene Plant. Brittle fracture failure of an exchanger in the acetylene converter system resulted in two fatalities, multiple serious injuries and extensive equipment damage. This paper presents an overview of the subsequent efforts pursued by LyondellBasell, the parent company of Equistar Chemicals, to mitigate ARBF hazards via a process involving awareness and response training, hazard analysis, and process modification. This is a process that has evolved, and continues to evolve, over the 25 years passing since the 1989 incident.

DESCRIPTION OF THE 1989 MORRIS AUTO-REFRIGERATION INCIDENT

In September, 1989, the Morris Ethylene Plant was being restarted after an extended shutdown. The unit was more than 24 hours into startup, a process that normally takes 24 to 48 hours. The three main process compressors, the ethylene and propylene refrigeration, and the charge gas compressor were on line and stable. The unit operations personnel had initiated process gas forward flow from the charge gas compressor shortly after the beginning of the night shift on September 11th. Process forward flow had been established into the deethanizer. As the deethanizer became inventoried, the C3 and heavier tower bottoms stream started flowing to the depropanizer. The tower overhead C2 stream flow was initiated through the acetylene conversion preheat and drying system, while bypassing the acetylene converters. This stream was being flared at the outlet of the Acetylene Converter Effluent Knock-Out Drum. See the figure 1 below for a simplified process flow diagram.

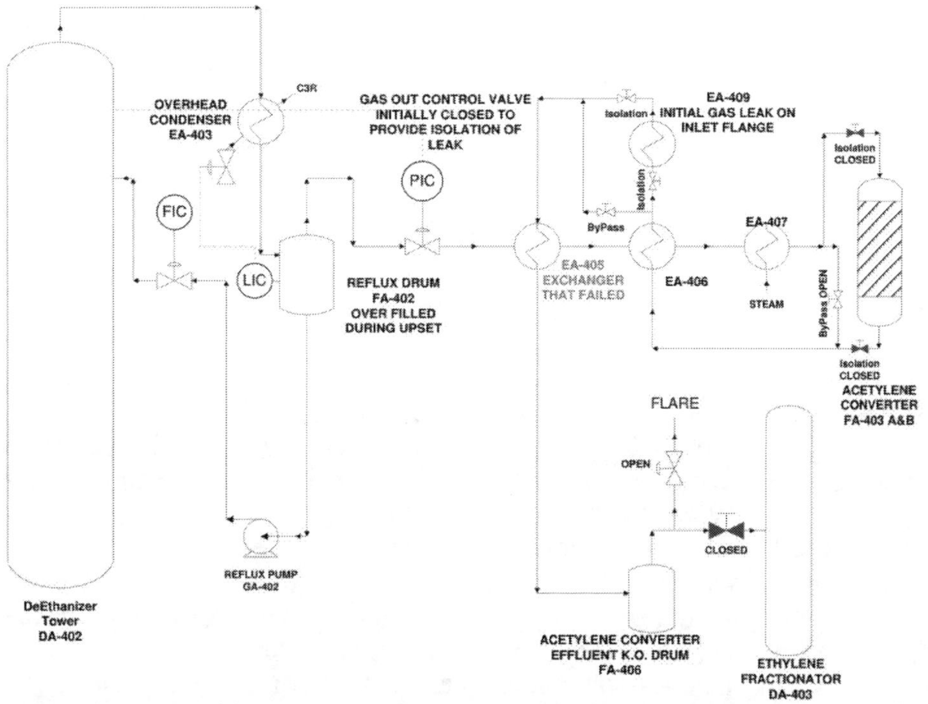

Figure 1 – Morris Olefins Unit Acetylene Converter Train

Deethanizer and Acetylene Converter Event Conditions:

The deethanizer reflux drum overhead control valve was controlling feed to the acetylene conversion system. Pressure on the acetylene conversion preheat system was being controlled by the control valve to the flare. The automatic block valve that allows flow to the ethylene fractionator was in the closed position. At approximately 0100 hours on September 12th, shortly after starting forward flow to the acetylene converter system, a significant process gas leak developed on the inlet flange of EA-409, a cooling water exchanger in the system. Emergency response was activated and fire water spray was initiated on the area of the exchanger flange leak. To stop the leak, as an emergency response action, the deethanizer overhead pressure control valve on the vapor outlet of the reflux drum was closed and the acetylene conversion preheat and dryer system pressure was dropped to flare pressure. Subsequent investigation determined that the deethanizer overhead pressure control valve was not a tight shutoff valve and flow of material continued through the system.

A period of approximately 30 minutes passed from the time the flow to the acetylene conversion preheat system was discontinued until flow was reinitiated. During this time, based on local level gauge indication, the deethanizer reflux drum was determined to have a high level. Due to the changing process conditions, the deethanizer pressure increased rapidly. This resulted in a substantial inventory of overhead process gas condensing in the overhead condenser (EA-403) and overfilling the reflux drum (FA-402). The continued forward flow (now liquid ethylene/ethane) leaking though the closed overhead gas-out control valve auto-refrigerated the downstream acetylene conversion preheat system. It is estimated that temperatures were as low as minus 100°F.

The exchanger which developed the leak was equipped with a bypass and block valves to isolate the exchanger. After the leaking exchanger had been successfully bypassed for repair, Operations attempted to open the control valve on the outlet of the reflux drum to reinitiate flow to the acetylene conversion system. The valve failed to respond as the board operator called for opening from the valve's computer controller. Operations personnel were sent to the valve in the field to investigate why it wasn't opening. Actions by Operations personnel did successfully open the valve after lowering the pressure of the deethanizer tower. The valve did begin to respond and opened rather quickly. As the valve opened, the acetylene converter process equipment pressurized from near flare header pressure to a pressure of 270 psig. As the acetylene converter system pressure increased, the first exchanger in the system downstream of the deethanizer overhead (EA-405) failed in a brittle and explosive manner. The ensuing process gas release ignited instantaneously with the heat exchanger failure. See Figure 2 for EA-405 post fracture incident condition. The subsequent conflagration resulted in two fatalities and seven serious burns to individuals who were in the area. Additionally, the accident caused major damage to the olefins unit and extensive business interruption. The Olefins unit was not restarted until May, 1990.

Heat Exchanger Failure:

The Acetylene Converter Pre-Heater (EA-405) brittle fractured was a result of an auto-refrigeration excursion to conditions less than the vessel's minimum allowable temperature (MAT). This event was the first documented auto-refrigeration – brittle fracture failure in the Olefins industry. The vessel's material-of-construction selection lacked adequate low-temperature ductility. The potential for ductile-to-brittle transition in pressure vessels had only recently been identified as a potential failure mechanism in process unit designs. The use of plain carbon steels in light hydrocarbon services was a common material selection given the Plant's age of construction (late 1960's).

The heat exchanger shell which failed during re-pressurization of the system was fabricated from carbon steel, SA515-70. Charpy V-Notch impact testing on the fragments of the failed shell demonstrated that the ductile-to-brittle transition temperature for the shell material was higher than the design temperature of the exchanger. The original specification did not require a normalizing heat treatment step of the shell plates, subsequent post weld heat treatment (stress relief) or Charpy V-Notch impact test qualification of the materials of construction. The shell was nominally one inch thick and had a Charpy V-Notch impact test value of 15 Ft-Lb. at +58°F. Figure 3 below shows the API-579/ASME FFS-1 Brittle Fracture Analysis of the exchanger made of SA515-70 relative to the process service (ethane and ethylene). It should be noted that upon re-pressurization, the vessel was in excess of 100°F below the Minimum Allowable Temperature (MAT). Post incident forensic examination postulates that the initial de-pressurization and the resulting auto-refrigeration caused a small ½" long weld flaw to grow to an approximately 12" long through-wall crack. This crack was not detected due to the low vessel pressure and vessel being insulated. Upon re-pressurizing, this new crack grew catastrophically in a brittle manner resulting in loss of pressure containment of the shell.

Figure 2 – Brittle Fracture of EA-405 Exchanger Shell

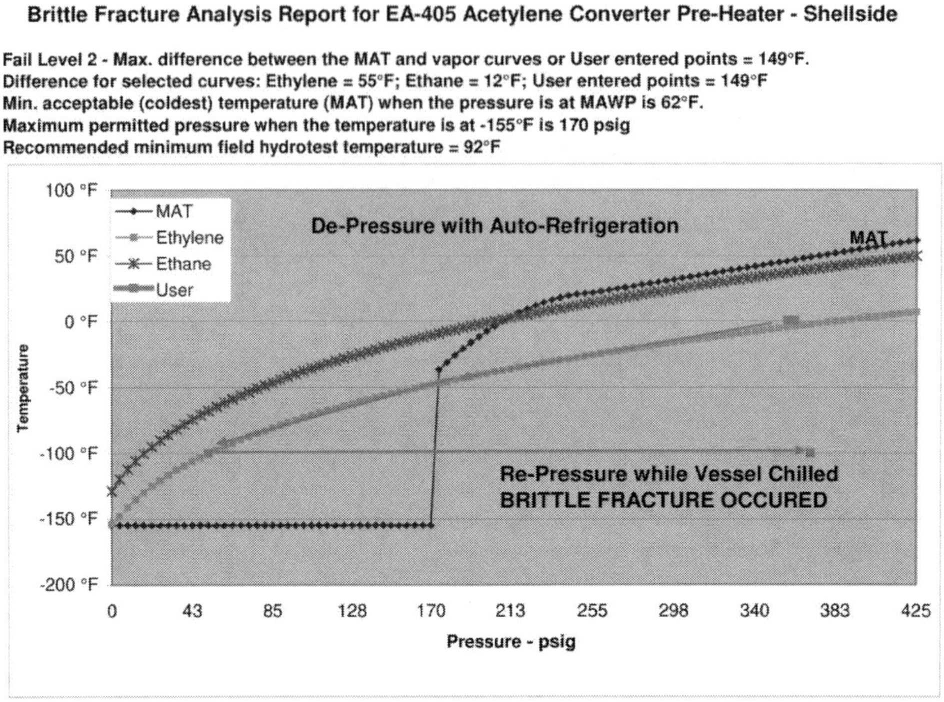

Figure 3 - EA-405 Acetylene Converter Pre-Heater – Shell Side

Post-incident Investigation of Other Equipment Metallurgy:

Following the incident, and with understanding of the cold metal embrittlement mechanism of failure, an evaluation of other process equipment subject to possible auto-refrigeration excursions was undertaken. Emphasis was placed on equipment made of SA515-70 and SA285 Gr. C non-normalized steel (the same as the failed EA-405 exchanger) and also SA516-70 grades carbon steel.

A large number of vessels were replaced, including the deethanizer distillation column, prior to the restart of the unit. The deethanizer overhead process was modified to include flaring capabilities ahead of the acetylene converter system. All deethanizer overhead control valves were design reviewed and replaced with new control valves suitable for potential upset condition temperatures. The Morris Ethylene Plant was successfully restarted in May 1990.

Morris 1989 Auto-Refrigeration Incident Key Findings:

- Recognition of auto-refrigeration and the potential for brittle fracture did not fully exist at the site of the incident or within the Olefins industry prior to this event. Recognition of the hazard did not exist.
- Material-of-construction selection for existing plants does not always fully account for abnormal situations (upsets); particularly auto-refrigeration. Existing vessels are not always designed to be inherently safe under auto-refrigeration conditions or during normal recovery sequence actions while chilled.
- The emergency response to isolate the exchanger flange leak and to de-pressure the system, while potentially causing auto-refrigeration, was the correct response given the potential consequence of the gas release. The subsequent recovery, without the knowledge of the potential for auto-refrigeration, resulted in vessel conditions under the MAT with catastrophic brittle fracture.

AUTO-REFRIGERATION AND BRITTLE FRACTURE

Brittle fracture basics, auto-refrigeration phenomena, and application of API's/ASME's Fitness-For-Service standard (API 579/ASME FFS-1) are well documented within other papers covering this topic (references 1-4). Those unfamiliar with these principles are encouraged to review these references as this fundamental information is not repeated within this paper.

There are two basic fundamental scenarios which can culminate in brittle fracture failure of piping and equipment. The first scenario involves auto-refrigeration in the conventional sense. Uncontrolled depressurization of a system containing a light hydrocarbon liquid (C1s through C4s typically) results in process and equipment chilling as liquid temperature follows equilibrium conditions. Turbulent boiling action of the liquid results in rapid heat transfer and subsequent chilling of the surrounding metal. With system repressurization, heat transfer rate is slow (boiling stops), and consequently both liquid and metal temperatures remain relatively unchanged as system pressure increases. As represented in the following Minimum Allowable Temperature (MAT) curve, stresses increase with increasing pressure until a point is reached in which the metal fails catastrophically. This scenario is classified as a two-step ARBF scenario since it first requires depressurization with associated chilling of vessel contents and therefore equipment. Then, secondly, with equipment in a brittle condition, repressurization provides the stresses necessary to cause equipment failure.

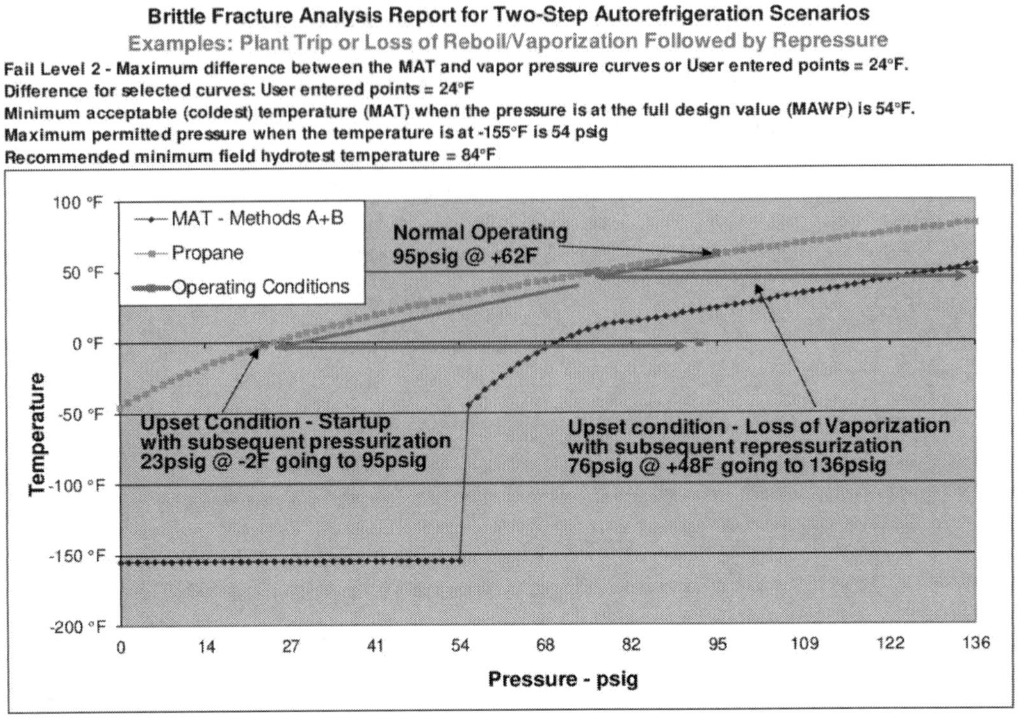

Figure 4 – Two-Step Auto-Refrigeration

The second scenario that can result in brittle fracture equipment failure is not truly an auto-refrigeration phenomenon. While at some sustained elevated pressure, equipment temperature steadily drops losing ductility until a point is reached at which the equipment fails catastrophically. Temperature reduction may be caused by unintentional introduction of a cold liquid, loss of a heat source (Demethanizer for example), ambient chilling of standby equipment, or other possible mechanisms. This scenario is classified as a one-step ARBF scenario since a single change in conditions can result in equipment failure.

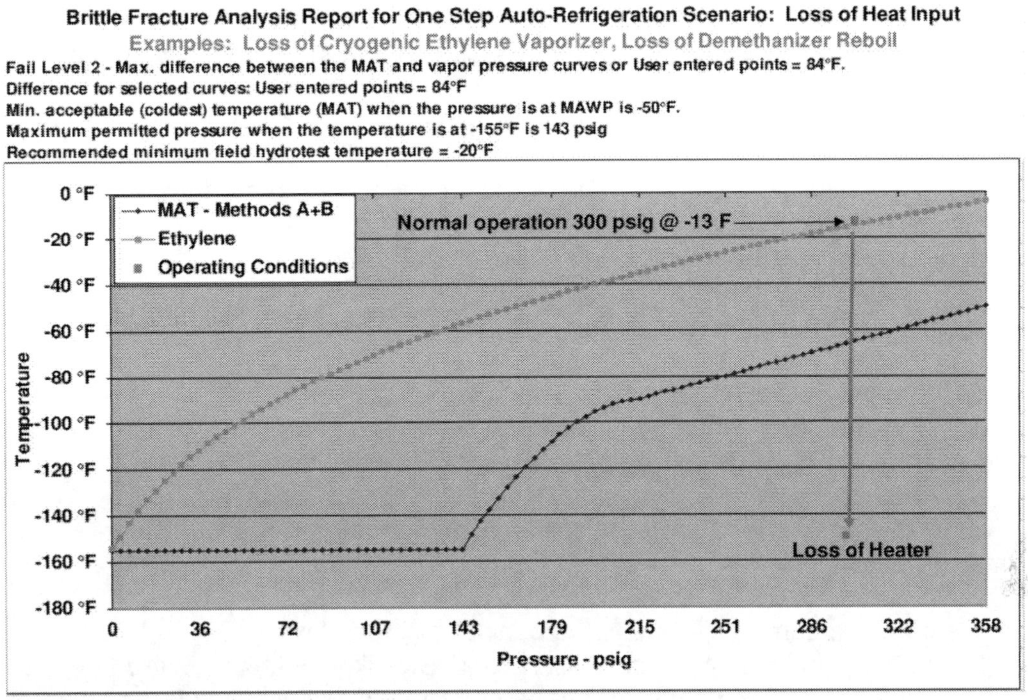

Figure 5 – One-Step Auto-Refrigeration

Ensuring operational and technical personnel understand these scenarios, as well as the fundamentals of brittle fracture failure, is critical to properly training these individuals on the principles of brittle fracture failure hazard recognition/identification, prevention, response and mitigation.

ENGINEERING GUIDELINE DEVELOPMENT

Comprehensive efforts to develop uniform engineering standards and training programs for the purpose of preventing and mitigating auto-refrigeration hazards took some time to materialize. Prior to their development, associated standards were primarily site specific and developed by predecessor companies. Coincident with API 579 Fitness-For-Service standard development efforts (1st edition released January 2000), the company's efforts to develop a corporate standard received renewed emphasis.

A multi-disciplined team was assembled for the purpose of developing engineering standards and training materials encompassing the following specific objectives:

- Develop PHA guidelines for identifying auto-refrigeration susceptible equipment.
- Develop methodology to determine equipment resistance to auto-refrigeration.
- Develop administrative and engineering control guidelines for safe operation of susceptible equipment.
- Develop recovery guidelines for equipment subjected to auto-refrigeration conditions.
- Develop post-incident equipment inspection guidelines.
- Develop an auto-refrigeration training package based on the above elements.
- Generate best practice and engineering documents to prevent auto-refrigeration.

A comprehensive engineering guideline including all of these elements was completed.

Engineering standard application criteria includes the following:

- Equipment constructed of carbon steel and low alloy metallurgy, and
- Equipment containing liquefied petroleum gas (C4 and lighter LPG) at normal operating conditions, or
- Equipment that can be exposed to LPG due to liquid carry-over from upstream or otherwise connected equipment, or
- Equipment operated in standby mode which can be at risk due to cold ambient conditions, or
- Equipment at risk during shutdown or start-up conditions during cold ambient conditions, or
- Other equipment identified by Operational or Technical staff as at risk due to other unique factors.

TRAINING PROGRAM DEVELOPMENT AND IMPLEMENTATION

Following the Morris Ethylene Plant incident, auto-refrigeration training was developed and delivered to all Morris Ethylene Plant personnel prior to the May 1990 restart of the plant. The training included;

- Auto-refrigeration phenomenon basics
  - All personnel trained on the use of Mollier pressure-enthalpy diagrams for pure components
- Brittle fracture and susceptible metallurgies
- Methods to avoid auto-refrigeration
- Review of new safety procedures warning of the potential for auto-refrigeration
- Review of new Standard Operating Procedures containing equipment MDMT's
- Special procedural requirements
  - Procedure with requirement to ensure SA515-70 equipment was at +50°F before pressurizing to 1/3 of MAWP
  - Procedure for "Cold Service Pumps Preparation for Maintenance"
  - Procedure for adding stainless steel bleed valve to carbon steel when deliquifying equipment through tubing to flare

As part of the effort of developing a corporate auto-refrigeration engineering standard, a comprehensive Auto-Refrigeration Training package was created. The Auto-Refrigeration Training package included;

- Review of historic company and industry auto-refrigeration events and near misses
- Auto-refrigeration basics
- Brittle fracture and susceptible metallurgies
- Process mechanisms that cause auto-refrigeration
  - Lowering pressure with liquid LPG present
  - Introduce dry inert gas in intimate contact with LPG
    - Disturbing liquid/vapor interface (bubble through or stir)
    - Increasing surface area
- Methods to avoid auto-refrigeration
- Proper response to auto-refrigeration events
- What not to do when confronted with an auto-refrigeration event

The last section of the training, "what not to do when confronted with an auto-refrigeration event", was a significant addition and enhancement to the company's auto-refrigeration training and knowledge. It established clear and concise course of action which includes;

- **STOP!** Compose - Evaluate.
  - Take actions **SLOWLY** (order of 10's of minutes).
  - Take re-warming actions very slowly (order of hours).

- **STOP** source of auto-refrigeration.
  - Determine the cause of auto-refrigeration.
    - De-pressure with LPG present.
    - Leaking valve.
    - "Dry" Inert Gas in intimate contact with LPG.
    - LPG material in the wrong place.
    - Uncontrolled process swing (like loss of reboil on column).
  - Take action to stop the cause or source of auto-refrigeration.
    - De-inventory liquid LPG - DO NOT INCREASE PRESSURE.
    - Close leaking valve or isolate upstream.
    - Stop source of "Dry" Inert Gas - Purge with warm LPG vapor if available.
    - Stop source of LPG - de-inventory liquid LPG - DO NOT INCREASE PRESSURE.
    - Reestablish heat input - DO NOT INCREASE PRESSURE.

- Where the Minimum Design Metal Temperature (MDMT) or Minimum Allowable Temperature (MAT) curves exist, check and determine if the temperature is below the safe operating range for the vessel.
- Determine if the vessel pressure is < 40% of Maximum Allowable Working Pressure (if vessel was built before 1998) or <33% (if vessel was built after 1998).

  IF NOT: DROP PRESSURE TO <40% (OR <33%) OF MAWP.

- De-inventory the equipment of liquid LPG if possible or wait until all LPG vaporizes. Begin the re-warming process by purging with a warm LPG vapor (gas). If no vapor is available, allow ambient conditions to re-warm the vessel. **Re-warm very slowly with temperature increase not to exceed 50ºF/hr.**
- Initiate required actions per Post Auto-Refrigeration Incident Inspection Guidelines. The guidelines establish post incident inspection requirements that are dependent upon both the rate of auto-refrigeration cooling (gradual, rapid, or fast) and the rate of vessel temperature recovery (gradual, rapid, or fast). In most cases, a full mechanical integrity inspection and evaluation will be required before the equipment can be placed back into service.

Auto-refrigeration training, using the comprehensive package, was delivered following development. Refresher auto-refrigeration training is now routinely included at all LyondellBasell ethylene plants as part of pre-turnaround training.

AUTO-REFRIGERATION PROCESS HAZARD ANALYSIS

Although a detailed PHA was completed following the 1989 Morris Plant incident including focus on auto-refrigeration hazards, with knowledge gained since that event, a need was recognized for a much more comprehensive and structured evaluation of ARBF hazards.

Focused PHA Guidelines

A critical element of the company's engineering standard is the Auto-Refrigeration Process Hazard Analysis Guidance document and associated procedures. The objective of this section of the standard is to provide guidelines for identifying, evaluating and mitigating potential risk of process equipment exposure to auto-refrigeration and brittle-fracture events. This guidance document was developed with several goals in mind which include:

- Providing a consistent basis for PHA Teams to identify potential process equipment auto-refrigeration and brittle-fracture scenarios.
- Providing a consistent basis for PHA Teams to assess whether process equipment is potentially at risk of brittle-fracture failure for the worst-case scenario identified by the Team.
- Providing a consistent format for documenting PHA Team findings and recommendations.

Considering processes are particularly susceptible to auto-refrigeration hazards during non-routine operation, the PHA guidance needed to address all modes of operation. The modes of operation addressed within the PHA Guidance document include:

- Normal Operation
- Upset Conditions
- Normal Startup
- Normal Shutdown
- Inventory
- De-inventory
- Emergency Shutdown
- Air Freeing / Nitrogen Freeing
- Not in Operation / Stand-by / Maintenance-in-Progress
- Commissioning / Leak Testing

The Auto-Refrigeration PHA Guidance document includes specific questions addressing each of these operating modes. The guidance document (or Guide-List) is completed for each piece of susceptible covering all of the modes of operation listed above. Included within the document is a discussion of observed and potential auto-refrigeration scenarios with the expectation that knowledgeable individuals reviewing the Guide-list who did not participate in the PHA come away with a clear understanding of the rational applied and the basis and justification for the contained recommendations. For each ARBF scenario identified, a frequency analysis is developed and documented, often supported by an event tree document. Typically, the Guide-List is completed while referencing a Minimum Allowable Temperature (MAT) curve developed specifically for the piece of equipment under evaluation. This curve represents the vessel's minimum allowable temperature for all pressure conditions. Process conditions are also represented on the MAT curve and the curve is used to document potential auto-refrigeration scenarios. For a complete review of MAT curve development and use, see references 1 through 4.

The PHA and Project Scope Development Process

There are several factors important to successful implementation of a process hazard analysis focused on auto-refrigeration hazards. Some of these factors are summarized below:

Management Support – Before undertaking this program, both Corporate and Site management needs to appreciate the commitment in resources and funding necessary for this comprehensive effort to identify and mitigate ARBF hazards. Preliminary resource, time and funding estimates were generated prior to initiating the program. The AR focused PHA process was initially piloted at a selected plant, resource and cost estimates updated and communicated to management. Central engineering staffing and resource commitment plans were developed to support this timetable.

PHA Team Membership – As with any PHA, the qualifications and experience of the PHA team members directly impacts the adequacy of the analysis. The mechanical lead is an ARBF subject matter expert (SME) knowledgeable with API 579/ASME FFS-1 content and governing fundamentals. A limited number of mechanical and process engineers trained in ARBF principles and mitigation methods served as technical SMEs. In addition to providing the needed expertize, this helped to maintain some level of consistency in the analysis, mitigation option assessment and documentation. Other core team members include site process engineers, site mechanical/inspection engineers and site operations specialists. All members of the PHA team are trained in ARBF principles. An experienced operational SME fully versed in all modes of plant operations, all operational systems and operational procedures (written AND practiced) is important. This individual is capable of, and responsible for, developing and communicating methods and procedure changes to the Unit shift organization.

MAT Curve – The equipment's Minimum Allowable Temperature (MAT) curve captures all critical equipment parameters defining susceptibility to brittle fracture over the range of possible operating conditions. The graphical representation, including process parameters, supports both the analysis process and mitigation option identification effort. When incorporated into a report or procedure it also becomes a very effective tool for communicating the hazard scenarios, findings and a basis for mitigation recommendations.

Historian – Included within the PHA Guide List is a requirement to review historical conditions to identify low temperature excursions and assist in the identification of operational scenarios creating ARBF risks. The unit historian trending feature is used to quickly screen process temperatures and pressures to identify abnormal excursions. Selected incidents are then examined in further detail to develop an understanding of the causal factors. In the majority of cases the event cause is readily understood and comes as no surprise. However, in a number of instances, a scenario has been uncovered that was not previously understood and required extended analysis to fully understand the initiating events. Additionally, the historian becomes a useful tool used to identify situations where instrumentation ranges are inadequate preventing the determination of actual minimum temperature excursions.

Process Simulation – Simulating process conditions under upset, start-up, shut-down as well as normal operations provides valuable information supporting the assessment. MAT curves are set-up providing the capability of selecting and plotting one or more pure component saturation curves along-side equipment MAT curves supporting scenario analysis.

Mitigation Scope Development – The properly assembled PHA team, in addition to having the necessary expertizes to perform the ARBF risk analysis, is also best positioned to identify, evaluate and recommend associated hazard mitigation options. Rather than merely identifying the hazard scenario, the PHA team provides specific scope recommendations for mitigating hazards in compliance with corporate risk standards. Options are developed, the selected mitigation scope defined, and critical design parameters specified to the extent possible without performing detailed engineering. A layers of protection analysis (LOPA) is performed. Both interim and long term solutions are developed and analyzed. All of this is documented within the auto-refrigeration PHA guide list. Although extending the PHA duration, this significantly expedites mitigation efforts and improves the efficiency of the project development process.

AUTO-REFRIGERATION RESPONSE AND RECOVERY GUIDELINES

Critical components of the auto-refrigeration engineering guideline are the auto-refrigeration incident recovery and equipment inspection guidelines. Incident response and recovery has been addressed previously within this paper as part of the training program description. The guideline also includes detailed direction defining the specific type of equipment inspection that must occur as a function of the conditions to which the equipment was subjected. Variables impacting inspection requirements include the magnitude and rate of temperature change, the peak pressure relative to MAWP, as well as the warming media fluid state and temperature. Resulting inspection requirements range from simple leak checks to comprehensive vessel inspection.

AUTO-REFRIGERATION BRITTLE FRACTURE HAZARD MITIGATION

Mitigation Approaches

Most companies follow a corporate risk assessment standard and procedure. A number of factors are taken into consideration when developing and accessing risk mitigation alternatives. The alternative selected depends in part on whether associated independent protection layers (IPLs) are determined adequate and valid while insuring operational reliability and performance are not compromised.

Operator Intervention/Alarms – As with other hazards, operator intervention in response to an alarm is a frequently used IPL which at least partially mitigates the hazards. Auto-refrigeration scenarios are often associated with abnormal operating conditions in which board operators must respond to a high volume of alarms. Consequently, alarms used to mitigate ARBF hazards are annunciated via an independent critical alarm system. Via this system, excursion response instructions can be rapidly accessed providing operator guidance. Additionally, current and recent operating conditions are presented graphically on the affected equipment's MAT curve enabling the operator to track the process response relative equipment limitations as a result of the process upset and mitigation actions.

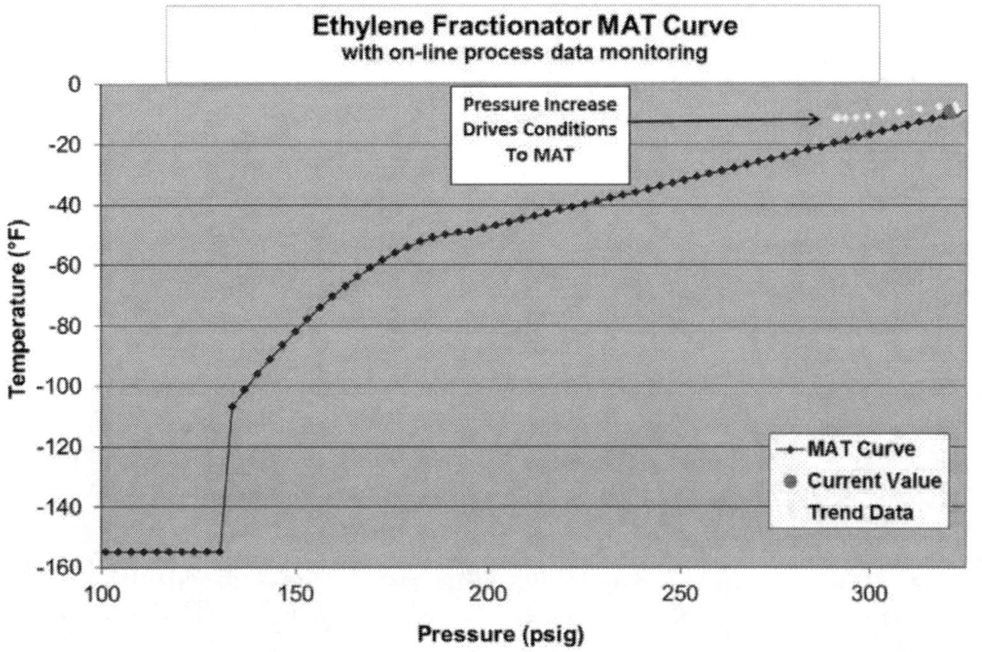

Figure 6 – Control Room MAT Monitoring

<u>Interlocks and Process Overrides</u> can effectively mitigate many ARBF scenarios which result from depressurization/repressurization cycles. In these cases, hazardous conditions are avoided if system pressure is maintained which can be accomplished via an isolation or "bottle-up" interlock. Interlock initiation may involve detection of reduced pressure, low temperature, MAT curve approach margin, forward flow loss, compressor shutdown or a combination of these measurements. Interlock initiator redundancy, i.e., selecting the two highest or two lowest transmitter signals from a group of three transmitters (2oo3 voting) to activate an interlock, is important for both reliable interlock action during hazardous conditions as well as prevention of false trips. In addition to redundant instrumentation, dual coil solenoids, and short duration time delays (0.5 to 2 seconds) are commonly incorporated into the interlock design to improve reliability. When including time delays, it is particularly important to analyze scenario dynamics to insure risk mitigation actions are not compromised. This may necessitate an adjustment in the interlock trip set-point.

Brittle Fracture Analysis Report for Reflux Drum

Example of delayed interlock response: Reboiler condensate isolation interlock

Fail Level 2 - Max. difference between the MAT and vapor pressure curves or User entered points = 8°F.
Difference for selected curves: User entered points = 8°F
Min. acceptable (coldest) temperature (MAT) when the pressure is at MAWP is 4°F.
Maximum permitted pressure when the temperature is at -155°F is 162 psig
Recommended minimum field hydrotest temperature = 52°F

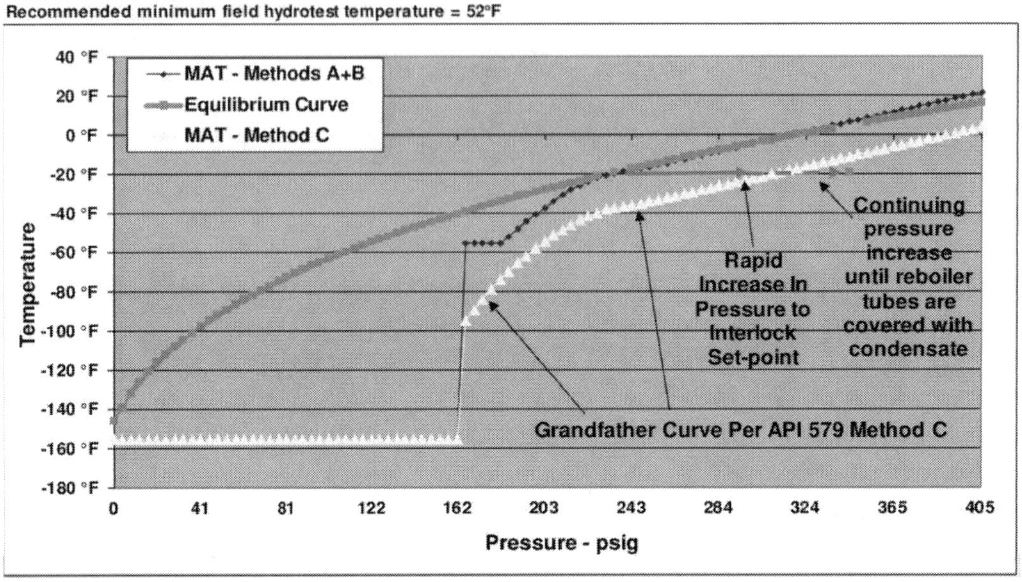

Figure 7 – Mitigation Via Interlock Example

Although prevention of the hazardous condition is the preferred course of action, this is not always possible. If depressurization has created hazardous conditions, catastrophic equipment failure risk can be mitigated by preventing system repressurization until system temperatures recover. Interlock action involves isolation of all energy sources that can result in an unacceptable rate of repressurization. Action to reset the interlock can be based on system temperature, a combination of temperature and pressure, or deviation from MAT temperature. A compressor system restart permissive can be effectively used where applicable.

Brittle Fracture Analysis Report for Compressor Discharge Drum

Example of use of compressor start-up permissive based on limiting vessel temperature

Fail Level 2 - Maximum difference between the MAT and vapor pressure curves or User entered points = 10°F.
Difference for selected curves: User entered points = 10°F
Minimum acceptable (coldest) temperature (MAT) when the pressure is at the full design value (MAWP) is 44°F.
Maximum permitted pressure when the temperature is at -155°F is 70 psig
Recommended minimum field hydrotest temperature = 74°F

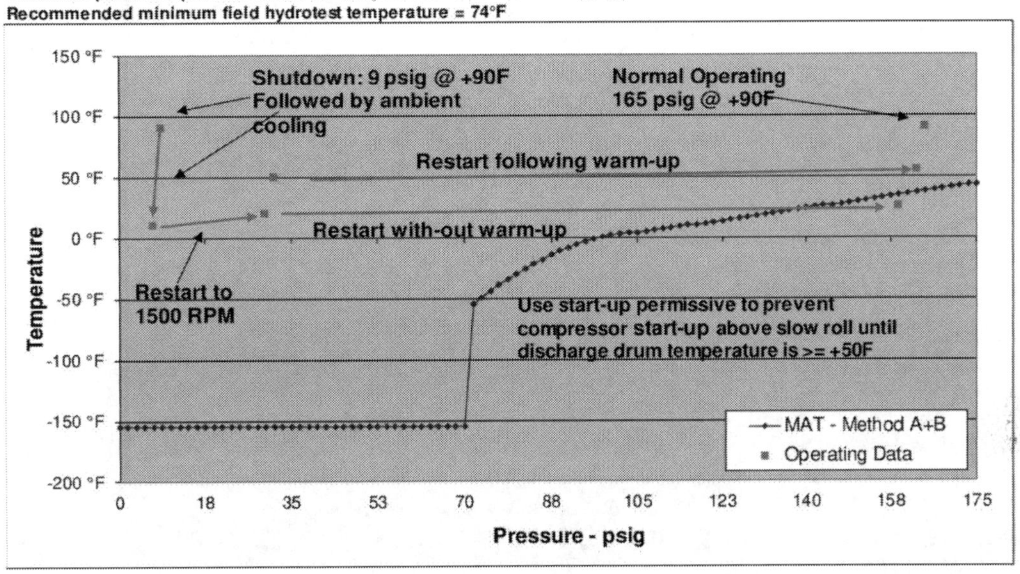

Figure 8 – Compressor Start-up Permissive Example

Of course when designing any interlock it is important to fully analyze interlock response under all possible operating scenarios. It is particularly important to evaluate the actions needed to clear interlock conditions and methods engaged to reset the interlock. Ideally the need for a manual bypass is avoided but this is not always possible. Risks associated with manual bypasses should be minimized by implementing auto-reset functionality when viable, including limited time delays or conditional resets. Operational sustainability under normal process variations, start-up conditions and shutdown conditions requires detailed review to assess the viability of the mitigation method. Often, modifications in start-up and shutdown procedures are required to avoid interlock initiation under those conditions.

Equipment Modification – Equipment modification may prove to be a more cost effective mitigation solution than a solution involving interlocks. The process of developing an equipment's MAT curve identifies the minimum allowable temperature associated with each vessel component. In some cases the equipment's MAT can be substantially reduced by addressing the limitation created by a single vessel component. Solutions may involve man-way blind replacement, individual nozzle replacement, individual component post weld heat treating (PWHT) or post weld heat treating the entire vessel.

Vessel Component MAT As Is

Component Description	Material	Governing Thickness or Bolt D.	PWHT?	MAT from Curve	Limiting Comp. at MAWP?	Field PWHT MAT
MH-1 - 18" Top Head at nozzle			No	27 °F		-3 °F
MH-1 - 18" nozzle neck	SA-181 Gr. II	1.500 in.	No	51 °F	Yes	21 °F
MH-1 - 18" flange	SA-181 Gr. I	NA	No			-50 °F
Bottom Head	SA-516-70	0.938 in.	No	27 °F		-3 °F
B-02 - 10" Bottom Head at nozzle	SA-516-70	0.938 in.	No	27 °F		-3 °F
B-02 - 10" nozzle neck	SA-106-B	0.593 in				-28 °F
B-02 - 10" repad	SA-516-70	0.938 in				-3 °F
B-02 - 10" flange	SA-181 Gr. I	NA	No			-50 °F

Limiting component MAT

MAT if PWHT

IMPACT OF REPLACED COMPONENT

Component Description	Material	Governing Thickness or Bolt D.	PWHT?	MAT from Curve	Limiting Comp. at MAWP?	Field PWHT MAT
MH-1 - 18" Top Head at nozzle	SA-516-7			°F		-3 °F
MH-1 - 18" nozzle neck	SA-350-LF1	1.500 in.	Yes			-30 °F
MH-1 - 18" flange	SA-181 Gr. I	NA	Yes			-50 °F
Bottom Head	SA-516-70	0.938 in.	Yes	27 °F	Yes	-3 °F
B-02 - 10" Bottom Head at nozzle	SA-516-70	0.938 in.	Yes	27 °F	Yes	-3 °F
B-02 - 10" nozzle neck				°F		-28 °F
B-02 - 10" repad				7 °F	Yes	-3 °F
B-02 - 10" flange	SA-181 Gr. I	NA	Yes			-50 °F

New Component MAT

These components now limit MAT and establish vessel MAT

Figure 9 – Vessel Component PWHT and Replacement

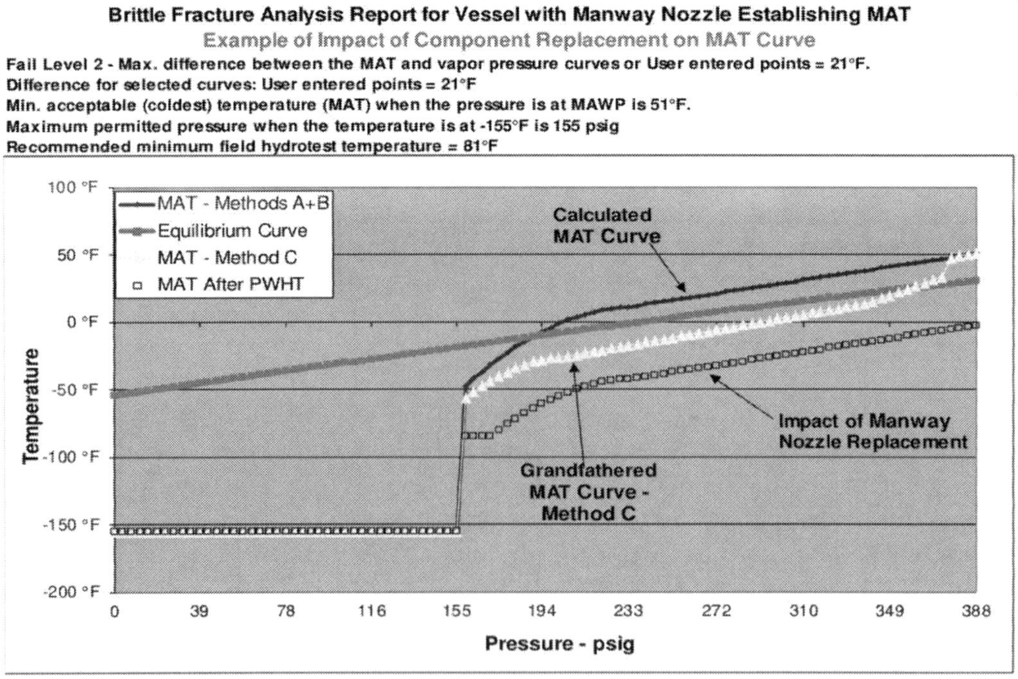

Figure 10 – MAT Curve – Vessel Component Replacement Impact

Equipment Replacement – Sometimes the only viable mitigation option involves equipment replacement. This may either driven by the lack of an alternative method which adequately mitigates the hazard or due to operability considerations associated with other possible solutions. New equipment minimum design metal temperature (MDMT) should be specified equal to, or less than, the equilibrium temperature of the vessel liquid contents at atmospheric pressure. Liquid composition during upset conditions needs to be taken into consideration. In the case of dryers, as well as catalyst containing reactors, when purged with dry gases the minimum temperature can drop substantially below the atmospheric equilibrium temperature. This should be taken into consideration when defining the MDMT of the equipment.

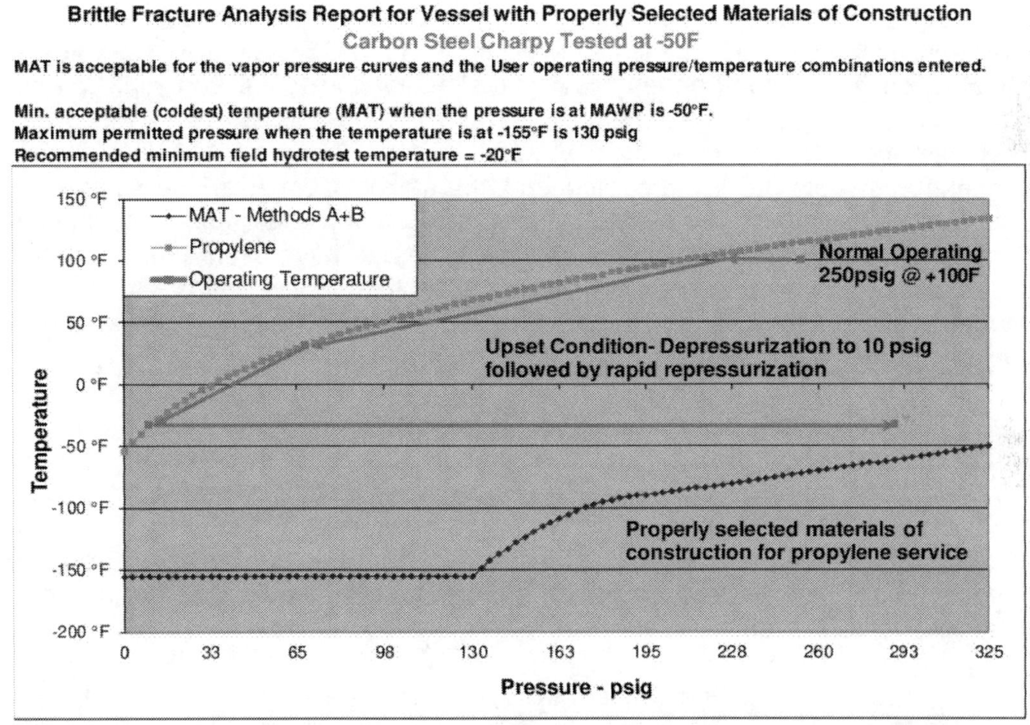

Figure 11 – Vessel With Properly Selected Materials of Construction

UNIQUE OR CHALLENGING AUTO-REFRIGERATION BRITTLE FRACTURE SCENARIOS

Identification and discussion of many ARBF scenarios can be found in several of the references listed at the end of this paper. Rather than repeat that information here, the following includes a few of the more unique or challenging scenarios. All of the following scenarios were either not identified or else inadequately mitigated via original ARBF PHA recommendations and thus required additional scope implementation to properly mitigate. As a colleague of ours is fond of say: "we reserve the right to get smarter".

Dryer Applications

Equipment containing molecular sieve or other porous media pose unique problems. Following equipment deliquification, up to 25% of the bed volume contains "sponge" liquid trapped within the sieve by capillary action. Unassisted, it takes a very long period of time for this trapped liquid to dissipate. Initiation of inert gas flow through the bed results in very low bed and equipment temperatures since the liquid component's partial pressure near the surface of the liquid is extremely low. Temperatures approaching the liquid's equilibrium temperature near vacuum conditions are possible and have been demonstrated. Note that the definition of an 'inert gas" is relative and is not limited to nitrogen or methane. For example, in propylene dryers, nitrogen, methane, ethane and ethylene purge gases behave as inert gases.

The minimum temperature expected contacting an inert gas with a liquid hydrocarbon can be estimated based on the liquid's equilibrium temperature at 2.5 psia. Although lower temperatures are possible, a point is reached in which the sensible heat of the inert gas flow stream prevents any further reduction in temperature. As the inert gas flow is increased, the temperature begins to rise. Expected minimum temperature for various hydrocarbon liquids are summarized within the following table.

Table 1 – Inert Gas Impact on Temperatures

Equilibrium Temperatures, DegF			
Component	**@ 14.7 Psia**	**@ 2.5 psia**	**@ 1 psia**
Methane	-259	-290	-303
Ethylene	-155	-200	-218
Ethane	-127	-176	-196
Propylene	-54	-112	-136
Propane	-44	-104	-128
I-Butane	11	-57	-85
Butene	21	-48	-75
N-Butane	31	-38	-66

Although there is risk of catastrophic brittle fracture failure when repressuring a chilled dryer vessel, this would only occur if the dryer's pressure increased sufficiently prior to regeneration. This certainly is possible and could occur via a valve misalignment error, due to a leaking valve, or due to a problem with the on-line dryer requiring necessitating an unplanned dryer swap. However, the more probable risk associated with this scenario is vessel cracking due to secondary stresses. The consequence of this specific failure mechanism is a leak rather than catastrophic equipment failure. The consequence and probability of both scenarios should be assessed to define appropriate hazard mitigation methods.

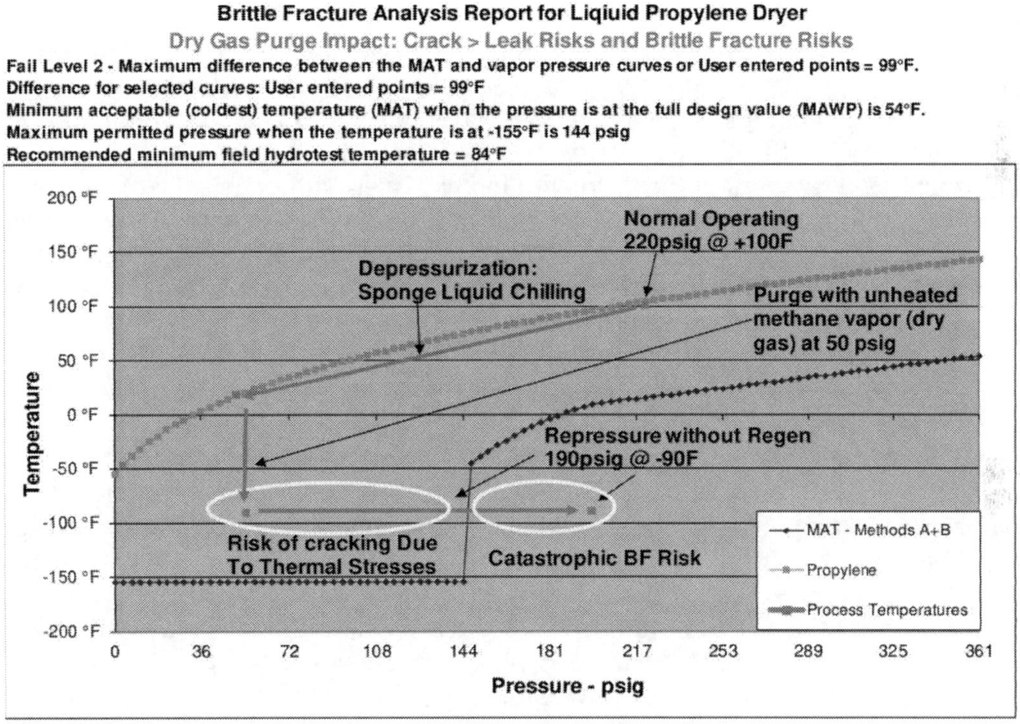

Figure 12 – Dryer Example - Dry Gas Purge Impact

Ideally, clearing the vessel and sieve of free liquid and "sponge" liquid is accomplished via purge with a hot regeneration gas of similar composition to the application fluid. In the case of the propylene dryer, either hot propylene or propane vapor is the best choice. However, this is not always feasible necessitating development of alternative approaches. Removing all free liquid is a critical first step. Dependent on fluid, environment and application, the liquid's own vapor pressure may be used to push liquid out. Application of vessel heating in a controlled manner, i.e., heat tracing/jacketing, is a possible approach to expedite liquid removal, however, there are specific precautions that must be taken with this approach. A qualified engineer must review the specific method/design for applying heat to the vessel to verify acceptability and, in particular, to verify that the temperature differentials created don't create unacceptable stresses. Additionally, an acceptable pressure control methodology (not dependent on manual adjustments or relief valve actuation) must be implemented. If free liquid is pushed out with an inert gas, the flow rate of the gas must be controlled and blow-through prevented. A means of DCS temperature monitoring and alarming should be provided.

Once free liquid is removed, the alternative methods described above will be ineffective in removing "sponge" liquid. Use of an inert gas to remove "sponge" liquid requires sufficient heat and volume to prevent low temperature excursions of unacceptable magnitude. Prior to initiating inert gas flow through the dryer, the gas stream must be heated requiring the stream to be diverted upstream of the vessel until targeted temperature is reached. The hot purge stream through the vessel must then be introduced at a rate and temperature adequate to offset the chilling which occurs as trapped liquid vaporizes. Providing DCS monitored temperature instrumentation with alarming enables verification of acceptable purge conditions. The sieve or catalyst supplier should be consulted when developing these procedures. If temperature drops below acceptable limits or differentials, at minimum, vessel leak checks should be performed prior to reapplying pressure to the equipment.

Distillation Towers

Both one step and two step ARBF scenarios can be associated with distillation tower applications. The one step scenario can be particularly challenging to adequately mitigate without major equipment modification and/or replacement. One step scenarios specifically associated with ethylene plant demethanizers have been the subject of prior publications (reference 1 and 7). Considering the significant temperature differential present in demethanizer towers, the presence of this hazard is not particularly surprising. However, this hazard can also be present in towers operating with much smaller differential temperatures such as ethylene fractionators. In the case of a typical high pressure splitter, the upper tower, lower tower, reboiler tube-side components and reboiler shell-side components may all have different minimum design temperatures. The upper tower section will be typically be constructed from normalized carbon steel specified and certified (Charpy tested) for a MDMT of -50°F. The lower tower section and reboilers may have been specified with MDMTs of -20°F and consequently, particularly in the case of pre-1990 plants, actual MDMT may be much higher. The authors are aware of actual MDMTs as high as +47°F associated with this equipment

Although grandfathering equipment per procedures defined within API 579/ASME FFS-1 may produce a grandfathered minimum allowable temperature (GMAT) curve deemed acceptable for continued operation, inevitably by the nature of the grandfathering process, the delta between the operating temperature curve and the GMAT curve is relatively small. This necessitates reliance on energy source (feed and reboiler heat media as well as reflux) isolation interlocks to adequately mitigate ARBF hazards. Avoiding interlock trips can create start-up challenges and will likely necessitate start-up procedure modifications. Interlock activation during significant process upsets can be expected. Energy source isolation of reboilers using a condensing heat media such as steam or propylene vapor may have a delayed response that must be taken into consideration. If the reboiler's condensate outlet control valve (or separate trip valve) is used for this purpose, process vaporization does not cease until the reboiler's tube area is fully covered with condensate. This may represent an energy source isolation delay of several minutes.

Interlocks may provide inadequate protection for reboilers. In the case of a C2 splitter reboiler using propylene refrigerant vapor as the heat media, dependent on system design, continuing reboiler heat input following a propylene refrigeration compressor trip may be sufficient to prevent conditions from crossing the MAT curve. Heat input may be sustained until compression system pressures equalize. On the other hand, failure of inlet or outlet heat media control valves reduces the duration heat input is sustained compared to a compressor trip conditions. In either event, consideration must be given to the resulting temperature of the tower bottoms liquid inventory as the colder tray liquid drops to the bottom of the tower following loss of reboil. Also recognize that changes in operating conditions and rates can shift tower operation creating a colder tower profile than experienced previously. For example, as other plant constraints are addressed, or feedstock is increasingly shifted toward ethane, the C2 splitter may become the predominant plant constraint. Increasing allowable ethylene content in ethane recycle in order to maximize ethylene production shifts the column temperature profile down the tower. Loss of reboil then results in what can be a significantly lower tower sump temperature not previously experienced. Lack of equipment failure in the past does not insure failure will not occur in the future.

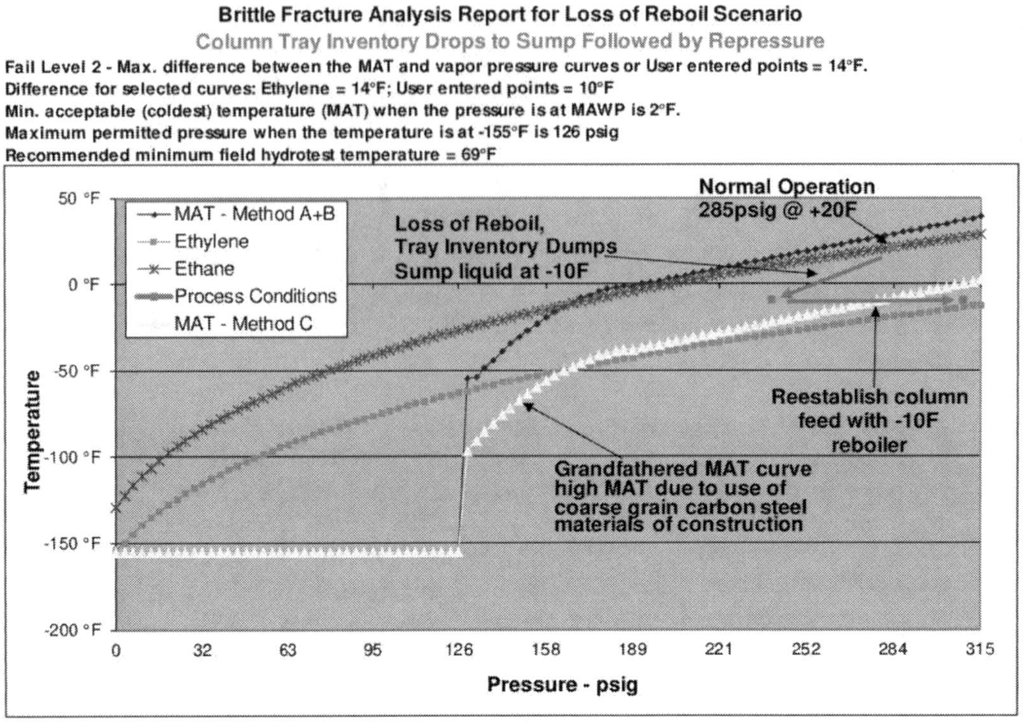

Figure 13 – Distillation Column Reboil Loss

Overpressure Conditions

Attributable to the significant safety margins included in vessel design standards, catastrophic vessel failure is not expected even with overpressure magnitudes up to 300% or more of MAWP if vessel integrity has not been compromised (reference 11). However, this is not true if brittle fracture failure risks are present. API 579/ASME FFS-1 defines the procedure for extending a vessel's MAT curve beyond its MAWP. The vessel's MAT continues to increase as pressure increases above MAWP. Compression systems in particular are at risk of overpressure well in excess of MAWP in the event of a check valve failure. Risk of overpressure in excess of 300% of MAWP has been identified with some compression systems (reference 12). However, it doesn't necessarily require a significant overpressure magnitude to create a brittle fracture failure risk. First stage suction equipment within an Ethylene Plant's ethylene refrigeration system is a case in point. Often, this equipment has been constructed from 3-1/2% nickel alloy steel (SA 203). Although this material has a MAT of -150°F at MAWP, the MAT increases to -144°F at 110% of MAWP and -138°F at 120% of MAWP. With this equipment normally running below -150°F (ethylene saturation temperature at 3 psig), following compressor trip the equipment rapidly pressures up while liquid and metal temperatures remain at pre-trip operating temperature. If required relief loading exceeds relief system capacity either due to design inadequacies or check valve failure, vessel conditions can be driven well across the MAT curve as illustrated below.

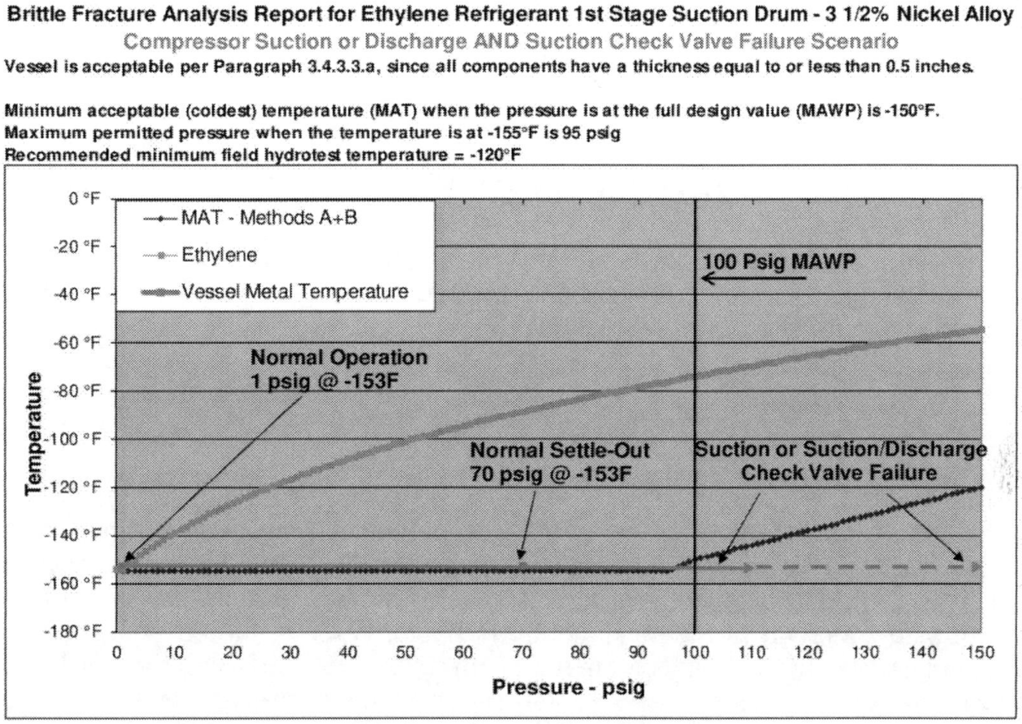

Figure 14 – Compressor Suction Drum Overpressure

Surplus Equipment

Awareness of all aspects of the API 579/ASME FFS-1 Fitness-For-Service standard is certainly warranted when evaluating internal or external surplus equipment for alternate applications. API 579/ASME FFS-1 Part 3 addressing assessment of brittle fracture of existing equipment is no exception. Older equipment predating development of industry knowledge captured in API 579/ASME FFS-1 may have a U1 form certifying a minimum design temperature of -20°F, however, in reality the MAT based on API 579/ASME FFS-1 assessment methods may be much higher, possibly well in excess of +50°F. A simple material of construction review can quickly identify equipment unsuitable for low temperature (process or ambient) service. Coarse grain carbon steel as well as steel which has not been heat treated/normalized will have elevated MATs with magnitude a function of the specific material type/grade and thickness. When considering surplus equipment for C2, C3 or C4 liquid service, or even operation in cold ambient conditions, equipment with material lists including SA-105, SA-181, SA-212, SA-283, SA-285 or SA-515 should be of particular concern. Although hydro-testing equipment per procedures defined within API 579/ASME

FFS-1 can reduce MAT (Method 2C); there is the potential for equipment failure during the hydro-test. Additionally, the MAT improvement may be insufficient to provide an adequate margin between operating temperatures and MAT, particularly when taking into consideration shutdown, start-up, and process upset conditions.

CONTINUOUS IMPROVEMENT

Auto-refrigeration knowledge and hazard recognition, incident and near-miss reporting, incident prevention, incident response, and post-incident inspection have been a continual evolution since the Morris Ethylene Plant incident.

Focused Auto-Refrigeration Brittle Fracture PHAs

Auto-refrigeration focused PHAs have been completed at facilities where light hydrocarbons are processed and the potential for auto-refrigeration with brittle fracture exists. The ARBF PHAs generated MAT curves for all susceptible equipment. The curves are available and used whenever there is a question about safe operation of the equipment, or should there be an auto-refrigeration incident or near miss. After completion of the focused ARBF PHAs, the identified ARBF scenarios are subsequently merged with the operating unit PHAs and revisited during PHA revalidation. During PHA revalidation, the PHA team continues to maintain focus on ARBF scenarios taking into consideration additional scenarios that potentially may not have been recognized during the original ARBF focused PHA. ARBF hazard identification and mitigation continues.

New Employee Training

Auto-refrigeration training is included in appropriate new-hire training curriculums.

Turnaround Training and Startup Monitoring

Auto-refrigeration refresher training is included in pre-turnaround training. Additionally, the Morris Ethylene Plant has developed real-time MAT graphics for all vessels protected by auto-refrigeration SIS interlocks. These graphics are displayed and monitored during Ethylene Plant startups to ensure that no vessels approach their MAT (typical for SIS to alarm at the MAT +10 °F and interlock activation at the MAT +5°F).

Lessons Learned Library

LyondellBasell has an incident and lessons learned sharing process called "Learning From Incidents" (LFI). One facet of the LFI process includes a "Lessons Learned Library", which contains information on significant historical events (sometimes multiple incidents of the event type). Auto-refrigeration is one of the LyondellBasell Lessons Learned Library modules. Each lesson in the library contains a text document detailing the incident, the causes determined by the incident investigation, and the lessons learned. Each lesson also includes a training PowerPoint and a guideline for assessing comprehension of the lesson's learnings.

Process Safety Monthly Topic

Process Safety Monthly Topic presentations are prepared and offered for shared use in plant safety huddles or start-of-shift safety toolbox discussions. An Auto-Refrigeration safety topic presentation has been created and distributed.

Abnormal Situation Overview Screen

The Morris Ethylene Plant Control Systems Engineering Group has created an Abnormal Situations Overview screen for the Olefins Process, which includes key equipment real-time Minimum Allowable Temperature (MAT) graphics. The purpose of the Abnormal Situations Overview graphic is to provide a one-view summary of key process safety variables during significant plant upsets or shutdowns. The graphic can be accessed at any DCS control station. The graphic is a first-check item for Technical Engineers responding to provide support during a unit upset. As control board Operators respond to an upset and direct field Operators on required actions, Technical Engineers provide oversight assistance in monitoring the Abnormal Situations Overview graphic. The real-time MAT curves show where key pressure vessels are currently operating with respect to their MAT. The graphic also color-highlights any equipment that has crossed its MAT, even if the equipment is currently in the acceptable range of its MAT pressure-temperature curve.

Auto-Refrigeration and Near-miss Responses & Reporting

Perhaps the most important outcome of organizational training and education on auto-refrigeration is that personnel are knowledgeable on preventing auto-refrigeration incidents, but also on appropriate response when incidents do happen. This is reflected by the following events which have occurred since the implementation of the Auto-Refrigeration Engineering Guideline. Employee training was an important factor resulting in the appropriate response to these incidents.

Propane Feed Drum Auto-Refrigeration During Post-Turnaround Startup - During the startup of an Ethylene Plant, following a turnaround, a Propane Feed Drum was being nitrogen purged to the flare in preparation to receive feed. Simultaneously, the unit's C2 Splitter column was being inventoried with liquid ethane and vapor ethylene. A liquid drain to flare was opened on the C2 Splitter column bottom. The liquid drain flare line was common to the liquid drain flare line from the Propane Feed Drum. The liquid drain flare line filled with ethane from the C2 Splitter and backed into the Propane Feed Drum. Operators observed frost had formed on the exterior of the Propane Feed Drum and investigated. They determined the cause of the frosting to be material backing into the drum from the liquid flare drain line and blocked in Propane Feed Drum drain line. Operations took steps to ensure that the drum pressure was not above 40% of the MAWP, and was not allowed to increase. Investigation determined that the drum likely experienced localized temperatures which would have been well below the MAT had the drum been allowed to pressurize. The drum was allowed to gradually warm to ambient temperature and was subjected to a full post-incident internal inspection. Liquid drain lines have since been segregated to prevent reoccurrence.

Ethane Feed Drum Auto-Refrigeration During Unit Shutdown - During an unplanned Ethylene Plant shutdown, the unit's ethane feed drum lost vaporization. As a result, the drum pressure began to fall. The drum was equipped with a low pressure override control which closed a pressure control valve in the vapor outlet from the drum. The low pressure override activated, however the pressure control valve failed to close fully, remaining approximately 5% open. The vessel pressure decreased to a pressure which corresponded to an ethane temperature of -70°F. This would have been well below the vessel MAT had the vessel been allowed to repressurize. The Operations Group recognized the issue and took appropriate steps to avoid an auto-refrigeration event by draining off liquid from the vessel. Technical assistance was requested to determine how to warm up the vessel before bringing the vessel back to operating pressure. The vessel was never brought above 40% of MAWP with the temperature below the allowable MAT curve.

Liquid Propylene Purification Column Auto-Refrigeration during Regeneration - A liquid propylene purification column, which normally operates at 650 psig and approximately 40°F, had been taken offline and was in the process of being regenerated. The first step in the regeneration process is to deliquify the purification column vessel by opening a flare drain line on the bottom of the column. Once all of the liquid has been removed from the vessel, the flare vent is continued until the vessel pressure has decreased to 60 psig. In this particular incident a Supervisor was reviewing the status of the column at the beginning of his shift and determined that there was the potential that the vessel had undergone auto-refrigeration. Upon further investigation the Supervisor determined that the vessel had been depressured to 60 psig via a flare line from the top vapor-space of the vessel, resulting in the auto-refrigeration of the liquid in the vessel. The temperature of the vessel would have been well below the vessel MAT had the vessel been allowed to repressurize. The Supervisor called for an immediate stop to the regeneration procedure and summoned Technical assistance. The vessel was safely deliquified and warmed without incident.

Propylene Dryer Auto-Refrigeration Incident - While making routine rounds, an Operator observed a Propylene Product Dryer Vessel to be frosted up, which was unusual. Operations investigated and determined that the vessel, which had been pressurized and was in standby, had been inadvertently partially depressured to flare. This resulted in vessel temperatures of -40°F. The vessel flare line was blocked in and the pressure was maintained at the same pressure (less than 40% of MAWP). The vessel was allowed to warm very gradually to ambient temperature. The vessel was pressure tested with low pressure nitrogen before being returned to service.

CONCLUSIONS

Industry knowledge concerning auto-refrigeration brittle fracture fundamentals, scenarios identification/analysis, as well as mitigation options and safety system design has grown considerably over the 25 years since the ARBF incident at the LyondellBasell Morris, IL Ethylene Plant in 1989. Available references on the subject from EPC proceedings alone are extensive as evident from the list of references included with this paper. API 579/ASME FFS-1 captures much of that knowledge as it applies to equipment assessment practices. However, this standard is only effective if adequately understood and used in conjunction with a comprehensive program to identify, understand, prevent and mitigate ARBF scenarios under all possible modes of operation. A thorough training program, including refresher training for both technical and operational personnel, is fundamental to this effort. This process has evolved over time and continues to evolve. Engineering and operations personnel continue to seek out ARBF risks, identify mitigation requirements and implement protective systems. After 25 years, we continue to identify and mitigate ARBF risks. We reserve the right to get smarter.

REFERENCES

1. King, R.E., Auto-Refrigeration / Brittle Fracture Analysis of Existing Olefins Plants - Identification of Potential Excursions, 2004 AIChE Spring National Meeting, 16th Ethylene Producers Conference, New Orleans, LA April 25-29.

2. McLaughlin, J.E., Sims, J. R., Findley, M. and Jones, J.P., Assessment of Older Cold Service Pressure Vessels For Risk of Brittle Fracture, 1995 AIChE Spring National Meeting, 7th Ethylene Producers Conference, Houston, TX, March 19-23.

3. King, R.E., Workshop on Auto-Refrigeration/Brittle Fracture Analysis, 2006 AIChE Spring National Meeting, 18th Ethylene Producers Conference, Orlando, FL, April 24-26.

4. King, R.E., Tutorial on Auto-Refrigeration and Brittle Fracture Analysis, 2014 AIChE Spring National Meeting, 26th Ethylene Producers Conference, New Orleans, LA, March 30 - April 3.

5. API 579/ASME FFS-1 Recommended Practice For Fitness-For-Service, Section 3 Assessment of Existing Equipment for Brittle Fracture.

6. Scego, J.P., Cooke, D.L., DeBose, M.E., Guinn, J.D., Polito, C.T., Ethylene Fractionator Auto-Refrigeration Incident, 2007 AIChE Spring National Meeting, 19th Ethylene Producers Conference, Houston, TX April 22-27.

7. Messinger, J.R., Findlay, M., Cold Service Risk Assessment of Existing Bimetallic Demethanizers, 2001 AIChE Spring National Meeting, 13th Ethylene Producers Conference, Houston, TX April 22-26.

8. Kennedy, M.J., Auto-refrigeration Near-Miss Incident at Equistar Chemicals, 2005 AIChE Spring National Meeting, 17th Ethylene Producers Conference, Atlanta, GA April 10-14.

9. Mullenix, D., Brittle Failure of a Carbon Steel Flare Line, 2000 AIChE Spring National Meeting, 12th Ethylene Producers Conference, Atlanta, GA April 5-9.

10. Kuo, A., Pitt, R., Flare Line Failure Case, What Have We Learned?, 2004 AIChE Spring National Meeting, 16th Ethylene Producers Conference, New Orleans, LA April 25-29.

11. Williams, P., Reliability For Safety Instrumented Systems, Chemical Engineering Progress, pp. 27-32, September 2004

12. Thompson, C., King, R., Compression System Check Valve Failure Hazards, 2010 AIChE Spring National Meeting, 22nd Ethylene Producers Conference, San Antonio, TX, March 22-25.

Auto-Refrigeration / Brittle Fracture Prevention

Process Hazard Management

Date: March 2014

Prepared by: Ralph E. King, P.E.
Senior Staff Consultant
Plant Services Practice

Taking on your toughest technical problems

STRESS ENGINEERING SERVICES INC.

an employee-owned company

Limitations

Limitations of This Training

- This training report is limited to the matters expressly covered. This report is prepared for the sole benefit of our industry clients. In preparing this report, Stress Engineering Services, Inc. has relied on information provided from various public sources. Stress Engineering Services, Inc. has made no independent investigation as to the accuracy or completeness of such information and has assumed that such information was accurate and complete. Further, Stress Engineering Services, Inc. is not able to direct or control the operation or maintenance of our clients equipment or processes.
- All recommendations, findings, and conclusions stated in this report are based on facts and circumstances as they existed at the time that this training report was prepared. A change in any fact or circumstance upon which this report is based may adversely affect the recommendations, findings, and conclusions expressed in this report.

NO IMPLIED WARRANTY OF MERCHANTABILITY OR FITNESS FOR A PARTICULAR PURPOSE SHALL APPLY. STRESS ENGINEERING SERVICES, INC. MAKES NO REPRESENTATION OR WARRANTY THAT THE IMPLEMENTATION OR USE OF THE RECOMMENDATIONS, FINDINGS, OR CONCLUSIONS OF THIS REPORT WILL RESULT IN COMPLIANCE WITH APPLICABLE LAWS OR PERFECT RESULTS.

THIS TRAINING REPORT CONTAINS INFORMATION, MATERIALS, FORMS, METHODS, AND PROCESSES WHICH ARE CONFIDENTIAL AND PROPRIETARY TO STRESS ENGINEERING SERVICES, INC. AND IS BEING PROVIDED TO OUR CLIENTS FOR THE SOLE PURPOSE OF GENERAL ENGINEER TRAINING . THE CONTAINED INFORMATION MAY NOT BE COPIED, REPRODUCED, ALTERED, DISTRIBUTED OR DISPLAYED IN WHOLE OR IN PART WITHOUT THE PRIOR WRITTEN CONSENT OF AN AUTHORIZED REPRESENTATIVE OF STRESS ENGINEERING SERVICES, INC.

2

STRESS ENGINEERING SERVICES INC.

an employee-owned company

Agenda

- Overview
  - Nomenclature
  - Incidents
  - Brittle Fracture Basics
- Analysis Methodologies and Examples
  - API 579-1 / ASME FFS-1 Brittle Fracture Analysis
  - Piping Brittle Fracture Analysis
  - Process Hazards Analysis (PHA)
  - Pressure Tests
- Example Scenarios
- Operator Training
- Post-excursion inspection
- Conclusions and Closing Remarks

3

STRESS ENGINEERING SERVICES INC.
an employee-owned company

Nomenclature

- Auto-Refrigeration (AR)
  - The UNINTENTIONAL and UNCONTROLLED phase change from a liquid to a vapor of a hydrocarbon that results in refrigeration
    - Refrigeration when you don't want it – inadequate matl. of construction
    - When you least expect it
    - You are NOT controlling – if you were, you would not do it
- Brittle Fracture (BF)
  - The sudden and often catastrophic fracture of a material with little or no plastic deformation
    - Relatively low levels of energy required to initiate or propagate a flaw
    - Associated with rapid crack propagation through the material
    - Break-before-leak scenario

4

STRESS ENGINEERING SERVICES INC.
an employee-owned company

Nomenclature (cont'd)

- CET - Critical Exposure Temperature
  - The lowest temperature experienced by the piece of equipment as a result of:
    - Auto-Refrigeration
    - Joule Thomson (J-T) effect
    - Exposure to cold (sub-cooled) liquid not designed for exposure
    - Low ambient
- MAT - Minimum Allowable Temperature
  - The lowest temperature allowed at a coincidental pressure to ensure the vessel remains ductile
- MAWP - Maximum Allowable Working Pressure
  - The maximum pressure the vessel may experience so as to keep within code-allowable stress
- MDMT - Minimum Design Metal Temperature
  - Specified value by owner or value determined as part of Code design (designer default – ASME BPVC Sect. VIII Div. 1 UCS-66)
  - Equals the MAT at MAWP - The lowest temperature allowed at MAWP to ensure the vessel remains ductile
- PWHT - Post Weld Heat Treatment
  - Post welding thermal heat treatment of the vessel to reduce residual stresses

5

STRESS ENGINEERING SERVICES INC.
an employee-owned company

Industry Auto-Refrigeration / Brittle Fracture Incidents

- Morris Exchanger Failure 1989
  - 2 killed, 7 burned, 5 disabled
- Lake Charles Ethylene Plant Flare Drum 1991
  - No injuries
- Longford Exchanger Failure 1998
  - 2 killed, 8 burned, $1.2B financial impact
- Beringen HDPE Plant Loop Failure 1998
  - No injuries
- Westlake Flare Piping 2002
  - No injuries
- Numerous pressure (hydro) test brittle fracture failures every year

6

STRESS ENGINEERING SERVICES INC.
an employee-owned company

HERALD-NEWS

Tuesday, September 12, 1989

One dead, 17 injured in explosion at Quantum

They saw fellow work 'disappear into flames

STRESS ENGINEERING SERVICES INC.

an employee-owned company

7

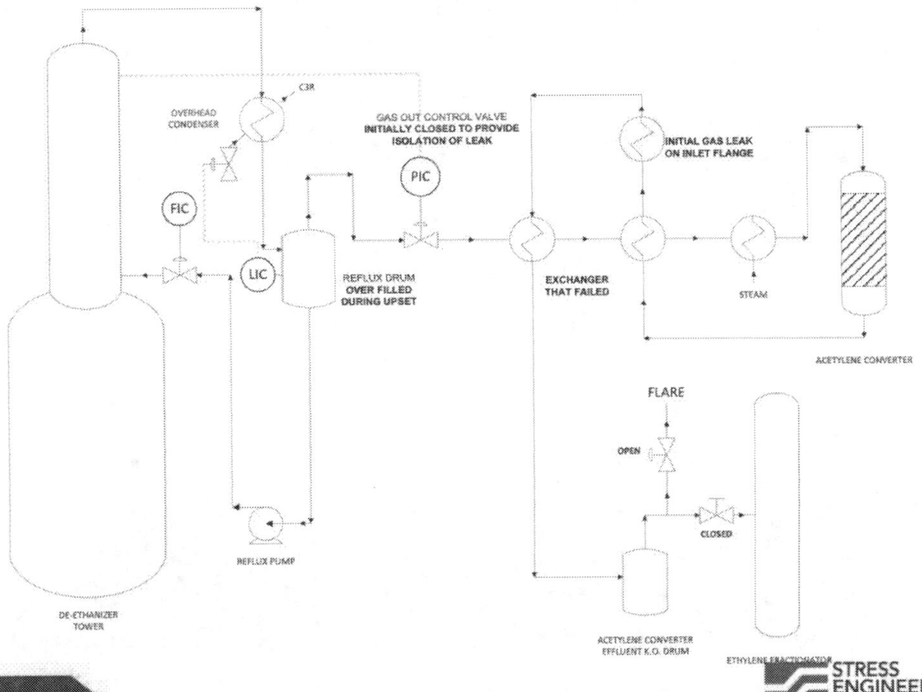

STRESS ENGINEERING SERVICES INC.

an employee-owned company

8

Auto-Refrigeration / Brittle Fracture

9

STRESS ENGINEERING SERVICES INC.
an employee-owned company

Lake Charles Flare Drum Failure

10

STRESS ENGINEERING SERVICES INC.
an employee-owned company

Lake Charles Flare Drum Failure

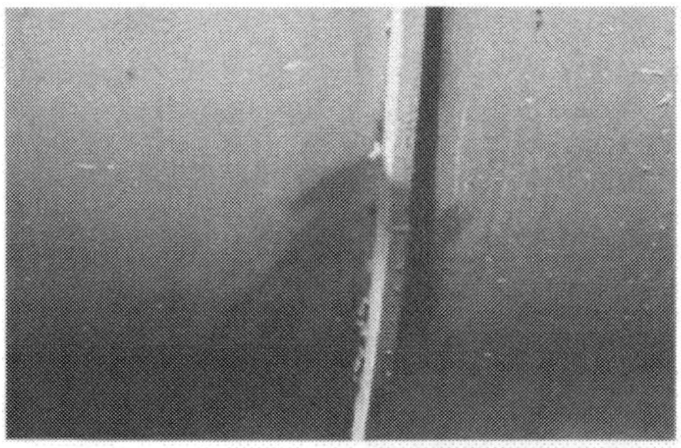

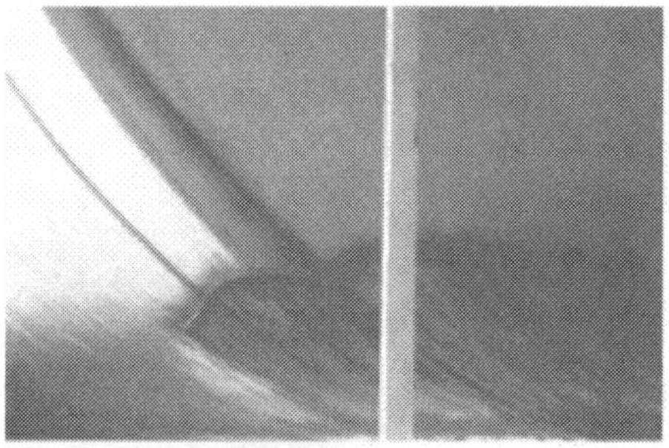

STRESS ENGINEERING SERVICES INC.

an employee-owned company

11

Longford Exchanger Auto-Refrigeration / Brittle Fracture

STRESS ENGINEERING SERVICES INC.

an employee-owned company

12

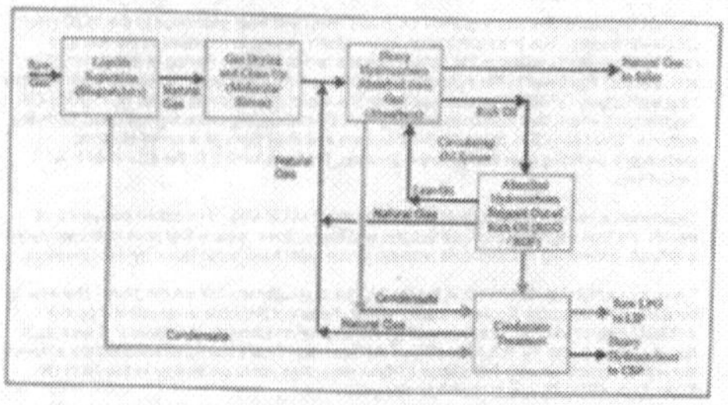

The process flow diagram (Figure 2) shows only the oil circulation for clarity, not the gas and condensate streams. In broad terms the process passes chilled natural gas containing condensate (ethane, propane and butane) into the Absorber vessels and in counter-current flow absorbs the condensate in Lean Oil at a temperature of about -20C. The saturated Lean Oil leaving the Absorbers is termed Rich Oil.

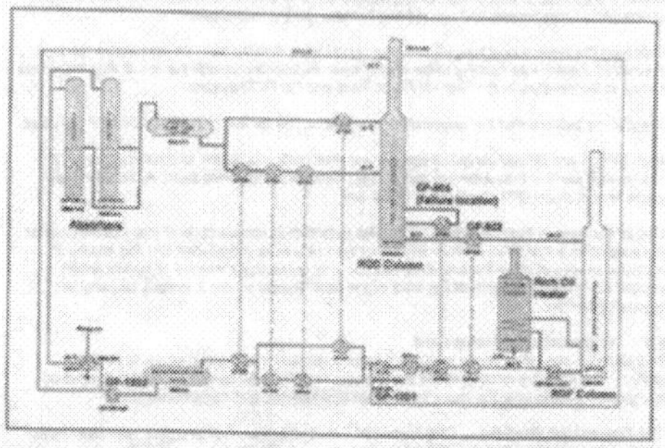

Figure 2. Process Flow Diagram of Absorber Oil Recirculation Loop

STRESS ENGINEERING SERVICES INC.
an employee-owned company

13

Flare Line Brittle Fracture

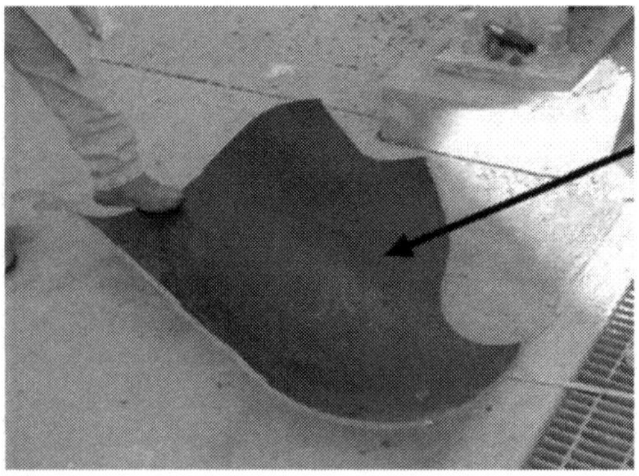

STRESS ENGINEERING SERVICES INC.
an employee-owned company

14

Auto-Refrigeration / Brittle Fracture Incidents

- Never occurs during normal operation
  - Except where an environmental-embrittlement mechanism is underway
- A process upset of some sort is often the initiator
- Subsequent actions by Operations can cause the chain of events to progress to a higher consequence
- Risk of brittle fracture is transitory with potential only occurring within a window of vulnerability
  - Often this window passes without notice
  - Doing the wrong thing during the vulnerability window can be catastrophic

15

STRESS ENGINEERING SERVICES INC.
an employee-owned company

Brittle Fracture Basics

16

STRESS ENGINEERING SERVICES INC.
an employee-owned company

Brittle Fracture

- Requires three elements to be present for failure to occur
  - A susceptible steel (susceptible at cold temperature)
  - A crack or a notch (as often found in weld defects)
  - Sufficient applied stress (above some minimum, ~7,000 psi [48 MPa] for carbon steels)
    - This is NOT operating pressure but vessel stress

17

STRESS ENGINEERING SERVICES INC.
an employee-owned company

Brittle Fracture Triangle

Need all 3 legs:

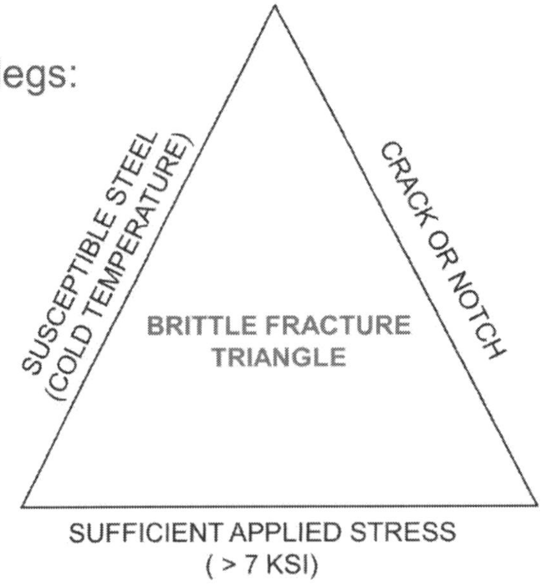

18

STRESS ENGINEERING SERVICES INC.
an employee-owned company

Brittle Fracture Failure: Liberty Ships

USS Esso Manhattan, 3/29/43

Fracture at entrance to NY harbor.

John P. Gaines, 11/43

Vessel broke in two off the Aleutians (10 killed).

USS Schenectady, 1/16/43

Liberty tanker split in two while moored in calm water at the outfitting dock at Swan Island, OR.

http://www.uh. edu/liberty/photos/liberty_summary.html

Coast Guard Report: USS Schenectady Without warning and with a report which was heard for at least a mile, the deck and sides of the vessel fractured just aft of the bridge superstructure. The fracture extended almost instantaneously to the turn of the bilge port and starboard. The deck side shell, longitudinal bulkhead and bottom girders fractured. Only the bottom plating held. The vessel jack-knifed and the center portion rose so that no water entered. The bow and stern settled into the silt of the river bottom.

The ship was 24 hours old.

Official CG Report attributed fracture to welds in critical seams that "were found to be defective".

STRESS ENGINEERING SERVICES INC.
an employee-owned company

Brittle Fracture Failure: Pipe Failure

Ductile failure:
- One piece
- Large deformation

Brittle failure:
- Many pieces
- Small deformation

STRESS ENGINEERING SERVICES INC.
an employee-owned company

Effect of Temp. and Thickness on Impact Energy

Ductile - Brittle Transition of Carbon Steel

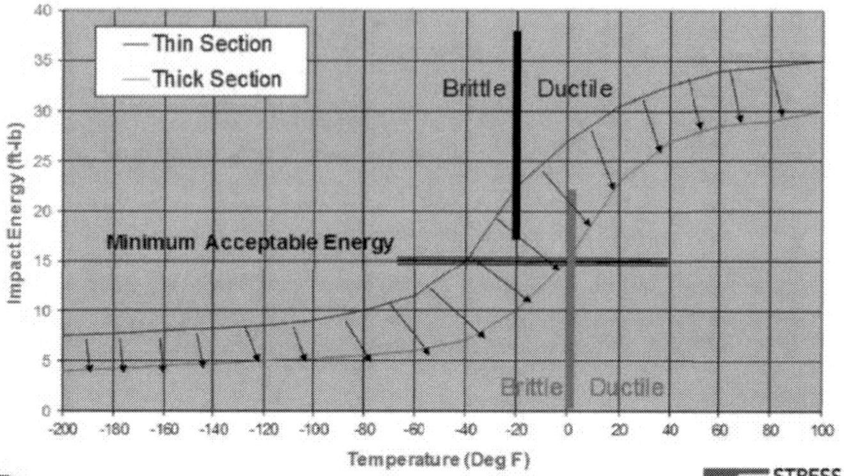

21

STRESS ENGINEERING SERVICES INC.
an employee-owned company

Ductile-Brittle Transition Curves

Effect Of Nickel On Transition Temp.

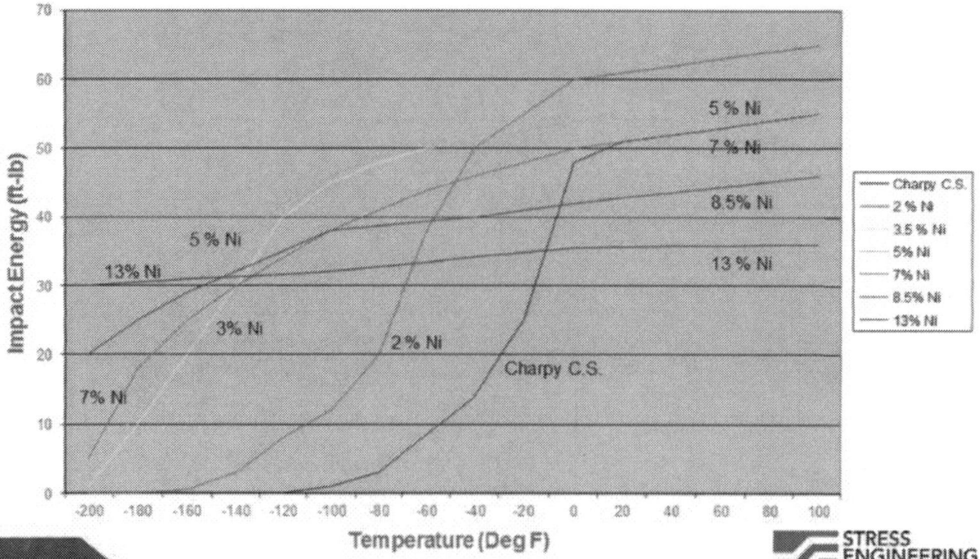

22

STRESS ENGINEERING SERVICES INC.
an employee-owned company

In Design: Stay Above DBTT

WWII: Liberty Ships   Pre-WWII: The Titanic

Problem: DBTT ~ Operating Temperature

STRESS ENGINEERING SERVICES INC.
an employee-owned company

23

Processes Subject To Potential AR/BF

- Olefin Plants
- Gas Processing
- Refineries
- Polymer Units
- Ammonia Units
- Any light liquid processing unit
  - C4 to C1 liquids
- Any Joule - Thompson (J-T) let-down
  - Top-side gas processing

STRESS ENGINEERING SERVICES INC.
an employee-owned company

24

Primary Program Elements

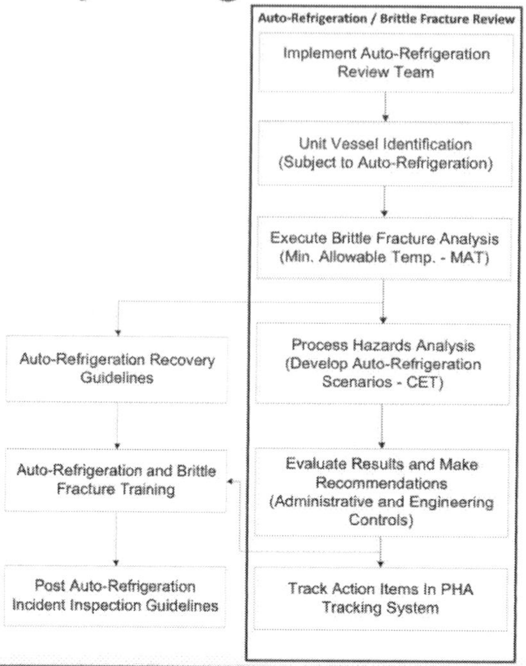

25

STRESS ENGINEERING SERVICES INC.
an employee-owned company

Brittle Fracture Analysis

- Utilizes API 579-1 / ASME FFS-1 Methods
- Generates a MAT curve (2a method) for each pressure containing envelope of the vessel (exchangers - shell side / tube side)
- MAT results are used to compare to developed PHA scenario CETs (how cold at what pressure?)
  - Sequence of events
  - Post chilling response or actions

26

STRESS ENGINEERING SERVICES INC.
an employee-owned company

API 579-1 FFS Assessment Procedures

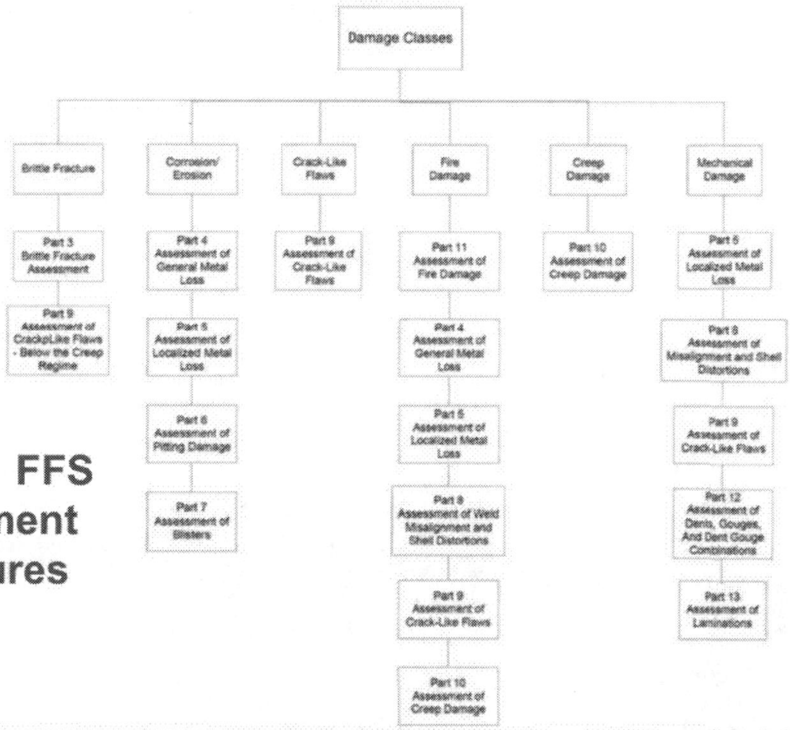

27

API 579-1 Overall Brittle Fracture Assessment Procedure for Pressure Vessels and Piping

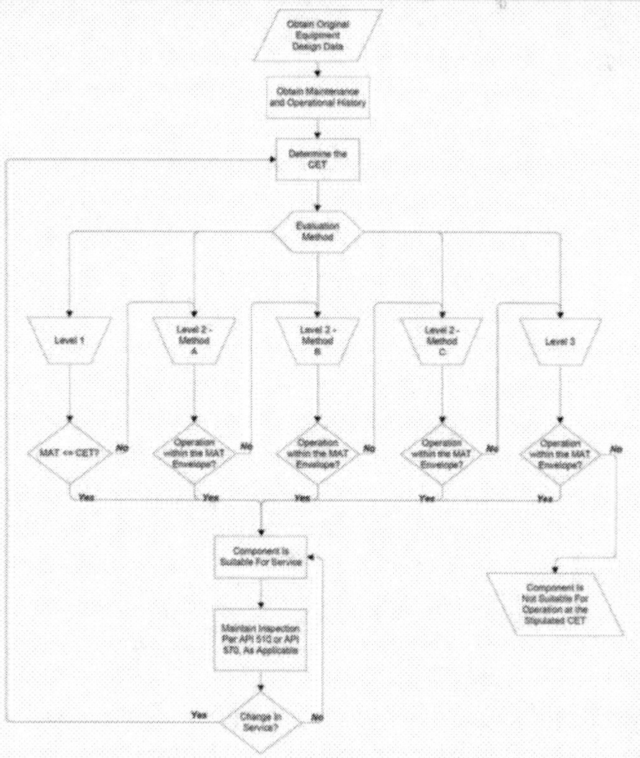

28

API 579-1 / ASME FFS-1 MAT

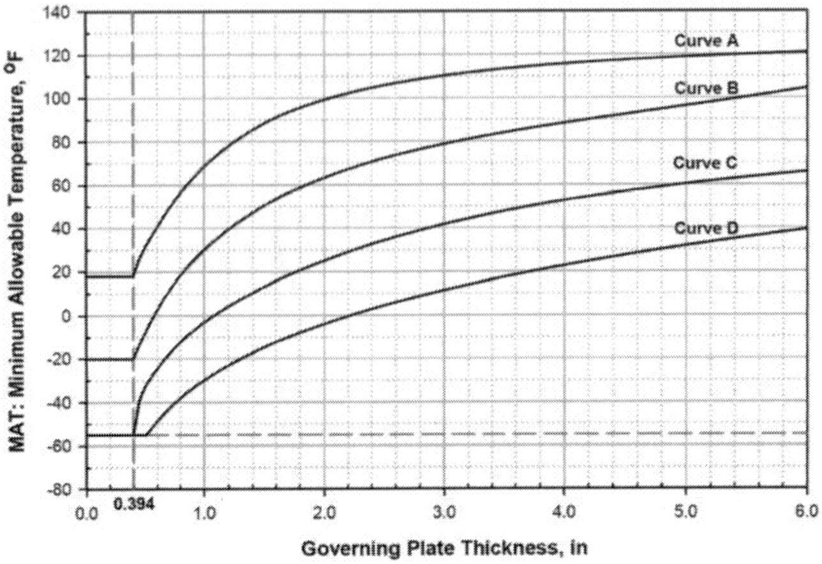

STRESS ENGINEERING SERVICES INC.
an employee-owned company

Governing Thickness – by component

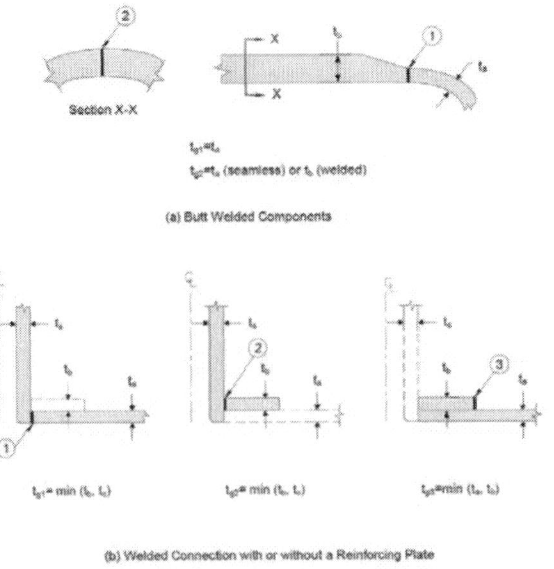

Figure 3.5 – Some Typical Vessel Details Showing the Governing Thickness

STRESS ENGINEERING SERVICES INC.
an employee-owned company

Temp. Reduction with Stress

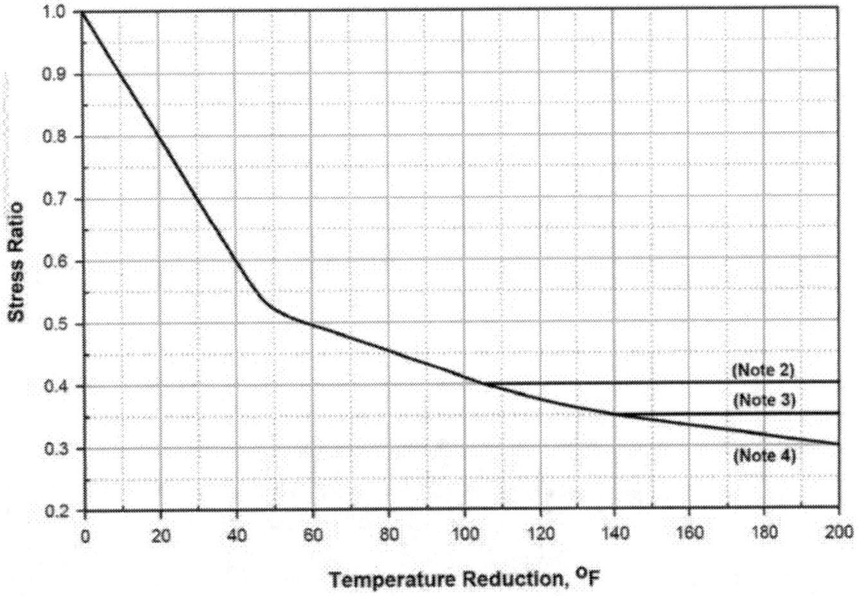

STRESS ENGINEERING SERVICES INC.
an employee-owned company

Analysis and Mitigation

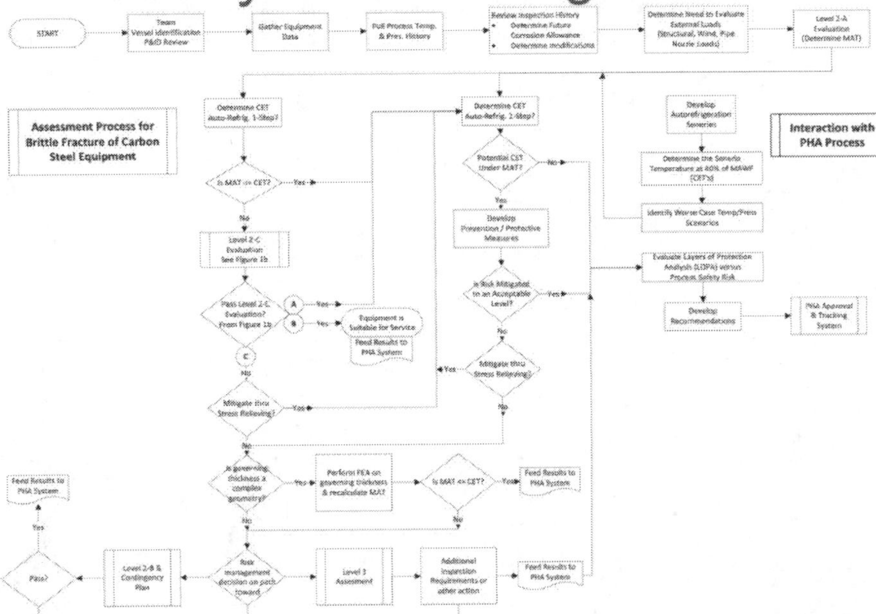

STRESS ENGINEERING SERVICES INC.
an employee-owned company

Brittle Fracture During Pneumatic Pressure Test

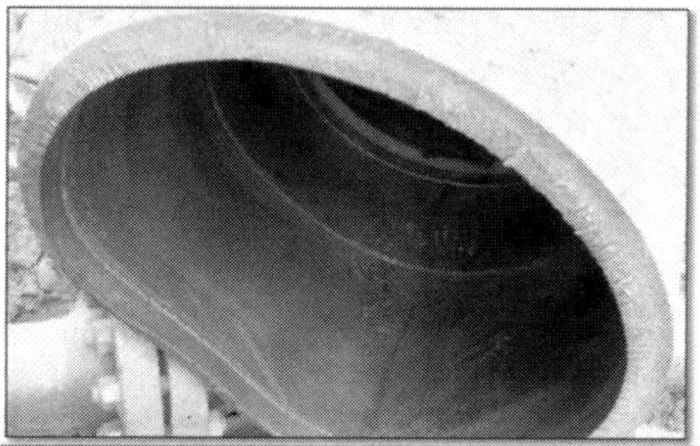

33

STRESS ENGINEERING SERVICES INC.

an employee-owned company

Documented Failures

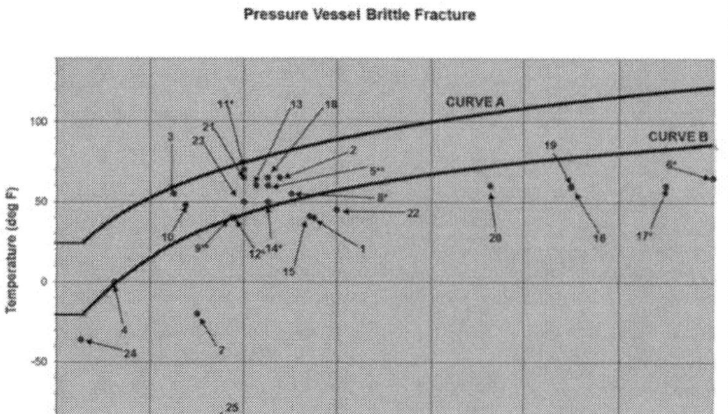

Num	Thickness (in)	Failure Temp (F)	Curve	Hydro Fail. (Y/N)
1	1.625	40	A	N
2	1.438	65	A	Y
3	0.875	55	A	N
4	0.563	0	A	N
5**	1.375	60	C	Y
6*	3.750	65	B	N
7	1.000	-20	A	N
8*	1.500	55	B	Y
9**	1.188	40	C	Y
10	0.938	48	A	N
11*	1.250	70	B	Y
12*	1.200	40	B	Y
13	1.313	60	A	Y
14*	1.375	50	B	Y
15	1.600	41	A	N
16	3.000	59	A	Y
17*	3.500	60	B	Y
18	1.375	65	A	Y
19	3.000	60	A	Y
20	2.563	60	A	Y
21	1.250	65	A	Y
22	1.750	45	A	Y
23	1.250	50	A	Y
24	0.380	-36	A	N
25	1.000	-100	A	N

34

STRESS ENGINEERING SERVICES INC.

an employee-owned company

Example Vessel MAT

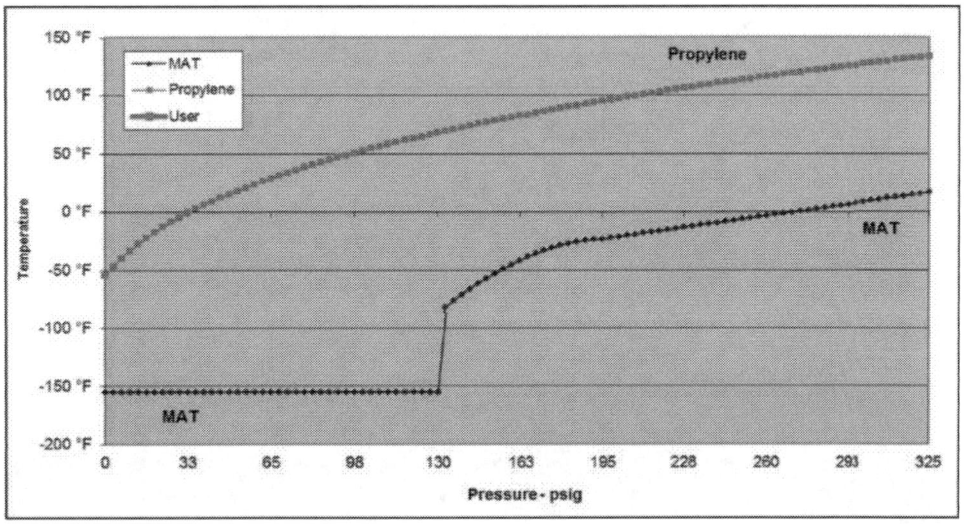

35

STRESS ENGINEERING SERVICES INC.
an employee-owned company

Inherently Safe Vessel

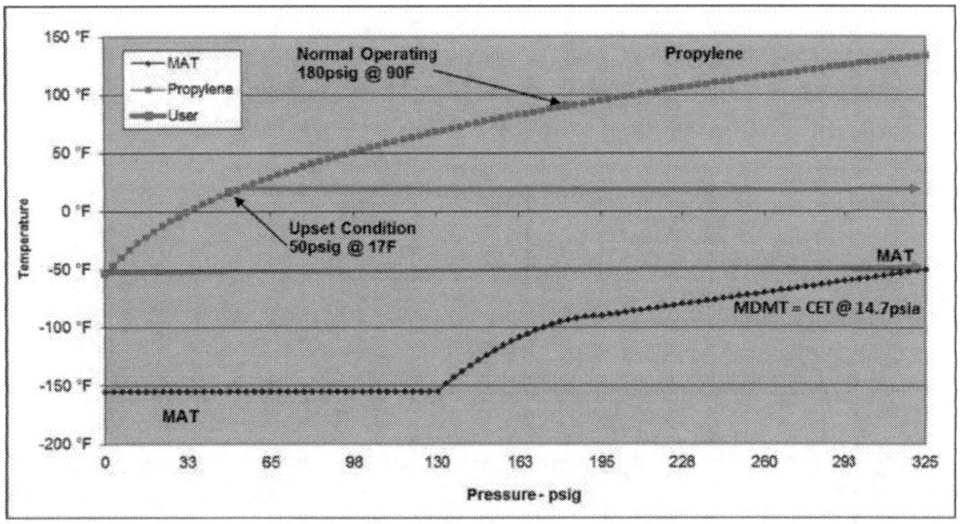

36

STRESS ENGINEERING SERVICES INC.
an employee-owned company

Acceptable Vessels

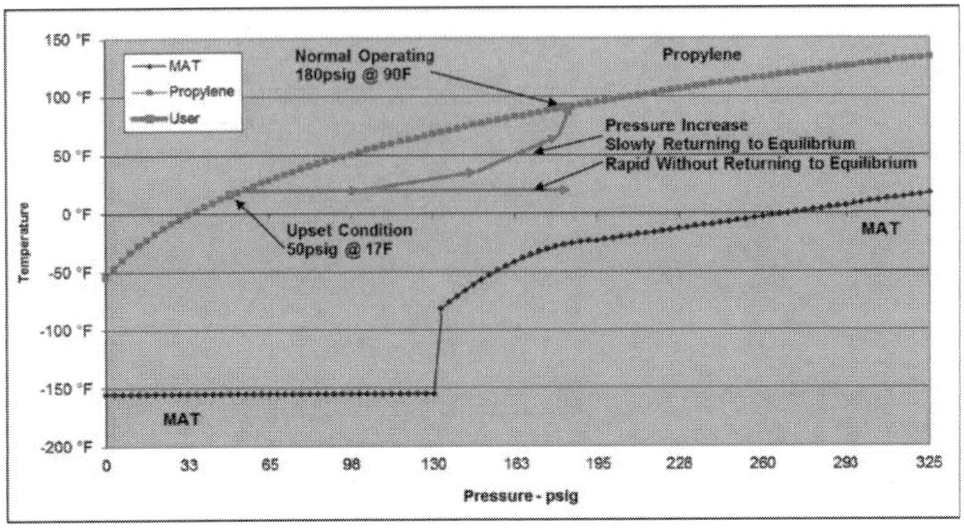

STRESS ENGINEERING SERVICES INC.
an employee-owned company

37

Example Vessel MAT – One Step

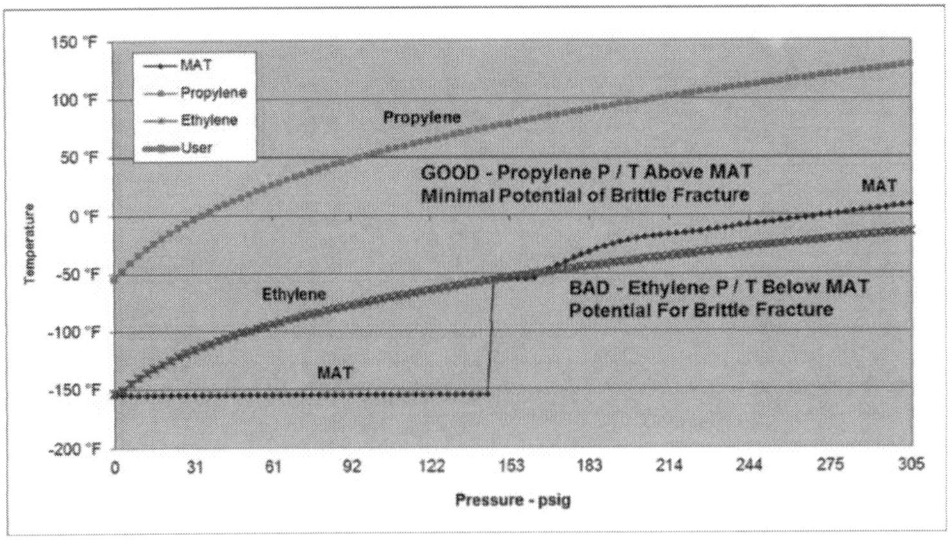

STRESS ENGINEERING SERVICES INC.
an employee-owned company

38

Example Vessel MAT – Two Step

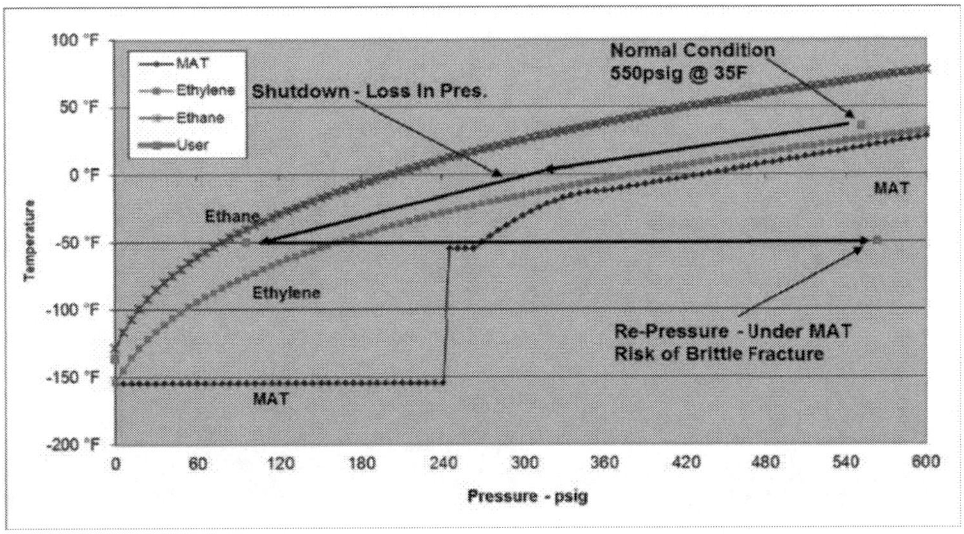

39

STRESS ENGINEERING SERVICES INC.
an employee-owned company

Existing Vessel – Operates Below Calc. MAT

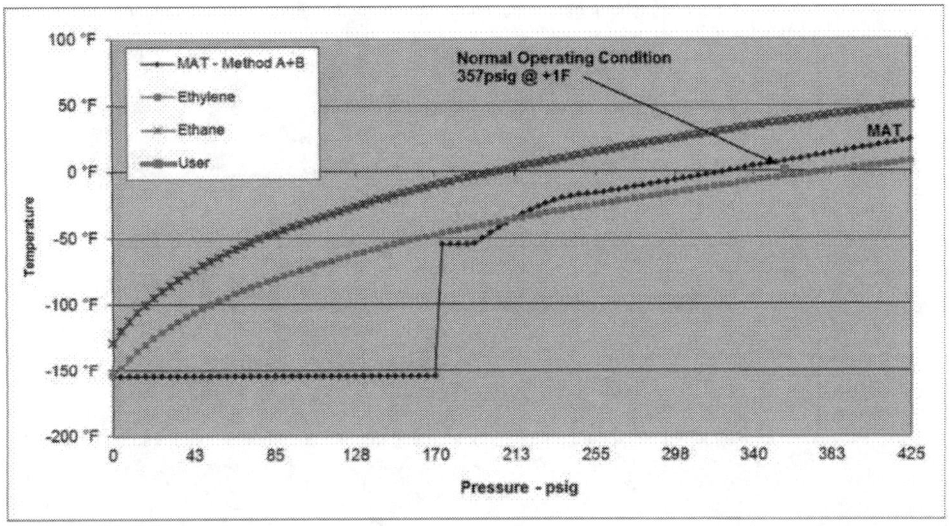

40

STRESS ENGINEERING SERVICES INC.
an employee-owned company

Level 2C Grandfathering

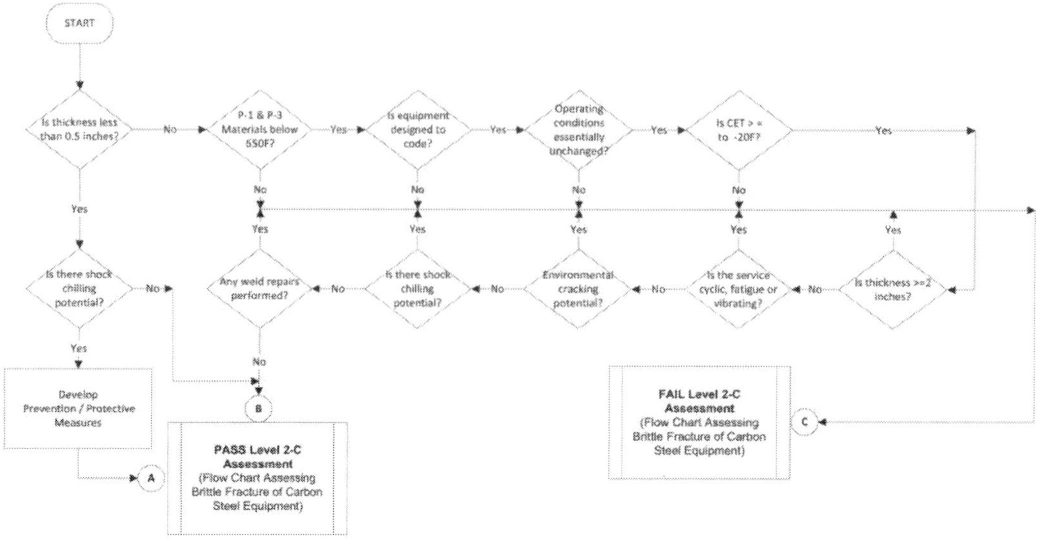

41

Level 2-C Evaluation

STRESS ENGINEERING SERVICES INC.

an employee-owned company

Level 2C Grandfathering Example

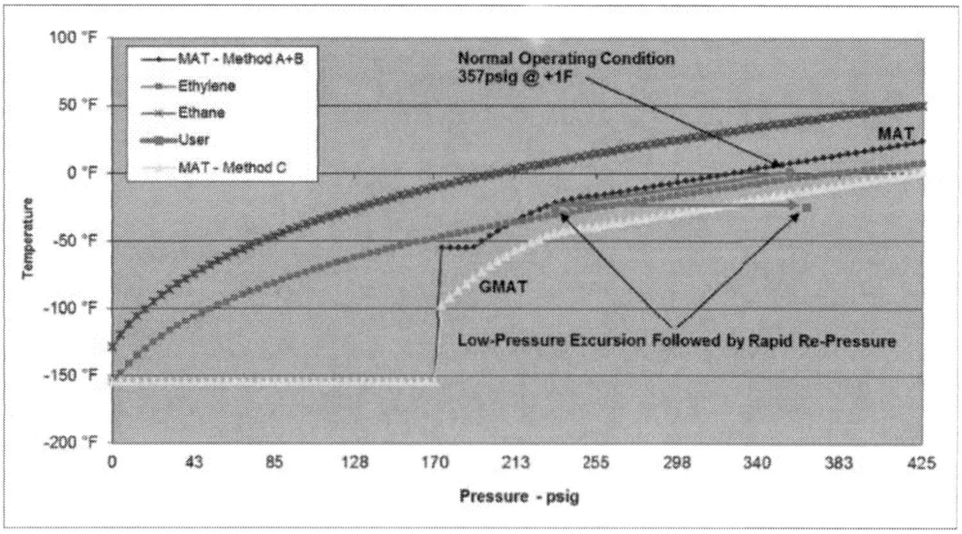

42

STRESS ENGINEERING SERVICES INC.

an employee-owned company

150# Class Piping – Pres. Stress MAT

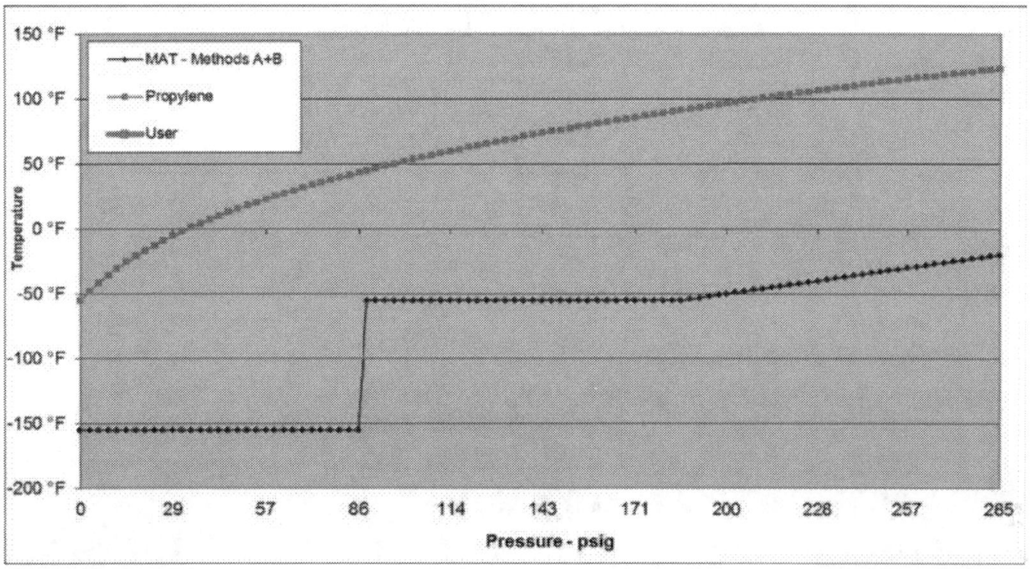

43

STRESS ENGINEERING SERVICES INC.
an employee-owned company

600# Class Piping – Pres. Stress MAT

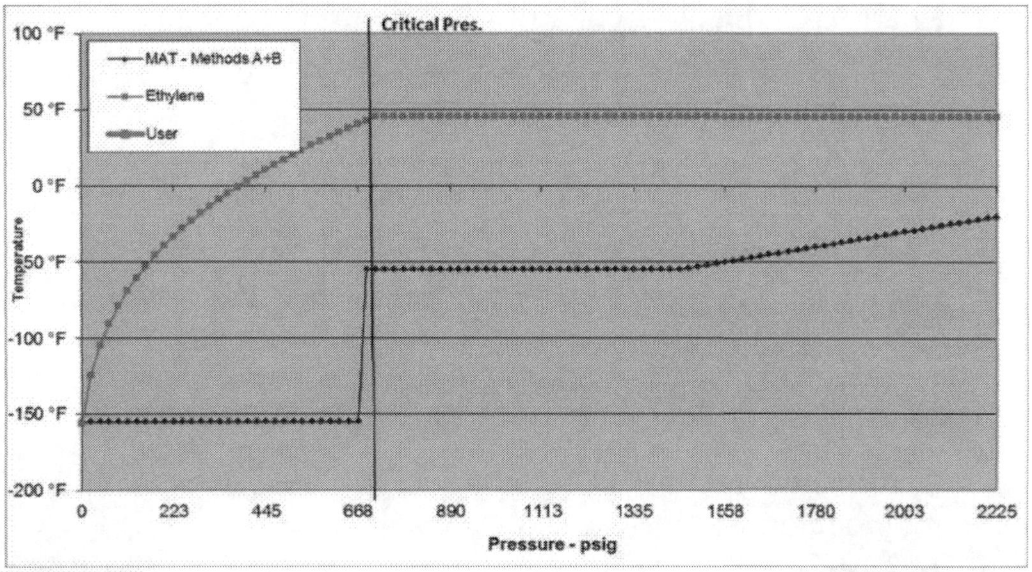

44

STRESS ENGINEERING SERVICES INC.
an employee-owned company

Piping Brittle Fracture Analysis

- Screening utilizing MAT curve based upon 2A analysis using piping design pressure
  - EXCEPT
    - Long straight lines without expansion / contraction provisions in the design
    - CET very close to the MAT – insufficient clearance
- Piping stress analysis using ratio of stress for sub-normal pressure conditions
  - Accounts for contraction stresses
  - Accounts for dead-load stresses

45

STRESS ENGINEERING SERVICES INC.
an employee-owned company

Critical Exposure Temperature - CET

- Determined based upon potential operating conditions
  - Startup (lowest one-day mean ambient temp.)
  - Normal operation (lowest metal temp.)
  - Shutdown
  - Upset (failure of system components [pumps, control valves, etc.], colder than expected warming streams, reboiler failure or stall, etc.)
  - Standby – not in operation
  - Pressure tightness testing – hydrotest
  - Auto-refrigeration and Joule-Thompson cooling
  - Shock Chilling
- Determined using PHA or other comparable assessment methodologies
  - Process change of particular concern

46

STRESS ENGINEERING SERVICES INC.
an employee-owned company

Hydrocarbon CETs

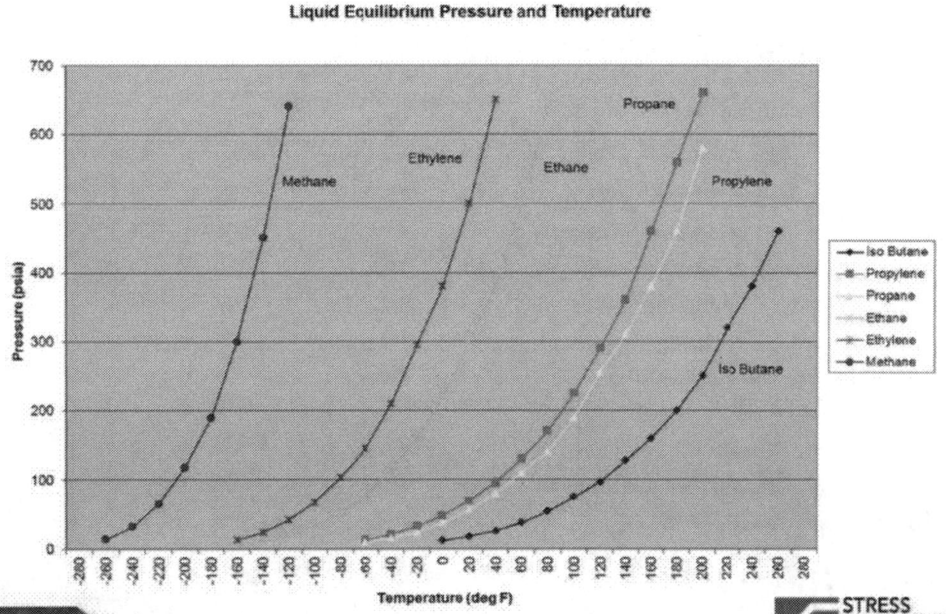

STRESS ENGINEERING SERVICES INC.
an employee-owned company

47

PHA Analysis Approach

- Process Hazards Analysis (PHA)
  - Use adapted PHA techniques (Event Tree, Guide-List, Black-Box)
  - Evaluate different operating modes
    - Normal Operation
    - Upset Conditions
    - Normal Startup / Shutdown
    - Emergency Shutdown
    - Air Freeing / Nitrogen Freeing
    - Inventory / De-inventory
    - Not in Operation / Maintenance-in-Progress
    - Commissioning / Leak Testing
  - Result of evaluation is to determine Critical Exposure Temp. (CET)
  - Compare MAT to CET and make recommendations
    - Execute LOPA of mitigation to ensure compliance with acceptable risk profile

STRESS ENGINEERING SERVICES INC.
an employee-owned company

48

PHA Methodology Approach

- Use "Event Tree" Method
  - Fault Tree without probabilities
  - Describe exact chain-of-events required to top event
    - Auto-Refrigeration Excursions
      - Auto-Refrigeration one-step
      - Auto-Refrigeration two-step
    - Builds Team understanding
      - Importance of sequence
      - Automatic control responses
      - Operations actions during recovery / startup
  - Determine how close to top-event did past historical excursions reach
  - Evaluates potential mitigation options
    - Interrupt chain-of-events as near the top as possible

STRESS ENGINEERING SERVICES INC.
an employee-owned company

Example Event Tree

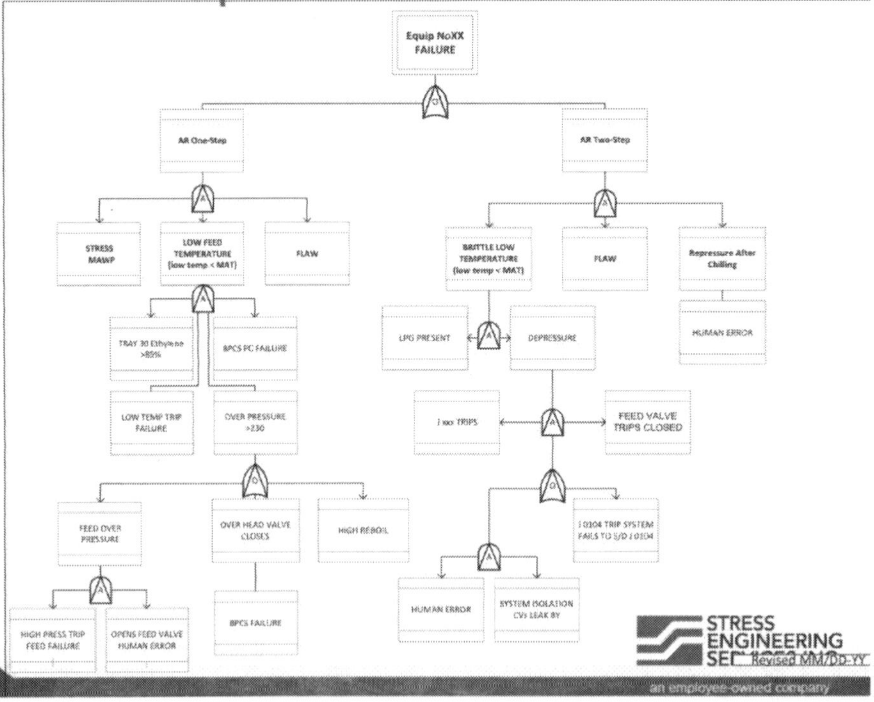

STRESS ENGINEERING SERVICES INC.
Revised MM/DD-YY
an employee-owned company

Guide-List Method

- Developed a Auto-Refrigeration Specific Guide-List
  - "What-If Checklist"
  - Asks specific process excursion questions
    - For each possible operating state
      - Transitions between operating states
    - Documents search results from "Black-Box" review
  - Emphasis on comparing actions that result in auto-refrigeration – prevention
  - Evaluates normal and abnormal responses
    - Control actions
    - Operator actions
  - Covers non-operating conditions, e.g., pressure testing

51

STRESS ENGINEERING SERVICES INC.
an employee-owned company

Example Guide-List Questions (1/3)

- Normal operating
  - What is the normal operating conditions CET relative to the MAT?
  - Are there any changes in process composition?
    - If so, what are the impact on process CET conditions relative to the MAT?
  - Are there process states (regeneration, etc.)?
    - If so, what are the process condition CET relative to the MAT?
- Upset conditions
  - What historical upsets scenarios have occurred and what were the process conditions relative to the MAT?
  - What scenarios might inadvertently introduce a light hydrocarbon liquid into a system that is NOT designed to contain liquid?
    - What is the upset CET relative to the vessel MAT?
  - What scenarios might cause loss of heat input?
    - What is the loss in heat input CET relative to the MAT?
  - For heat exchangers, does an auto-refrigeration excursion on one side of the vessel result in cross-exchange to the other side?
    - What is the cross-exchange CET relative to the MAT?

52

STRESS ENGINEERING SERVICES INC.
an employee-owned company

Example Guide-List Questions (2/3)

- Startup
  - Do process procedures document minimum pressurization requirements (and pressurization vapor source) before admission of liquid hydrocarbons?
  - As process flow progresses forward, what possible equipment / instrument failures could occur that have the potential for auto-refrigeration?
    - What is the CET relative to the MAT?

53

STRESS ENGINEERING SERVICES INC.
an employee-owned company

Example Guide-List Questions (3/3)

- Shutdown
  - What is the equipment's settling-out pressure for fail-safe condition?
    - What is the shutdown CET relative to the MAT?
  - If system is de-pressured prior to draining, what is the resulting CET relative to the MAT?
    - Do existing procedures require draining of equipment BEFORE de-pressuring?
  - For vessels with internal media (sieve, catalyst, etc.), is there a potential to purge the wet beds with a dry, inert gas?
    - What is the purge CET relative to the MAT?
  - Is there potential during decommissioning steps for a leaking isolation valve to manually let-down a light hydrocarbon liquid into section of process that is out-of-service?
    - What is the let-down CET relative to the MAT?
    - What subsequent actions could increase pressure resulting in conditions approaching the MAT?

54

STRESS ENGINEERING SERVICES INC.
an employee-owned company

FAA – "Black-Box" Data Analysis

- Concept – don't wait for the plane crash to start identifying safety issues (Started in 2009)
- Routinely Pull "Black-Box" Data
- Execute data analysis (bin-analysis) to identify potential safety issues
- Integrate all flight information from multiple carriers to identify "outliers" to the normal operation
- Goal is to reduce "near-hit" events

Do you really need the references?

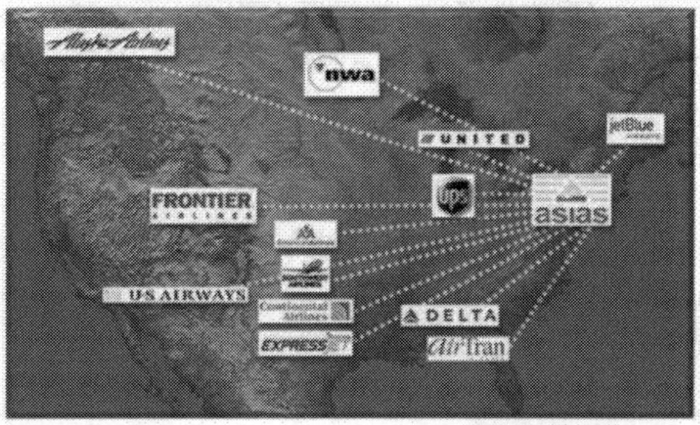

Aviation Safety Information Analysis and Sharing (ASIAS—pronounced "a-sigh-us") initiative.
http://www.mitre.org/news/digest/aviation/04_09/data.html
http://web.mit.edu/newsoffice/2011/black-box-analysis-0912.html

STRESS ENGINEERING SERVICES INC.
an employee-owned company

"Black-Box" Data Dump

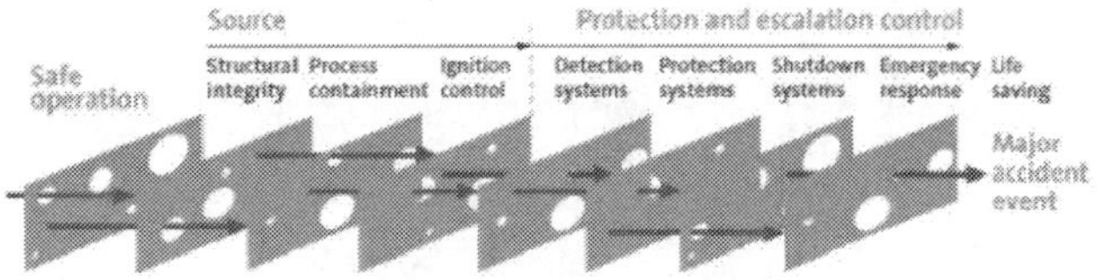

- Execute query (via spreadsheet) for:
  - Parameter conditions outside (above or below) or nearly outside (90% of) system limits
    - Pressure, temperature, flow, composition limits, etc.
    - Identifies "events" where data indicates "at-risk" deviation
      - Known initiating events or
      - Key enabling events
- Utilize root-cause analysis methods in determining the cause for the excursion including the chain of events to the parameter deviation
- How close to the top-event did excursion reach?

http://www.bg-group.com/sustainability11/ManagementSystems/Pages/HSSE.aspx

STRESS ENGINEERING SERVICES INC.
an employee-owned company

"Black-Box" Review

- Data Historian Search
  - Locate historical results where
    - Temperature approached MAT
    - Low-pressure excursions
    - Unit or major equipment trips or upsets
  - Used to determine sequence of events
    - Enabler and initiator to auto-refrigeration excursion
    - Excursion recovery
      - Control actions
      - Operations actions
  - How close to "top-event" did chain-of-events reach?
- Technology to automate data query is available

STRESS ENGINEERING SERVICES INC.
an employee-owned company

57

"Black-Box" Review (cont'd)

- Data Historian Search
  - Generate a query of the historian using unit instrument tags
  - Evaluate process variables for
    - Exceedance of limits
    - Sequence of events
    - Root-cause for excursion

  NOTE: Requires discipline to sift through data to locate and understand scenario – root cause analysis methods
- Utilized as part of Layers of Protection Analysis (LOPA) to establish unmitigated event frequency
- Example – Ethane Feed System
  - High-pressure scenario under auto-refrigeration excursion
  - System pressure at 200psig against a Relief Valve (RV) set-point of 141psig
    - Relief scenario not properly identified
    - RVs significantly undersized
    - Auto-refrigeration excursion due to liquid feed during loss of vaporization

STRESS ENGINEERING SERVICES INC.
an employee-owned company

58

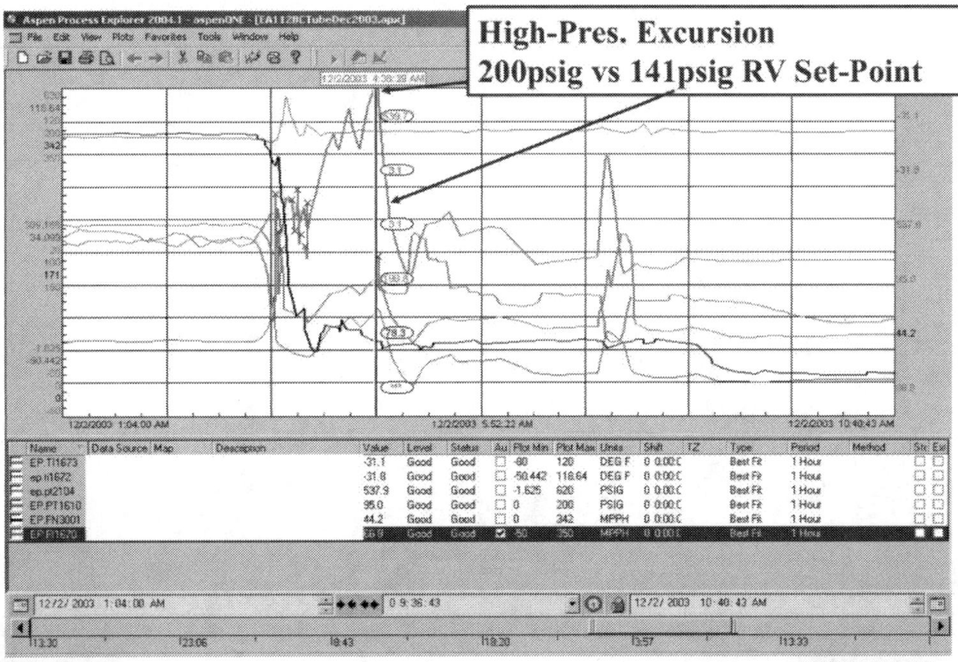

59

STRESS ENGINEERING SERVICES INC.

an employee-owned company

“Black-Box” Root Cause Analysis

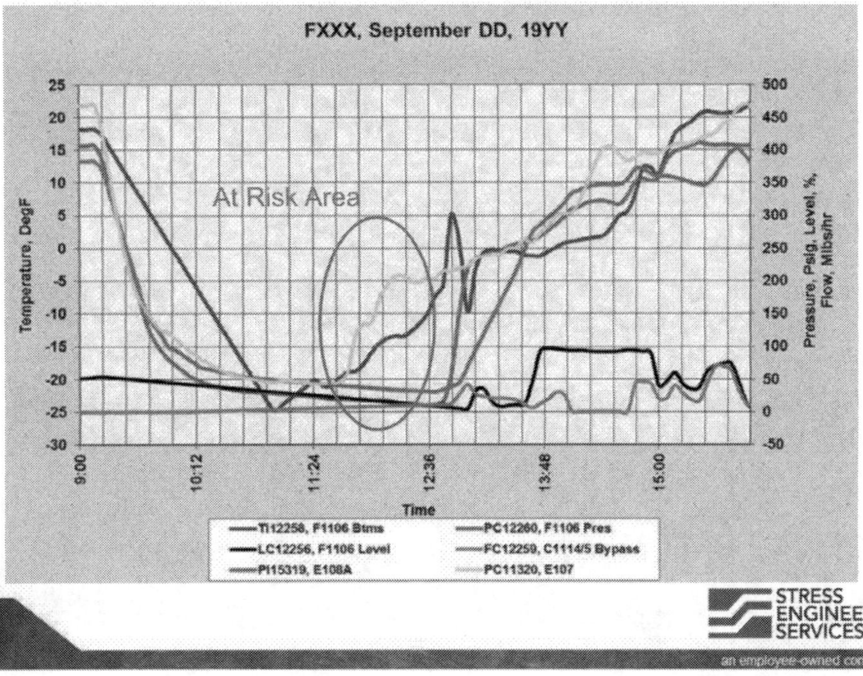

60

STRESS ENGINEERING SERVICES INC.

an employee-owned company

Non-Normal Operation

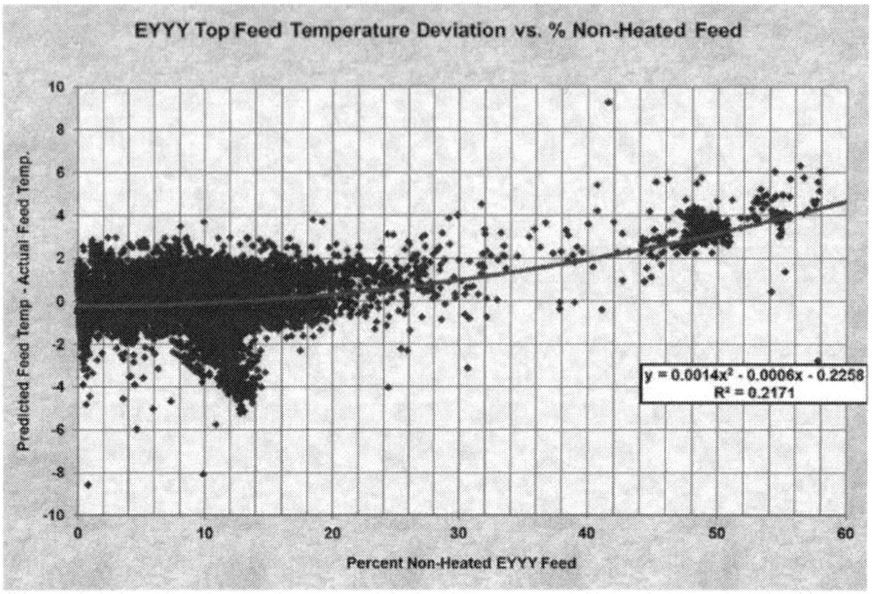

STRESS ENGINEERING SERVICES INC.
an employee-owned company

Automated Data Historian Search

1. **Launch the Envoy-Stress Form**
   After installing the add-in launch the form by going to Envoy-Stress → Find Auto Refrigeration

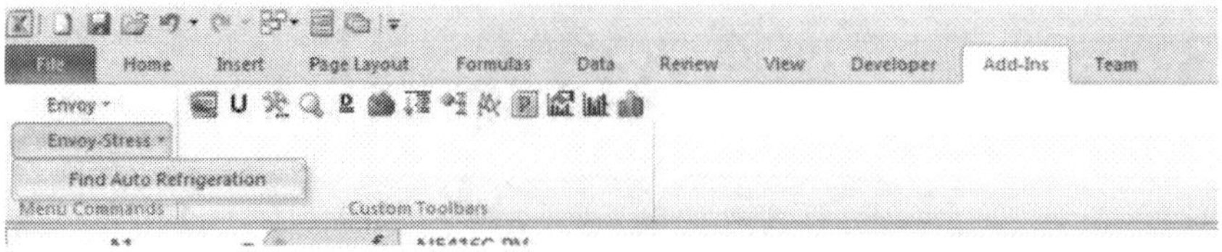

STRESS ENGINEERING SERVICES INC.
an employee-owned company

Automated Data Historian Search

Fill in the Form

Follow the directions on the form to fill it in. Ref edits are used for the tag list, limit dates, start date, and end date. It is helpful to have the information for the ref edit already on a sheet. The user may specify via the start and end dates if the application will look forwards or backwards in time. The limit dates should be configured by low limit then high limit for each tag.

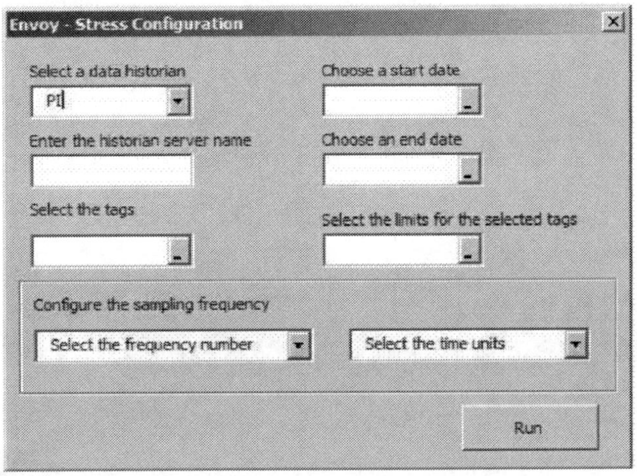

STRESS ENGINEERING SERVICES INC.
an employee-owned company

63

Automated Data Historian Search

Form configuration with example:

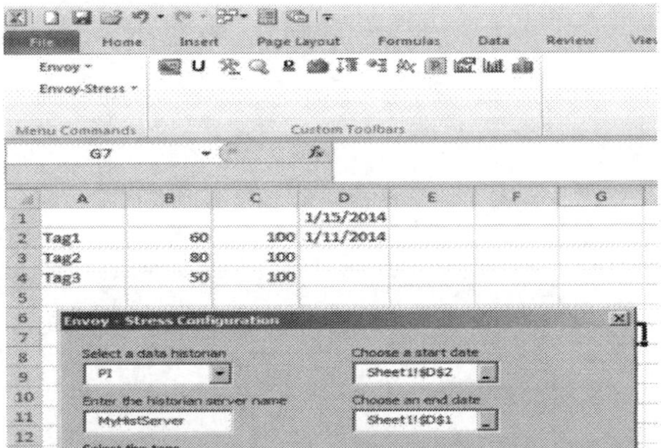

STRESS ENGINEERING SERVICES INC.
an employee-owned company

64

Automated Data Historian Search

Find times exceeding the given limits

Click the Run button to find times when auto refrigeration occurred as determined by the limits. The findings will be output on a new sheet.

	A	B	C	D	E	F	G
1		Tags	Tag1	Tag2	Tag3	Duration	
2		Low Limit	60	80	50		
3		High Limit	100	100	100		
4							
5	AR Start	1/11/2014	70.55475	53.3424	57.95186		
6	Time OK	1/11/2014 10:00	82.98016	82.46021	58.9163	10	
7	AR Start	1/11/2014 11:00	98.60931	91.09643	22.6866		
8	Time OK	1/12/2014 3:00	69.44853	91.37176	83.48172	16	
9	AR Start	1/12/2014 4:00	2.26292	54.33606	91.6164		
10	Time OK	1/12/2014 21:00	70.43958	92.8786	53.02125	17	
11	AR Start	1/12/2014 22:00	8.964139	75.77293	40.1842		
12							
13							
14							

Process Simulation

- For identified scenarios often process simulation is required
  - To accurately indicate the process temperature/pressure
  - To provide heat transfer rate and heat capacity – Level 3 analysis
  - Cases are often non-normal so compositions are often markedly different than standard operating conditions
- Can provide critical insight into understanding "black-box" identified excursions

Process Simulation

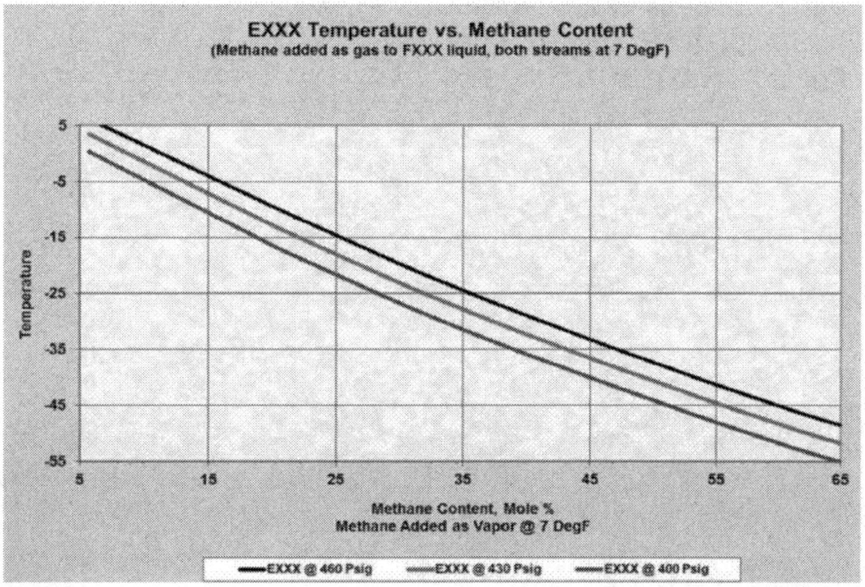

Pressure Tests

Pressure Test Failures – Thick Wall

(b)

Figure 5.92. (continued)

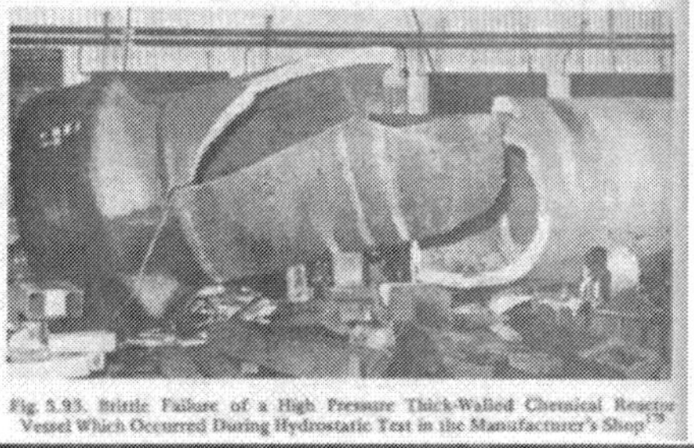

Fig. 5.93. Brittle Failure of a High Pressure Thick-Walled Chemical Reactor Vessel Which Occurred During Hydrostatic Test in the Manufacturer's Shop[79]

69

STRESS ENGINEERING SERVICES INC.

an employee-owned company

Pressure Test Gone Bad!

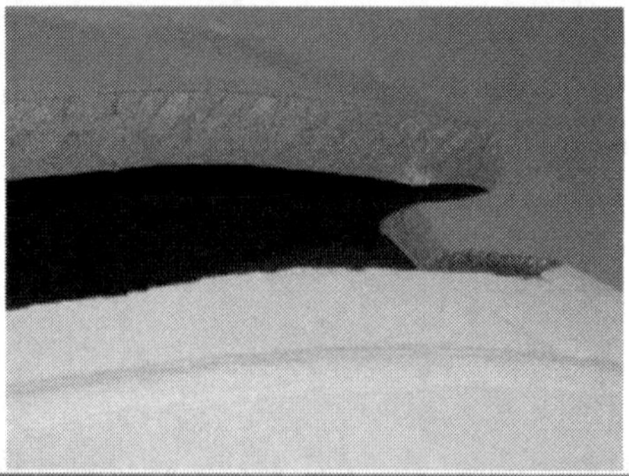

70

STRESS ENGINEERING SERVICES INC.

an employee-owned company

Pressure Test Mitigation

- Must ensure pressure and tightness testing accounts for
  - Minimum required temperature at the increased stress level to ensure vessel remains ductile
    - Code recommends +30°F above MAT at MAWP or MDMT
  - Accounts for any environmental embrittlement mechanisms (e.g., hydrogen)
- Appropriate personnel protection is taken
  - Treats all vessels as at-risk of failure until proven otherwise
  - Stand-off distance when pressure above MAWP

71

STRESS ENGINEERING SERVICES INC.
an employee-owned company

Hydro-Test Brittle Fracture

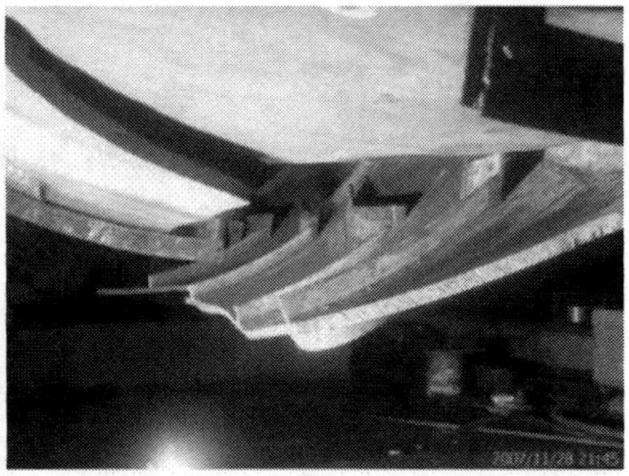

72

STRESS ENGINEERING SERVICES INC.
an employee-owned company

API-579 / ASME FFS-1 Remediation

- 3.6.2.1 Limiting Operation
  - Limit operating conditions to within P-T MAT envelope
  - Control startup via operating procedures
- PWHT of non-PWHT components that limit the MAT
- Execute a field pressure test at appropriate temperature for equipment not subject to a pressure test previously

Hierarchy of Remediation Actions

- Avoid / prevent the excursion
  - Process Change
- Administrative / Procedural changes
- Alarm with administrative response
- Alarm / Interlock to protect
- Upgrade Equipment
  - PWHT
  - Component upgrade (change material)
- Level 3 Analysis
  - Model transient conditions of process and equipment
  - Utilize material testing to quantify toughness under low-temp. conditions
- Total Upgrade of Equipment
  - Vessel replacement (change material)

Example Scenarios

Ethylene Plant (Hot Section)

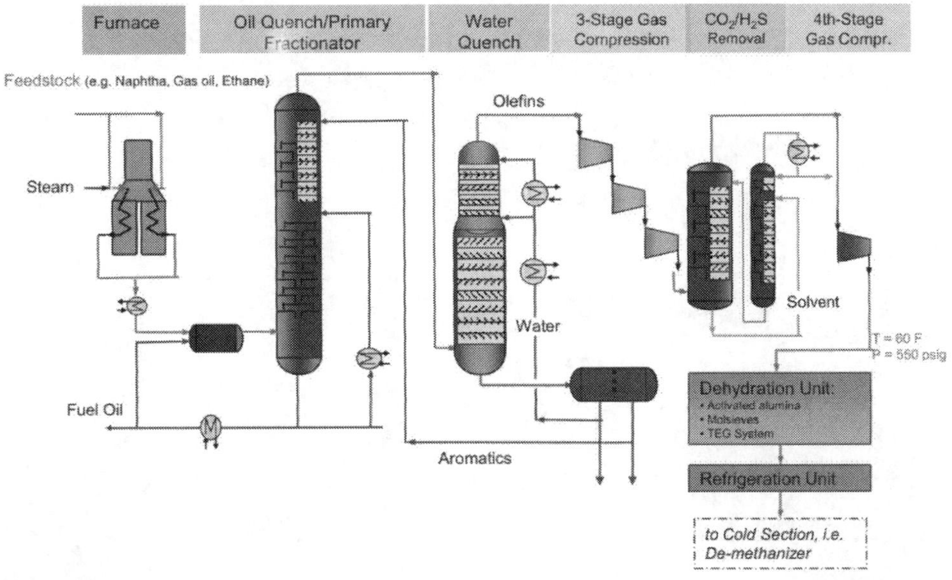

Slide Courtesy of Sulzer Chemtech

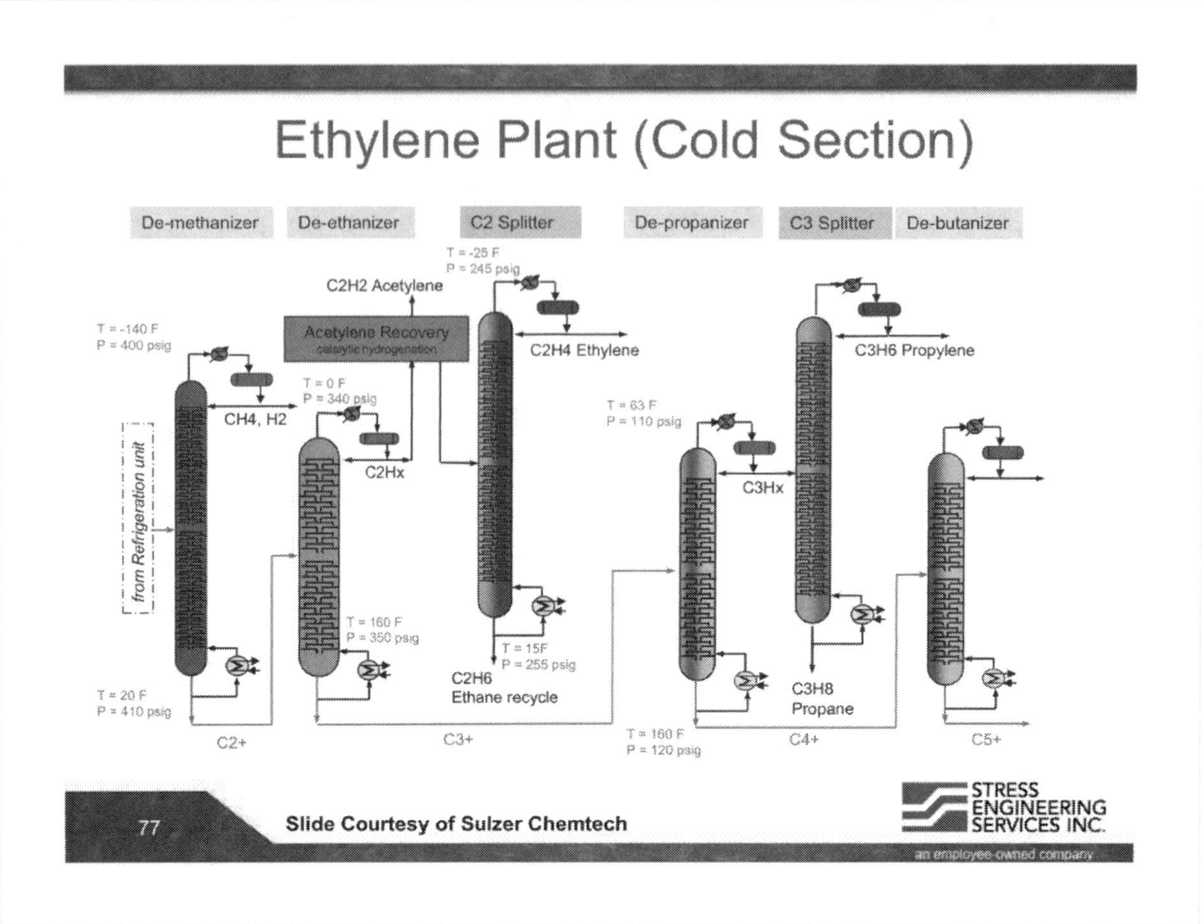
Ethylene Plant (Cold Section)
De-methanizer
De-ethanizer
C2 Splitter
De-propanizer
C3 Splitter
De-butanizer
T = -25 F
P = 245 psig
C2H2 Acetylene
Acetylene Recovery
catalytic hydrogenation
C2H4 Ethylene
C3H6 Propylene
T = -140 F
P = 400 psig
T = 0 F
P = 340 psig
CH4, H2
C2Hx
T = 63 F
P = 110 psig
C3Hx
from Refrigeration unit
T = 160 F
P = 350 psig
T = 15F
P = 255 psig
C2H6
Ethane recycle
C3H8
Propane
T = 20 F
P = 410 psig
T = 160 F
P = 120 psig
C2+
C3+
C4+
C5+
77
Slide Courtesy of Sulzer Chemtech
STRESS ENGINEERING SERVICES INC.
an employee-owned company

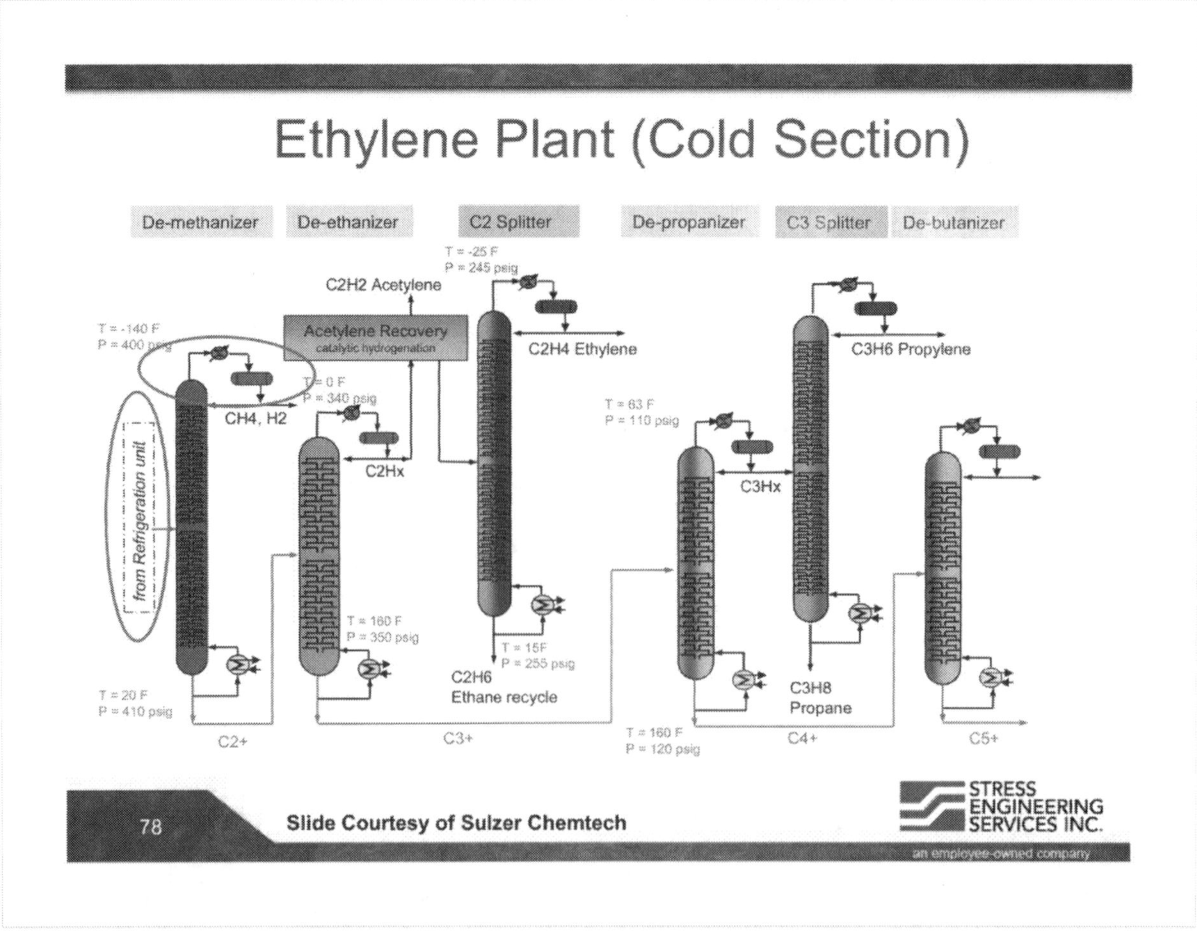
Ethylene Plant (Cold Section)
De-methanizer
De-ethanizer
C2 Splitter
De-propanizer
C3 Splitter
De-butanizer
T = -25 F
P = 245 psig
C2H2 Acetylene
Acetylene Recovery
catalytic hydrogenation
C2H4 Ethylene
C3H6 Propylene
T = -140 F
P = 400 psig
T = 0 F
P = 340 psig
CH4, H2
C2Hx
T = 63 F
P = 110 psig
C3Hx
from Refrigeration unit
T = 160 F
P = 350 psig
T = 15F
P = 255 psig
C2H6
Ethane recycle
C3H8
Propane
T = 20 F
P = 410 psig
T = 160 F
P = 120 psig
C2+
C3+
C4+
C5+
78
Slide Courtesy of Sulzer Chemtech
STRESS ENGINEERING SERVICES INC.
an employee-owned company

Three-And-One-Half Nickel Vessels

in Liquid Methane Service

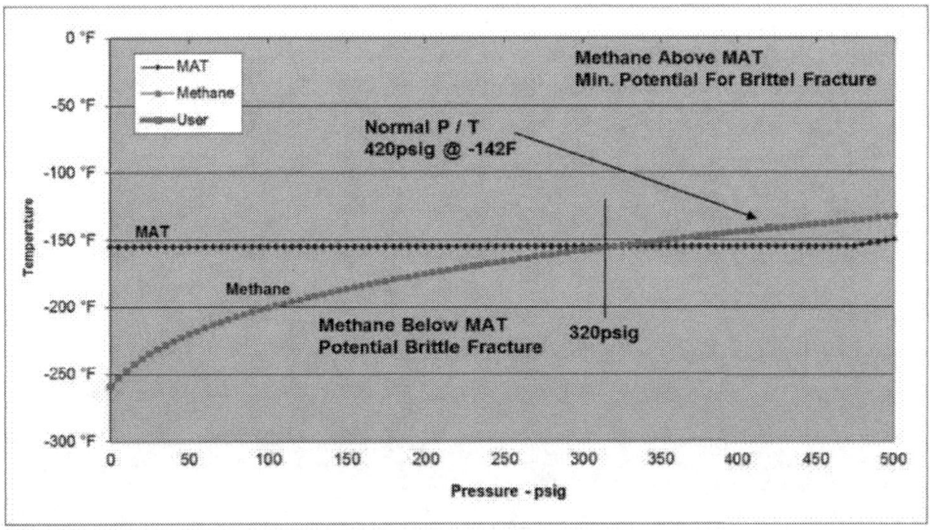

STRESS ENGINEERING SERVICES INC.
an employee-owned company

79

Chilling Train Vessels

Cross Heat Exchange with Methane/Hydrogen Streams

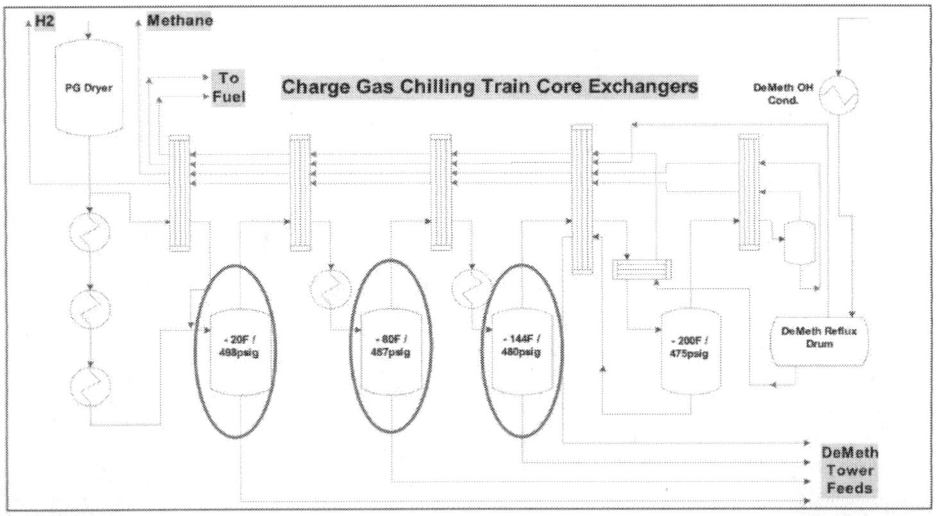

STRESS ENGINEERING SERVICES INC.
an employee-owned company

80

Chilling Train Vessels

Continued Chilling Streams Through Core Exchangers with Condensing and Liquid Accumulation in Chilling Train Drums

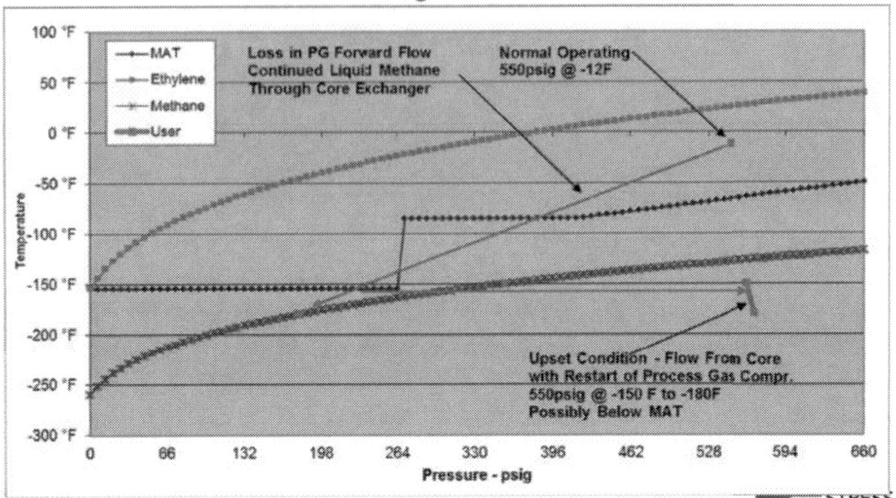

81

STRESS ENGINEERING SERVICES INC.
an employee-owned company

Ethylene Plant (Cold Section)

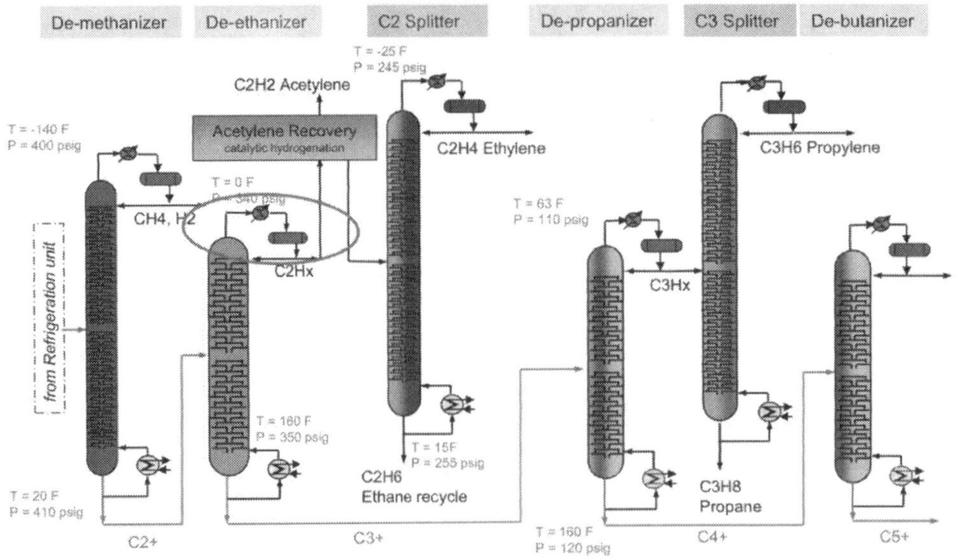

82

Slide Courtesy of Sulzer Chemtech

STRESS ENGINEERING SERVICES INC.
an employee-owned company

Distillation Towers and Peripherals

De-Ethanizer Reflux Drum MAT

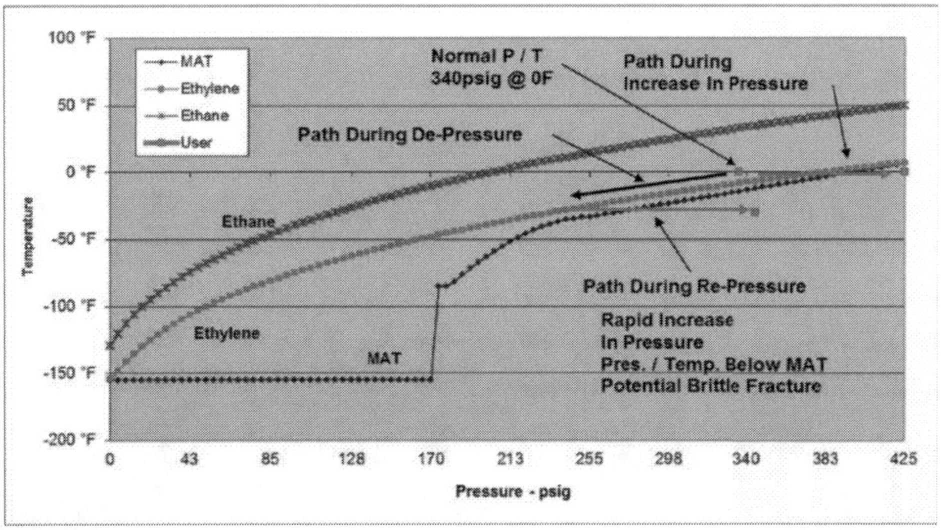

STRESS ENGINEERING SERVICES INC.
an employee-owned company

Ethylene Plant (Cold Section)

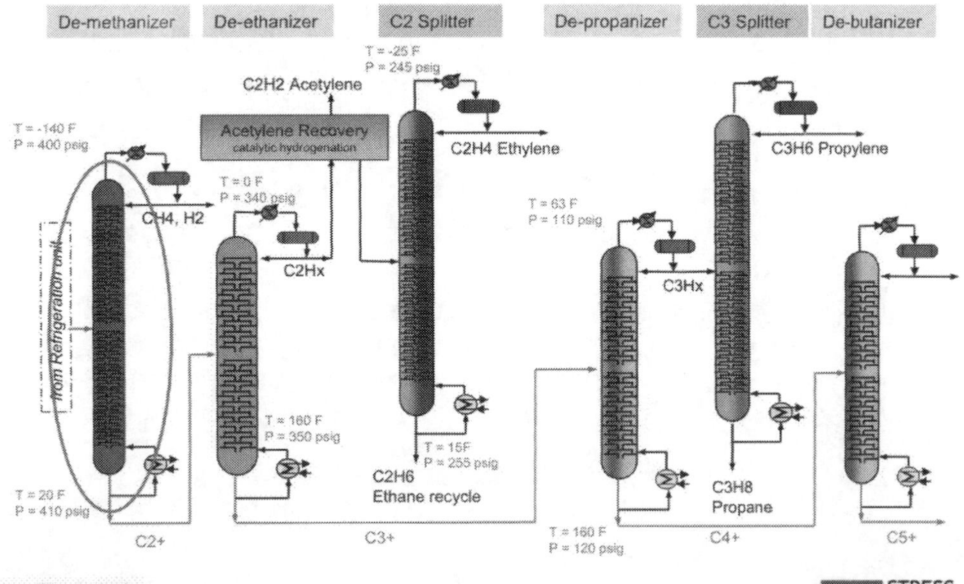

Slide Courtesy of Sulzer Chemtech

STRESS ENGINEERING SERVICES INC.
an employee-owned company

Bimetallic De-Methanizer

Carbon Steel Bottom Section

Loss of Reboil with Slump of Tower

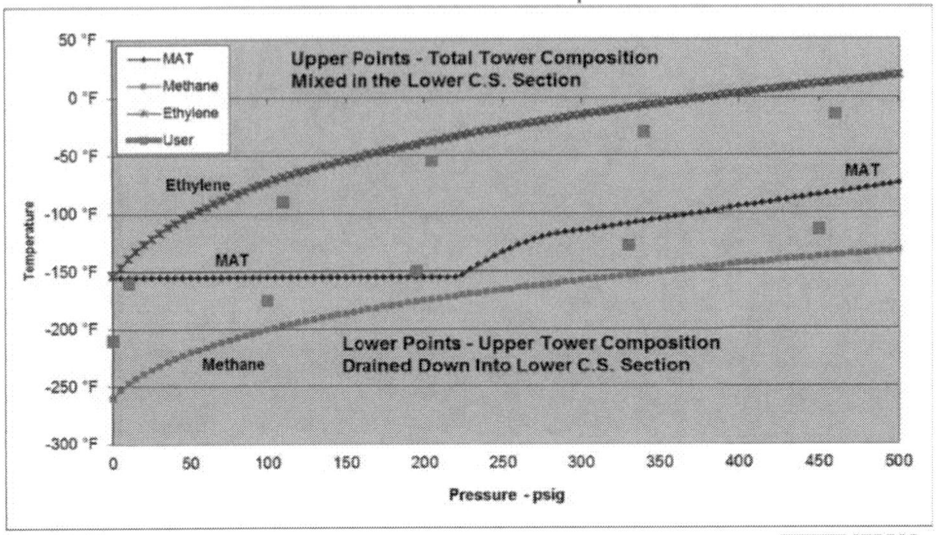

85

STRESS ENGINEERING SERVICES INC.

an employee-owned company

Nitrogen Freeing of Equipment

Nitrogen-Filled Vessel During Inventory

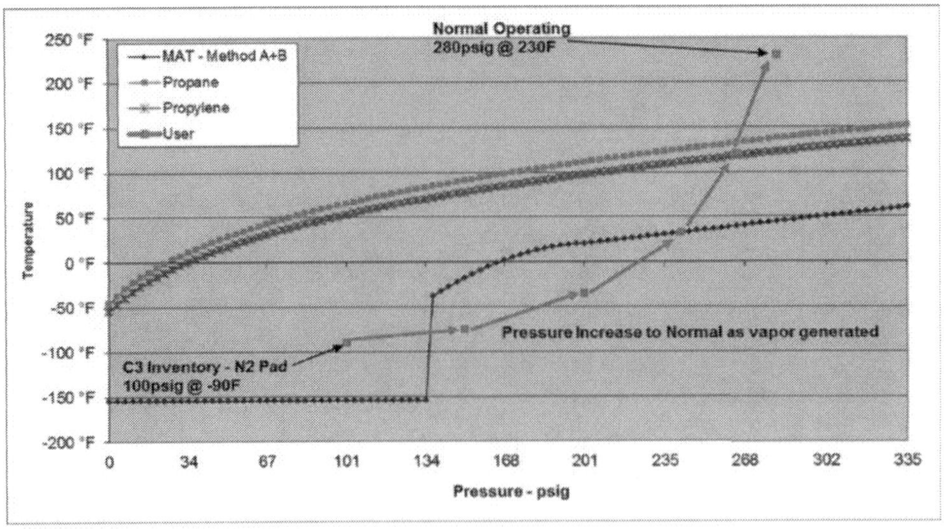

86

STRESS ENGINEERING SERVICES INC.

an employee-owned company

Compressor Discharge Drums

Vessel MAT - Loss in Pressure with Restart

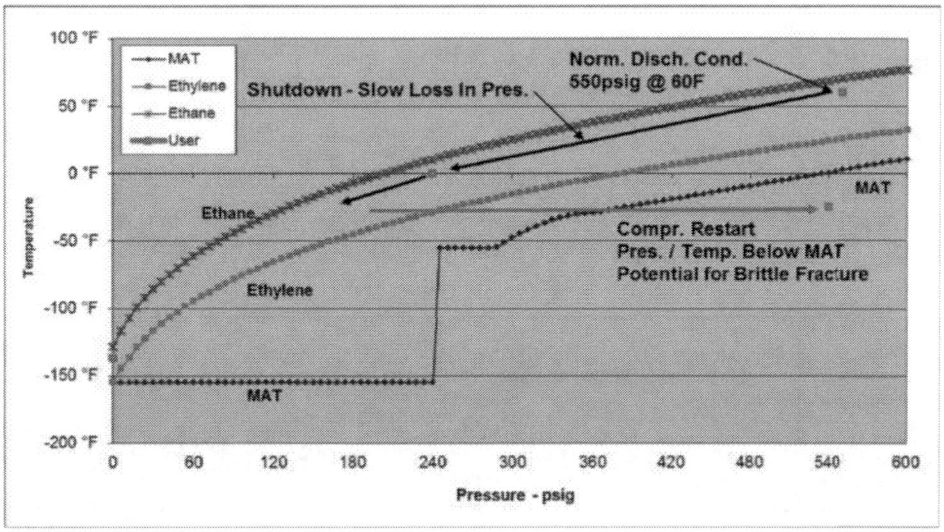

87

STRESS ENGINEERING SERVICES INC.

an employee-owned company

Compressor Suction Drum

Vessel MAT – compressor trip with vapor return

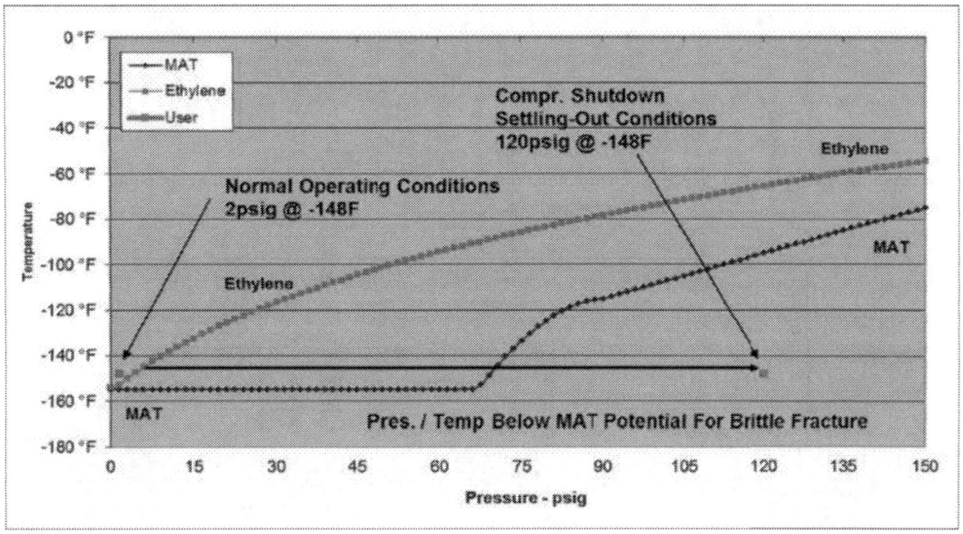

88

STRESS ENGINEERING SERVICES INC.

an employee-owned company

Loss in Heat Input

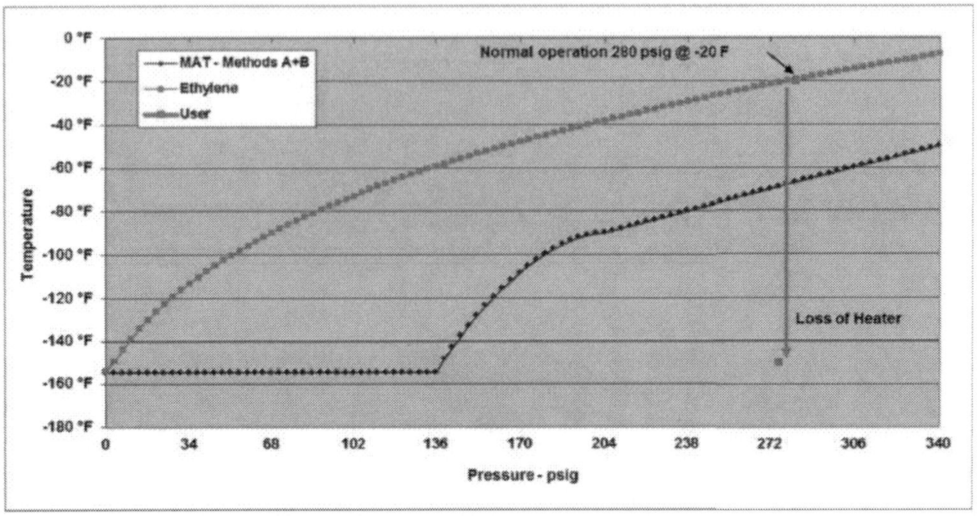

89

STRESS ENGINEERING SERVICES INC.
an employee-owned company

Overflow/Transport Liquid

Normally Vapor Only Vessels Downstream of Separator Overhead

- For Brittle Fracture potential to exist
  - Increase level in drum, causing overflow to downstream equipment
    - Inventory error
    - Level control failure
  - The pressure must be lowered, causing auto-refrigeration of the liquid
    - Unit trip (automatic or manual) due to liquid impact downstream
    - Pressure let-down valve
  - Through operator actions, normal process pressure is restored while vessel is chilled

90

STRESS ENGINEERING SERVICES INC.
an employee-owned company

Overflow/Transport of Liquid

Acetylene Converter Preheat Exchanger MAT

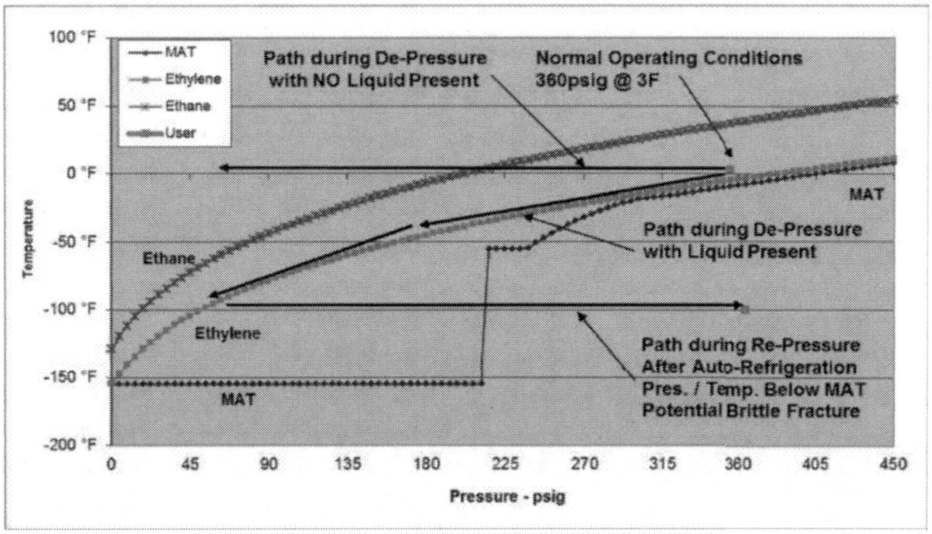

91

STRESS ENGINEERING SERVICES INC.
an employee-owned company

Non-Hydrogen Charged As-Designed

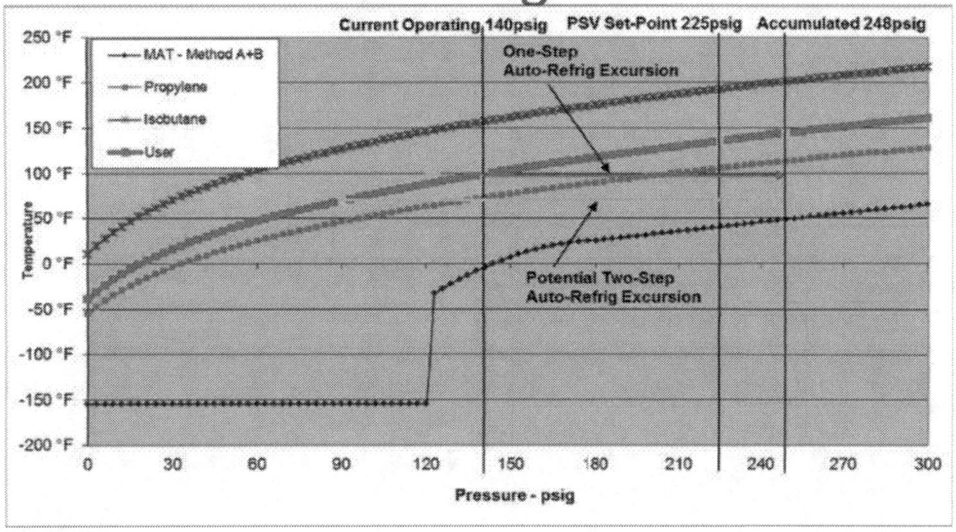

92

STRESS ENGINEERING SERVICES INC.
an employee-owned company

Hydrogen Charged – Loss of Ductility

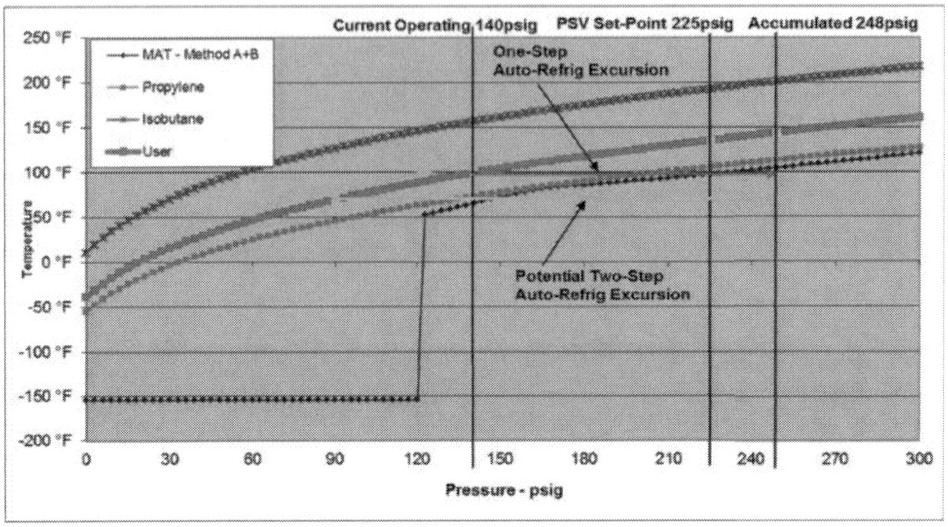

93

STRESS ENGINEERING SERVICES INC.
an employee-owned company

Operator Training

94

STRESS ENGINEERING SERVICES INC.
an employee-owned company

Operator Training

- Must have defined training content
- Subject-focused on auto-refrigeration / brittle fracture prevention only
- Teaches what must be done and what must NOT be done
- Covers highest risk operating states
  - Inventory / Startup
  - Upset
  - Shutdown

Operator Training

- Training Goals
  - Basic understanding of equipment limits
    - Brittle fracture basics
  - Fully understand the phenomenon of auto-refrigeration
    - How to avoid it
    - If it occurs
      - What to do
      - What NOT to do
  - Know and understand best practices in controlling and mitigating potential auto-refrigeration

Indications of an Auto-Refrigeration Excursion

- Abnormally frosted or iced steel surfaces such as vessel supports, vents, drains or manways
- Abnormally heavy atmospheric condensation and sweating of carbon steel surfaces
- Temperature indicators which have gone "down scale" and are "flat lined" at the bottom-of-range
- Sounds (rumbling) of boiling (flashing) liquid or equipment or piping shaking

Auto-Refrigeration Recovery Guidlines

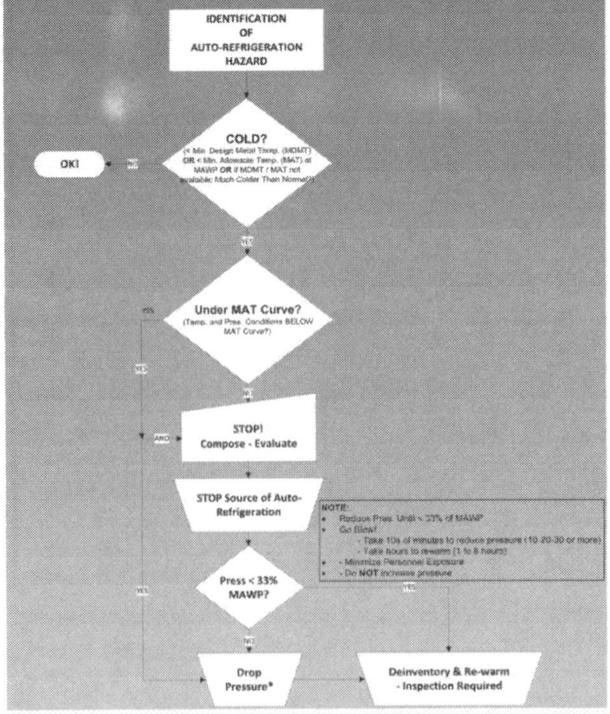

Auto-Refrigeration Recovery

Actions if AR occurs

- STOP! Compose - Evaluate
  - Take Actions Slowly (10s of minutes: 10-20-30 or more)
  - Re-warm VERY Slowly (Hours: 1 – 8 hours or more)
- Conditions under MAT curve?
  - Immediately begin pressure reduction
- Never increase pressure!
- Determine and stop source of Auto-Refrigeration
- Is pressure < 33% MAWP? (Hoop stress < ~ 7,000psi)
  - Drop pressure to <33% MAWP!
- Minimize personnel exposure
- Gradually
  - De-inventory liquid
  - Use warm vapor if available to heat equipment
    - NEVER use steam to heat the equipment – thermal shock potential
- Inspect as required – must demonstrate fitness-for-service

99

STRESS ENGINEERING SERVICES INC.
an employee-owned company

Important Reminders

- If you just stop, isolate and let the boiling liquid continue, the refrigeration will stop BUT the pressure will increase.
  - Don't let this happen!
- Dropping pressure to <33% MAWP is the fail-safe mode. Equipment may develop a leak due to strain limited fracture, but it should not catastrophically fracture.
- De-inventory the liquid, stops auto-refrigeration and minimizes potential spill inventory.

100

STRESS ENGINEERING SERVICES INC.
an employee-owned company

Example Vessel MAT – Two Step

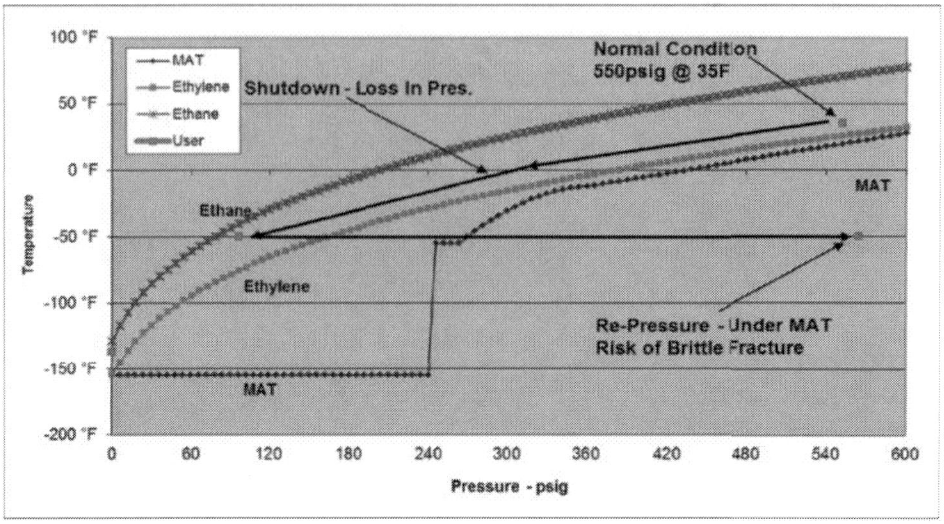

101

STRESS ENGINEERING SERVICES INC.
an employee-owned company

What NOT to Do – A Critical Safety Note!

The following must NEVER be done when a vessel is cold

- Apply or increase pressure when vessel is COLD
  - Stresses the equipment when it is least able to handle it
- Add a "DRY" gas or vapor in an attempt to push liquid out
  - Lowers partial pressure making the boiling (flashing) occur at a lower temperature
- Put steam or other heating media on equipment in attempt to warm the equipment
  - Increases thermal stresses

102

STRESS ENGINEERING SERVICES INC.
an employee-owned company

Training Focus – Key to Mitigation

AR/BF mitigation of existing units fundamentally

- Changes the operation of the unit
  - Establishes NEW operating limits (Safety Critical Variables - SCVs)
  - Sequence control
  - Prevention steps
- Changes the process
  - NEW alarms and interlocks
  - NEW hardware and systems

TRAINING OF THE OPERATING STAFF IS CRITICAL TO THE SUCCESS OF AUTO-REFRIGERATION / BRITTLE FRACTURE PREVENTION

103

STRESS ENGINEERING SERVICES INC.
an employee-owned company

Pressure Test Gone Bad!

104

STRESS ENGINEERING SERVICES INC.
an employee-owned company

Post Excursion Inspection

105

STRESS ENGINEERING SERVICES INC.
an employee-owned company

Post Excursion Inspection

- Excursions to the "lower-left" of the MAT can
  - Result in cracking of vessel from thermal stresses
  - Grow existing flaws which are under critical flaw length (strain limited growth)
    - The next excursion or subsequent excursion results in reaching critical flaw size
- Fitness-For-Service must be established prior to returning the equipment to service
- Inspection must take into account
  - Thickness of components and possible stress risers
  - Rate of chilling
  - Vessel condition relative to MAT
    - Low-temperature excursion without going under MAT
    - Excursion to the MAT
    - Excursion under the MAT
      - By how much?
  - Rate of warming during recovery
  - Contraction restraints – locations of high thermal contraction stress

106

STRESS ENGINEERING SERVICES INC.
an employee-owned company

General Observations

- Plants were designed for steady-state operating conditions:
  - Many built prior to AR/BF identification
  - Operating modes may present AR/BF hazard
    - Materials of Construction limitations
    - Startup / Shutdown / Upset conditions were not fully evaluated in the initial plant design
- Single Event Block
  - Operating mode sequence can result in auto-refrigeration
  - Often only one event block separates conditions from potential risk of brittle fracture
  - For many client risk profiles; additional protection required

107

STRESS ENGINEERING SERVICES INC.
an employee-owned company

General Observations (cont'd)

- Materials of construction selection:
  - Lack adequate separation between CET and MAT
  - Inherently Safe Vessel Design:
    - Make lowest possible CET (CET @ 14.7psia = MAT (MDMT))
  - Large Vessel Design:
    - A detailed process hazards analysis required
    - Provide adequate separation between CET and MAT
    - Allow vessel wall to return to equilibrium with process without reaching MAT

108

STRESS ENGINEERING SERVICES INC.
an employee-owned company

General Observations (cont'd)

- Full management support is required for a successful Auto-Refrigeration / Brittle Fracture Prevention Program
  - Substantial commitment for brittle fracture analyses
  - Time and resource commitment for PHA
  - Mitigation costs can be substantial
- Use Process Data Historian – "Black-Box Review"
  - Identify Temps = / < Vessel MAT @ MAWP
  - Identify inadequately ranged temp. measurement (bottom of scale during excursion)
  - Identify past potential AR/BF excursions
  - Important tool for evaluation of PHA identified scenarios
  - Highly useful in LOPA in establishing event frequency

109

STRESS ENGINEERING SERVICES INC.
an employee-owned company

General Observations (cont'd)

- Operator Training
  - First line of defense against possible AR/BF excursions
    - Understand what causes auto-refrigeration - avoids the excursion
  - Knows what to do if an AR/BF excursion occurs
    - Drain liquid
  - Knows what NOT to do if an AR/BF excursion occurs
    - NEVER increase pressure
    - NEVER add a "Dry" gas
    - NEVER heat with steam

110

STRESS ENGINEERING SERVICES INC.
an employee-owned company

Concluding Remarks

- Do you have MATs for all of your equipment subject to Auto-Refrigeration or Brittle Fracture?
- Have you utilized the equipment MATs in your PHAs?
- Have you utilized your process data to identify scenarios and verify unmitigated frequency of occurrence?
- AR/BF is a real process safety hazard and can have significant (actionable) process safety risk.
- Past process and equipment designs did not fully account for the potential for auto-refrigeration during non-steady state operation
- Making appropriate process changes and installation of alarms and interlocks to avoid the auto-refrigeration excursion may provide the needed levels of protection

111

STRESS ENGINEERING SERVICES INC.
an employee-owned company

Concluding Remarks (cont'd)

- Replacing equipment may be needed if sufficient protection cannot be gained by other means
- Operator training is key to avoiding and properly responding to auto-refrigeration excursions
- Inspection Guidelines should be provided based on the excursion CET compared to the vessel MAT and the rate of chilling and subsequent warming recovery

112

STRESS ENGINEERING SERVICES INC.
an employee-owned company

113

STRESS ENGINEERING SERVICES INC.
an employee-owned company

353247 Tutorial on Auto-Refrigeration and Brittle Fracture Analysis (Part 2)

Abstract:

The tutorial is aimed at entry level to seasoned veteran engineers that desire to have a better understanding of this multidiscipline process safety risk.

Part 2 of the Tutorial will Cover:

- Example Scenarios
- Operator Training
- Post Incident Inspection

351215 Low Temperature / Auto-Refrigeration Incidents Panel Discussion

Abstract:

A number of ethylene producers will share their experiences with low temperature / auto-refrigeration (near miss incidents, PHAs, mitigation, etc.). Conference attendees are encouraged to ask questions, and to engage with the panelists in this session.

AUTHOR INDEX

AUTHOR INDEX